Booker Tropical Soil Manual

Principal Contributors

C Berryman

R P Bower

C J Chartres

H B Davis

R A Davison

B W Eavis

P F Goldsmith

J R Landon

A M Lavelle

D R Mellors

G Murdoch

P L Searl

J S O Suttie

J A Varley

A J West

R A Yates

Booker Tropical Soil Manual

A handbook for soil survey and agricultural
land evaluation in the tropics and subtropics

Paperback Edition

Edited by
J R Landon

Longman
Scientific &
Technical
Copublished in the United States with
John Wiley & Sons, Inc., New York

BOOKER TATE

Longman Scientific & Technical,
Longman Group UK Ltd,
Longman House, Burnt Mill, Harlow,
Essex CM20 2JE, England
and associated companies throughout the world.

Booker Tate Limited
Masters Court
Church Road
Thame, Oxon OX9 3FA
England

Copublished in the United States with
John Wiley & Sons, Inc., 605 Third Avenue, New York, NY 10158

First Published 1984, paperback edition 1991

British Library Cataloguing in Publication Data

Booker tropical soil mannual.
  1. Tropical regions. Soils. Surveying
  I. Landon, J. R. (John Richard) 1947-
  631.4713

  ISBN 0-582-00557-4

Library of Congress Cataloging-in-Publication Data

Booker tropical soil mannual : a handbook for soil survey and
  agricultural land evaluation in the tropics and subtropics / edited
  by J.R. Landon.
       p.   cm.
    Reprint. Originally published: London : Booker Agriculture
  International Ltd. ; New York : Longman, 1984.
    Includes bibliographical references and index.
    ISBN 0-470-21713-8 (pbk.)
    1. Soils--Tropics--Handbooks, manuals, etc. 2. Soil surveys-
  -Tropics--Handbooks, manuals, etc. 3. Land capability for
  agriculture--Tropics--Handbooks, manuals, etc. I. Landon, J. R.
  (John Richard), 1947-   . II> Booker Agriculture International
  Limited.
  (S599.9.T76B66 1991)
  631.4'7113--dc20   90-19251
                                                               CIP

Produced by Longman Group (FE) Limited
Printed and bound by Antony Rowe Ltd, Eastbourne
Transferred to digital print on demand, 2003

Contents

CONTENTS (cont)

(vi)

List of Annexes

List of Figures

List of Tables

List of Abbreviations

A	annual soil loss (in the USLE) or ampere		exch	exchangeable
ac	acre		f	following (or force, in force and pressure units)
ADAS	UK Agricultural Development and Advisory Service		FAO	Food and Agriculture Organisation of the United Nations
adj	adjusted		FAP	fifteen atmosphere percentage
ads	air dry soil		FC	field capacity
ADS	Agricultural Development Service, World Bank		fS	fine sand
AFP	air-filled porosity		fSL	fine sandy loam
am	ammonium		ft	foot
AOAC	Association of Official Analytical Chemists		g	gram
API	aerial photographic interpretation		gal	gallon
ASTM	American Society for Testing and Materials		GWT	groundwater-table
atm	atmosphere		h	hour
AWC	available water capacity		ha	hectare
b	bar		HC	hydraulic conductivity
b & w	black and white		hp	horsepower
BAI	Booker Agriculture International Limited		Hz	hertz
BD	bulk density		ID	internal diameter
BOD	biochemical oxygen demand of effluent and sludge		ie	that is
BSI	British Standards Institution		ILRI	International Institute for Land Reclamation and Improvement
BSP	base saturation percentage		IMM	Institute of Mining and Metallurgy
C	clay (or crop management factor in the USLE)		imp	imperial (measure)
cc	cubic centimetres		in.	inch
CCTA	Conseil pour Cooporation Technique en Afrique		INEAC	Institut National pour l'Etude Agronomique du Congo
CEC	cation-exchange capacity		$\overline{IR}$	average infiltration rate
cf	compare		IR	infra-red, or instantaneous infiltration rate
CL	clay loam		ISRIC	International Soil Reference and Information Centre
cm	centimetre		ITCZ	Inter-Tropical Convergence Zone
cS	coarse sand		IUPAC	International Union of Pure and Applied Chemists
cumec	cubic metres per second		J	joule
cusec	cubic feet per second		K	saturated hydraulic conductivity (or soil erodibility factor in the USLE)
CV	curriculum vitae		KE	kinetic energy
cwt	hundredweight		kg	kilogram
cZ	coarse silt		km	kilometre
DHSS	Department of Health and Social Security (in UK)		kPa	kilopascal
dia	diameter		ℓ	litre
dm	decimetre		L	loam (or length of slope factor in the USLE)
DOS	Directorate of Overseas Surveys		lb	pound
dS	decisiemens		LcS	loamy coarse sand
DTA	differential thermal analysis		LF	leaching fraction
DTPA	diethylene triamine pentacetic acid		LfS	loamy fine sand
EC	electrical conductivity (subscripts: p = paste, e = saturation extract, 2.5 = 1:2.5 soil:water suspension)		liq	liquid
			LmS	loamy medium sand (infrequently used)
EDDHA	ethylene diamine-di (ortho-hydroxyphenylacetic acid)		ln	natural logarithm
EDTA	ethylene diamine tetracetic acid		LRD	Land Resources Division/Department
eg	for example		LRDC	Land Resources Development Centre
Eh	redox potential		LS	loamy sand (or slope factor in the USLE)
EPP	exchangeable potassium percentage		LvcS	loamy very coarse sand
ESP	exchangeable sodium percentage		LvfS	loamy very fine sand
ET	evapotranspiration (subscripts: o = reference (eg Penman; Class A pan), a = actual; m = maximum)		LWB	long wheel base
			m	metre
et seq	and the following		M	million or moles per litre
EWt	equivalent weight			

(x)

MAFF	Ministry of Agriculture, Fisheries and Food	TSAU	Tropical Soil Analytical Unit (in LRD)
mb	millibar	UK	United Kingdom
MC	moisture content	UNDP	United Nations Development Programme
me%	milliequivalent (also meq)	Unesco	United Nations Educational, Scientific
mg	milligram		and Cultural Organisation
mi	mile	USBR	United States Bureau of Reclamation
min	minute	USDA	United States Department of Agriculture
mℓ	millilitre	USDI	United States Department of the Interior
mm	millimetre	USLE	Universal Soil Loss Equation
mmho	millimho (obselete; now mS)	UV	ultraviolet
mol	mole or molar	v	coefficient of variation
mph	miles per hour	V	volt
mS	millisiemens	VAT	value added tax
MSS	multispectral scanner	vcS	very coarse sand
mV	millivolt	vcSL	very coarse sandy loam
N	normal (of chemical solutions) or newton	vfS	very fine sand
N/A	not available	vfSL	very fine sandy loam
NF	Norme francais	viz	namely
No	number	v/v	volume for volume
ods	oven-dry soil	W	watt
ODA/ODM	Overseas Development Administration/Ministry	WHO	World Health Organisation
OM	organic matter	WP	wilting point
ORSTOM	Office de la Recherche Scientifique et	wt	weight
	Technique Outre-Mer	w/w	weight for weight
OSP	ordinary superphosphate	yd	yard
oz	ounce	Z	silt
p	page (plural pp)	ZC	silty clay
P	Conservation Practice factor (in the USLE)	ZCL	silty clay loam
Pa	pascal	ZL	silty loam, silt loam
pF	measure of soil water tension	ZSL	silty sandy loam (infrequently used)
PF	permeable fill	°C	degree Celsius (Centigrade)
pH	measure of acidity/alkalinity	°F	degree Fahrenheit
pp	pro parte (in part or part of) or pages	°K	degree Kelvin
ppm	parts per million	4WD	four-wheel-drive
pto	power take-off	>	greater than
PWP	permanent wilting point	<	less than
qv	which see	≥	greater than or equal to
R	rainfall erosivity factor (in the USLE)	≤	less than or equal to
rad	radian or radius	ρp	particle density
RBV	return beam vidicon	ψ	thermodynamic potential (of soil water)
rev	revolution		(subscripts: h = hydraulic, m = matric,
RH	relative humidity		p = pressure, s = solute, t = total,
RNa	sodium adsorption ratio		z = gravitational)
RSC	residual sodium carbonate	µg	microgram (10^{-6} g)
s	second	µm	micrometre formerly micron (10^{-6} m)
S	sand (or gradient in the USLE, or siemens)		
SAR	sodium adsorption ratio		
sat ex	saturated extract		
SC	sandy clay		
SCL	sandy clay loam		
SE	standard error		
SG	specific gravity		
Si	silt (see also Z, which is the abbreviation		
	BAI prefers)		
SI	International System (of units and measures),		
	from Système International d'Unités		
SL	sandy loam		
sp	species (plural spp)		
SP	saturation percentage		
t	tonne		
TDS	total dissolved solids		
TEB	total exchangeable bases		
TM	thematic mapper		
TOR	terms of reference		
TRAWC	total readily available water capacity		

Chemical elements are abbreviated to their standard symbols.   For SI unit abbreviations, additional to
those listed above, see pp 339-340.

Preface to the Paperback Edition

This is a more portable version of the Booker Tropical Soil Manual, in which the format (and weight) of the first edition have been reduced whilst retaining as much as possible of the original clarity. Since the Manual is written by professional soil scientists in busy commercial practice, constraints of cost and time have not allowed production of a fully revised second edition. However, this is not merely a re-issue of the old contents. We have taken the opportunity to add two new Appendices. These latter, and appropriate entries in the main text, cover the revised FAO publications on soil classification and on water quality for agriculture. We have also corrected minor errors, and changed some obviously dated items, such as the costs in Annex A and remote sensing data in Annex K.

Since our objective has been to keep the bulk down, the additions are strictly limited, and we have not included material that is too long and detailed to summarise usefully. This applies, for example, to the series of FAO Soils Bulletins on land evaluation guidelines for specific applications, such as irrigated agriculture, forestry etc and to newer methodologies, such as soil science linkages within geographic information systems. However, some of the more useful references are included in an expanded bibliography, page 411 ff, which cites 190 more works than the hard cover volume.

This edition has again seen valuable contributions from a range of professional and administrative colleagues. In particular, we should like to thank Alan Bird of Mott MacDonald International Ltd for editing and updating the remote sensing Annex, Philip Beckett and Roger Mills of Oxford University for assistance with bibliography additions and the following Booker Tate Ltd staff: John West, George Murdoch and Richard Bower for supervision of alterations through to the camera-ready copy stage, Brian Plant for updating costs and Johanna Claudel for word-processing this edition for publication.

Thanks are due to the Food and Agriculture Organisation of the United Nations for permission to reproduce copyright material additional to that which was acknowledged previously - see also page (xiv) - ie excerpts and adaptations from Soils Bulletin 55, World Soil Resources Report 60 and the 1985 edition of Irrigation and Drainage Paper 29.

Finally, we are grateful to the managements of Booker Tate Ltd (the successors in 1988 to Booker Agriculture International Ltd) and of MacDonald Agriculture Services Ltd (the current employers of the Editor) for facilities and support in the production of this edition.

Acknowledgements (1984 Edition)

Almost all the members, past and present, of the Land Resource Services unit in Booker Agriculture International Limited (BAI) have contributed to this manual, which forms an amalgamation of several in-house monographs and discussion papers. Peter Searl was the prime mover behind the early editions, beginning the compilation over 10 years ago under the company's Technical Director, Bob Innes. Subsequent work has been carried out with the encouragement of the Managing Directors of BAI, successively Jonathan Taylor and Ed Robinson, of Consultancy Division Managers Alan Yates, Jim Evans, Laurence Cockcroft and Martin Evans, and with the technical assistance of George Murdoch, the current Director of BAI Land Resource Services.

This edition has been comprehensively revised, so that most of the individual contributions cannot be precisely identified. However, large parts of Peter Searl's papers appear in Chapters 2, 3, 7 and 8, and parts of Carl Berryman's Soil Physics Manual are retained in Chapter 6 and Annex B. Tony Lavelle made substantial contributions to Chapters 4 and 5 and Annex C, as did Harold Davis to Chapter 7 and Annexes N and P, Rob Davison to Annexes F and K, and Peter Goldsmith to Annex G. Jim Suttie, Ronny Osborne, Bill Mitchell, Gerry Wood and Alan McEwan contributed useful information to the Annex sections dealing respectively with mechanisation, bananas, coffee, camping and medical topics. All these individuals are warmly thanked for their contributions, as are the many people who commented on earlier drafts. Amongst these latter particular thanks for their observations are due to George Watson, Gerry Wood, Ian Hill, Jim Harbord, Maurice Keech, Orville Seaton, Tony Smyth, Peter Thomas and Tony Young. Special thanks are also due to Richard Bower, George Murdoch and Alan Yates for their detailed comments on the pre-publication volumes of the manual, and to Stuart Donald for helping both to check the word processor output and to compile the Index. George Murdoch and Tony Lavelle are also thanked for their invaluable assistance in tracing copyright holders and preparing the final copy ready for printing.

Two additional contributors are advisers to BAI but have not been permanent staff members and their assistance has been especially welcome. Bryan Eavis of the ODA Land Resources Development Centre, Tolworth, made many useful additions and corrections to Chapters 6 and 8 and Annex B, whilst John Varley of the Commonwealth Development Corporation, London, corrected and added to Chapter 7 and Annexes J, N and P. Four members of the Soil Science Department at Reading University, John Dalrymple, Peter Gregory, Chris Mott and David Rowell are also thanked for the specific advice and lists of references that they have provided.

Despite the helpful suggestions received from all the above sources, however, responsibility for any errors in this edition remains with the present editor.

An immense amount of secretarial assistance has been provided over the years for which Lynda Beacon, Irene Kittel, Doris Lobo Wendy White and Debbie Dawson are heartily thanked, the last named in particular for the difficult task of preparing the present edition for publication. John West and Mary Davies are also thanked for drawing the diagrams and checking the camera-ready copy.

Thanks are due to the following for permission to reproduce copyright material:

Academic Press (London) Ltd for the table on pp 60-61 of 'Environmental chemistry of the elements' by H J M Bowen, 1979; Academic Press (New York) Inc and D Hillel for the table on pp 326-327 of 'Applications of soil physics', 1980; George Allen and Unwin Ltd for the tables on pp 31 and 43 of 'Geology for Civil Engineers' by A C McLean and C D Gribble, 1979; B T Batsford Ltd for the figure on p 187 of 'Soil conservation' by N W Hudson, Second Edition 1971; Cambridge University Press for the tables on pp 301 and 310 of 'Tropical soils and soil survey' by A Young, 1976; Centre d'Etude de l'Azote for tables on pp 97 and 419-420 of 'Fertiliser guide for the tropics and sub-tropics' by J G de Geus, 1973; Chapman and Hall for the tables on pp 327-328 of 'Soil chemistry' edited by F E Bear, 1964; Elsevier Scientific Publishing Company, the Dutch Ministry of Agriculture and Fisheries and ILACO BV for the table on pp 569-570 of 'Agricultural compendium for rural development in the tropics and sub-tropics', 1981; the Food and Agriculture Organisation of the United Nations (FAO) for excerpts from pp 75 and 203 of 'Irrigation, drainage and salinity - an international source book', pp 18-21 and 54 of Soils Bulletin 32, pp 42 and 67 of Soils Bulletin 42, pp 26-31 of Irrigation and Drainage Paper 29 and pp 6-7 of Irrigation and Drainage Paper 33; W H Freeman and Co Ltd and G L Ashcroft for the table on p 17 of 'Physical edaphology' by S A Taylor and G L Ashcroft, 1972; Granada Publishing Ltd for the tables on pp 323, 324, 327, 329 and 330 of 'Site investigations' by C R I Clayton, 1982; Hamlyn Publishing Group Ltd for the figure on p 147 of 'The Hamlyn Guide to minerals and fossils' by W R Hamilton et al, 1974; Hutchinson Ross Publishing Company for tables on pp 168 and 174 of 'The encyclopedia of soil science' by R W Fairbridge and C W Finkl, 1979; the International Institute for Land Reclamation and Improvement for the figure on p 6 of 'The auger hole method' by W F J Van Beers, Fourth Edition 1976, and the extracts on hydraulic conductivity in Annex B, taken from Volume III of 'Drainage principles and applications', 1974; Iowa State University Press for the table on pp 204-205 of 'Soil genesis and classification' by S W Buol et al, Second Edition 1980; the MacMillan Publishing Co Inc for the figure on p 348 of 'The nature and properties of soil' by N C Brady, 1974; McGraw Hill Book Company for tables on pp 56-57, 83 and 136 of 'Soil physics' by H Kohnke, 1968; the National Academy of Science Washington, for the tables on pp 21-24 of 'Soils of the humid tropics' edited by M Drosdoff, 1972, and also for the Environmental Studies Board data quoted on p 70 of FAO Soils Bulletin 42; Oxford University Press Ltd and Blackwell Scientific Publications Ltd for the tables on pp 137 and 138 of 'Charge distribution and the cation exchange capacity of an iron-rich kaolinitic soil' by R G Barber and D L Rowell, 1972; Rothamsted Experimental Station for the table on p 57 of 'Soil survey field handbook' edited by J M Hodgson, 1974, the land use capability class definitions from 'Land use capability classification' by J S Bibby and D Mackney, 1969, and the soil drainage classes and soil profile diagram from 'Soils in Norfolk, Sheet TM49 (Beccles North)' by W M Corbett and W Tatler, 1970; Sceptre Books Ltd, London for the figure on p 9 of 'The Cambridge encyclopaedia of earth sciences' by D G Smith, 1981; the Soil and Irrigation Research Institute, Pretoria, for the table and figures on pp 21-23 of 'The influence of initial moisture content on field measured infiltration rates' by D P Turner and M E Sumner, 1978; the Soil Bureau of New Zealand for tables on p 173 of 'Methods of chemical analysis for soil survey samples' by A J Metson, 1961; the Soil Conservation Society of America for the soil erodibility nomograph from 'A soil erodibility nomograph for farmland and construction sites' by W H Wischmeier et al, 1971; the Soil Science Society of America for the tables on pp 281-282, 313 and 332 of 'Micronutrients in agriculture' edited by J J Mortvedt et al, 1972; Springer-Verlag in Heidelberg and A P A Vink for the table on pp 148-149 of 'Land use in advancing agriculture', 1975; Times Books and John Bartholomew and Son Ltd for the map on p 24 of the Times Atlas, Sixth Comprehensive Edition, 1980; University Tutorial Press for excerpts from pp 86-94 of 'Climate, soils and vegetation' by D C Money, 1974; Alan Wild, editor of the Journal of Soil Science, for reproduction of its guide to SI units and SI related units; John Wiley and Sons (New York) Inc for the figures on pp 56, 63 and 290 of 'Remote sensing and image interpretation' by T M Lillesand and R W Kiefer, 1979, and for tables on pp 56, 58-60 and 69 of 'Properties and management of soils in the tropics' by P A Sanchez, 1976; Williams and Wilkins for excerpts from 'Lime in relation to availability of plant nutrients' by E Truog, in Soil Science 65, 1948, and with F T Bingham, for the table on p 94 of 'Chemical tests for available phosphorus' in Soil Science 94, 1962; and Carl Zeiss (Oberkochen) Ltd for their stereo-vision test. Whilst we have attempted to identify the sources of all the material quoted in this manual, we have been unable to trace some copyright holders, and should be grateful for any information that would enable us to do so.

Chapter 1

Introduction

1.1   Aims and scope of the manual

This manual stems from many years' work by BAI on agricultural
consultancy and management assignments and reflects the need felt
by staff members for a concise, comprehensive – and above all
readable – summary of soil-related methods and terminology, with
clear guidelines for interpretation of soil and land evaluation
data used in project design, costing and management. The manual
therefore has two main aims: it is intended firstly to be a
practical source-book for soil scientists in the field, and
secondly to be a reference work on soil and related studies for
members of other disciplines who require information for purposes
such as proposal compilation, project planning, survey
implementation and practical interpretation of soil and land
capability reports. It is not intended as a replacement for the
standard soil survey manuals, but rather as an expansion of
particular topics of most immediate use for consultancy assignments
in the tropics and subtropics. The contents also reflect the
working practices of BAI, and are thus fairly strictly confined to
those aspects of development studies handled by soil scientists as
members of multidisciplinary teams. The manual does not,
therefore, cover the wider aspects of multidisciplinary investigations, which are treated in, for example, ILACO (1981),
Shaxson et al (1977) and Ministère de la Coopération (1974).

1.2   Selection of contents

The contents of the manual are not intended to form a 'balanced'
work on soils and land evaluation. Instead, the main emphasis is
placed on those items which receive scant attention in published
works (such as numerical values for critical levels of soil
constituents) and on those which are treated in too many sources
and need amalgamation to be of practical value (such as methods in
soil physics). The presentation also reflects the preferences of
BAI soil scientists in the level of detail accorded to different
systems and methods. In most cases a recommended practice is
covered in some depth, whereas less suitable or less convenient
methods are only summarised, although full references have normally
been included. Items such as soil morphological descriptions are
treated only superficially in the manual, since they are dealt with
more fully in well-known references that would usually be available
on site during projects.

Other inclusions or omissions arise as a result of the division of responsibilities between specialists of different disciplines within BAI, or are a reflection of the range of projects that the company normally undertakes; specific omissions, with some suggested references, are as follows:

Quantitative and numerical methods, including soil information systems, automated mapping, spatial and statistical analyses
(de Gruijter, 1977; Webster, 1977; Burrough, 1982, 1986)

Land surveying for soil studies
(Wright, 1982)

Soil and land evaluation for tree crops or forestry
(Chan Huen Yin, 1980; Pritchett, 1979; World Bank, 1980; Laban, 1981)

Soil pollution
(D'Itri, 1977; Harmsen, 1977; HMSO, 1979; Davies, 1980)

Soil engineering
(Terzaghi and Peck, 1948; Jumikis, 1967; Lambe and Whitman, 1969; USBR, 1974; BSI, 1981; Brink et al, 1982)

Economic evaluation
(Bergman and Boussard, 1976; Irvin, 1978; Gittinger, 1981)

Climatic analysis
(Troll, 1966; Money, 1974; Doorenbos and Pruitt, 1977)

Hydrological studies
(FAO series of Irrigation and Drainage Papers)

Soil biology and microbiology
(Kononova, 1966; Phillipson, 1971; Wallwork, 1976)

Ecology and environmental impact
(Odum, 1964; Walter, 1971; Golley and Medina, 1975; Unesco series of Man and the Biosphere Papers)

Land management techniques
(Hudson, 1975; Schwab et al, 1966; Greenland and Lal, 1977; in Africa Allan, 1965 and Upton, 1973)

Wherever possible the subjects covered have been quantified and numerous rules of thumb are included in the manual. The numerical values are mostly for use at the prefeasibility or feasibility levels where other data, and particularly crop trial data, are lacking. Most of the values are described as 'tentative' or 'indicative', since it would be impossible in a book of this size to summarise all the effects and interactions, even if they were all known. The figures do, however, serve a useful purpose in helping to warn of potential problems which could appear during project implementation, and which should therefore be allowed for in project designs and costings. There is a certain amount of repetition in the book, and frequent cross-references are included; this is deliberate, to facilitate use of the manual as a reference book for specific topics.

1.3   Layout of the manual

The main text of the manual is aimed primarily at the non-specialist land resource assessor, and the Annexes at the specialist soil scientist. The main text is compiled roughly in the order the information is needed in most consultancy projects, but with general aspects preceding the more detailed and specialised sections, a format which is paralleled within individual chapters. Discussions of survey planning and organisation thus form the opening sections followed by an outline of soil classification and land evaluation. These are followed in turn by chapters on specific soil physical and chemical measurements and their interpretation, before a final chapter on reports and maps. Great emphasis is placed on the collations of physics and chemistry interpretive data (Chapters 6 and 7 take up 100 pages, more than half of the manual's main text) because the BAI experience is that these topics tend to be scattered amongst references and sources, or treated cursorily or from viewpoints other than the soil surveyor's in the general textbooks.

The set of 15 Annexes contain more specialist information, mainly of interest to practising soil scientists; they include numerous checklists and worked examples for planning and execution of soil surveys and related studies.

1.4   The need for soil surveys

The importance of soil and land suitability studies for agricultural development projects has long been recognised by their proponents, such as soil surveyors and land use planners, but even today many workers in other disciplines still tend to regard such studies as being, at best, of only peripheral value to their work. All too often only a fraction of the potentially important information contained in a soil report is used, the remainder being considered too academic, too riddled with incomprehensible jargon or too remote from the practical decisions and actions which are needed for successful development and management of agricultural enterprises. In general, this under-use appears to be the result of a 'communications gap' between producers and users of soil reports, since the importance and relevance of soil-based studies have often been stressed. Storie (1964), for example, quoted in FAO Soil Bulletin No 42 (1979a), lists 11 major areas in which soil and land evaluation studies can contribute significantly to irrigation developments. In most projects, soil and related studies should form an indispensable part of the basic planning process (see Chapter 2), and without them very costly mistakes can be, and have been, made.

All too often the perceived expense of soil survey and land evaluation means that such work is skimped or even omitted from project planning, but the costs need to be seen in relation to the overall development of a project. Young (1973) indicates that a survey can pay its way if it prevents planting of unsuitable land

3

over as little as 10% of a project area, and Nieuwenhuis (1975) suggests that soil surveys typically take up only 10 to 20% of planning costs which themselves form only about 10% of total project development costs. As a rough guide to the quantities involved, the figures of Dent and Young (1981) are illustrative: reconnaissance field surveys can be made relatively cheaply at costs of the order of £10 km^{-2}, and semi-detailed surveys for about 5 to 10 times that, at £50 to £100 km^{-2}. Depending on soil complexity and the intensity of field-work required, more detailed surveys down to scales of about 1:25 000 can be about three times more expensive, and 1:10 000 work some three to four times more expensive still, with costs of about £500 to £1 000 km^{-2}.

Using again the example of Dent and Young (1981) with slightly amended figures, a survey for rainfed development at 1:25 000 could cost about £2 ha^{-1}, an area with a yield potential of some 2 t of grain valued at £200 on the world market. Assuming the effects of the survey raised net gains by a modest 5%, the produce value would increase by about £10 ha^{-1}, or five times the survey cost, in a single year. In appropriate circumstances, therefore, soil surveys can be highly cost effective, provided a careful choice is made of scale and intensity relative to the development envisaged.

Chapter 2

Types of Land Resource Field Studies

2.1 <u>Introduction</u>

The types of land resource surveys undertaken by soil surveyors attached to commercial companies can be categorised under one or more of the following headings:

a) Exploratory;

b) Reconnaissance;

c) Semi-detailed;

d) Detailed.

All of these terms are in common use, although there are wide differences in the definitions of the scales and intensities included under the individual headings. To avoid confusion, therefore, the purpose, intensity and scale of a survey should always be stated (see Tables 2.1 and 2.2). The descriptions that follow are particularly relevant to the projects undertaken by BAI; general definitions are given in FAO (1979a, p 88).

The more intensive the survey, the more detailed are the mapping units, the more direct are the measurements of the parameters mapped and, normally, the more specific is the purpose of the survey. The use of aerial photos or satellite imagery is correspondingly reduced with increasing scale and intensity of survey.

In small-scale surveys, and where soil/landform/vegetation relationships can be established, a soil surveyor can use free survey techniques, which involve the location of representative sites based on his own professional judgement. At larger scales, grid surveys are usually preferable, making the sampling more objective and, if the grid is randomly aligned, capable of statistical interpretation.

2.2 <u>Exploratory surveys</u>

The purpose of an exploratory survey is to obtain a rapid general appraisal of an area. Normally this is either to determine what further studies are required or to locate suitable sites for a

Types of land resource field studies

Summary of types of land resource surveys

Table 2.1

Type of survey [1]	Nearest FAO equivalent nomenclature and final map scale [2]	Aim and level	Site intensity and survey method	Approximate proportion of time input (%)			Preferred scales	
				API	Literature	Field work and sampling	Aerial photos	Final maps [3]
Exploratory	Exploratory to low intensity ≤ 1:1 000 000 to 1:100 000	Resource inventory Project location Prefeasibility	Free survey of variable intensity usually much < 1 per 100 ha	60	20 (Probable averages, very variable)	20	≤ 1:60 000 ≤ 1:100 000	Variable
Reconnaissance	Medium intensity 1:100 000 to 1:25 000	Prefeasibility Regional planning Project location	Free survey of variable intensity usually < 1 per 100 ha	50	25	25	1:40 000 to 1:20 000	≤ 1:50 000
Semi-detailed	High intensity 1:25 000 to 1:10 000	Feasibility Development planning	Flexible or rigid grid. Intensity 1 per 15 to 50 ha	20	20	60	1:25 000 to 1:10 000	1:25 000 to 1:10 000
Detailed	Very high intensity > 1:10 000	Development Management Special purpose	Rigid grid. 1 per 1 to 25 ha	5	20	75	1:10 000 to 1:5 000	1:10 000 to 1:5 000

Notes:
[1] These terms are loosely used for a wide variety of intensities and final map scales: see Young (1973), Stobbs (1970) and Western (1978, Chapter 3).
[2] See FAO (1979a, p 88).
[3] For many integrated projects the final map scale may be chosen to conform to civil engineering or project development requirements, rather than to the most appropriate scale for the survey intensity and complexity of the soil pattern (see Subsection 9.5.1).

Summary of mapping units used in land resource surveys

Table 2.2

Types of survey	Final mapping unit	Landscape components 1/			
		Geomorphology	Soil	Vegetation	Land use
Exploratory	1. Physiographic units/ land systems 2. Potential development areas	Major relief units	Orders 2/ to associations	Soil/climate-related types	Agro-ecological groups
Reconnaissance	1. Physiographic units/ land systems 2. Soil associations 3. Land capability units 4. Potential development areas	Relief units, major landforms	Associations	Soil/climate-related types, plant associations	Land use systems, cultivation density
Semi-detailed	1. Geomorphic units 2. Soil series/ associations 3. Land suitability/ management classes 4. Major constraints or development parameters	Detailed landforms and elements, slope units	Series, complexes or associations; soil phases and selected parameters	Plant associations and distribution	Land use and farming systems, specific parameters, cropping patterns
Detailed	1. Soil phases and/or land parameters 2. Land management units	Slope units	Soil phases and selected parameters	Specific crop or natural vegetation variables related to soil parameters – eg drainage or salinity effects	

Notes: 1/ After Baulkwill (1972).
       2/ ie the highest level of soil classification (see Annex C); not to be confused with the USDA term 'order of soil survey' which refers to the kind of survey (see Orvedal, 1977).

specified development.   The soil and land capability inputs to an exploratory survey will usually, therefore, form part of a broadly based study at the prefeasibility level.

The intensity, scale and method of operation in an exploratory survey vary greatly, depending on the specified development, the nature of the terrain, and the data already available.   Because large areas of land have to be examined in limited time, the surveyor must concentrate on those constraints and beneficial properties which appear critical for selection, deletion and/or further assessment of the land, making the fullest use of all the available data.   Aspects of pedological interest should only be pursued if they have direct relevance to the aims of the survey.

2.3   Reconnaissance surveys

Reconnaissance surveys are usually mounted at the prefeasibility or feasibility stage for one of two purposes:

a)   to locate and give initial assessments of areas suitable for more intensive studies; or

b)   to locate and investigate areas which are unlikely to justify the costs of more detailed studies, such as those suitable only for non-intensive agriculture, or rough grazing.

The scale and intensity of a survey will depend on the variability of the soils and landforms, and on the proposed land uses. Maximum use should be made of aerial photographic interpretation (API), but field investigations should also be as intensive as possible, since cost will often preclude any follow-up survey being undertaken.

Where reconnaissance surveys are designed to locate areas for further study, detailed definition and delineation of the selected areas are not required, since these will be covered by the later work.   A broad classification scheme should be used to summarise as many land qualities as possible, subject to the range of envisaged land uses and the costs of providing the relevant data.

An important aspect of all reconnaissance surveys is the identification of major development constraints and the deletion of unsuitable areas (eg those suffering from rock outcrops, severe flooding, excessive relief etc).   This is a comparatively rapid process, often only requiring API, and further time should not be wasted in detailed characterisation of such unsuitable land.

In most areas physiographic differentiation is sufficient to distinguish major variations in soils and land capability at a reconnaissance scale.   The most effective field approach therefore usually involves API (ideally at scales of about 1:20 000 to 1:40 000), followed by free survey investigations to characterise the geomorphology, land use patterns and soil associations/complexes of each physiographic unit.   The investigations should normally include check-line transects across the larger mapped units.   Grid surveys will only be required in those areas of exceptionally high soil variability that cannot be correlated

8

with changes in visible features such as landform or vegetation (see Subsection 3.5.1).

Almost all the mapped soil or land capability boundaries at this survey level will be drawn from API, and as much information as possible about each unit should be gathered from pre-existing data such as published articles, reports and research station records. A proportionally large part of the survey time must therefore be spent on collecting this information, much of which may well be available only outside the survey area.

Reconnaissance surveys are generally regarded as those with final map production at scales of about 1:50 000 or smaller. However, mapping scale is often dependent on the scale of available air photographs or topographic base maps, since additional provision of maps or photographs at different scales is seldom economic at this stage. As a result, a wide latitude in the relationship between inspection site intensity and map scale has to be accepted. The final map scale also depends to some extent on the scale of any more detailed maps to be produced later. In practice, reconnaissance map scales range from about 1:300 000 to 1:50 000. Since map scales can be readily altered, a more useful criterion for indicating a survey level is the intensity of inspection sites on which the mapped boundaries are based. Except in areas with very simple soil patterns, reconnaissance surveys can be taken to be those with an inspection site intensity of less than one per 100 ha (1 km^2), and are frequently of the order of one per 200 to 400 ha.

The mapped units normally consist of physiographic areas with correspondingly broad soil associations or complexes and land use patterns. Derivative maps showing priority development areas may also be presented.

2.4 Semi-detailed surveys

The purpose of surveys at this level is either to assist with feasibility assessments, or to contribute to the implementation of a development programme, or both.

Feasibility surveys are essentially concerned with the economic evaluation of development options. Thus the investigation of soil and land should be confined largely to those properties which determine development and operating costs, and statements about the 'technical feasibility' of an option will chiefly refer to the feasibility with respect to specified economic criteria.

Although semi-detailed surveys are mainly employed at the feasibility stage, they may also be valid for development planning, provided that the soils are simple to categorise and map. At the development level of investigation, the soil survey may be adequate not only to determine the precise nature and distribution of the land and soils, but also to establish an outline of the required land management techniques related to the mapping units.

Feasibility surveys should therefore be designed to distinguish variability of land and soils within major physiographic units.

9

For this purpose, interpretion of air photographs, preferably at a scale of about 1:20 000, should be complemented by a system of regular inspection sites (see Subsection 3.5.1).

Depending on the nature of the terrain, boundaries will in places be defined from aerial photographs and in others from field observations. The final accuracy or reliability of the boundaries should be indicated in the written report.

Map scales for feasibility surveys generally range from about 1:10 000 to 1:25 000 or occasionally 1:50 000, and site intensities are usually of the order of one site per 15 to 50 ha, depending on the complexity of the land and soil and the purpose of the survey. The scale may often be chosen to coincide with that used for the accompanying engineering or agricultural studies.

The maps produced at this level usually include:

a)   soil series/associations;

b)   land capability or suitability classes and subclasses.

Additional maps may be prepared of geomorphic units and of major development constraints or individual parameters (eg, salinity).

2.5   Detailed surveys

Detailed surveys are usually designed to provide land and soil data for use at the project development stage, although they also include very-high-intensity surveys to characterise soils on research farms, to identify specific soil features over a limited area or to investigate soil variability in small sample blocks.

Development surveys at this level of intensity are concerned with management of soil and land resources, usually in order to achieve optimum yields of specified crops. The investigations should therefore be directed towards the accurate definition and location of the characteristics relevant to project planning and implementation. The finer subdivisions of a natural soil classification may be valid in such a survey if they can be used to predict some useful crop behaviour or land management practice. However, more important is the establishment of distinct 'management units'. Land suitability classifications based on, say, repayment capacity (see Chapter 5) are not applicable at this level; for example, Land Class 3 in the USBR system may include both sands and clays and, whilst their repayment capacities may be similar, their management is not.

A detailed survey should be implemented on a grid system (see Subsection 3.5.1), although full use should also be made of those vegetation and topographic features that are related to soil and land management characteristics. API is of more limited value, but may be included in the studies to assist, for example, with locating drainage lines or topographic features; scales as large as 1:5 000 may be used if available. Mapped boundaries, however, will be drawn mainly from the results of the field survey.

As increasingly large-scale surveys are employed, the variability of the parameters used to define soil and management units becomes more apparent. Thus, for example, detailed investigations within a given soil phase might indicate important differences in the depth of indurated subsurface horizons, the degree of aggregate stability in the topsoil, or the depth of the GWT. All of these characteristics could be crucial for project development, but their incorporation into soil-phase definitions would inevitably lead to highly complex maps which would be difficult to interpret and cumbersome to use. The information at this level of survey is, therefore, best presented in a series of single parameter maps. These parameters may include directly observable phenomena such as soil depth or subsoil permeability, or interpreted units such as optimum drain spacing, or restrictions on irrigation methods.

Final mapping scales of detailed surveys vary from about 1:15 000 to 1:5 000. The intensity of inspection sites usually varies between one per 1 ha and one per 15 ha. This can be much greater in very-high-intensity specific-purpose surveys, but these are rare and have not been considered here.

2.6   The role of land resource surveys in integrated projects

2.6.1   Interaction with other disciplines   Land resource surveys are increasingly being implemented as part of integrated projects involving engineers, agriculturalists, ecologists, economists, sociologists and extension workers. It is therefore vital on these projects that a soil surveyor should organise his work as part of a team exercise and not as an isolated entity. As far as possible, he should arrange for a free flow of information between himself and the representatives of other disciplines, both at the beginning of a project and as a continuing process as methods and objectives are clarified. In particular, other disciplines must be involved in discussions about quantitative limits (eg engineering or economic) to be placed on soil and land features. Since these are often not known until comparatively late in the survey programme, the soil surveyor must devise a flexible system of assessment in order to incorporate new or modified data as they become available.

One of the commmonest problems in an integrated survey is the organisation of the data required from, and by, the different members. The team leader is the key person in this liaison process, and he needs to clarify early in the study what information is required by whom, and when it will be needed. He must also ensure that each team member sees his work in the context of the overall study and appreciates the overlap and interactions of the different disciplines; some examples are illustrated in Table 2.3 and a general discussion is given in Russell (1966). As soon as possible during such a project the soil surveyor should find out exactly what he is expected to produce and by what deadline, and he should also make known his own requirements to those involved. In order to ensure that information is exchanged when required, the soil survey should be designed to fit into the overall project plan using, if possible, a logic flow diagram (see Section 3.3) to ensure that the inputs of different disciplines dovetail. A checklist of the main items of information on the

11

Types of land resource field studies

Simplified interdisciplinary relationships in feasibility studies 1/   Table 2.3

A. List of basic background data

    SOIL/LAND SUITABILITY

    Topographic/mapping details

    Climatic records

    Socio-economic data

    Hydrological data

    Existing agricultural data and practices

    Existing infrastructure

    Equipment and construction costs

B. Crop production estimates

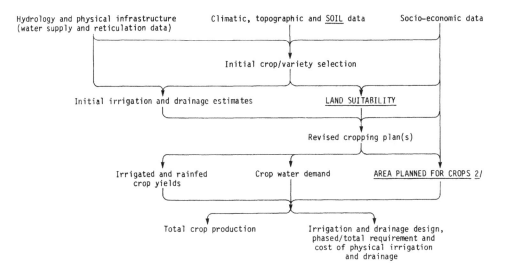

C. Land clearance and development

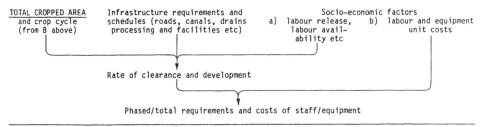

Notes: See page 13.   cont

D. Routine cultivation

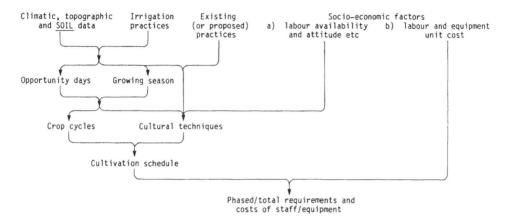

E. Harvesting and transport

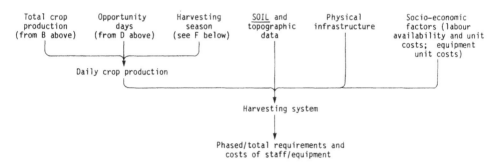

F. Crop processing (using sugarcane as an example)

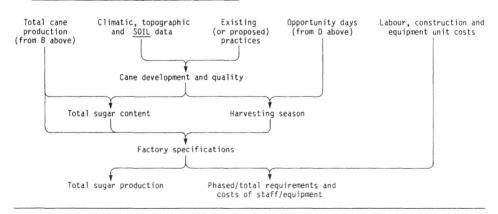

Notes: 1/   Items in CAPITALS underlined indicate main inputs from soil surveyors.
         2/   cf Figure 9.1.

Source: R A Yates (1981).

'interface' between soil survey and other disciplines is given in Table 2.4.

2.6.2   Data collection and interpretation   It should be noted that the soil surveyors (and topographic surveyors) are very often the only members of a team to make a systematic traverse of an area. They are, therefore, normally the best qualified to make generalisations about a range of features based on direct observations. Before the start of a survey, a soil surveyor should find out what additional field information he can usefully contribute to a project, subject to the time available. As a general rule, soil surveyors should always note information on:

a)   groundwater table (GWT) depths;

b)   crop patterns and quality;

c)   natural vegetation;

d)   ecology.

The information from soil and land suitability studies is normally required immediately for the design or implementation of integrated projects. This information is most efficiently transmitted if other team members are closely involved in the discussions to determine the critical soil and land properties. The actual interpretation and mapping should, however, be made by the soil surveyor, since only he is intimately acquainted with the full range of soil and land characteristics and can achieve the correct emphasis.

2.7   Limitations of land resource surveys

A number of misconceptions about the information that can be supplied by soil surveys are common amongst members of other disciplines, the main misunderstandings arising from ignorance of the level at which a survey is being conducted (see Gibbons, 1961 and Davidson, 1980). Amongst the more commonly misunderstood limitations are the following:

a)   Limitations of survey data - Soil physical data in relation to engineering aspects (eg infiltration rates, permeability, land grading requirements) can only be supplied by detailed soil, and/or topographic, surveys. Surveys at the feasibility or prefeasiblity levels can usually give ranges or estimates, but no more.

Standard soil survey results cannot be used for regional drainage designs, since the latter rely on much deeper soil observations than are usually made. Similarly, information on GWT levels is liable to be patchy, except where the GWT is encountered within auger depth.

Crop yields cannot be accurately predicted from soil survey results alone, although extrapolations from experience on similar soils nearby can form a good basis.

Major interdisciplinary overlaps involving soil surveyors in integrated studies

Table 2.4

	Information	Required by	Required from	Stage of project/degree of detail
1.	**Project planning information**			
a)	Background assumptions/scope of project	All	Terms of reference or decisions of project leader	Premobilisation stage. Project concept and emphases; delineation of individual responsibilities
2.	**Data for preliminary decision on location/extent of study area**			
a)	Land conditions (topography, rock outcrops, dissection)	All	Soil surveyors	Initial estimates – as soon as possible after start of field-work
b)	Soil conditions (depth, texture, distribution, microtopography)	Agronomists Field engineers	Soil surveyors	As 2a above
c)	Engineering considerations (flooding, topography)	All	Field engineers	As 2a above
d)	Crop requirements (climate, soils, economics)	All	Agronomists	As 2a above
e)	Limits of study area (if not defined in TOR)	All	All	As 2a above
3.	**Specific project details**			
a)	Location, extent of development areas and suitability classification	Agronomists Field engineers Economists	Soil surveyors	Revised estimates – throughout field-work. Final definition and measurements – as soon as possible after end of field-work
b)	Soil types and areas	As above	Soil surveyors	As 3a above, for yield calculations and layout design(s)
c)	Specific soil characteristics 1/	As above	Soil surveyors	As 3a above. Specific measurements required as soon as possible for engineering calculations, yield estimates etc
d)	Quantified limits to soil/land features 2/	Soil surveyors	Agronomists Field engineers	As soon as possible after (1) to start classification system
e)	Input costs for, and benefits of, development	Soil surveyors	Economists	a) Preliminary estimates for provisional land classification – as 3a above b) As soon as possible after end of field-work – final estimate for land classification (ie accurate enough for comparative purposes) c) Final figures – for report
4.	**Report and map preparation**			
a)	Standard project data: place names, base map, climatic data, project areas	All	Team leader (as co-ordinator)	As soon as possible
b)	Recommendations on soil and land management	Agronomists Field and civil engineers	Soil surveyors	Co-ordination meetings needed before report finalised

Notes: 1/ Such as available water capacity, infiltration rate, hydraulic conductivity, bulk density, depth, salinity and chemical data.

        2/ Such as soil texture and depth; land levelling requirements; available water content and infiltration rate as related to irrigation practices; soil nutrient/toxicity levels related to crop yields; land management criteria.

Subsequent management experience will almost certainly modify the conclusions of a soil survey. A specific management problem (eg a nutrient deficiency) may arise at a particular location, but such a deficiency cannot always be assumed to be coextensive with the soil unit on which it occurs. This is because:

i)   the accuracy of the soil survey may not be sufficient;

ii)   there may be considerable variation of the problem within the soil unit;

iii)   the soil properties used for mapping may not correspond to all those that subsequently prove to be significant in agriculture;

iv)   specific details may be lacking in the soil survey results.

The criteria selected for soil classification and mapping are inevitably confined to features that are:

i)   observable (directly or easily);

ii)   mappable, ie with distinct clustering distributions at the scale used.

For specific purposes, the most important criteria chosen for soil classification, especially 'general purpose' classification, may well not coincide with those that are critical for project development.

b)   <u>Limitations of soil maps</u> - Soil mapping units are rarely, if ever, completely uniform and, as noted by Young (1976, Chapter 18), the variability within a unit is seldom completely explicable in terms of a rational and ascertainable variation in influences on the soil. Conventionally, soil mapping units are supposed to be about 85% pure, but in practice the figure is likely to be about 50 to 65% (Ragg and Henderson, 1980). Even in areas of comparatively simple, uniform soils the purity of mapped series, in terms of observed morphology, can vary between about 65 and 90% (BAI, 1979), and the use of soil associations can reduce the chances of predicting a soil at a given site to values well below 50% (McRae and Burnham, 1981).

Much of the total variability in soil properties over a large region can occur within areas as small as 1 ha, or even 10 m^2, for which the corresponding coefficients of variation of individual soil properties are rarely less than 20%, and sometimes are as high as 70%. Variation tends to be highest for chemical properties and lowest for physical properties, and these variations can even be found within soils mapped as a single series (see Beckett and Webster, 1971; Nielsen et al, 1973; Topp et al, 1980; Areola, 1982). It should be noted, however, that measured values reflect both variations inherent in the soil and those ascribable to the methods of measurement. Further aspects of variability are discussed in Subsection 6.1.2 and Section 7.3.

Soil boundaries are sometimes inferred by API, are sometimes determined by ground observation and are sometimes merely drawn between observation sites. The accuracy and significance of boundaries can therefore be highly variable, even on a single map, since some boundaries represent narrow zones of rapid soil change between uniform areas, whilst others merely approximate to the mid-points of gradual changes.

Finally, it must be remembered that soil management techniques are also critical; based on their studies in Malawi, Young and Goldsmith (1977) quote a fivefold difference in some crop yields due to management effects, whereas changes in soil types accounted for only twofold or threefold yield variations amongst the same sampled plots (farmers' fields) in the same harvest seasons.

Chapter 3

Survey Organisation and Practice

3.1   Terms of reference

The first reaction of a soil surveyor upon receiving a survey brief
must be to examine the operational details of the study in relation
to the information which the client requires.   It is the
surveyor's responsibility to ensure, before commencing a survey,
that the programme of work he is about to carry out will be
sufficient to conform to the final terms of reference (TOR) agreed
with the client.   The TOR should also be critically examined in
order to ensure that they will lead to the kind of information the
client needs, and that any ambiguities or impossibilities are
resolved.

In some cases the surveyor will be involved at the inception of the
project and, in discussion with colleagues of other disciplines,
will be able to indicate both the nature and intensity of survey
best employed, and the likely time required.   This is the most
satisfactory approach, and should be followed whenever possible.
On other occasions, however, a surveyor may only be involved after
a contract has been signed and a fixed input for soil studies
agreed.   In these circumstances, he must of necessity tailor the
survey to fit the time constraints, but at the same time he must
clearly indicate to the Project Director what can, and cannot, be
achieved in the time available.   If he fails to do this at the
onset of the work, he can blame no one but himself if he is then
held responsible in the event that the soil report is inadequate
for the purposes of the project.

A selection of checklists involved in survey organisation are
presented in Annex A, and further information is contained in
Mohrmann (1966) and Western (1978);   comments on compilation of
proposals for land resource surveys are given in Annex E.   Note
that when translation is required a good technical dictionary of
agricultural or soil terms is essential;   some of the more common
languages are covered in Haensch and Haberkamp (1975) and
Jacks et al (1960).

3.2   Background data

Often the best source of background information is the client,
particularly if the project is for a government department.
Background literature relating to soil, vegetation and proposed

land use for each project should always be carefully studied, wherever possible including data from the area itself or at least from local investigation sites. Bibliographic searches in libraries specialising in the area/region concerned are often worth while, as is consultation of specialist libraries – for example, that at the Land Resources Development Centre (LRDC) of the Ministry of Overseas Development, and the Directorate of Overseas Surveys (DOS) should both be consulted. In some cases a literature search, such as that offered by the Commonwealth Agricultural Bureaux, may be justified. General background books on pedology and soil properties include Bridges and Davidson (1982), Brady (1974), Bridges (1978), Russell (1973) and Duchaufour (1982); tropical soils are discussed in Buringh (1979). Informal contact with other private consultants, including topographic surveyors, civil engineers, foresters and agriculturists, may also prove useful.

The main items of background data may be summarised under the following headings:

a) Air photography – If available, what area is covered? What are the scale and date? Do mosaics exist? Colour, false colour, side-look radar, and Landsat satellite imagery should all be checked. DOS is a good starting point. Clients often underestimate the importance of aerial photography – and the need for it to be available as soon as the survey starts.

b) Topographic maps – Again, scale and date required; also the contour interval and reliability of information on roads and tracks. Note any areas with restricted, or prohibited, access.

c) Previous soil studies – Both in the general region of the project and any broad land resource inventories, descriptions of geomorphology, land use or natural vegetation. Any national or regional classification systems. Climatic data, usually atmospheric only but sometimes soil temperature records too.

d) Geological maps and reports – Less vital where the soils are developed on alluvial sediments, but still useful. Mineral or oil companies may have useful data.

e) Crop yield and land management data – Both locally and from areas with similar climate or soils.

3.3   Survey logistics and planning

After determining the nature and intensity of the survey (see Chapter 2), the surveyor should make estimates of his on-site requirements for labour, transport, local equipment, office space, and accommodation. These estimates should be submitted to the Project Manager as soon as possible, in order to facilitate overall project planning. The main items required are summarised below; detailed checklists are given in Annex A.

a) Equipment – Should be sufficient for independent survey work by each surveyor, plus some spares. If possible, local equipment availability and costs should be obtained in advance to determine which items are to be shipped out with the UK surveyors, and which purchased on site.

b) Labour – Estimates should be based on the following:

   i) amount of trace cutting and density of bush;

   ii) necessity for trace cutters to remain ahead of surveyor's programme;

   iii) need to measure and lay out transect lines;

   iv) assistance required to auger and/or dig pits;

   v) assistance required in a field laboratory;

   vi) persons required to assist with field tests.

Whenever possible the estimated labour requirement should be given on a weekly basis rather than as a gross figure for man-days over the field period, since labour inputs are likely to fluctuate widely at different stages of the survey.

Working rates will vary greatly from region to region; examples from BAI experience are summarised in Table 3.1.

Approximate working rates on different terrain types   Table 3.1

Project location	Type of land/land use and soil complexity	Vegetation clearance required (and daily length of trace cut by 6-man team)	No of 150 cm auger holes per team in working day (and ha covered)
Coastal lowlands, Ecuador	Fairly flat, banana plantations/regrowth thicket. Average soils	Machete transects cut concurrently on 300 m grid (1 to 1.5 km)	9 (approx 80 ha)
Sudan savanna, Nigeria	Flat, open cultivated parkland. Simple soils	No clearance needed 500 m grid (4 to 6 km in grassland)	16 (approx 400 ha)
Juba River, Somalia	Fairly flat, open bushland	Bulldozed and hand traces pre-cut on 250 m grid (2 to 3 km)	20 (approx 112 ha)
San Pedro, Paraguay	Primary forest	As Somalia above, but 500 m grid (1 to 1.5 km)	10 (approx 250 ha)

Note: Survey timings are also highly dependent on the means of access to the survey area; time must be allowed in project planning for travel between the base and the survey sites, including individual sectors of the journey by air, road, river or foot, as appropriate.

c) <u>Access and transport</u> - Transport should be sufficient for the following:

   i)   the surveyor;

   ii) the labour;

   iii) additional equipment (such as 50 gal drums for in-filtration measurements).

On rare occasions one vehicle may be sufficient for all three purposes. More commonly two or more vehicles are required from time to time, normally 4WD. A tractor and trailer may be needed for heavy loads (eg several water drums).

Access should be checked as early as possible; traces may need clearance by bulldozer or by hand cutting.

d) <u>Local equipment</u> - As a rule, this is a very small item in the overall cost of the survey. Normally included under this heading are picks, shovels, auger handles, machetes, ranging poles (locally cut), rope for the chain-men, infiltration rings and various other small pieces of equipment. Local equipment needs to be mentioned in as much detail as possible at the planning stage, so that local funds can be allocated before the arrival of the surveyor.

e) <u>Office space</u> - Sufficient space should be requested for a field laboratory and soil-drying area, if they are required; storage space for equipment and tools, and desk facilities for written work, map production and API.

f) <u>Timing of survey</u> - Allowance should be made for adverse climatic conditions during rainy and/or hot seasons (indications of these are given in good atlases - eg Times Books, 1980) and for periods of local holidays. In Muslim countries field-work should be avoided or at least kept to a minimum during the month of Ramadan, when labourers cannot be expected to do a full day's work. For the months affected up to the year AD 2000, see Freeman-Grenville (1963).

Planning is particularly important in order to obtain the optimum balance between various activities and to ensure that the work is completed within the time and economic budget allowed. This is best achieved by the construction of a critical path network diagram and/or a bar-chart before any active involvement in the field. Using a network presentation the critical times (eg for sample collection, production of draft maps and liaison with other team members) can be easily determined; an example is shown in Figure 3.1. Further discussions of planning, and the use of systems analysis, are set out in Beek (1978).

3.4   <u>General guidelines for survey planning</u>

The types of survey in which BAI is commonly involved are given in Table 3.2, together with optimum inspection intensities, office and field activities and normal rates of progress.

21

Indicative flow diagram for soil survey planning

Figure 3.1

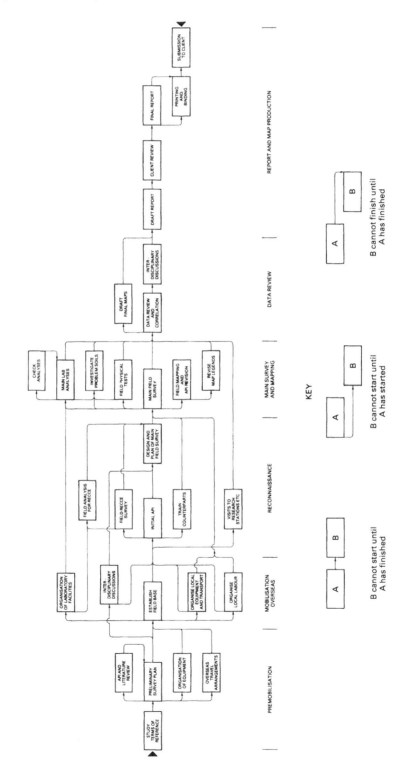

Time inputs for different types of survey

Table 3.2

Nature of survey	Degree of difficulty 1/	A Pre-field-work (days)	B Intensity (ha/site)	C Auger sites/day	D Auger survey	E Pits and sampling	F Field tests	G Field lab	H Office time	I Contingency	J Area/man-month 2/ (ha)	K Report time (weeks)	L Contingency % added
Exploratory survey	Unknown	3-10	Will vary widely	Will vary widely	Together 15		-	-	5 +	2	Will vary widely	2-4	10
Reconnaissance survey for prefeasibility objective	Simple Average Complex	3 5 8	400 200 100	8 9 10	15 15 15	2 2 2	- - -	- - -	3 3 3	2 2 2	48 000 27 000 15 000	2-4 2-5 3-6	10 10 10
Semi-detailed survey for feasibility objective	Simple Average Complex	3 5 10	100 50 25	10 11 12	14 12 11	2 1.5 1	1 3.5 5	- - -	3 3 3	2 2 2	14 000 6 600 3 300	4-7 4-8 5-10	12.5 12.5 12.5
Detailed survey for development objective	Simple Average Complex	5 8 12	25 10 4 or less	12 14 15	11.5 9.5 7.5	1 1 1	3.5 4 6	- 1.5 1.5	4 4 4	2 2 2	3 450 1 330 450	6-9 6-10 6-12 +	15 15 15

Field activities schedule (days per 22 day man-month)

Notes:
A) Will vary depending on data available; if two or more surveyors used, the time may well exceed that shown.
B) An appropriate value based on field experience.
C) Note that this rate should normally permit some office work each day.
D) Note that basic auger survey drops from 70% of allocated time on prefeasibility study to less than 35% on a complex development project; the auger survey speed controls the overall rate of progress.
E) This refers to representative pits only; others may be included as part of the auger survey.
F) Chiefly infiltration and hydraulic conductivty, bulk density.
G) For electrical conductivity and pH.
H) Note the proportion in an exploratory survey for collecting and reviewing previous data. Draft maps should be produced during the survey, not after.
I) To include some visits outside the area to institutes etc.
J) An approximate average only – particularly at the reconnaissance level.
K) Presumes the completion of draft maps in the field. Not valid for prolonged surveys with > 3 or 4 months' field-work.
L) Add on more contingency if local conditions are likely to be very poor, access very difficult, or rains/floods/sandstorms likely to be encountered.

1/ Definition of soil complexity
a) Simple – Generally older landscapes where the soils have been developed directly from the underlying rocks. The designation presupposes that the geology is itself simple or straightforward. This group also includes soils developed from uniform aeolian or marine sediments. Examples of simple soil distributions include many northern Nigerian soils, the Sudan Gezira and many of the older desert soils.
b) Complex – Flood plain soils, except where they are very extensive and uniform; many recent alluvial sediments and soils with particular salinity and/or drainage problems.
    Examples include the levee soils in the Juba Valley in Somalia, wadi soils at some sites in Saudi Arabia and soils of the Southern Ghors in Jordan.
c) Average – Soils which fall into neither of the categories above. For example, most of the Guyana front-land soils, the basin soils in Somalia and those of the southern coast of Ecuador.

2/ See Table 2.1.

From Table 3.2 a broad estimate can be made of the total time likely to be required for a survey of a specific kind over a given area. Assuming UK work inputs of five days per week, and 4.3 weeks per month overseas, the following empirical formula can be used to estimate the overall timing:

$$\text{Soil survey input (man--weeks)} = \frac{A}{5} + K + \frac{4.3 \times \text{area in ha}}{J} \times \frac{(100 + L)}{100}$$

(See Table 3.2 for key to symbols).

Worked example

A feasibility study is required for an area of 10 000 ha; the soils are of average complexity.

From Table 3.2   A = 5
                            J = 6 600
                            K = 4 to 8 (say 6)
                            L = 12.5

From equation above:

$$\text{Time required} = \frac{5}{5} + 6 + \frac{(10\ 000 \times 4.3)}{6\ 600} \times \frac{112.5}{100}$$

$$= 15.2 \text{ man--weeks}$$

Or for a two-man team = 7.6 weeks

Values obtained using this formula must be treated with caution, but they should represent reasonable order-of-magnitude estimates of the input times required.

3.5   Survey techniques

3.5.1   Survey methods   Three principal methods of survey, as described by Beckett (1971) and Young (1976), may be summarised as being:

a)   General-purpose API survey – Used for low-intensity surveys in which mapped boundaries (soils or land resources) are largely or entirely inferred from API with a free survey at low intensities of field observations to characterise physiographic units. Compound mapping units (land systems, soil associations) are common.

b)   General-purpose free survey – This method is employed for medium-intensity surveys using API, but also with a relatively high intensity of field observations. These observations are again sited by judgement to sample representative areas.

c)   Special purpose grid survey – In this method, individual soil properties are recorded on a grid pattern and may be mapped parametrically. Observation points may commonly be located on an elongated grid sited to cross the 'grain' of the land. Such systems, with observations along each line spaced more closely than the transects themselves, is particularly suited to land with readily identifiable soil catenae, or where dense vegetation makes trace cutting difficult. However, where soil variation is more random as, for example, in recent alluvial sediments, a regular grid becomes more efficient. A formal grid is best laid down at an intensity slightly less that the calculated optimum, since insertion of additional sites is invariably required to give more accurate information in boundary areas. Grid surveys may also be used where there is too little surface evidence and/or too few landmarks to allow free survey of representative areas.

3.5.2   Sequence of survey operations   A number of activities that are related to survey techniques have already been discussed, in particular the different advantages of grid and free survey and the importance of relating soils as far as possible to landforms, vegetation or topography (see Chapter 2). Further practical details are given in, for example, Handbook 18 (USDA, 1951) or Hodgson (1974, 1978).

The following general plan is recommended for most soil and land surveys at whatever level (see also Figure 3.1).

a)   Initial reconnaissance - To identify the major survey problems, establish methods of working and to familiarise the team with the area and form a consensus on priorities, special features etc in the light of survey purpose.

The reconnaissance should include: preliminary identification of major soil/landform/vegetation relationships, and the study of all relevant existing data.

b)   Main survey - To establish distribution of soil and land units. Must be flexible to accommodate changes in definitions based on additional information produced during survey; preliminary classification and mapping to proceed concurrently.

c)   Sampling - To sample soils of major mapped areas for characterisation (see Section 3.7).

d)   Consolidation - 'Filling in' in areas of complex soils or difficult boundaries; testing of mapped areas. Revisions of API; incorporation of soil analytical results; final classification.

e)   Reporting - Production of final report and accompanying maps and diagrams.

3.5.3   Practical hints   Much survey technique is a question of experience and may be acquired to a reasonable level quite rapidly. However, there are a number of points that are often forgotten, even by experienced surveyors. These include:

a)   Knowing site locations - As far as basic survey work is concerned, it is usually pointless to collect data if you are unsure of your location. This is so simple and self-evident that everybody forgets it at some time or other.

b) <u>Using soil exposures</u> – Road cuts, ditches, wells, sawpits and construction sites often have exposed soil and should be checked. There is no point in augering a site if you have a clean dry ditch only a few metres away; second-hand profiles are very useful for soil observations, provided that the soil is not too disturbed and that the exposed surface is cleaned before description. Beware of disturbed and artificial profiles along main roads, railways, power-line cuts etc; these soils often occupy a width of several tens of metres.

c) <u>Using a flexible approach</u> – Do not settle on one method/intensity/transect spacing and apply it rigidly to the whole of a varied area. Some parts may well need a different approach, with higher or lower intensity.

d) <u>Careful scheduling of field-work</u> – Spend an adequate time, preferably one day per week on a long job, in the office keeping up to date with maps and report drafts. Workloads should, as far as possible, be spread evenly over the project time span or the quality of the work will decline.

Large survey areas can be split into blocks for separate surveying and sampling. This helps to break up the work and ensures a steady flow of samples to the laboratory, with constant feedback of results.

Do as much office work as possible concurrently with field survey; keep maps up to date, incorporate field and laboratory data as available and write up sample site descriptions ready for final report.

e) <u>Checking mapping unit variability</u> – Mapping units can vary considerably in the uniformity of their features. Variability should be quantified by detailed survey in small sample areas, preferably sited by a statistically random procedure.

The accuracy of mapping unit boundaries should also be estimated, by using check transects on all feasibility and development level surveys. Boundaries identified both by API and by field-work should be tested.

f) <u>Double checking survey data in 'public' places</u> – Survey data must be correct at sites likely to be used subsequently as check or reference points. Take extra care at major sampling sites, and make sure the data are right where mapped boundaries cross known sites or easy access points, because these places are where the information will be checked. Arguments about 'intrinsic variability' will not convince a non-technical client if he finds out you cannot even get it right when next to the local highway.

g) <u>Allowing correlation time</u> – On-site discussions between surveyors about soil and land categories are essential to ensure survey uniformity. Polaroid photos are very useful for recording land and vegetation types for comparison between team members.

h) Consulting other team members - Ensure full consultations with agronomists, engineers, economists and others as appropriate. Involve them in all decisions on quantitative limits, and the information that they (and you) most need, as summarised in Table 2.3.

i) Using standard notations - Use the 'Guidelines for Soil Profile Description' (FAO, 1977a) for auger and pit descriptions, so that all nomenclature is standardised. Avoid imprecise or undefined terminology.

j) Observing crops and natural vegetation - Try to identify 'indicator' plants reflecting changes in local soil conditions and soil-related problems in vegetation (Sprague, 1964).

k) Using duplicate data sheets - These allow a (comparatively) clean copy to be preserved for the client and obviate copying time and expense. Keep duplicates in separate places so that if one set is lost, stolen or destroyed the survey does not have to be repeated!

l) Respecting local customs - Find out any unexpected local customs as early as possible and abide by them when dealing with labourers, drivers, landowners, local government officials etc. Time apparently wasted in observing local formalities is often recouped several times over by the development of a positive local reaction to a survey. Soil surveyors are often the first team members in the field, and have a correspondingly important 'diplomatic' role to fulfil.

A key requirement in organising survey work is the rapid establishment of efficient technical and administrative routines to ensure that daily operations proceed smoothly. A checklist of operations that need to be established early in the programme is given in Table 3.3.

3.6   Data recording

Standard data sheets should be used for recording soil profile characteristics (see Annex H). Any additional data too lengthy to be included may be recorded separately, but should be cross-referenced to the relevant data sheet. Standard abbreviations should be used throughout.

Special attention should be paid to those characteristics that are of direct relevance to crop growth. Landform characteristics, soil drainage, profile permeabiliy and crop rooting often receive only a cursory appraisal because they are less easy to quantify than soil colour, depth or texture. Nevertheless, the former characteristics are always of equal, and often of greater, importance from the point of view of agricultural development, and must therefore be carefully considered and recorded at each inspection site. A brief checklist of soil and related properties described and/or measured in most soil surveys is given in Annex H; further details of procedures and nomenclature are given in USDA (1951) and FAO (1977a).

Checklist of routine survey operations   Table 3.3

	Operation	Comment
A.	**Technical procedures**	
	Depth and intensity of auger/pit sites	Establish criteria for depth and location on basis of TOR and reconnaissance survey
	Soil sampling	Decide on method (eg by horizons, or standard depths; frequency of sampling within uniform horizons etc) and criteria for selection of sites
	Soil and land capability/suitability	Decide on characteristics and quantitative limits to use in descriptions and definitions
	Soil physical tests	Decide on types, methods, numbers and depths
	Aerial photos and field sheets	Decide on method and symbols for filling in and filing. In particular, establish foolproof site/sheet numbering system to prevent repetitions; standardise methods of indicating location; standardise descriptions of local details (eg size/density classes for tree cover)
	Daily routines	Establish, as soon as possible, procedures such as: correction and correlation of field sheets (eg drainage classes, soil and land suitability units); plotting points/boundaries/other data (eg GWT depths) on field maps; discussions of technical amendments/additions
	Mapping	Decide on sizes/layout/contents of base maps, and subjects/symbols for thematic details (eg soil series, phases, parameters) to be added
B.	**Administrative procedures**	
	Note:	As far as possible the survey should be organised to allow each surveyor and his team to operate independently
	Team responsibilites	Assign responsibilities for specific technical/ administrative operations to individuals and make sure individual TOR are carefully designed to avoid overlaps
	Daily routines	Establish, as soon as possible, such items as: departure and return times; vehicle refuelling and maintenance procedures; drying, labelling and bagging of samples; preparation of photos, field sheets and equipment for following day; field and office timetables; allocation and checking of tasks for draughtsmen, typists etc
	Logistic support	Food, fuel and maintenance supplies; communications; office assistance; labour payments; local purchases; information flow to and from head office

All field information should be presented in a neat and legible form, because data sheets are not normally retyped, but are submitted to the client and/or filed in their original form. Arrangement and presentation of data for final reports are covered in Chapter 9.

3.7 Soil sampling

3.7.1 **Introduction** With the present accuracy and reproducibility of chemical and physical analyses, the weakest link in the whole soil analytical chain is often the sampling procedure. Whereas most reputable laboratories would regard duplicate variation of greater than 10% as unacceptable in most of their chemical analyses (see Table 7.2), it would not be unusual to encounter differences in specific chemical characteristics of more than twice this magnitude between comparable horizons taken from two profiles of the same soil type in the same field (see Section 2.7b). Moreover, this variation need not be restricted to the immediate topsoil; subsoil clay contents from similar horizons in adjacent profiles of, say, 17 and 22%, or exchangeable sodium values of 4 and 6 me/100 g, would not be uncommon, even though differences of 29 and 50% respectively are involved. Therefore, when analyses are being performed in order to assess the feasibility of a project, or to assist in the implementation of agricultural development, the surveyor must take great care to ensure that the samples collected are as fully representative as possible of the defined soil types, and that any interpretation of the results takes into account the constraints imposed by the sampling procedure. The most accurate soil analyses are of limited value if their interpretation is valid only over a very small area immediately around the sampled site, and for most survey purposes it is better to have a large number of 'approximate' analytical figures than a few 'precise' ones.

The time of collection and processing of samples must also be carefully controlled. All too often results arrive late and are slotted into the final report more or less as independent annexes without adequate interpretation.

The subject of soil sampling has been dealt with in detail in several publications (Cline, 1944; Hammond et al, 1958; Petersen and Calvin, 1965; Hodgson, 1974, 1978) which, between them, consider both the theory and practice of obtaining representative material for analysis. Six important aspects are emphasised in the following subsections.

3.7.2 **Judgement and random sampling** Judgement samples are those selected by the surveyor on the basis of field or API studies as being representative of the soils in a particular area, or of a particular type. Random samples are collected in a statistically random manner and, in general, the fewer the number of samples actually collected, the greater the possible representational errors. With judgement sampling, the possible errors also decrease with sample numbers, but they do so at a slower rate than with random sampling. At some point, therefore, random samples become more statistically representative than judgement samples from 'typical' sites. In practice, random sampling (in its purest form) is rarely used in soil surveys or land suitability studies.

This is usually because the number of soil samples are cut to the barest minimum on grounds of cost.

Provided that great care is maintained in the selection of 'typical' sites, judgement samples are acceptable. These sites should be chosen to represent those soils occupying the greatest areas regardless of their nature. There is a tendency, which must be resisted, to sample soils which are modal only in terms of a standard classification, or which represent the clearest examples of particular pedogenetic processes, rather than sampling the most commonly occurring soils of an area.

3.7.3   Depth and intensity of sampling   In many regions of the tropics and subtropics, the soils have little horizon differentiation and, where this is the case, there is much to be said for the standardisation of soil-sampling depths. This permits rapid correlation between analysed profiles, and allows samples to be collected (if required) by persons having little or no detailed soil knowledge. The system becomes more difficult to apply when the soil profile is differentiated into distinct horizons, and is essentially unusable where profound differences occur between contiguous horizons; when this is the case, the profile must be sampled by horizons. The decision on which system to use, or whether to combine the two and, say, sample the surface horizons by morphologic characteristics and the remainder by arbitrary depths, is a decision best left to the surveyor himself.

In coordination with potential users such as agronomists or engineers, the soil surveyor should select appropriate depths to characterise both topsoils and subsoils, and those profile features most relevant to a particular project, such as the environment of seedling development, and the main soil materials providing nutrients and water during the crop cycle. The whole of the effective rooting depth, often arbitrarily taken as about 1 m, should be included in the sampling programme, and thick horizons should be sampled at more than one depth, at intervals of about every 40 cm. Actual depths selected will obviously vary, depending on the soils and crops involved; two possible systems might be as follows (depths in cm):

a)   0-15,   15-30,   30-60,   60-90,   90-120

b)   0-10,   40-50,   90-100,   140-150

Deep borings (of about 3 to 5 m) are often required to characterise substrata - particularly with reference to permeability and salinity, and for groundwater sampling. The observations, depth and intensities of sampling should again be discussed with other team members.

3.7.4   Composite samples   Despite the interrelated pressures of time and costs that seem to accompany most soil and land suitability studies, the collection of composite samples from all soil profile horizons should be a standard objective where fertility, rather than pedogenetic, characteristics are being examined.

Since the topsoil is generally that part of the profile most subject to variation over short distances, it is recommended that composite samples be obtained from this source as a matter of routine. This might, in practice, involve the collection and mixing of about five or six equal weight subsamples over a small area (perhaps 0.5 ha or less). Care should be taken, however, to avoid collecting subsamples from locations having a different history of land use or recent fertiliser application. In this case, additional representative surface samples should be taken for separate analysis. In all cases, the history of land management, and particularly of fertiliser application, should be recorded for sites sampled.

3.7.5   Correlatory samples   The remarks made above refer specifically to those samples collected for detailed laboratory analysis. In addition to these, large numbers of correlatory samples may be collected, on which only a very limited number of tests are made – such as pH and electrical conductivity. These tests are usually carried out on site or in a field laboratory. They can be used to monitor particular characteristics of the soil, such as salinity, but more commonly are used as a continual check on the apparent uniformity or otherwise of the soils, and help to reduce the number of samples requiring detailed analysis.

3.7.6   Undisturbed samples   Undisturbed samples are sometimes taken for soil physical tests, such as pF and bulk density (see Chapter 6) and, in most respects, the previous comments are equally applicable. Note also the requirements for adequate replication discussed in Chapter 6, pp 58-59.

3.7.7   Samples for nematode analysis   Whenever coarse-textured soils are being surveyed, consideration should be given to the desirability of sampling for nematode analysis. This is particularly the case where the proposed agricultural development involves high-value crops susceptible to nematode attack, such as sugarcane, sugarbeet, tobacco, citrus, bananas, tomatoes and onions. The sampling procedure differs from that normally used in that the soil must be collected in a moist (but not wet) condition, and must reach the test laboratory not more than three days after sampling. The soil should be kept in a sealed polythene bag to prevent drying. Samples for nematode analysis are normally taken from depths of 0 to 15 cm and 15 to 45 cm; a method is illustrated in Crops and Soils (1979).

3.7.8   Duplicate analyses   As a matter of routine, suitably 'disguised' samples should accompany all samples sent for outside analysis. These duplicates should be obtained as subsamples of carefully mixed and evenly divided samples to ensure uniformity. The temptation to take 'duplicates' from adjacent parts of a horizon should be firmly resisted, since spatial variability of the soil will inevitably lead to some differences between the samples. Proposals should always include costings for duplicate analyses. Normally about 5 to 10% of the total number of samples should consist of duplicates, although the proportion should be much higher in the initial batches. As a further check, if there are suspicions about reliability of results, a limited number of duplicate samples should be sent to a laboratory with a sound

31

international reputation. Duplicate analyses should continue to be sent throughout an analytical programme, as many analyses are subject to 'operator error'. The results of duplicate analyses should be carefully checked, and any inconsistencies investigated. Acceptable limits of error and cross-checks for consistency are given in Chapter 7.

3.7.9 <u>Sample preparation, recording and despatch</u> In order to minimise problems of handling and recording, the following points should be borne in mind when sending samples for analysis:

a) If possible, the samples should be air dried before being sent to the laboratory (this eliminates one stage in the laboratory process and problems of mislabelling during the drying process).

b) Each sample should be labelled twice; once with a tie-on label and once with a label inside the sample bag.

c) Labels should be short and simple, since soil laboratories occasionally make mistakes in transcribing numbers. The project name should be clearly marked on each label. The 'laboratory register' kept by the soil surveyor should show for each sample where applicable the date, site number, depth and replicate number (see Annex A).

d) The laboratory should be requested to keep any unused samples, carefully labelled, until about six months after completion of the project in case of any queries.

e) About 1 kg of air-dried soil is sufficient for complete analysis; for a restricted range of chemical analyses only 150 g may be needed. Weights required for different analyses are given in Annex A; just over twice the minimum amount should be collected to allow for repeat analyses.

3.8 <u>Field laboratories</u>

On many surveys, particularly at the semi-detailed and development level, it is advantageous to set up a field laboratory in regions having salinity hazards, in order to carry out routine tests whilst the work is in progress.

Normally the tests are restricted to EC, pH, bulk density and field capacity. The amount of apparatus will clearly depend on the size of the project and the number of samples to be tested. As a general guideline, a list of equipment which is considered adequate to accompany a development soil survey over 5 000 to 10 000 ha is given in Annex A. For larger projects, the disposable items such as chemicals and filter papers should increase in proportion, but the basic equipment would be unlikely to be materially different unless it is intended that more than one technician should be involved on a full-time basis.

Equipment for a complete laboratory is listed in Golden et al (1966) and Cottenie (1980); laboratory designs are given in Ferguson (1973).

Chapter 4

Classification and Mapping of Soils

4.1   Introduction

Several standard references exist on classification and mapping of
soils and land, so the notes which follow are intended only to
highlight a number of important points.   Full references are given
to the standard works, and systems used by BAI are treated in
detail, but no attempt has been made to repeat or summarise all the
literature.   Soil morphological properties are discussed in detail
by USDA (1951) and FAO (1977a);   some of these properties and their
quantitative limits are summarised in Annex H, whilst specific
earth science terms (including geology, mineralogy, geomorphology
and soils) are given in Annexes L and N.   A recent review of soil
classification   appears   in   Butler   (1980)   and   a   historical
perspective is provided by Finkl (1982).

4.2   Soil classification

Only   two   systems   of   soil   classification   enjoy   very   wide
international recognition: Soil Taxonomy (USDA, 1975b, 1982) and
Soil Map of the World (FAO-Unesco, 1974, revised 1988).   The Soil
Map of Africa (D'Hoore, 1964) was until recently much quoted by
soil surveyors working in that continent.   The Francophone tropics
are served by Aubert and Tavernier (1972) and Duchaufour (1982).

These   formal   taxonomic   classifications   are   of   limited   use   as
mapping tools for BAI investigations, for the following reasons:

a)   they tend to be based on inferred pedogenesis and/or detailed
      laboratory results.   This makes them difficult to apply in
      the field;

b)   their choice of criteria for differentiating between soils may
      not   allow   immediate   adaptation   to   specific   development
      objectives, particularly on irrigated agricultural projects;

c)   they   are   often   not   suitable   for   practical   use   at   the
      feasibility and more detailed levels, where for example local
      values of diagnostic criteria for a single soil series may
      straddle the values differentiating the taxonomic units.

In addition, each system suffers from other inherent drawbacks (see
Young, 1976, Chapter 13).   The USDA Soil Taxonomy - formerly the

Seventh Approximation - has been particularly heavily criticised (Webster, 1968a, 1968b, but contrast Mitchell, 1973b).

For many feasibility projects, the main use of soil classification at the higher levels is to group the similar soil series for discussion, to make correlations with previous work and to act as an aide-mémoire of general soil properties. For these purposes the FAO system is recommended since it is designed for world-wide use, is more applicable in the field than Soil Taxonomy and, although it is an 'artificial' system, the parameters have been chosen in an attempt to define natural classes. The 1974 system and the associated mapping units are summarised in Tables 4.1 to 4.3; further details and a recommended procedure for soil classification are presented in Annex C. Key tables for the 1988 revision appear in Appendix I. Where necessary, correlation of different systems can be made using the comparison tables in Annex C.

For many commercial soil surveys, the client will specify the classification system to be used, and problems can arise if this is an international - or even a national - hierarchical system. In particular, the soil unit divisions of the system may not coincide with the classification or mapping unit divisions that the surveyor wishes to make at the series, or phase, level to reflect local relationships of, say, soils and landscape. Since a local system is likely to be more relevant to planning and implementation of a specific project, it is best to adopt this system as the basic classification and then make the best possible correlations with the national or international schemes, even if there is a consequent 'hiatus' (Butler, 1980, p 40). If an unfamiliar classification is required by the client, or is in use locally, the surveyor should examine its structure very carefully to decide what, exactly, are the properties and values used in its formulation, and whether they are relevant to the project being considered. In practice the basis of a classification can often be rather different from that formally stated, and may even be reducible in essence to just one or two key parameters.

4.3   Soil mapping units

The notes that follow highlight a number of points that need to be borne in mind when selecting or using mapping units.

a)   Distinguishing criteria for mapping units - With one or two exceptions, the criteria used should be visible in the field. They should have a morphological, or in some cases a physiographic, basis which can be directly recorded as the survey proceeds. A surveyor cannot use criteria such as percentage of organic matter, exchange capacity or some fertility index in order to separate the mapping units, unless this characteristic is clearly and directly related to a morphological property that can be directly observed. Thus, although the definition of a xerosol depends partly on the A horizon having an organic matter content of more than 1% if the weighted average sand/clay ratio is one or less, such a characteristic clearly cannot be used in the field as a diagnostic criterion for a mapping unit.

Simplified key to FAO—Unesco soils (1974 Classification)   Table 4.1

Soil name	Symbol	Brief description 1/
Acrisols	A	Acid low base status lessivés; more strongly leached than luvisols, but insufficiently weathered for ferralsols. Base saturation of B horizon < 50% (cf luvisols)
Andosols	T	Soils derived from recent volcanic deposits, high in volcanic glass
Arenosols	Q	Coarse, weakly developed soils with an identifiable B horizon; clay content < 18%
Cambisols	B	Brown earths or brown forest soils, with cambic B horizons as major feature
Chernozems	C	Black earths of the temperate steppes; deep A horizons rich in organic matter
Ferralsols	F	Strongly weathered soils of the humid tropics with oxic horizons. CEC < 16 me/100 g clay
Fluvisols	J	Recent alluvial soils (include fluviatile, marine, lacustrine and colluvial sediments). Depositional rather than pedogenetic profiles
Gleysols	G	Hydromorphic properties dominant above 50 cm
Greyzems	M	Grey forest soils of cool temperature zones (cf chernozems, phaeozems)
Histosols	O	Organic soils – peats
Kastanozems	K	Chestnut or brown temperate steppe soils – similar to chernozems, but with shallower mollic and cambic horizons, and with carbonate/gypsum horizons (cf phaeozems, greyzems)
Lithosols	I	Shallow soils of < 10 cm depth over hard rock
Luvisols	L	Lessivés – soils having argillic B horizons, with high base status. Base saturation of B horizon > 50% (cf acrisols)
Nitosols	N	Tropical soils with merging argillic B horizons (prominent shiny clay skins); formed on basic rocks usually
Phaeozems	H	Prairie soils – lighter than chernozems; chernozem—kastanozem intergrade (cf greyzems)
Planosols	W	Soils with an albic E horizon with hydromorphic properties, and a slowly permeable B horizon
Podzols	P	Soils with spodic B horizons (note: some lowland tropical podzols (bleached sands) are albic arenosols – Young, 1976, p 245)
Podzoluvisols	D	Soils intermediate between podzols and luvisols
Rankers	U	Shallow soils with umbric A horizons overlying rock
Regosols	R	Weakly developed soils from unconsolidated materials
Rendzinas	E	Shallow soils with mollic A horizons immediately overlying calcareous material ($CaCO_3$ equivalent > 40%)
Solonchaks	Z	Saline soils
Solonetz	S	Sodic soils with natric B horizons; alkaline reaction
Vertisols	V	Dark montmorillonite—rich cracking clays
Xerosols	X	Semi—desert soils with weak ochric A horizons (usually 0.5% to 1.0% OM)
Yermosols	Y	Desert soils with very weak ochric A horizons (usually < 0.5% OM)

Note: 1/   These descriptions are highly abbreviated;   the full definitions are given in FAO—Unesco (1974, p 32f).

A key to Soil Unit names is given in Table 4.2;   diagnostic horizons and properties, possible horizon sequences and soil phase names are presented in Annex C, Tables C.1 to C.3.

For 1988 Classification see Appendix I, Table I.2

Key to descriptive adjectives used for FAO-Unesco soil units (1974 Classification) Table 4.2

Adjective	Symbol	Major property [1]	A Acrisols	B Cambisols	C Chernozems	D Podzoluvisols	F Ferralsols	G Gleysols	H Phaeozems	J Fluvisols	K Kastanozems	L Luvisols	M Greyzems	N Nitosols	O Histosols	P Podzols	Q Arenosols	R Regosols	S Solonetz	T Andosols	V Vertisols	W Planosols	X Xerosols	Y Yermosols	Z Solonchaks
Acric	a	Weathered; CEC < 1.5 me per 100 g clay in B horizon	x				x																		
Albic	a	Containing albic material										x					x								
Calcaric	c	Accumulation of CaCO3		x				x		x								x							
Calcic	k	Having a calcic horizon									x	x											x	x	
Cambic	c	Having a cambic B horizon		x													x								
Chromic	c	High chromas		x								x									x				
Dystric	d	Low base status – low fertility BSP < 50%		x		x		x		x				x	x			x				x			
Eutric	e	High base status – moderate/high fertility BSP > 50%		x		x		x		x				x	x			x				x			
Ferralic	f	Showing ferralic properties (high sesquioxides)		x													x								
Ferric	f	Showing ferric properties	x									x				x									
Gelic	x	Containing permafrost		x		x		x				x			x	x		x				x			
Gleyic	g	Showing hydromorphic features	x	x		x	x	x	x	x		x	x			x			x						x
Glossic	g	Having an umbric A or histic B horizon			x																				
Gypsic	y	Having a gypsic horizon																					x	x	
Haplic	h	Unit with normal horizon sequence			x				x		x												x	x	
Humic	h	Having an umbric A horizon	x	x			x	x						x		x				x		x			
Leptic	l	Weakly developed; see FAO-Unesco, p 39														x									
Luvic	l	Having an argillic B horizon			x				x		x	x					x						x	x	
Mollic	m	Having a mollic A horizon						x											x	x					x
Ochric	o	Having an ochric A horizon																		x					
Orthic	o	Most commonly occurring; 'typical' unit	x				x					x	x			x			x						x
Pellic	p	Low moist chromas of < 1.5																			x				
Placic	p	Thin iron pan over spodic B horizon														x									
Plinthic	p	Containing plinthite	x				x	x				x													
Rhodic	r	Red to dusky red oxic B horizon					x					x													
Solodic	s	> 60% exchangeable Na in slowly permeable horizon																				x			
Takyric	t	Barren, showing takyric features																						x	x
Thionic	t	Having a sulphuric horizon								x															
Vertic	v	Showing vertic properties		x								x													
Vitric	v	Lacking a smeary consistence																		x					
Xanthic	x	Yellow to pale yellow oxic B horizon					x																		

Notes:
[1] Highly abbreviated; full definitions are given in FAO-Unesco (1974, pp 32-41).
[2] Lithosols (I), Rendzinas (E) and Rankers (U) are not listed, since they are not subdivided.

For 1988 Classification see Appendix I, Table I.3

FAO–Unesco mapping legend (1974 Classification)   Table 4.3

Property and order referred to	Description of map symbol, as used on the World Soil Map sheets at 1:5 000 000
1. Soil association	Symbol of dominant soil unit, or if lithosols predominate the symbol I followed by 1 or 2 associated units
2. Description of association	Given by a number referring to the descriptive map legend
3. Soil texture	Top 30 cm of dominant soil referred to by numeral symbol separated by hyphen:
	–1: Coarse < 18% clay, > 65% sand
	–2: Medium > 35% clay, < 65% sand (or ≤ 82% sand if clay ≥ 18%)
	–3: Fine > 35% clay
	Where textures cannot be separated, two symbols may be used separated by an oblique
4. Relief 1/	Slope classes indicated by small letter:
	a: level to gently undulating 0 to 8%
	b: rolling to hilly 8 to 30%
	c: steeply dissected to mountainous > 30%
	Two symbols may be used in areas of complex topography

Examples

a)	Fx 1 – 2ab	Xanthic ferralsols, medium textured, level to rolling, described as Unit 1 on legend
b)	Wm 5 – 1/2b	Mollic planosols, coarse and medium textured, rolling, and ferric luvisols, described as Unit 5 on legend
c)	I – Lc – To 3 – c	Lithosols, chromic luvisols and ochric andosols, steeply dissected (no information on texture); Unit 3 on legend

Note: 1/   These classes are too broad for most development work, particularly irrigation planning.

Source: Adapted from FAO–Unesco (1974, p 8f).   For 1988 Classification see Appendix I, Table I.4.

b) <u>Salinity surveys</u> - These provide the chief exception to the use of morphological criteria for mapping. Sites usually need to be sampled at regular intervals on a grid or transect basis. Sampling by horizons is rarely satisfactory since the data will not be directly comparable between sites. Composite samples (see Section 3.7.4) are often useful.

c) <u>Units</u> - Wherever possible, the classic format of soil association, complex, series, type and phase should be employed (see USDA, 1951). Care should be taken to correlate identified soil units with any already notionally established and to ensure that names given to new soil series or associations will be acceptable according to local practice.

d) <u>Parameter mapping</u> - When the soils are developed on variable recent alluvial sediments and only very limited pedogenesis has occurred, the use of 'series, phase and type' designation is rarely satisfactory. In these circumstances, arbitrary criteria have to be used in order to separate out different soil units, and it is generally best if the properties selected are likely to reflect differences in crop response or land management requirements.

e) <u>Size and numbers of soil mapping units</u> - Surveyors commonly tend to identify too few mapping units at the reconnaissance level and too many at the detailed level. The latter is particularly the case on alluvial soils where arbitrary criteria, rather than major morphological changes, are used to separate soil units. A balance needs to be maintained amongst what can be identified as 'soil units', what can be mapped with an acceptable degree of accuracy and what the project/client needs for the purposes for which the survey was implemented. Some aspects of deciding on numbers of mapping units are discussed by Valentine and Chang (1980).

The size limits of soil mapping units are discussed further in Chapter 9, but note that the minimum size is often determined by the minimum management area (ie the minimum area that can be treated differently) and/or by cartographic considerations involving map scale and the shape of individual units, and/or by other user considerations (eg preferred scales for engineering designs).

f) <u>Mapping legends</u> - These should be designed with careful attention to the purpose of the survey and needs of the client. Descriptive information on the map should avoid specialist jargon and should emphasise two kinds of soil properties:

   i)   those enabling the non-specialist to recognise the soil in the field (especially colour and texture);

   ii) those of agricultural significance.

g) <u>Grouping of units</u> - Where soil units are grouped into classes of a higher category, this is commonly done in terms of a formal, extant soil classification. It is worth noting, however, that alternative groupings (by geomorphic units, or by agricultural significance) would often be of more practical use, particularly for map legends.

h) <u>Compound mapping units</u> - Where compound mapping units are used, the soil report and map legend should indicate the proportions of the soils included and their relative positions in the landscape.

Chapter 5

Land Evaluation

5.1 Introduction

5.1.1 <u>Approaches to land evaluation</u>   Soil surveys usually include interpretations of their findings for land-use planning purposes in the form of land capability or suitability classifications.   The aim of these classifications is to guide planning decisions, ideally in such a way that the resources of the environment are put to the most beneficial use, whilst at the same time conserving them for the future.   The role of soil surveyors is normally concerned with the physical assessment of land resources, although this should always be seen in the broader context of a land evaluation which takes into account the technological and socio-economic consequences for the people of the area and the country concerned.   Note that most agronomists and soil surveyors tend to use physical measures, such as potential yields per unit area, to build up an evaluation system, but such an approach may be inappropriate in some circumstances (eg under shifting agriculture, where much of the surface may be occupied by tree roots, fallen tree trunks etc).   A more realistic measure, often overlooked in textbooks, is the return in yield or economic terms per person per unit of time worked.   Such evaluations are however usually outside the scope of soil surveyors' inputs to projects.

The following sections describe some of the more common approaches in land evaluation;   specific crop requirements useful for land evaluation are given in Annex F.   Summaries and reviews of the more important land evaluation systems are contained in Young (1976), Dent and Young (1981) and McRae and Burnham (1981);   the last-named give particularly detailed comparison tables of the advantages and disadvantages of the main methods.   Terrain evaluation is covered by Mitchell (1973b).

5.1.2 <u>Land evaluation terminology</u>   There is considerable confusion over the terms used and the methodology of various interpretations, the most basic of which is the distinction between 'soil' and 'land'. Both soil and land interpretation may sometimes be restricted to soil properties alone, but land evaluations in particular more commonly refer to land in a broader environmental sense.   These broader systems incorporate (implicity or explicitly) such factors as topography, climate, vegetation and perhaps socio-economic features of an area and/or combinations of their effects that form constraints on the ease of agricultural development (eg land grading or drainage requirements, flood or erosion risks). Such systems may be general purpose, with an implied list of priorities, or may be specific to a particular crop and/or system of management.

The nomenclature for the different systems is often used very loosely, with corresponding confusion over the aims and methods employed; for most purposes, the following terminology is recommended:

a) Land evaluation - A general term embracing all forms of interpretation, and not implying any particular method of evaluation, or classification or final land use. There is regrettable but inevitable ambiguity because of the 'monitoring and evaluation' concept employed by economists, financial analysts and planners, restricted to their post hoc assessments and reviews. The contrast is brought out by Murdoch and Lang (1978) and Cracknell (1978).

b) Land capability classification - A rather more specific term derived from the USDA (see Section 5.3) and used for a ranked system based on the severity of land limitations for general agricultural use (eg slope, flood or erosion risk).

c) Land suitability classification - A term which relates to specific use(s); the USBR system (see Subsection 5.2.3) is an example - it classifies land by its economic suitability for irrigated agricultural development. These systems are usually more detailed, with more specific quantitative limits to the classes.

A variety of other terms are also found in the literature, such as land use capability classification (in the UK), soil capability classification (Canadian Land Inventory) and soil suitability classification; a careful check on these systems will usually indicate that they fall into one of the three broad categories outlined above. Note that 'land use classification' is a matter of the present actual use to which the land is put and not necessarily a reflection of the land's potential and limitations. 'Terrain evaluation' is a broader, more geographical, term covering analysis of all aspects of land with regard to its natural features and configuration (Mitchell, 1973b). 'Land classification' is too vague and should be avoided except in very general statements.

Land evaluation may be qualitative or quantitative. In a qualitative system, environmental factors are compared and ranked subjectively, whilst for quantitative comparison the factors must be measured in common numerical terms, usually (quasi) economic, whereby costs and returns for a specified land use are calculated or implied for a given type of land. Most BAI land evaluations are aimed at specific project developments, for which the more quantitative land suitability approaches (Section 5.2) are preferred. The qualitative systems (Section 5.3) are more often used in wider studies, such as regional planning exercises, where the quantitative data, particularly the economic information, may be sparse or even entirely lacking.

5.1.3 <u>General problems of land classification</u>   The use of standard national or international land evaluation schemes suffers from the same problems as previously noted for standard soil classifications (Section 4.2).   In particular, the quantified limits for distinguishing features of subclasses (defined, for example, as part of a standard system, or considered as optimum for a given land use) may not coincide with the values measurable or mappable in practice.   Sometimes the limits can be modified appropriately for local use, or complex land suitability mapping units can be used.   Alternatively, the standard limits to the land suitability classes or subclasses can be redefined to coincide with the naturally occurring, mappable limits of soil and land features;   in these cases, appropriate alterations in overall yield or economic estimates for each class or subclass should be made to reflect the different proportions of the 'standard' classification units that occur within each mapping unit.   Particular care must be taken in such heterogeneous areas when smallholder development is being considered;   individual holdings are often very small (down to 3 ha, or less) and it is essential to ensure that they are not sited on unsuitable land.

Additional problems may occur in surveys of large regions where different land evaluations apply to different areas (eg hot, lowland plains suitable for irrigated field crops, and adjacent cool, upland plateaux suitable for grazing).   The use of different diagnostic criteria in class definitions within the two (or more) systems may prevent simple interrelation of classes, and will very often preclude simple recommendations on development priorities for the region based on land suitabilility criteria alone.   At this level however, a whole range of outside factors must be taken into account, including government policy, sociology, demography, market forces, financial and economic analyses etc and the exercise necessarily becomes one involving a multidisciplinary approach.

5.2   <u>Land suitability classification</u>

5.2.1 <u>Scope of land suitability classifications</u>   By narrowing down the range of land uses considered, it is possible to be more specific about the fitness of the land for a given use, this being implied by the word 'suitability' rather than 'capability'.   Thus, for instance, it is possible to map land suitability for rainfed maize, or for surface irrigation in general, or for sprinkler-irrigated sugarcane.   The physical factors themselves can only be weighed against one another subjectively, unless one or other is clearly limiting, but by careful and explicit economic assessment of a given development at a given time, the factors and possible uses can be quantitatively compared.   This requires relating the physical factors to comprehensive information on yields, market prices and cost of inputs, both for land preparation and for recurrent expenditure.   Normally the land is classified separately for more than one kind of use, for comparison of alternatives.

5.2.2 <u>The FAO framework</u>   The Framework for Land Evaluation (FAO, 1976a) is a standard set of principles and concepts on which national or regional land evaluation systems can be constructed.   It emphasises in particular the importance of explicitly stating the intended land use and the level of management envisaged, and that land evaluation may be either on current suitability or, as for irrigation/drainage schemes, on potential suitability. It should be emphasised that the system is only a framework, and for most projects it will need quantifying with detailed specifications, as discussed in FAO (1979a) and Smyth et al (1979); an example, based on the USBR format, is given in Subsection 5.2.3.

Using the concepts of the FAO framework, reconnaissance surveys are concerned with 'major kinds of land use' such as rainfed agriculture, irrigation, forestry; they are usually qualitative, and incorporate economics only as broad background assumptions. Semi-detailed or detailed surveys include comparisons of costs and returns quantitatively for a more specific 'land utilisation type', such as subsistence-level, rainfed maize production by smallholders. The degree of suitability for a given use may be expressed in terms of net income per unit area or per standard management unit, or per unit of irrigation water applied to different types of terrain. Scarcity value may increase the 'suitability', for example as regards wildlife conservation, over more profitable uses. Social and environmental effects are difficult to evaluate in monetary terms and may be better dealt with in a qualitative system.

The Framework's structure, as shown in Table 5.1, is compatible with other systems but allows great flexibility. There are two orders, termed suitable (S) and not suitable (N). Conditionally suitable land (Sc) is a 'phase' of the order suitable, and approximates to Classes 4 and 5 of USBR, but its extent must be small with respect to the total study area; if not, the land utilisation types or potential suitability classification must be redefined. Land classes, subclasses and units are similar categories to those in the USBR and USDA systems (see also Section 5.3). Definitions of FAO land classes are given in Table 5.2, and some suggestions on presentation in Chapter 9.

The Framework employs several terms to define or describe land features, in particular 'land quality' and 'land characteristic', as described in Table 5.3, but care has to be exercised in the use of land qualities, some examples of which are shown in Table 5.4. Whilst these rightly emphasise the combination of land characteristics which affect crop growth, there is a danger of constructing an over-complex system (involving - say - moisture, oxygen and nutrient availability, soil erodibility, trafficability and opportunity days) when their effects could be simply indicated in terms of just one or two land characteristics such as soil texture or depth. Although these latter may not be the land qualities which directly affect crop growth, their effects are sufficiently well known by users of major land suitability classifications not to require further explanation and incorporation into a formal (and unfamiliar) system. In some cases - particularly as abbreviated symbols on land suitability maps - the use of land qualities rather than land characteristics may obscure the user's immediate appreciation of which lands he can and cannot improve, and by what means. For example, a 3mo designation (see Table 5.4) might indicate a compact or a poorly drained soil - or one which is both - whereas symbols such as 3d (d for drainage) or 3c (c for compaction) give a clearer indication

Structure of the FAO land suitability classification   Table 5.1

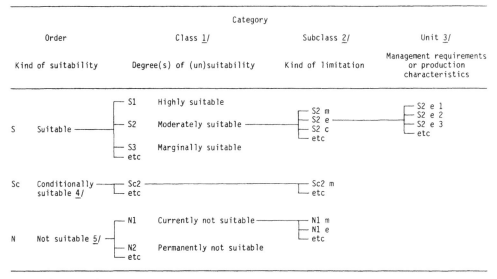

	Category		
Order	Class 1/	Subclass 2/	Unit 3/
Kind of suitability	Degree(s) of (un)suitability	Kind of limitation	Management requirements or production characteristics

S   Suitable
  ┌ S1 Highly suitable
  ├ S2 Moderately suitable ─── ┌ S2 m / S2 e / S2 c / etc ─── ┌ S2 e 1 / S2 e 2 / S2 e 3 / etc
  ├ S3 Marginally suitable
  └ etc

Sc Conditionally suitable 4/ ── ┌ Sc2 ─────────────── ┌ Sc2 m / etc
                                └ etc

N   Not suitable 5/
  ┌ N1 Currently not suitable ─── ┌ N1 m / N1 e / etc
  ├ N2 Permanently not suitable
  └ etc

Notes: 1/ The class names shown here are those recommended for a three-class system. Up to five suitable classes are permitted.

2/ The number of subclasses should be kept to the minimum necessary to distinguish land with significantly different management requirements or production potential. As few limitation symbols as possible should be used for each subclass. Note that S1 land is not divided into subclasses.

3/ Units are normally for use at the farm planning level, and are often definable by differences in detail of their limitation(s).

4/ For small areas of land where certain conditions additional to those specified for the 'suitable' classes must be fulfilled for successful land use; once these conditions are met, the land is included in the class or subclass indicated by the code following the Sc designation. NB: 'conditionally suitable' does not imply that the interpretation is uncertain, either because the land is only marginally suitable or because the relevant factors are not understood.

5/ The additional symbol NR (not relevant) is used for areas not being considered.

Source: FAO (1976a, p 21).

FAO recommended land class definitions   Table 5.2
(for a system with three suitable classes)

Class	Designation	Definition
S1	Highly suitable	Land having no significant limitations to sustained application of a given use, or only minor limitations that will not significantly reduce productivity or benefits and will not raise inputs above an acceptable level
S2	Moderately suitable	Land having limitations which, in aggregate, are moderately severe for sustained application of a given use; the limitations will reduce productivity or benefits, and increase required inputs to the extent that the overall advantage to be gained from the use, although still attractive, will be appreciably inferior to that expected on Class S1 land
S3	Marginally suitable	Land having limitations which, in aggregate, are severe for sustained application of a given use and will so reduce productivity or benefits, or increase required inputs, that this expenditure will be only marginally justified
N1	Currently not suitable	Land having limitations which may be surmountable in time, but which cannot be corrected with existing knowledge at currently acceptable cost; the limitations are so severe as to preclude successful sustained use of the land in the given manner
N2	Permanently not suitable	Land having limitations which appear so severe as to preclude any possibilities of successful sustained use of the land in the given manner

Notes: 1. In quantitative classifications, both inputs and benefits must be expressed in common measurable terms, normally economic.
2. Where additional refinement is needed, this should be done by adding classes (S4, S5) rather than by subdividing classes, since the latter procedure is reserved for subclass designation.
3. A particular class may sometimes not appear on a map of a given area, although it may be included in the classification, if such land occurs - or is believed to occur - in other areas relevant to the study.
4. Boundaries between suitability classes will be subject to revision with time in the light of technical developments and economic and social changes. However, the boundary of Class N2 is normally physical and permanent.
5. For 'conditionally suitable' land, see footnote 4 to Table 5.1.

Source: FAO (1976a, p 18).

Land evaluation

Table 5.3

FAO term	Meaning	Examples
Land quality	Complex attributes which directly affect specific kinds of land use; derived from land characteristics	Erosion resistance, water availability, flood hazard. Crop yield levels are aggregate land qualities
Land characteristic	Measurable or estimated parameter, use for land resource mapping	Slope angle, rainfall, soil texture/type/salinity
Diagnostic criterion	Land characteristic or quality which determines suitability of land for a particular use. Expresses the limitations of land for the given use, and the requirements of that use	Usually a combination of land qualities and/or characteristics 1/

Note: 1/ For example, a plant requirement for 'available oxygen in the root zone' could be specified as the period during which the redox potential (E_h) is less than + 200 mV. Alternatively, seasonal depths to the GWT, mottling or even vegetation types could be used as increasingly crude diagnostic criteria for the same requirement.

Examples of FAO land qualities   Table 5.4

Land quality	Symbol
Crop growth	
Water availability 1/)	m
Oxygen availability) pore-size distribution and climatic effect	o
Nutrient availability 1/	n
Toxicities	t
Land and crop management	
Erodibility	e
In-field trafficability and accessibility	a
Opportunity days for land preparation and harvest	h
Compaction (and resistance to)	c
Land improvement requirements	
Land levelling or grading	ℓ
Vegetation clearance	v
Installation of drainage and irrigation	i

Note: 1/ Note distinction between these land quality terms, which apply to the actual uptake a crop is able to make, and the land characteristic terms 'available water capacity' and 'available nutrient levels', which are measurements made as part of the land quality assessment; the latter also includes properties such as soil hydraulic conductivity, drainage status and rooting pattern.

Adapted from FAO (1976a, p 54).

of the land management practices required and practicable. Table 5.5 lists some suggested land characteristics and qualities likely to be of practical use for subclass designations. Note that symbols used for subclass designation are not standardised and should always be clearly specified; to avoid confusion of too many symbols a useful system is that of Comerma and Arias (1971) whereby capital letters indicate major subclass designations (T topography, S soil, D drainage) with lower-case letters for specific indication of limitations. The capital letters should be carefully distinguished from USBR land restriction symbols (see Annex D).

Climatic contributions to land qualities are implicit in some of the above, but for projects where climatic variation is large, explicit consideration may also be needed of:

danger of frost or storms; evaporation rate; humidity; net radiation; precipitation (including frequency, distribution and intensity); temperature; wind speed and direction.

Combinations of these and other factors may also be used to derive terms such as opportunity days, and crop germination/growing/ripening/harvesting periods and requirements.

5.2.3 The USDI Bureau of Reclamation (USBR) land classification system
A specific land suitability classification, the USBR system (USDI, 1953), and modifications of it, are widely used by BAI for land evaluation. Unlike many other systems, they are (implicitly or explicitly) based on the economics of land development although physical features are used as a basis for the economic rating. Depending on the nature and scale of the surveys, such systems can have varying degrees of quantitative assessment built in to both the physical and the economic criteria used.

The basic system is intended to reflect the 'payment capacity' of the land to support agricultural production. Ideally, as is illustrated in FAO (1979a, p 136f), farm budgets are used for evaluating the costs and benefits, initially for the best land (Class 1). Less suitable land is then downgraded according to the economic effects of a number of physical land deficiencies acting singly or together; these are considered under the basic headings 'soil', 'topography' and 'drainage'.

There are four main classes, as follows:

```
Class 1 )
        )      Arable (suitable      ↑   Increasing payment
Class 2 )      for irrigation)           capacity from Class 3
        )                                 to Class 1
Class 3 )

Class 6        Non-arable
```

For detailed studies, two additional special classes are recognised:

Class 4: Suitable for special use or restricted crop range
Class 5: Temporarily non-arable (pending further study of problems)

Recommended symbols for land characteristics and qualities used in subclass designation   Table 5.5

Land characteristic/quality	Suggested symbol 1/	Land characteristic/quality	Suggested symbol 1/
Drainage		Soil hydraulic conductivity	p
– surface drainage/flooding	f		
– subsurface drainage/GWT	w	Soil fertility	y
– drainage outlet	o		
		Soil salinity/toxicity	x
Erosion hazard	e		
		Soil sodicity	a
Flood risk	f		
		Soil structure	
Infiltration rate	i	– including pans, crusting, compaction	c
Soil acidity	j	Soil texture	
		– rock	r
Soil AWC	q	– gravel or stones	g
		– coarse (sands, loamy sands)	v
Soil consistency	n	– moderately coarse (sandy loams, loams)	l
		– moderately fine (silt loams, clay loams)	m
Soil depth (effective for plant growth)		– very fine (clays)	h
– to coarse sand, gravel or cobbles	k		
– to relatively impervious substrata	b	Topography/slope	t
– to carbonate/gypsum accumulations	z		
		Vegetation cover	u

Note: 1/   Based on the USBR system for indicating deficiencies (USBR, 1953, and see Annex D).   There are, however, no universally agreed symbols so these must always be carefully specified in reports and on maps.   Note that the following symbols in particular are often used for two or more land qualities/characteristics or deficiencies and may be particularly confusing:
    c = climate or soil compaction or brush or tree cover
    d = soil depth or drainage
    e = erosion or soil structure
    s = general soil deficiency, or salinity, or stoniness, or even slope.
Capital letters may be suffixed to indicate recommended restrictions to particular land uses (eg P = restriction to non-intensive pasture production).

Sources: USDI (1953) and Purnell (1978);   see also McRae and Burnham (1981, p 24f) and Dent and Young (1981, Table 10.5).

It should be noted that the payment capacities of these last two classes are <u>not</u> necessarily intermediate between those of Classes 3 and 6, but will depend on the particular economics of a given project. The crops included in the restricted range of Class 4 may have a higher or lower payment capacity than those of the other 'arable' classes, and the problems of Class 5 lands (eg behaviour of regional GWT) may prove to be anything between negligible and completely unsurmountable at reasonable cost.

The general class descriptions of the USBR system (see Annex D) are invariably adapted for particular projects. In South-East Asia two separate classes for rice land, which would otherwise be Class 4, are recognised, Classes 1R and 2R, and for most surveys the USBR survey scales (reconnaissance, semi-detailed etc - see Annex D) are usually not adhered to at the intensities stated. In projects with little or no economic data, the classes may be defined on the basis of physical criteria, with implied economic consequences. One such modified USBR system used by BAI is summarised in Tables 5.6 to 5.8.

Each classification on the USBR system is thus usually a relative classification depending on the specific use(s) or crop(s) envisaged, the economic circumstances of the proposed development project at the time of survey, and the inputs required to bring the land to full productive capacity. A distinction needs to be drawn, therefore, between land preparation costs (eg of land reclamation, clearance and grading; installation of irrigation or drainage systems) and operational costs (eg fertiliser applications, tillage) and also between 'correctable' and 'non-correctable' land deficiencies (eg minor undulations, salinity and pH problems, as against steep slopes, soil depth and texture problems). However, the scope of practicable changes will vary, depending on the economic feasibility of ameliorative processes within each project.

USBR terminology also distinguishes between 'arable' land and 'irrigable' land. The former is land that is mapped as suitable for irrigation, but the latter is land actually selected for development, so land out of command and small, isolated or odd-shaped tracts are excluded.

5.3   Land capability classification

Qualitative, general-purpose classifications such as that used in Britain (Bibby and Mackney, 1969) are usually based on the USDA system of Klingebiel and Montgomery (1961) and its predecessors. The system is designed primarily for soil conservation, rather than for economic purposes, and is intended for determining the maximum intensity of land use consistent with low erosion risks and sustained productivity. In its original form it is criticised specifically for being too generalised and unquantified (amongst other defects - see McRae and Burnham, 1981).

The basic USDA Land Capability Classification (see Figure 5.1) allocates land suited to cultivation to Classes 1 to 4, followed by land suited to grazing, Classes 5 to 7, and forestry, Class 7, leaving Class 8 for wildlife and recreation (see also Annex D).

Land evaluation

                                    Table 5.6

Class 1:   Irrigable

Lands that are the most suitable for irrigation farming, being capable of producing sustained and relatively high yields of a wide range of climatically suited crops at reasonable cost. They have gentle slopes.   The soils are deep and of medium to fairly fine texture with mellow, open structure allowing easy penetration of roots, air and water, and having free drainage – yet good available moisture capacity.   These soils are free from harmful accumulations of soluble salts.   Both soils and topographic conditions are such that no specific farm drainage requirements are anticipated, minimum erosion will result from irrigation and land development can be accomplished at relatively low cost

Class 2:   Irrigable

This class comprises lands of moderate suitability for irrigated farming, being measurably lower than Class 1 in productive capacity, adapted to a somewhat narrower range of crops, more expensive to prepare for irrigation or more costly to farm.   They may have a lower available moisture capacity, as indicated by coarse textures.   Any one of the limitations may be sufficient to reduce the lands from Class 1 to Class 2, but frequently a combination of one or more of them is operating

Class 3:   Irrigable

Lands that are suitable for development, but are approaching marginality for irrigation and are of distinctly restricted suitability because of more extreme deficiencies in the soil, topographic or drainage characteristics than described for Class 2 lands.   They may have good topography but, because of inferior soils, have restricted crop use, require larger amounts of irrigation water or special irrigation practices, and demand greater fertilisation or more intensive soil improvement practices.   They may have uneven topography or restricted drainage, amenable to correction but only at relatively high costs.   Generally, greater risk may be involved in farming Class 3 lands

Class 4:   Restricted irrigable

Lands are included in this class when a specific excessive deficiency or deficiencies indicate that the land is irrigable only for particular types of crops, or by restricted methods of irrigation.   The deficiency may be very slow profile permeability and poor drainage, leading to a restriction to rice; steeply sloping land whose utility seems best served by tree crops, or very coarse soils with high infiltration rates which are recommended for irrigation by overhead methods.   The Class 4 lands have a range of payment capacity which, in some circumstances, may be as high as or higher than in the irrigable land classes

Class 5:   Provisionally non-irrigable

Lands in this class are not considered economically irrigable pending further study.   They consist, for example, of land underlain by laterite at depths between 1.0 and 1.5 m.   Final designation of land class depends on detailed assessment of the hazard presented by the laterite, particularly as related to regional drainage

Class 6:   Non-irrigable

Lands in this class include those failing to meet the minimum requirements of the classes described above.   They comprise land with excessive slopes or dissection; with inadequate drainage and very variable soils, or with a liability to flooding.   Under the envisaged development programme, these lands are not considered to have sufficient repayment capacity to warrant irrigation

Source: Adapted from USDI (1953).

Example of modified USBR land suitability class specifications   Table 5.7

Land characteristics	Class 1 – Irrigable	Class 2 – Irrigable	Class 3 – Irrigable
Soil			
Topsoil texture (0 to 30 cm)	Porous fine sandy loams to fine sandy clay loams	Fine sand to loamy fine sand	Fine sand to loamy fine sand
Subsoil texture (30 to 80 cm)	As topsoils	Porous fine sand loams to fine sandy clay loams	Fine sand to loamy fine sand
Depth (minimum)	150 cm	130 cm	110 cm
Available water capacity (minimum)	150 mm m^{-1} soil	120 mm m^{-1} soil	90 mm m^{-1} soil
IR after 4 h	0.7–5.0 cm h^{-1}	5.0–12.0 cm h^{-1}	12.0–15.0 cm h^{-1}
Topography			
Slopes	$\leqslant 0.5°$	$\leqslant 0.5°$	0.5° to 1°
Levelling requirements for surface irrigation	$\leqslant 350$ m^3 ha^{-1}	350–750 m^3 ha^{-1}	750–1 000 m^3 ha^{-1}
Vegetation cover	Moderate to low clearing costs	Moderate clearing costs	Moderate to high clearing costs
Drainage			
GWT	Normally > 10 m	7–10 m	5–7 m
Drainage	No immediate farm drainage required; profiles well drained	No immediate farm drainage required; profiles well drained	Minor farm drainage required in places. Good to moderate profile drainage

Class 4:   Restricted irrigable or special use

Includes lands with coarse soils (fine and medium sands, loamy fine sands) to depth; high IR rates of > 15.0 cm h^{-1}; low AWC values; slopes between 1° and 3°; land levelling requirements > 1 000 m^3 ha^{-1}; GWT levels within 5 m of the surface; poorly drained profiles. These soils are considered suitable only for overhead or drip irrigation systems, although small basin irrigation may be possible on a small scale.

Class 5:   Provisionally non-irrigable

Includes lands underlain by laterite within 150 cm of the soil surface; additional economic and engineering studies are required to determine whether drainage is required or is practical.

Class 6:   Non-irrigable

Includes lands with excessive topographic, flooding or drainage problems which are considered to be non-correctable at an economic rate.

Note: 1/ In practice, a selection of diagnostic properties is often used to define the Classes; additional properties may then be used in the overall descriptions.

Source: Adapted from USDI (1953), see Annex D, Table D.2; several other examples are given in FAO (1974, and 1979a, p 146f) and McRae and Burnham (1981, p 134f).

Land evaluation

Subclass 3s

Land with very flat microtopography and very gradual slopes of less than 1%. The soils consist predominantly of the Vindu and Clarence soil series and are deep ($\geqslant$ 150 cm), friable and well drained. The topsoils consist of fine sand or loamy fine sand to a depth of between 30 and 80 cm, and overlie medium-textured subsoils of porous fine sandy loam to sandy clay loam

Average steady-state infiltration rates are moderately rapid to rapid, ranging from 12 to 15 cm h^{-1} with a mean of about 12 cm h^{-1}. AWC values are moderate to low, varying between 90 and 150 mm m^{-1}, and available nutrient levels are very low. Dry season GWT levels lie between 5 and 10 m. Levelling requirements for surface irrigation are low, of the order of 350 m^3 ha^{-1} or less

These lands are recommended for overhead irrigation. They are considered as only marginally suitable for surface irrigation because of the rapid infiltration rates, low AWC values and potentially high percolation losses. Leaching of nutrients is liable to be severe and drainage problems are likely to develop quickly in response to inefficient water use. Lined supply canals and very small basins would have to be used for surface water applications, but would lead to a rapid, although as yet unquantifiable, rise in GWT levels

Subclass 3sd

Land and soil similar to Subclass 2s but with imperfect drainage, as indicated by slight hydromorphic features below a depth of 70 cm. The soils are often noticeably compacted. The GWT fluctuates throughout the year, but normally lies within 5 m of the surface. Levelling requirements are similar to those of Subclass 3s

This subclass corresponds to imperfectly drained soils of the Zangaro complex which border major depressions and water courses. Minor drainage works are required for crops sensitive to waterlogging. The lands are suitable for either overhead or surface irrigation methods, although in view of the fragmented distribution of the subclass the former system may prove to be easier to manage if the surrounding areas are already served by sprinklers

Subclass 3st

Land with similar soils to Subclass 2s but with moderate microtopographic defects usually consisting of low undulations. The high levelling requirements of between 750 and 1 000 m^3 ha^{-1} render these lands extremely marginal for development for surface irrigation, and they are therefore recommended for overhead irrigation

Subclass 3std

Land similar to Subclass 2st but with additional drainage problems similar to those of Subclass 3sd. Overhead irrigation methods are recommended

---

USDA land capability classification   Figure 5.1

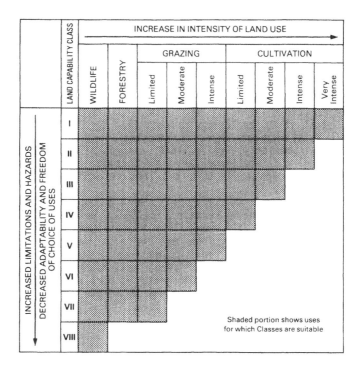

Notes: 1. The intensity with which each land capability class can be used
           with safety, and the limitations acting, increase as one moves
           from land capability Class I to Class VIII.

        2. An alternative land suitability presentation is given in Table 9.2.

Source: Brady (1974) after Hockensmith and Steele (1949).

These Classes may then be divided into subclasses based on the number and severity of several limitations which downgrade land from Class 1. These limitations include erosion risk (e), wetness (w), rooting zone (soil) limitations (s) and climatic limitations (c). The British system also specifically recognises gradient and soil pattern limitations (g) but has only seven classes. The letter notation is added as a suffix to the class to form a subclass, such as Subclass 2w or Subclass 7sc. Subclasses can be further divided into land capability units, the lowest category, which have uniform management requirements and yields that deviate not more than about 25% from norms.

The emphasis of the system is on environmental factors that cannot be modified practically. Thus physical soil features, such as texture, are accorded greater significance than chemical characteristics, such as pH. Poor soil drainage would lower the land class if it could not be rectified by field drainage, but if an arterial scheme were implemented the land class would be raised. Similarly, soils with a drought hazard need not be downgraded if supplementary irrigation becomes economically feasible. Economic factors are thus recognised implicitly rather than explicitly.

Apart from assuming a certain ranking of priorities for competing uses of land, the USDA capability classification also assumes that what is best land for one use is also best for another. In general of course this tends to be true. For instance most crops are said to do best on a 'deep, well-drained, fertile loam, on gentle slopes'. Urban development, airport sites and industry are likely to compete for the same good arable land. Nevertheless, particular crops, notably rice, have different soil/land requirements from the majority.

Less obvious assumptions built-in are economic, such as the level of management expected and the limit at which land improvement is considered impractical.

5.4 Field evaluation of agronomic potential

Virtually all land capability systems employed by BAI are based on the repayment capacity of the land. It is always useful, therefore, and sometimes essential, to have quantitative data on crop yields, either in the immediate vicinity of the project area, or on similar soils in a comparable environment. Even if the available crops are not those proposed for development, their response to various standards of husbandry and land management should be carefully noted. As a matter of routine, all crops encountered in a survey should be checked for deficiency symptoms, pests, disease and nematodes. Wherever possible, these factors should be correlated with the detailed method of farming and yields obtained.

In survey planning (see Section 3.4), time should be specifically allocated for evaluation of agronomic potential, and allowance should be made for travel outside the immediate project area. As a general rule, between 5 and 10% of the total survey time is usually adequate in semi-detailed and detailed surveys. This period does not include the time spent describing crops and vegetation at each inspection site.

In addition to agronomic data, information on other economic aspects of land development is important, although this is often not available to the soil surveyor. The information includes land clearance costs, land grading specifications and costs, water supply and distribution costs (on irrigation schemes) and fertiliser application rates and costs. Whenever possible, such data should be included in any land suitability assessment, as well as any management recommendations.

5.5   Productivity indices or ratings

5.5.1   Scope and problems of indices   In several parts of the world attempts have been made to develop 'productivity indices' as aids to land evaluation. These systems usually consist of empirically derived equations for relating productivity to parametric values assigned to soil or land properties. For every property, the full spectrum of values is split into convenient ranges, each of which is then assigned a numerical value to insert into the equation (ILACO, 1981, pp 164ff). A simple example of such an equation is that of Riquier et al (1970):

Productivity index = H x D x P x T x N (or S) x O x A x M

where the nine parameters are respectively soil moisture, drainage, effective depth, texture/structure, base saturation or soluble salt concentration, organic matter content, cation exchange capacity/nature of clay, and mineral reserves.

Whilst such methods can, with suitable refinements, give useful and reproducible results in specific areas and circumstances (see Subsection 5.5.2 below), they are definitely not recommended for universal application, since wildly inaccurate predictions can arise when a productivity index is used outside the area where it was developed. In addition, amongst several drawbacks (see McRae and Burnham, 1981) there are three main objections to their use:

a) Use of several arbitrary procedures - Despite the appearance of objectivity afforded by the use of standard procedures and numerical values, rating systems involve subjective - and often largely arbitrary - processes at five stages: the selection of properties to be used; the formulation of the equation; the ranges of the properties to be differentiated; the numerical value to be assigned to each range of each property; and the translation into planning or operational terms of the productivity index obtained. The final results are themselves therefore highly arbitrary and, at least in the initial development of a scheme, do not necessarily accurately reflect the individual importance and mutual interactions of land characteristics.

b) The possibility of bias (especially unconscious bias) - The choice of parameters and values to be used may result in just a few, or even one, land characteristic being the principal factor determining the final index. This arises because the parameters chosen may not be independent variables and because it is difficult to decide what parameter value to choose for, say, a given depth range to ensure that the same effect on productivity is reflected by a similar value assigned to, say, organic matter content. The resulting bias may not be spotted by an inexperienced operator and may lead to spurious results. If the actual productivity of the land is highly dependent on just a few land characteristics, then it is better to base a system on those characteristics explicitly.

c)   The complexity and obscurity of the rating – The rating obtained for an area is not immediately translatable into management terms, since it gives no direct indication of the nature or severity of individual land deficiencies requiring specific management practices. Similarly, the complexities involved in the design (as opposed to the operation) of the system obscure how the physical properties of an area contribute to the final designation, and this may alienate many potential users.

In areas where soil surveyors have to use a local rating system, particular attention must be paid to interpretation of the values obtained if the formula is not a multiplicative one such as the Riquier example above. In multiplicative systems the low value of unfavourable properties automatically reduce the final rating, and in extreme cases where an individual value is zero, the final rating is also zero. In additive systems, however, high values of some properties may produce a satisfactory final rating and totally obscure the presence of a very low, and even crucially limiting, value for some other property.

5.5.2   The merits of judiciously designed indices   Although on the whole the predictive value of parametric methods is too erratic to permit their recommendation as standard land evaluation aids, the post hoc assignment of numerically based ratings and rankings to soils, terrain types and land classes can be a rewarding exercise. That was how the FAO/ORSTOM multiplicative system was pioneered, as regards groundnut performance on Madagascar soils by Riquier (1965) and cotton etc in Togo by Vieillefon (1967): see also Frankart et al (1972) and Sys and Frankart (1972). The Swaziland ratings for maize, soil series by soil series, based on more than 700 harvests from farmers' plots (Murdoch, 1970) proved satisfactory guides in Rural Development Area planning for several years subsequently. For land taxation purposes, the Bonitierungskala was established in Germany (originally Bavaria – Fackler, 1924) as a soil and general physical resource assessment with politically agreed methods, definitions and interpretations.

There will be diminishing applicability of such ratings with time, as husbandry methods, management standards and crop cultivars change. The built-in obsolescence of land evaluation ratings has been well recognised in respect of perhaps the most successful system, the Storie Index. Since its origins half a century ago (Storie, 1933) this classification of soil and land for citrus has been revised locally many times (the current edition being Storie, 1978) and modified, with rigorous testing, for use in field conditions far from California where it was developed – there are versions for Colombian and Spanish orchards. Without the revisions/extensions the Storie Index might well have faded into obscurity. Land evaluation has to be dynamic – the last word is never spoken. Where farmers' attitudes and aptitudes are too disparate or heterogeneous, land evaluation at a district or regional level fails (Zwerman and Prundeanu, 1958; Young and Goldsmith, 1977) but this does not invalidate the conclusion that, for a defined management, or on a single farm/estate/settlement with a unified method of using the land, a ranking and mapping of physical features will always be useful and will sometimes be essential for planning purposes – assuming that good basic data are to hand and crop performance trends, field by field and soil by soil, are known.

Chapter 6

Soil Physics

6.1   <u>Background to soil physical measurements</u>

6.1.1   <u>Introduction</u>   Soil physical measurements are of very great importance for project planning and design, not least because they form the basis for many of the more costly operations that are required in project development. This chapter describes the main tests that are routinely carried out, and the interpretation of results for project implementation.   Details of practical procedures are given in Annex B and descriptions of the basic theories in a number of soil physics textbooks (Baver et al, 1972; Childs, 1969;   Hillel, 1971;   James et al, 1982;   Kirkham and Powers, 1972;   Taylor and Ashcroft, 1972;   Yong and Warkentin, 1975).   Methodology is discussed in standard texts such as Black (1965, Volume 1), which also includes sections on observation errors, bias, operator variation and sampling.

The study of soil physics is often hindered by the division of the subject into broad theoretical considerations on the one hand, and immediate practical considerations on the other.   Many of the former tend to concentrate on soil, or models of soil, as a basis for description and quantification of idealised behaviour, whilst the latter are often limited to technical specifications suitable only for very localised applications.   In either case, interpretation of the results for general project use is often lacking.   This chapter concentrates on a few simple tests that can be used for practical characterisation of soil physical properties.

6.1.2   <u>Interpretation of soil physical data</u>   Broad interpretations of soil physical data are often difficult to quantify accurately because of the nature of the soil and/or the tests employed.   As a general principle, as much information as possible should be used from projects already in operation, rather than the somewhat artifical results from the tests.   Difficulties in interpretation of test results arise for a number of reasons, which include:

a)   the degree to which the soil sampled represents the natural soil under consideration: soil spatial variation (see Section 3.7 and Warrick and Nielsen, 1980), poor sampling techniques,   and   disturbance   of   natural   soil   conditions (especially of structure) can all contribute to the production of unrepresentative test results;

b) the amount of external influence before or after sampling:
many physical properties can be substantially altered by soil
treatments such as cultivation practices or by factors such as
a change in soil-water content; test interpretations or
comparisons, unless made under standard conditions, are
therefore very difficult;

c) differences in methods used for testing: standard test
procedure varies considerably between countries and
organisations - and varies in particular between the use of
laboratory and on-site testing. Wherever possible the latter
is preferable, but again care must be taken in making
comparisons with other methods.

From the viewpoint of designing, as well as interpreting, a
programme of soil physical measurements, the variability of soil
properties is crucial. Table 6.1 lists some common physical tests
and examples of measured variation over short distances; the high
variabilities in many of the results serve to emphasise that it is
better to have several less precise, but carefully sited,
measurements than just a few highly accurate results. In
addition, most tests only provide 'order of magnitude' results for
broad project design; more precise information can only be
obtained by much more detailed work that is seldom economic for
routine surveys. Discussions of variability of some physical
parameters, and minimum sample numbers are given in Keisling et al
(1977) and Warrick and Nielsen (1980).

Table 6.2 summarises the main soil physical tests and their
interpretations and further information on each is given in the
following sections; as far as possible in situ tests are used to
minimise the effects of soil disturbance.

6.2   Infiltration

6.2.1   Background to infiltration measurements   Infiltration refers to
the vertical intake of water into a soil, usually at the soil
surface, and measurements of infiltration rates form a vital part
of many surveys involving irrigation development or soil
conservation. The results are normally used at the design stage
with other measurements (permeabilities, crop water demands,
evaporation data etc) for determining the most efficient method(s)
of application of irrigation water, and attendant details such as
furrow length and application rates. The results may also be used
in run-off calculations. Ideally, the measurements should be made
under conditions as similar as possible to those that will obtain
when the development is implemented. Thus the same irrigation
method should be used, the soil should have a similar moisture
content, and water of similar quality should be employed. In
practice, however, it is often difficult to meet all these
conditions and compromise methods are used; the results must
therefore be interpreted with great care.

6.2.2   Flooded basin infiltration   This involves the use of basins with
areas usually between 3 and 10 m^2. The soil surface is prepared
in a manner similar to that to be used when the land is developed,
and a number of graduated measuring staffs are located within the

Soil physics

Table 6.1

Property	Unit	Mean	Standard deviation	Coefficient of variation, v (%)	No of samples	Area (ha)	Comments
A. Low variation							
Bulk density	g cm^{-3}	1.3	0.09	6.9	64	15	5 combined depths
							5 series
		1.4	0.095	6.8	20	150	6 combined depths
		1.5	0.11	7.3	N/A	1.3	30–60 cm depth
Saturation percentage	vol %	40	4.5	11	20	150	30 cm depth
		45	4.8	11	20	150	6 combined depths
		47	4.8	10	N/A	225	15–30 cm depth
B. Medium variation							
Sand/silt/clay	%	53/28/19	15/9/7	28/32/36	64	15	30 cm depth
		59/29/12	22/18/6	37/62/53	64	15	5 depths
		26/27/47	11/6/8	42/22/17	20	150	30–45 cm depths
		24/30/45	14/8/10	58/27/22	20	150	12 combined depths
Moisture contents 0.1 bar	wt %						
		27	5.4	20	64	15	30 cm depth
		23	9.2	40	64	15	5 combined depths
15 bar		9.5	3.1	33	64	15	30 cm depth
		7.5	3.8	51	64	15	5 combined depths
		4.5	1.4	31	N/A	1.3	30–60 cm depth
Moisture contents 0.2 bar	vol %						
		32	5.4	17	20	150	30 cm depth
		32	7.7	24	20	150	6 combined depths
2.2 bar		34	4.1	12	N/A	225	15–30 cm depth
Infiltration rate 1/	cm h^{-1}	12.7	2.8	22	11	< 10 m^2	) Each value obtained
		1.4	1.1	8	11	< 10 m^2	) from 11 tests at 1
							) site. For several
							) sites, v varied
							) from 6 to 23%
C. High variation							
Saturated hydraulic	cm h^{-1}	14	26	190	64	15	5 combined depths
conductivity	cm day^{-1}	20	22	110	20	150	30 cm depth
		35	30	86	20	150	6 combined depths
Electrical conductivity (EC$_e$) 2/	mS cm^{-1}	9.9	9.0	91	232	150	) Average values for
		8.4	6.3	74	682	440	) 0–30, 30–60 and
		2.4	5.4	226	710	455	) 60–90 cm depth

Adapted from: Warrick and Nielsen (1980).

Notes: 1/ From Turner and Sumner (1978).
         2/ From Hajrasuliha et al (1980).

Summary of routine soil physical measurements and their interpretation   Table 6.2

Measurement	Recommended method(s)	Preferred units	Range of values and comments	
Infiltration rate	Basin, furrow or ring at site	cm h^{-1}	< 0.1	Rating for surface irrigation: Unsuitable (too slow) except for rice
			0.1-0.3	Unsuitable (too slow); marginal for rice
			0.3-6.5	Main suitable range (> 0.3 unsuitable for rice)
			6.5-12.5	Marginal (too rapid)
			12.5-25.0	Only suitable in special conditions (small basins)
			> 25.0	Only suitable for overhead irrigation
				Note: Values from ring infiltration may be high because of lateral seepage
Hydraulic conductivity	Auger hole or inverse auger hole	m day^{-1}	< 0.2	Very slow
			0.2-1.4	Slow to moderate
			1.4-3.0	Moderately rapid to rapid
			> 3.0	Very rapid
				Note: both horizontal and vertical components should be considered
Bulk density	Replacement at site and/or undisturbed core in laboratory	g cm^{-3}	0.9-1.2	Recently cultivated soil
			1.1-1.4	Main range uncultivated, uncompact soil
			1.6-1.8	Sands and loams) Ranges that
			1.4-1.6	Silts) restrict
			Very variable	Clays) roots
Porosity	From bulk density tests	% by vol	30-70	Usual range in soils
			10	Approx minimum value for aerobic soils
Field capacity (FC)	In situ tests	mm m^{-1} (% by vol)	100-450	High values for clays; low values for sandy soils
Permanent wilting point (PWP)	Pressure membrane method at 15 bar	mm m^{-1} (% by vol)	50-250	High values for clays; low values for sandy soils
Available water capacity (AWC)	FC - PWP	mm m^{-1} (% by vol)	50-230 but mostly 70-190	Approximate range for stone-free tropical soils. High values (> 180 mm m^{-1}) for soils with very fine sandy and silty textures; moderate values (120-180 mm m^{-1}) for clayey soils; low values (< 120 mm m^{-1}) for sandy soils
Water content	Gravimetric	% by vol or by mass	See ranges above	Tensiometers and neutron probe methods need careful calibration; former unsatisfactory on gravel soils
Structure	Water immersion	-	Class 1-8	Class 1 (least stable) to Class 8 (most stable); may be great variation within classes. Limited use in routine surveys
Strength	Penetrometer	kg cm^{-2}	Highly variable	Careful calibration and operation needed; highly dependent on water content; limited use in routine surveys
Stone content, and particle size distribution	Wet and dry sieving; sedimentation	Stones: % by vol Fines: % by wt	) See) Section 6.9	Note variation due to different pretreatments

basin. After construction of a bund of suitable height around the basin, water is introduced and the rate of intake into the soil is calculated from readings on the measuring staffs. In some circumstances, it may be desirable to maintain a constant head of water in which case the infiltration rate is obtained by relating the rate of water inflow to the surface area of the plot. Additional accuracy is achieved by surrounding the basin with an outer 'moated' area to minimise lateral seepage.

Although carried out under representative field conditions, the method suffers from the practical limitation that comparatively large areas of prepared land and large volumes of water are required. For this reason, the cylinder infiltrometer method (see below) is preferred in most surveys.

6.2.3   Cylinder infiltration   Cylinder, or ring, infiltration is usually the most practical and simplest method for the soil surveyor. Either a single or double ring infiltrometer can be used, but the latter method is preferred because it reduces errors due to effects at the edges of the inner ring.

The measurement, described in detail in Annex B, Section B.1, is usually performed in triplicate at any given site, and the three stations, not less than 10 m apart, should be located near to a described soil profile. At each pre-wetted station, a large- and a small-diameter steel cylinder are hammered concentrically into the ground to a depth of about 15 cm and levelled. Each ring is filled to an equal height about 15 to 20 cm above the surface of the ground, with water of the same quality as that to be used for subsequent irrigation. The height of water in the inner ring is allowed to fall about 5 to 10 cm between refills up to its original level, and the height of water in the outer ring is adjusted throughout to follow these changes.

Rates of flow are establised from measurements of the water-level at predetermined time intervals. In practice, the rate of inflow diminishes with time and the experiment is terminated when the rate has become constant, which is normally after 3 to 5 h of infiltration, depending on the soil.

6.2.4   Furrow infiltration   Infiltration rates may also be determined by passing water down a furrow of known length and wetted cross-section and measuring the inflow and outflow rates. The flow rates may be obtained empirically or by the use of a 'V' notch or rectangular weir sections associated with standard formulae (see Ven Te Chow, 1959). Ideally, the furrow selected for testing should be in the centre of an area being similarly treated. Where this is not possible, at least two or three furrows should be irrigated on either side of the one under test. For furrow irrigation the intake rate is usually expressed in m^3 min^{-1} per 10 m of furrow (or gal min^{-1} per 100 ft of furrow).

As with basin irrigation (Subsection 6.2.2), comparatively large volumes of water are required, but the method does approximate to an 'actual' irrigation practice, and can be useful in irrigation planning (see Borden et al, 1974).

An example of the graphical determination of infiltration advance rate from simple field tests is given by Finkel and Nir (1960) and is shown in ILRI (1972, p 537).

6.2.5   Sprinkler infiltration   This may be determined by the use of portable sprinkler infiltrometer equipment (see Tovey and Pair, 1963), or by utilisation of irrigation sprinklers with different sized nozzles on the site to be tested. The soil is normally brought to field capacity by use of the sprinklers and then rain gauges are strategically placed throughout the area (see Pair et al, 1969). The sprinklers are restarted and observations made adjacent to the rain gauges as to whether the application rate is slower than, faster than or equal to the infiltration of the water. At the conclusion of the test, the application rates determined from the rain gauges are related to the observations made in different areas of the site under test.

6.2.6   Calculation and presentation of infiltration data   The data from ring infiltration tests can be treated in a great many ways, but for most investigations the quantities required can be derived from a linear plot of the cumulative intake or instantaneous intake rate against elapsed time (see Figure 6.1). The readings from the three replicates are plotted separately on the same sheet of paper, and the individual values are calculated. Although infiltration results are often skewed, with a greater chance occurrence of relatively high rates (see Turner and Sumner, 1978) it is usually sufficient to take the arithmetic mean of the individual values. If several measurements are available, the geometric mean may be slightly more useful, but more complex estimates (Slater, 1957; Bouwer, 1969) are not usually worth while.

The quantities most commonly required for design purposes are:

a)   the basic infiltration rate, which is the relatively constant infiltration rate that develops after some 3 to 5 h;

b)   the time required to reach the constant infiltration rate;

c)   the time required for a given depth of water to infiltrate.

For most soil reports, it is normally sufficient to quote the basic infiltration rate, with additional tables giving cumulative intakes and times, and average and immediate infiltration rates (see Annex B). Further information, for example, profile textures, can also be shown in the tables as an aid to subsequent data interpretation.

It should be noted that deviations from ideal behaviour in homogeneous soils can result from delays in the first readings when the water has already started infiltrating, and from the presence of cracks and holes which are filled up in the first minute or so. In heterogeneous soils, some anomalous results are to be expected. If the subsoil has a different permeability from that of the topsoil, the curve will often show a change of slope at the point when the wetting front passes the boundary in the soil.

Figure 6.1

Infiltration rate: accumulated and instantaneous intakes

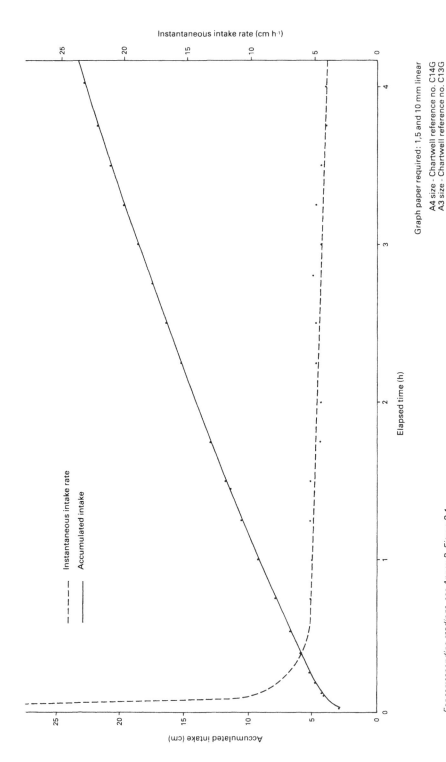

Graph paper required: 1,5 and 10 mm linear

A4 size - Chartwell reference no. C14G
A3 size - Chartwell reference no. C13G

For corresponding readings, see Annex B, Figure B.1.

6.2.7   Alternative forms of treatment of infiltration data   In some circumstances, soil surveyors may be required to produce or interpret more detailed analyses of infiltration results.   Among the more common quantities used are the following;   details are given in Annex B.

a)   The cumulative infiltration - (F) - This quantity is the characteristic which is measured in the field.   The measurements of F are plotted on log-log paper against time.

b)   The instantaneous intake rate - (IR) - This is the volume of water infiltrating through a horizontal unit area of soil surface at any instant.   It shows, in general, a rapid decrease at the beginning followed by a more stable, very slow decrease after some 3 to 4 h of infiltration (see Figure 6.1).

c)   The average infiltration rate - (ĪR) - This quantity is the cumulative infiltration F divided by the time from the start of infiltration.

d)   The Philip equation - A commonly employed equation for extrapolating infiltration results to the steady state is the Philip equation (Philip, 1954), which refers to the isothermal movement of water into homogeneous soils in one dimension. The equation can be stated as follows:

$$F = at^{0.5} + bt$$

where F = cumulative depth (cm), t = time (s)

and a and b are constants, respectively soil sorptivity (or storage capacity, depending on soil properties and ability to transmit water) and initial moisture content.

Limitations to the Philip equation are discussed in Taylor and Ashcroft (1972) and Turner and Sumner (1978);   in particular, the equation may not be valid when the water flow is not linear, which occurs notably when:

a)   soils are not homogeneous (and note possibility of changes during the test in swelling and shrinking clays);   and

b)   water is infiltrating in more than one dimension (eg in the early stages of furrow irrigation).

However, deviations from linearity may be small in many cases, particularly in the later stages of infiltration, and the equation can still yield sufficiently accurate results.   Direct measurements are to be preferred wherever possible, however.

Additional complexities can be allowed for by use of other equations (see Haverkamp and Vauclin, 1981, and Hachum and Alfaro, 1980), but where these are needed it is best to rely on field measurements alone.

6.2.8   <u>Sources of variation in infiltration rate measurements</u>   Ring infiltration is probably the most common infiltration measurement made during soil surveys.   However, the method can suffer from a number of errors, which often arise because the conditions of measurement do not accurately reflect the conditions under which water will be applied to the soil during project operations (see Parr and Bertrand, 1960, and Hills, 1970).   The more common errors include:

   a)   <u>Lack of adequate pre-wetting</u> – Very dry, expanding clay soils, for example, may need weeks of pre-wetting before major shrinkage cracks are closed.

   b)   <u>Trapping of air under the cylinder</u> – This can reduce the measured rate considerably, although some air entrapment may occur during irrigation and thereby also reduce the 'real' rate (cf Whisler and Bouwer, 1970).

   c)   <u>Soil disturbance effects</u> – The effects of cultivation during project development may alter the soil-water intake characteristics; pans and surface capping, for example, may be broken up.

   d)   <u>Irregular wetting effects</u> – With only a small area wetted, as under a cylinder, water penetration may proceed largely along a major crack or hole and give an unrepresentatively high measurement.   Sideways seepage above a less permeable horizon may also account for an unrealistically high proportion of the flow.

   e)   <u>Use of different quality water to that to be used for the development</u> – Both suspended and dissolved material can have very marked effects.   Silt suspended in the water can almost completely seal a soil surface, whilst higher infiltration rates are obtained if the electrolyte concentration of the applied water is increased (see Quirk, 1957).

Other reasons for variability include the nature of water flow in a soil and its initial water content (Turner and Sumner, 1978). Poiseuille's Law of flow in capillaries relates the rate of flow to the fourth power of capillary radius (Hillel, 1971), and so a small difference in capillary radius can account for correspondingly much larger differences in flow rate.   Although flow conditions in soils are obviously far more complex than in simple capillaries, the relationship helps to explain the high variation in infiltration rates which can occur over short distances within apparently uniform soils.   Examples of such variability, and additional differences produced by changes in the initial soil-water content, are shown in Table 6.3.   A further example of the effect of initial soil-water content is illustrated in Figure 6.2, which emphasises the need to standardise, or at least state, this water content for all infiltration tests; the variation is, however, slightly lower for the higher moisture contents, so pre-wetting a site will help to reduce such errors.

Example of variation in basic infiltration rate with initial   Figure 6.2
soil-water content

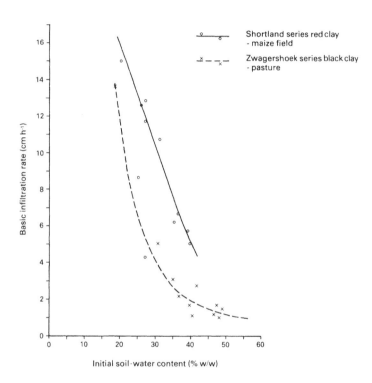

Source: Adapted from Turner and Sumner (1978).

Illustrative variations of basic infiltration rate with   Table 6.3
initial soil moisture content

	Initial moisture content (% W/W)					
	54.1	56.7	26.7	40.4	36.9	49.1
Infiltrometer	Measured basic infiltration rate (cm h^{-1})					
ring No 1/	Balmoral series 2/		Shortland series 2/		Zwagershoek series 2/	
1	18.0	16.2	12.4	4.4	1.5	0.8
2	12.0	10.2	11.9	3.8	0.2	0.0
3	18.0	8.4	12.1	5.5	0.6	0.1
4	15.6	9.0	10.1	4.1	2.4	1.6
5	9.0	3.6	12.1	6.1	5.3	4.1
6	8.0	9.0	11.0	5.2	0.8	0.2
7	8.4	15.6	15.2	5.4	1.7	1.2
8	11.4	4.2	12.1	4.9	3.7	2.0
9	12.0	8.2	8.8	3.0	3.2	1.8
10	21.6	10.2	18.7	7.4	2.7	1.5
11	7.5	3.0	15.0	6.0	1.1	0.6
12	7.7	12.6	–	–	3.6	1.4
Mean	12.4	9.2	12.7	5.1	2.2	1.4
SE	4.68	4.26	2.72	1.22	1.52	1.09
CV	38	46	22	24	69	78

Notes: 1/   Infiltrometers located 1 m apart on each soil series.
          2/   For each series, measurements in the second column
                were made 2 days after those in first column.

Source: Turner and Sumner (1978).

Because of all these variations in infiltration rate, it is most
important to ensure adequate replication of infiltration tests,
preferably in numbers sufficiently large to allow statistical
analysis.   Individual measurements are obviously of very limited
use on their own, and it is not worth wasting time to obtain
extremely accurate values except, possibly, in detailed management
studies.   Ring   infiltration   measurements   can   only   give
order-of-magnitude results, and interpretations must take fully
into account the conditions of measurement.

6.2.9   Evaluation   of   infiltration   results   The   evaluation   of
infiltration rates is at best a subjective exercise.   Different
agricultural policies and methods of irrigation, or cropping or
farm management, can all affect the use of land and the practical
limits represented by particular infiltration rates.   Similarly,
the applicability of the method of measurement to the conditions
expected during project development must be considered.   Ring
infiltration, for example, may give some broad indications of soil
behaviour for basin irrigation but, under sprinkler irrigation or

natural rainfall, rates are likely to be quite different, not least because different theoretical boundary conditions apply (see Rubin, 1966), and effects of raindrop impact may be important. It is vital, therefore, to view any standard infiltration criteria in relation to local conditions and specific project objectives, and the following section should be regarded as only a general guide to interpretation of results.

Optimum basic infiltration rates for surface irrigation are considered to be in the range of 0.7 to 3.5 cm h^{-1}, although acceptable values normally range from about 0.3 to 6.5 cm h^{-1}. This is reflected in the BAI system shown in Table 6.4; an alternative treatment, classifying rates on the basis of the time taken to infiltrate 10 cm of water is given by van Beers (1976).

Suggested infiltration categories   Table 6.4

Class	Infiltration category	Basic infiltration rate (cm h^{-1})
1	Very slow (non-irrigable)	< 0.1
2	Slow	0.1 – 0.5
3	Moderately slow	0.5 – 2.0
4	Moderate	2.0 – 6.0
5	Moderately rapid	6.0 – 12.5
6	Rapid) (overhead methods	12.5 – 25.0
7	Very rapid) preferred)	> 25.0

Source: BAI (1979).

a)   Slow infiltration rates – When basic infiltration values are less than about 0.3 cm h^{-1}, surface water losses during irrigation are liable to be excessive; deep percolation for leaching is liable to be difficult to achieve, and crops may be damaged in hot weather by scalding action; in some circumstances, yields could be depressed by lack of aeration in the root zone due to temporary excess moisture conditions. Therefore, soils which have average rates < 0.1 cm h^{-1} are usually considered non-irrigable for crops other than rice. For rice, infiltration rates < 0.1 cm h^{-1} are generally required, and this is usually achieved by puddling the soil during land preparation. Ideally, this should reduce infiltration rate to 0.02 cm h^{-1} or less.

b) <u>Fast infiltration rates</u> – When basic infiltration rates become faster than about 6.5 cm h^{-1}, soils are considered increasingly unsuitable for surface irrigation, and overhead methods become more applicable. This is due to poor uniformity of application, deep percolation losses, excessive leaching of crop nutrients and low irrigation efficiency. There is, however, no specific limiting value of infiltration rate for surface irrigation. The upper limit is frequently taken as between 10 and 12.5 cm h^{-1}, although it is acknowledged that rates of up to 25 cm h^{-1} may occur in special circumstances in the irrigation of some sandy soils. FAO (1979a, p 32) quotes 12.5 cm h^{-1} as the upper limit for gravity irrigation except in small basins. Similarly van Beers (1976) classifies as 'marginal' soil infiltration rates of 10 to 20 cm h^{-1}.

c) <u>Rating of IR values</u> – Table 6.5, compiled from the various sources mentioned in this subsection, summarises the suitability for surface irrigation of soils with different infiltration rates. The rating applies to most development projects, but there may be schemes – particularly those involving smallholders and/or highly labour-intensive methods – where different criteria apply. As a very general guide, the representative rates given in Table 6.6 can be used as indicative of values applying to different textural categories.

Rating of infiltration rates for surface irrigation   Table 6.5

Basic infiltration rate (cm h^{-1})	Suitability for surface irrigation
< 0.1	Unsuitable (too slow) but suitable for rice
0.1 – 0.3	Marginally suitable (too slow); marginally suitable for rice
0.3 – 0.7	Suitable; unsuitable for rice
0.7 – 3.5	Optimum
3.5 – 6.5	Suitable
6.5 – 12.5	Marginally suitable (too rapid). Small basins required
12.5 – 25.0	Suitable only under special conditions. Very small basins required
> 25.0	Unsuitable (too rapid); recommended for overhead methods only

Infiltration rates related to soil texture   Table 6.6

Soil texture	Representative $\overline{IR}$ (cm h^{-1})	Normal range of $\overline{IR}$ (cm h^{-1})
Sand	5	2 – 25
Sandy loam	2	1 – 8
Loam	1	0 1 – 2
Clay loam	0.8	0.2 – 1.5
Silty clay	0.2	0.03 – 0.5
Clay	0.05	< 0.01 – 0.8

Sources: Israelsen and Hansen (1962), quoted in FAO (1979a, p 34).

6.3   Hydraulic conductivity

6.3.1   Background to hydraulic conductivity measurements   The hydraulic conductivity (or permeability) of a soil, K, in cm h^{-1} or m day^{-1}, defines the volume of water which will pass through unit cross-sectional area of a soil in unit time, given a unit difference in water potential.   Due to rapid draining of the larger pores there is a rapid decrease in K with decrease in water content in unsaturated soils.   When the water content is reduced to the field capacity moisture content K commonly decreases to 1/100 or 1/1000 of its value at saturation 1/.

In practice, the saturated hydraulic conductivity is the property of most interest, measurements being made for two main purposes:

a)   for comparison of hydraulic conductivity rates of different soil horizons, particularly as a guide to water movement and possible drainage problems within soil profiles; and

b)   as a basis for in-field drainage design.   It should be noted that, for this purpose, measurements may often be required to about 2 m depth and regional drainage studies may call for even deeper measurements.   In either case measurement is needed below the proposed drain depth, since flow into drains develops a radial pattern perpendicular to the drainage line.

Subsurface flow in both vertical and horizontal directions is therefore important in, for example, the functioning of subsurface drains.   In anisotropic soils the differences can be highly significant, but many simple methods of measurement represent predominantly vertical or horizontal flows – a fact which must be remembered in interpretation of values.   Large differences in the

---

1/   For unsaturated hydraulic conductivity see Taylor and Ashcroft (1972, p 188) and Black et al (1965, pp 234ff).

two component flows may also occur in otherwise apparently uniform soils, because of the effects of pore/capillary sizes on water flow (see Subsection 6.2.8).

Both field and laboratory determinations of soil hydraulic conductivity can be made, but correlation between the two methods is often very poor, and obtaining genuinely undisturbed cores for laboratory tests can be exceedingly difficult. For these reasons, it is recommended that only field measurements should be undertaken; these are discussed in the following sections which cover measurements above and below the water-table. A comparison of field methods is given in Talsma (1960), Sillanpää (1959), Bouma et al (1982) and Bouma (1983). Indications of the large variations between field and laboratory tests are presented in Lal (1981, p 144).

The comments on water supply and soil conditions made for infiltration rates (Subsection 6.2.1) also apply to these hydraulic conductivity tests, detailed descriptions of which appear in Annex B, Section B.2.

6.3.2   Theoretical background to hydraulic conductivity measurements Saturated hydraulic conductivity is a constant referring to the flow of a fluid through a saturated conducting medium, derived from an empirical relationship established by Darcy (1856) between the rates of flow of water through saturated columns of sand and the hydraulic head loss; this may be expressed as follows:

$$q = KA \frac{h}{L}$$

where: q = volume rate of flow across a plane normal to the direction of flow

K = hydraulic conductivity, which is the volume rate of flow through a sample of unit cross-sectional area under the influence of a unit hydraulic potential or head gradient

A = cross-sectional area through which the flow takes place

h = hydraulic head expended in moving water from one side of the sample to the other

L = length of the sample in the direction of flow

Darcy's Law only applies to truly laminar flow, which is likely to occur in natural soil-water movement, but problems can arise in laboratory determinations.

6.3.3   Hydraulic conductivity measurement below the water-table: the auger-hole method   The simplest approach is described in Annex B, and bearing in mind the variations of hydraulic conductivity which occur in the field, it is probably the most effective method; experimental details are given in Annex B, Subsection B.2.2.

Sites and stations are located as for infiltration tests (Subsection 6.2.3), and at each station a hole of about 8 cm diameter is augered to a depth below the water-table, care being taken to eliminate smearing of the sides (Smitham, 1970). The hole is finished off with a special cutting tool designed to produce a flat bottom. The water in the hole is pumped out, or removed with a bailer, and the rate at which the water flows back in is measured until the hole is about a quarter full; the hydraulic conductivity is then calculated as shown in Annex B. If the hole penetrates two or more horizons below the water-table, the hydraulic conductivity of each horizon can be determined approximately using the method described by van Beers (1976). In such heterogeneous profiles, however, the piezometer method may be preferable (see Johnson et al, 1952; Luthin and Kirkham, 1949; review by Luthin, 1957; and ASAE, 1962).

Care should be taken to complete the measurements before 25% of the volume of water removed from the hole has been replaced by inflowing groundwater. After that, a considerable funnel-shaped water-table develops around the top of the hole. This increases resistance to the flow around and into the hole, and the effect is not allowed for in the formulae or flow charts developed for the auger hole method.

The auger-hole method is unsuitable for use in very stony or coarse soils because of the difficulty of augering and maintaining a uniform hole in such materials. It is also unusable when artesian conditions occur, or in soils containing small sand lenses within less permeable material, or in anisotropic soils, since the value measured by the method is mainly the horizontal component of the hydraulic conductivity. In general, however, the method is suitable for most agricultural soils.

Calculations for the one- and two-layer situations, employing the Ernst formula (Ernst, 1950), are presented in Annex B, Subsection B.2.5. A good rule of thumb is that the rate of rise in mm s^{-1} in an 8 cm diameter hole to a depth of 70 cm below the water-table approximately equals the K value of the soil in m day^{-1}. Other checks are given in the calculations in Annex B.

6.3.4 Hydraulic conductivity measurement above the water-table: the inverse auger-hole method   The inverse auger-hole method (see Annex B, Subsection B.2.8) is an auger-hole test above the water-table, and is described in French literature as the Porchet method. It consists of boring a hole to a given depth, filling it with water and measuring the rate of fall of the water-level.

Due to the swelling properties of soil, a K value obtained by this method may differ from one obtained if the soil is saturated. If this change of structure is significant, it has to be taken into consideration when the measured K is evaluated. Similarly, the test should not be done on a dry site but only after saturating the test site, for example by conducting the test immediately after an infiltration measurement. As noted for infiltration tests, air entrapment within the soil can affect values (see also Subsections 6.3.5 and 6.3.8).

6.3.5   <u>Hydraulic conductivity measurements above the water-table: other
methods</u>   In some circumstances it may be more appropriate to use
a method which employs a constant depth of water in the auger hole,
such as the shallow well pump-in test, also called the well
permeameter test.   The experimental set-up is slightly more
cumbersome, but is not difficult to use (see FAO 1979a, p 173f).

Values of K determined by this method are usually lower than those
obtained by the auger-hole method (ASAE, 1962), this difference
possibly varying with the size of K (Talsma, 1960); some measured
values of the ratio of K (pump-in) : K (auger hole) are 0.5, 0.85
and 0.72 to 0.97 (Talsma, 1960; Winger, 1956; Sillanpää, 1959
respectively).

The cylinder permeameter test (Black et al, 1965) permits field
measurements of the vertical component of hydraulic conductivity of
restricting or slowly permeable layers that obstruct percolation
and cause perched water-tables.   The other tests mainly measure
lateral rather than vertical permeability.

6.3.6   <u>Depth of hydraulic conductivity observations</u>   It is advisable to
obtain a rough assessment of the hydraulic properties of the soils
in an area and of the area's hydrologic conditions as early as
possible during a survey, so as to have a general idea of the
desired type of drainage system.   The anticipated depth of such a
system, for instance, is a factor to be kept in mind.

The K value of the soil layer in which the drain will be placed is
important because a large part of the energy flow will be expended
in this layer due to the radial contraction of stream lines to the
drain.   It may, for example, be preferable to design the drain
level a little deeper if the bottom of the drain then cuts through
layers that are more permeable.   One can then combine the
advantage of wider drain spacing with increased storage or, in arid
regions, it can mean a reduction in salinisation by capillary
rise.   Sometimes, however, the high excavation costs, the
instability of the deeper soil or seepage problems may prevent such
a design.

There is, furthermore, a close relationship between the spacing of
a drainage system and the effective depth of groundwater flow.
This depth may be taken as the maximum depth of the 90% flow line,
indicating that 90% of the flow takes place above this depth.   In
homogeneous soils the effective depth is around one-quarter to
one-eighth of the drain spacing.   If the soil becomes less
permeable with depth, the effective depth is less than one-eighth
of the drain spacing.   The converse is also true.   Consequently,
if a layer with a relatively low hydraulic conductivity is found at
a depth of one-twelfth of the drain spacing, the K value of this
layer need not be intensively investigated and the layer can be
considered impervious.   If, on the other hand, seepage problems
are being investigated, and if the layer of relatively poor
permeability is part of a semi-confined aquifer, its hydraulic
conductivity cannot be neglected.

For sound evaluation of drainage characteristics, investigations to a minimum depth of 3 m and occasionally more are recommended by the FAO (1979a, p 7).

6.3.7 Limitations and variability of hydraulic conductivity measurements   The methods described measure K values in various directions and for various sizes of soil bodies.   These measurements should be regarded as point determinations only with systematic errors of the order of 10 to 20%.

For many of the same reasons given for infiltration rates (Subsection 6.2.8), the variability of individual hydraulic conductivity values made by any field method can be large, even where measurements are made at points a short distance from one another on the same soil type: differences of greater than 100% are not uncommon (see Table 6.1).   This variability is more marked in the top 45 cm of soil, where management has a pronounced influence on soil properties, notably on structure.

Adequate numbers of replicates are therefore essential;   in practice the the number of replicates per site is about three or four, but in highly variable soil – and depending on the level of survey – up to about 20 may occasionally be necessary. Sophisticated methods to obtain highly accurate values of individual tests are seldom worth while.

Because of these limitations on the quoted hydraulic conductivity figures, most results obtained at the prefeasibility or feasibility levels will normally only be sufficient to give orders of magnitude for hydraulic conductivities of the soils.   Interpretation of the results must therefore be made with caution.

6.3.8 Evaluation of hydraulic conductivity results   Hydraulic conductivity values are related to textural and structural characteristics of a soil, as summarised in Table 6.7.   The FAO classification is summarised in Table 6.8.

Soils with hydraulic conductivity values below 0.1 m day^{-1} require excessively close drain spacings, and hence some artificial modification of subsoil water movement by moling or subsoiling is essential for a practical and economical field drainage system. Hydraulic conductivities of 0.1 to 1.0 m day^{-1} are the most critical for drainage design, and the greatest accuracy in measuring K is required in this range.   Above 1.0 m day^{-1}, in-field drainage is unlikely to be required unless such rapid rates are liable to lead to a rapid rise in GWT levels (eg above rock or an impermeable soil layer).   Great accuracy is not required in the measurement of such rapid rates.

In soils with abrupt horizon changes, corresponding changes in the hydraulic conductivity values can have serious effects on the movement of irrigation or drainage waters within the profile.   As a rule of thumb, a horizon with a K value less than 10% that of the overlying horizon should be regarded as effectively impermeable.

Approximate relationships between texture, structure and hydraulic conductivity

Table 6.7

Texture	Structure	Indicative hydraulic conductivity, K	
		(cm h^{-1})	(m day^{-1})
Coarse sand, gravel	Single grain	$\geqslant$ 50	$\geqslant$ 12
Medium sand	Single grain	25 - 50	6 - 12
Loamy sand, fine sand	Medium crumb, single grain	12 - 25	3 - 6
Fine sandy loam, sandy loam	Coarse, subangular blocky and granular, fine crumb	6 - 12	1.5 - 3
Light clay loam, silt, silt loam, very fine sandy loam, loam	Medium prismatic and subangular blocky	2 - 6	0.5 - 1.5
Clay, silty clay, sandy clay, silty clay loam, clay loam, silt loam, silt, sandy clay loam	Fine and medium prismatic, angular blocky, platy	0.5 - 2	0.1 - 0.5
Clay, clay loam, silty clay, sandy clay loam	Very fine or fine prismatic, angular blocky, platy	0.25 - 0.5	0.05 - 0.1
Clay, heavy clay	Massive, very fine or fine columnar	< 0.25	< 0.05

Source: Adapted from FAO (1979a, p 177).

Classification of hydraulic conductivity values   Table 6.8

| Hydraulic conductivity (K) | | Conductivity class |
$(m \ day^{-1})$	$(cm \ h^{-1})$	
< 0.2	< 0.8	Very slow
0.2 – 0.5	0.8 – 2.0	Slow
0.5 – 1.4	2.0 – 6.0	Moderate
1.4 – 1.9	6.0 – 8.0	Moderately rapid
1.9 – 3.0	8.0 – 12.5	Rapid
> 3.0	> 12.5	Very rapid

Source: FAO (1963).

The use of hydraulic conductivity measurements in calculations of drain spacing is given in, for example, ILRI (1972-74).

6.4   Bulk density

6.4.1   Background to bulk density measurements   Bulk density measurements are made in the course of many routine soil surveys as a guide to soil compaction and porosity (see Section 6.5).   The results are used as indicators of problems of root penetration and soil aeration in different soil horizons.   Bulk density values vary considerably with moisture content, particularly those of fine-textured soils;   samples should therefore be taken at or near to field capacity.

Bulk density refers to the overall density of a soil (ie the mass of mineral soil divided by the overall volume occupied by soil, water and air), and it should be distinguished from the density of the solid soil constituents, usually called the particle density, which is conventionally taken as 2.65 g cm^{-3} (see also Annex J).   The weight of soil solids in bulk density measurements is taken as the oven-dry constant weight at 105°C (BSI, 1975).

6.4.2   Replacement methods for bulk density measurements   The method of Avery and Bascomb (1974) is probably the simplest and most accurate.   A hole of about 20 cm x 20 cm area x 10 cm deep (about 4 ℓ volume) is carefully cut at the desired depth without causing compaction or shattering.   The removed soil is dried and weighed (or more commonly the removed soil is weighed moist and a subsample sent for drying and reweighing).   The original volume occupied by the soil is then determined by recording the number of spherical plastic balls of known packing density that are required to refill the hole;   the balls commonly used have a diameter of about 20 mm.

The only sizeable error which may occur is that due to difficulty in judging when the hole is exactly filled.   This error diminishes as the hole depth increases, but there are still liable to be differences between measurements by individual surveyors.   The number of balls required to fill a known volume of similar dimensions to the holes to be used in the field should therefore be made by each surveyor before the experiments are started. This will also act as a check on the dimensions of the balls, which may vary slightly between batches.

The bulk density of the removed soil is calculated thus:

$$\text{Bulk density} = \frac{M}{V} \quad \text{expressed as g cm}^{-3}$$

where   M = oven-dry weight of soil removed from hole (g)

V = volume of hole (cm^3)
= number of balls x packing volume per ball (cm^3)

The method is applicable on most soils and is accurate, although slow, since minimum disturbance of the soil surrounding the hole is achieved. It is a development of the sand replacement method (Zwarich and Shaykewich, 1969), in which sand is used in place of the plastic balls. This latter method is not recommended as it is slower, and in fissured soils the sand tends to run into the cracks, resulting in an over-estimate of soil volume.

6.4.3   Core sampling for bulk density measurement   This method consists of taking a core sample of soil using a coring cylinder of known volume (BSI, 1976) which is driven into the soil and then carefully dug out. The cylinder (often referred to as a pF ring, since it is also used for water tension measurements, see Section 6.6) is usually about 5 cm long with a 5 cm diameter, but the larger the better. For transporting purposes, it should be fitted with tightly fitting caps for the top and base, particularly when moisture-release measurements are also to be made on the sample. The method is described in Annex B, Section B.3.

In general, the core-sampling method has a somewhat limited applicability. When used in very wet soils, soil friction on the sides of the sampling cylinder causes viscous flow and compression, and in dry soils, the sample may shatter. The method cannot cope with gravelly or stony soils. Under suitable conditions, however, it has the advantage of producing an undisturbed sample which can be used for non-destructive physical tests.

6.4.4   Auger-hole method for bulk density measurement   Using a 10 cm diameter auger, a hole 15 cm deep is bored (Zwarich and Shaykewich, 1969), and the oven-dry weight of the extracted soil determined. The volume of the hole is calculated from the measurements of depth and cross-sectional area.

This method is more accurate than the core-sampling method, but a uniform hole has to be made and the soil removed is completely disturbed and cannot be used for other physical determinations requiring undisturbed soil.

6.4.5   Replication of bulk density measurements   At least three replicate determinations are normally made at each site, but for detailed work up to 10 may be required. The oven-dry mass of the replicates are summed and divided by the total volume to obtain the mean bulk density. Small errors in measuring the volume are counterbalanced by the increased accuracy of a bulk density value derived from a large sample. Bulk density is normally expressed to the second decimal place (eg 1.32 g cm^{-3}), so that accuracy of mass or volume need not exceed 1%.

Bulk density replicates at a site may vary considerably; variations of at least 15 to 20% are to be expected in most soils.

6.4.6 Interpretation of bulk density results   The bulk densities of clay, clay loam and silt loam topsoils may range between 1.00 and 1.60 g cm^{-3} depending on their condition.   Sands and sandy loams usually show variations between about 1.20 and 1.80 g cm^{-3}. There is very often a tendency for bulk density values to rise with depth, as effects of cultivation and 'organic matter content decrease.   Very compact subsoils, of whatever texture, may have bulk densities exceeding 2 g cm^{-3}.

Bulk densities above 1.75 g cm^{-3} for sands, or 1.46 to 1.63 g cm^{-3} for silts and clays, are quoted by de Geus (1973) as causing hindrance to root penetration;   these values correspond broadly with those shown in Table 6.9:

Typical bulk density ranges   Table 6.9

Material	Bulk density (g cm^{-3})
Recently cultivated soils	0.9 – 1.2
Surface mineral soils, not recently cultivated, but not compacted	1.1 – 1.4
Soils showing root restriction:	
Sands and loams	< 1.6 – 1.8
Silts	< 1.4 – 1.6
Clays	Extremely variable 1/

Note: 1/   But note deleterious effects of reduced air-filled pore space (see Subsection 6.5.4) when bulk density ≥ about 1.3 g cm^{-3}.

Source: Taylor et al (1966).

Even in horizons of similar texture lying at similar depths, there are usually great differences in bulk density values depending on organic matter levels, root penetration and soil structure. Increases in soil density impose the following stresses on a plant's root system:

a) The mechanical resistance to root penetration increases, so reducing the plant's ability to exploit its environment.

b) The air-filled porosity of the soil decreases, thus restricting the air supply to plant roots and facilitating the build-up of toxic products such as carbon dioxide and ethylene. As well as decreasing total porosity, compaction of soil decreases the volume of coarse pores relative to the volume of fine ones, and hence also increases the proportion of total porosity occupied by water at any given suction (Russell, 1973). Relationships between bulk density, AWC and air capacity are discussed by Archer and Smith (1972).

c) In general, permeability decreases with increasing density, making field crops more susceptible to the adverse effects of waterlogging. Note that under certain circumstances, such as where crops are direct drilled, permeability may be largely governed by the size and abundance of fissures and macropores (Cannell and Finney, 1973).

Using cotton seedlings, Taylor and Gardner (1963) found that, under laboratory conditions in which they considered aeration to be non-limiting 1/, root penetration was related linearly to soil strength as estimated with a penetrometer. The range of soil strength studies was achieved by a number of combinations of bulk density and soil-water tension values, but only poor correlation was reported between root growth and these two parameters individually. Mirreh and Ketcheson (1972) showed that soil strength increases with both density and tension (see Section 6.6).

6.5   Soil porosity

6.5.1   Introduction   The quantity of pores in a soil and their size distribution (as reflected in estimates of total pore space, coarse porosity and air-filled porosity) are useful general indicators of the physical condition of soils. However, pore characteristics change between seasons as well as within a cropping season and are not readily quantifiable in their influence on crop productivity. Apart from quantity and distribution, the tortuosity and continuity of pores are important features influencing aeration, water movement and root penetration in soils, but they are less easily measured, and in most surveys only qualitative observations are made.

6.5.2   Pore sizes   There are, unfortunately, several different systems of pore-size designation in use, some based on size and others on function of the pores. The variety of systems means that ambiguous terms like 'micropores' should always be explained when used, preferably in terms of the appropriate equivalent cylindrical diameters. A selection of systems is illustrated in Table 6.10.

1/ Fine sandy loam at moisture tensions in the range 200 to 660 cm of water.

Selected terms and size ranges for soil pores

Table 6.10

Equivalent cylindrical diameter (μm)	FAO (1977a) after Johnson et al (1960)	Jongerius (1957)	IUPAC (1972)	Russell 1/ (1966)	Greenland 1/ (1977)	Comments
5 000	Coarse pores					
2 000	Medium pores					
1 000	Fine pores	Macropores		Main root penetration	Fissures	Diameter of 18 gauge wire
500	Very fine pores		Macropores			Diameter of thin leads for propelling pencils
100				Main soil-water movement	Transmission pores	Diameter of 40 gauge copper or nichrome wire; Approximate visible limit without hand lens
75		Mesopores				
60				Root hair penetration		
50			Mesopores	Soil aeration	Storage pores	
30	Micropores					
10		Micropores				Approximate visible limit without microscope
0.1			Micropores		Residual pores	
0.05						
0.005					Bonding pores	

Note: 1/ Class divisions are indicated by a dotted line because there is no abrupt change in function with pore size. For convenience, limits can be taken at the following equivalent cylindrical diameters (in μm): 500, 50, 0.5 and 0.005 (cf Table 6.11).

Source: Adapted from Greenland (1979).

81

The simplest of these that retains some relation to processes in the soil is that of Jongerius (1957); using his classes, the following observations can be made:

a) Coarse (macro) pores — Macropores have diameters greater than 100 μm (0.1 mm), and their main function is aeration and drainage by gravity flow. They are also the pores in which roots proliferate. They are visible to the naked eye.

Rapid water movement through soil and aeration at field capacity takes place through the network of coarse soil pores. Measures of coarse porosity are, therefore, of some value as indirect estimates of likely waterlogging and consequent anaerobism within profiles. Estimates of coarse porosity involve measurement of total porosity by volume sampling of soil at field capacity (see Subsection 6.6.3) or equilibration of carefully taken undisturbed soil cores at an appropriate water tension. Coarse porosity (% by vol) equals the total porosity (% by vol) less the volumetric water content at field capacity.

Additional measurements of hydraulic conductivity (see Section 6.2) are necessary for characterising the spatial distribution of coarse pores as channels for drainage.

b) Medium (meso) pores — Medium pores have diameters from 30 to 100 μm and their main function is conduction of water by rapid capillary flow; they are visible at times 10 magnification.

c) Fine (micro) pores — Micropores have diameters less than 30 μm and their main function is water retention and slow capillary flow. Micropores are not visible, but their presence can be inferred from observation of the face of aggregates: when the aggregates have a rough surface, there are many micropores.

The measurement of fine porosity in soil and its significance are considered in the section on available water capacity (see Subsection 6.6.5).

In general, soils should contain an adequate number of pores of ⩾ 250 μm diameter to allow good root penetration; at least 10% by volume of the soil in the rooting depth should be composed of interconnected pores ⩾ 50 μm in diameter to allow free drainage, and at least 10% by volume of the soil should consist of pores with equivalent cylindrical diameter between 0.5 and 50 μm to allow for storage of available water.

6.5.3   Total porosity   The total porosity, or total pore space, of a soil is calculated from the dry bulk density and particle density as shown below; it is normally expressed as a volume percentage and is equal to the volume % water content at saturation:

$$\text{Total porosity (volume \%)} = \left\{ 1 - \frac{\text{Dry bulk density}}{\text{Particle density}} \right\} \times 100$$

The particle density of most soils is normally assumed to be 2.65 g cm^{-3} (but see Section 6.7).

Example: if the bulk density is 1.40 g cm^{-3} and the particle density 2.65 g cm^{-3}, the porosity equals:

$$100 \left(1 - \frac{1.40}{2.65} \right) = 47\%$$

The total porosity of soils usually lies between 30 and 70% and may be used as a very general indication of the degree of compaction in a soil in the same way as bulk density is used. For example, sands with a total pore space less than about 40% are liable to restrict root growth due to excessive strength (Harrod, 1975) whilst, in clay soils, limiting total porosities are higher, and less than 50% can be taken as the corresponding approximate value.

However, it must be stressed that values of total porosity should not be used as conclusive evidence for over-compaction problems in soil, but rather as indicators of likely risk. Moreover, the calculation gives only the overall volume percentage of the pore space and does not characterise the size of the individual pores.

6.5.4   Air-filled porosity   Pores that are not filled with water are filled with air, and an estimate of their volume can give an indication of the aeration and drainage status of a soil. A common measure is the air capacity, which is the air-filled porosity of a soil at field capacity. It corresponds to the volume of pores that have a diameter greater than about 30 to 60 μm (0.03 to 0.06 mm). Generally, the higher the air capacity, the better the drainage and aeration of a soil.

A limiting value for air capacity of 10% has been used as a general value above which anaerobic conditions are unlikely to occur (Greenwood, 1975). In practice, however, the limit depends on temperature, microbial activity, oxygen consumption by the plant and the continuity of the pores. In saturated soils, such as those used for rice production, plants receive oxygen through air-filled channels in stems and roots, as well as from oxygen dissolved in the water; the degree of oxidation/reduction in these soils is indicated by their redox potentials (see Subsection 6.6.10).

Eavis (1972) illustrated a way of distinguishing between the effects of mechanical impedance, anaerobism and soil moisture using root growth of pea seedlings in soils with various textures, compactions and water contents. Anaerobism occurred in soils with up to 30% air-filled porosity (AFP), but depended on local conditions around the roots and the presence or absence of intact water films; normally, however, the soils were anaerobic at AFP values below 10 to 15%.

6.6   Soil—water relations

6.6.1   Retention of water in soils   Water in unsaturated soil is held as thin films on soil particle or pore surfaces, or as 'wedges' where the particle or pore surfaces lie sufficiently close together.

The forces retaining soil water against the pull of gravity are essentially short-distance electrical forces which vary as the reciprocal of some power of the distance from the attracting surface (see Taylor and Ashcroft, 1972, p 141).

The idealised behaviour of soil-water retention can be visualised as follows: under saturated conditions, all the soil voids are filled with water, which can drain from the larger pores until, at field capacity (see Subsection 6.6.3), the retention forces are sufficient to prevent further drainage loss. The remaining water can still be depleted by evaporation and/or root action, but as these processes continue, so the water wedges disappear and the water becomes concentrated in smaller pores and thinner films. As a soil dries, therefore, the soil water becomes increasingly strongly held and increasingly less available to plants until, at the permanent wilting point (see Subsection 6.6.4), water uptake by roots ceases. Evaporation can continue to occur until, in air-dry soils, the films are so thin and so tightly held that further water loss takes place only if the soil is heated as, for example, in oven drying.

In practice, water uptake by roots is affected by the depth and density of rooting, by gradients in water potential and by the hydraulic conductivity of both soil and roots. In many soils it may also be affected by such properties as swelling and shrinking which influence contact at the soil/root interface.

Much of the water in soils is retained in small pores into which roots do not penetrate. The amount of available water is greatly influenced by the clay, silt and fine sand fractions and in tropical soils by the mineral composition of the clay (Lal, 1979, Table 1.10). In temperate soils there is a closer relationship between texture and available water capacity than in tropical soils, where the most important features are often the clay mineral type, the associated soil structure, and the amount of swelling and shrinking which occurs (Warkentin, 1974).

6.6.2   Soil water measurements and units   In most consultancy work, soil-water content is determined by gravimetric methods (ie from the weight of soil samples before and after oven drying for 24 h at 100 to 110°C); the loss in weight on drying is expressed as a percentage of the weight of the oven-dry soil (w/w%). If the volume of the original sample is also determined, the bulk density can be calculated by dividing this into the oven-dry weight. The gravimetric moisture content multiplied by the bulk density gives the volumetric moisture content (v/v%). Adjustments to volumes must be made for swelling and shrinking soils, and stony soils present special problems.

Various attempts have been made to determine moisture content by indirect but non-destructive methods. These include determinations based on nuclear, electromagnetic, hygrometric and remote sensing techniques. A discussion of these methods is given by Schmugge et al (1980) but only gravimetric determinations and measurements based on the neutron probe are the most widely used methods in the field.

The neutron probe is the most satisfactory instrument for determining volumetric soil-water content in situ. The probe is lowered into a 40 to 50 mm diameter access hole cased with aluminium tubing. A radioactive source in the probe emits fast neutrons which are slowed down, or thermalised, by hydrogen atoms in the soil before being counted by a detector. Since most of the hydrogen forms part of water molecules, the concentration of thermalised neutrons is related to the volumetric soil water content. Repeated readings can be taken at various depths down the profile to within about 15 cm of the depth of the access tube. Modern forms of the instrument have readout devices and a capability for computing the total moisture in a profile. The instrument does, however, have to be calibrated against gravimetric measurements and a separate calibration is necessary if the instrument is to be used in the upper 20 cm of soil because neutrons are lost into the atmosphere. Frequently a combination of the neutron probe for deeper soil layers and gravimetric sampling in the surface layers is used to measure the distribution of water in a soil profile. The neutron probe has considerable advantages where repeated measurements are required at a site but suffers the disadvantage of being expensive. Details of the most widely used UK probe are given by Bell et al (1979).

Most of the water in unsaturated soils occurs in the confined conditions of thin films or in soil pores, where it exhibits differences in physical properties (such as mobility, vapour pressure and internal energy) from the bulk liquid, which it only begins to resemble in the larger voids of nearly saturated soils. For unsaturated soils, energy relations have proved particularly useful in the study of the forces which bind water or induce it to flow. Discussions of these energy relations are, in modern terms, based on thermodynamic concepts of potentials (symbol ψ) which express the energy status of soil water, or its ability to do work, compared with pure free water. Unfortunately, there is some confusion over the various quantities involved, not least because of the past lack of standard terminology. However, the total potential of soil-water has recently been defined by the International Soil Science Society (Hillel, 1980, p 136) as 'the amount of work that must be done per unit quantity of pure water in order to transport reversibly and isothermally an infinitesimal quantity of water from a pool of pure water at a specified elevation at atmosphere pressure to the soil water (at the point under consideration)'.

One of the clearest expositions of the components of potentials is given in Taylor and Ashcroft (1972), the essential elements being as follows:

a) Matric potential ψ_m - This potential results from the attractive forces between the soil matrix and the water and always has a negative sign. The term capillary potential is sometimes used synonymously and the terms soil moisture tension, soil moisture suction and matric suction or tension are frequently used to express a similar idea except that their sign is positive.

b) <u>Solute potential</u> ψ_s - This potential results from the attraction of the solutes for water and always has a negative sign. The term osmotic potential is sometimes used synonymously, and the term osmotic pressure has a similar meaning except that its sign is positive.

c) <u>Pressure potential</u> ψ_p - This potential can result from either a hydrostatic pressure or a difference in air pressure between the soil and atmosphere. In unsaturated soils there is no hydrostatic pressure and no difference in pressure between air in the soil and air in the atmosphere; therefore the pressure potential is zero. In saturated soils, the sum of the atmospheric and hydrostatic pressures is usually greater than atmospheric pressure and the pressure potential is therefore positive.

d) <u>Gravitation potential</u> ψ_z - This potential takes account of differences in elevation between the soil water and a specified reference point. The sign can be positive or negative depending on the position of the reference point, but if the reference is the soil surface then all depths in the soil will be at a negative gravitational potential.

e) <u>Other potentials</u> - It is possible to specify other components of the soil-water potential (eg electrical potential, temperature potential) but these are rarely employed.

Since potentials are additive and some particular combinations are often convenient, they have been given specific names:

<u>Total potential</u> ψ_t - This is the sum of all the components of potential at a point
$$\psi_t = \psi_m + \psi_s + \psi_p + \psi_z + \ldots\ldots$$

<u>Hydraulic potential</u> ψ_h - A gradient of hydraulic potential causes liquid water flow in soils
$$\psi_h = \psi_m + \psi_z + \psi_p$$

The dimensions of potential are those of energy per unit quantity of water and the units depend on the way the quantity is specified. Common alternatives used are:

a) <u>Energy per unit volume</u> (pressure) - This is the most common method of expressing potential and can be written with units of either pascal, or bar, or atmosphere
$$1 \text{ Pa} = 10^{-5} \text{ bar} \approx 10^{-5} \text{ atmosphere}$$

b) <u>Energy per unit weight</u> (head) - This method of expressing potential is also common and has units of length. One bar pressure corresponds to 10 m water head.

c) <u>Energy per unit mass</u> - This method of expression is not widely used but has units of J kg^{-1}. One 1 bar pressure corresponds to 100 J kg^{-1}.

It is clear from the above examples that while a wide range of units for expressing potential is available, they are all easily interchangeable. One method of expressing potential peculiar to soil scientists is the use of pF (Schofield, 1935):

$pF = \log_{10}$ head (cm water), ie a head of 100 cm water has pF 2.0

Conversions of the more common units are given in Table 6.11 and in Annex J, Table J.2.1. Note that zero tension cannot be given exactly on the pF scale since it is a logarithmic function. Zero tension is conventionally taken as pF zero, or 1 cm water tension $(9.8 \times 10^{-4}$ bar).

Tensiometers can be used in the field for measuring water potential (usually $\psi_m + \psi_p$) from 0 to about -0.8 bar (800 cm of water tension). In drier soils, potentials can be measured using gypsum blocks (-1 to -10 bars) and thermocouple psychrometers (-1 to -50 bars), but the measurement is the sum of matric and solute potentials.

6.6.3   Field capacity   Field capacity (FC) is the term used to describe the maximum water content that the soil will hold following free drainage. It does not, therefore, correspond to a fixed soil-water potential, but instead represents the condition of each individual soil after the larger pores have drained freely under gravity. In practice, FC is usually taken as the moisture content of a soil which has drained freely for 1 or 2 days after saturation.

After saturation of free-draining sandy soils, the large pores soon empty and the drainage flow decreases rapidly, giving a fairly clearly marked transition. Beyond this point, the soil-water potential declines markedly even when only small volumes of water are removed by absorption or evaporation. In clay soils, however, there is a slower change and gradual drainage persists without a sharp cessation. In such soils there may be no specific change in drainage rate that can be correlated with the FC moisture content. Slowly draining water within the root zone is still available to plants.

Several attempts have been made to correlate FC with a particular soil-water potential, but none has been entirely satisfactory, because they fail to take into account the dynamic properties of the whole soil profile, such as transmission characteristics and hydraulic gradients, hysteresis 1/ and other factors. The FC is also affected by features such as soil stratification, soil swelling and shrinking, the presence of pans, and the occurrence of GWT levels < 2 to 3 m deep. With shallow groundwater levels, where capillary conductivity is maintained in the upper horizons, the water potential at a point is given simply by its height above the GWT, for example 80 cm water tension (8 kPa or pF 1.9) for a soil with GWT at 0.8 m, and the FC is the corresponding moisture content.

1/   Hysteresis is the lagging of effects behind causes: the curves relating soil-water potential and water content differ according to whether the soil is being wetted or dried.

Soil-water tension units

Table 6.11

Appearance of soil	pF[1]	Water tension units[2]				Soil air: Relative humidity at 25°C (%)	Soil pores			Soil moisture constants	Remarks
		cm of water	bar	atm (approx.)	Pa		Equivalent cylindrical diameter[3] (μm)	Size designation[3]	Dominant function		
Dry	7	10^7	9,800	10^4	$98,000 \times 10^5$	0				Oven dry	
	6.5					10					
	6	10^6	980	1,000	$9,800 \times 10^5$	50		Hygroscopic surfaces and residual pores	Residual soil water, unavailable to plants		
	5	10^5	98	100	980×10^5	93					
	4.5	31,623		31		98	0.05			Hygroscopic coefficient	Barely moist colour
Moist	4.18	15,340	15	14.8	98×10^5	99	0.2			15 bar Wilting point	
	4	10,000	9.8	10							
	3.7	5,168	5.1	5.0							Best moisture range for tillage
	3.5	3,100	3.0	3.0					Storage of plant-available water		
	3	1,000	0.98	1	9.8×10^5			Storage pores			Capillary adjustment sluggish
	2.53	341	0.33	0.3			5			⅓ bar "Field Capacity"	Approx. in situ Field Capacity range.
	2	100	0.098	0.1	0.98×10^5				Capillary conduction		
Wet	1.7	50	0.05	0.05			50			Aeration porosity limit	
								Transmission pores			
	1	10	0.0098	0.01	0.098×10^5				Aeration; rapid drainage and infiltration		
							500				
								Fissures			
Saturated	0	1	0.00098	0.001	0.0098×10^5		2000			Saturation	

Notes: (1) Zero tension cannot be shown on this logarithmic scale, so 1 cm water (pF = 0) is taken as saturation tension.

(2) For other conversions, see Annex J, Table J.2.1.

(3) Approximate values; see also Table 6.10.

Source: Adapted from Kohnke (1968, pp 56 and 136).

In the UK, Webster and Beckett (1972) reported that soil-water potentials 48 h after thorough wetting do not generally exceed -0.05 bar (-5 kPa, 50 cm of water tension or pF 1.7), and this figure is generally used to determine FC in the laboratory of the Soil Survey of England and Wales (Avery and Bascomb, 1974). Stakman (1974) in the Netherlands used -0.1 bar (-10 kPa, 100 cm of water tension or pF 2). In the tropics or subtropics significant evaporation losses in addition to drainage losses can take place during the 24 to 48 h period following wetting, and for medium-textured soils an FC moisture content corresponding to -0.33 bar potential (-33 kPa, 330 cm of water or pF 2.5) is commonly used if, for some reason, FC in the field cannot be determined.

Since FC values are used in calculations of available water capacity (Subsection 6.6.6), taking the water potential at FC as -33 kPa when it is actually nearer to -10 kPa can lead to underestimates of the volume of water available to plants following irrigation or rainfall (Hansen et al, 1980) and hence to costly over-specification of water supply systems. Conversely, assuming FC to occur at -10 kPa potential when the real value is closer to -33 kPa can lead to underestimates of irrigation frequency or volume, and hence to under-specification of irrigation supply equipment and water storage requirements. The FC should therefore be measured in situ after natural drainage, rather than in a laboratory at a predetermined water potential or various potentials (see Subsection 6.6.5).

Soils at -33 kPa potential have lost water in pores larger than about 10 μm diameter (Russell, 1973). A useful approximate relationship between tension and the size of pore is:

$$\text{diameter of pore (cm)} \approx \frac{0.3}{h}$$

where h (cm of water) = soil-water tension

Tensiometers can be used in the field for measuring water potential from 0 to about -80 kPa (-0.8 bar or 800 cm of water tension). For drier soils, samples must be taken into the laboratory for determinations using a pressure membrane or pressure plate apparatus. After equilibration at pressures equivalent to the appropriate tensions in these pressure chambers, the soil moisture content can be determined, and the relationship between tension and moisture content plotted.

6.6.4   <u>Permanent wilting point</u>   The permanent wilting point (PWP) is arbitrarily defined as the soil moisture content at which the leaves of sunflower plants wilt permanently, ie when they do not recover their turgor if subsequently placed in a saturated atmosphere. This test can be done on cores of soil taken into the laboratory and planted with sunflowers, but more commonly the moisture content at -15 bars water potential (pF 4.2 or 15 000 cm of water) is assumed to represent PWP. The PWP is taken as the lower limit of available water so that water in drier soils is assumed to be not available to plants. In fact, the value of PWP depends on the climatic and soil conditions, on the hydraulic conductivities at corresponding water potential levels in the soil/plant/atmosphere continuum, and on the plant species. Sykes (1969) found an interspecies variation in the water potential at PWP from -7 to -39 bars. However, in defence of the use of -15 bars as the water potential corresponding to PWP, many soil-water retention curves (see Subsection 6.6.8) tend to be flatter at low potentials, and unit changes in potential produce less variation in soil-water content than at the high potentials associated with FC values.

At the low potentials (high tensions) corresponding to the PWP, the soil water is held in micropores and on soil particle surfaces, and is less dependent on soil structural conditions than at FC. In general, therefore, although undisturbed samples are more satisfactory for -15 bar determinations, disturbed samples may be used provided the results are corrected for bulk density. They are not satisfactory for high bulk density soils.

6.6.5   Measurement of FC and PWP   The FC can be determined in situ or in the laboratory, but no laboratory method is a satisfactory substitute for the field determination. This arises from the difficulty in determining at what potential to measure FC values in the laboratory (see Subsection 6.6.3) and the fact that, for many soils, a small error in the potential can lead to a comparatively large change in moisture content of the soil because of the shape of the moisture retention curve (see Subsection 6.6.8). In addition, considerable care is needed in laboratory techniques to ensure that the soils are uniformly and consistently drained at the high potentials corresponding to the FC and that they are protected from evaporation during the test.

For all these reasons, FC measurements are normally made in the field where, except for slowly draining fine-textured soils (Avella and Rodriguez, 1973), fairly reproducible and reliable results can be obtained although, because of hysteresis effects, the values depend on previous history of wetting and drying and will not be the same on all occasions, especially for clay soils. The recommended method is as follows:

The soil is fully wetted to at least 30 cm below the proposed maximum sampling depth, and covered with a sheet of polythene to reduce evaporative losses to a minimum. Moisture content samples are taken at different depths every 24 h until the moisture contents at successive samplings agree to within 1% at each depth. These final moisture contents represent field capacity. Where circumstances do not allow such frequent sampling, the soil should be sampled 48 h after wetting, and the moisture content recorded as the FC.

Moisture is determined on a weight basis, and the associated bulk density and stone content determinations are needed to convert the result to a volume basis:

MC (% by volume v/v) = MC (% by weight w/w) x BD

where   MC = moisture content
                          BD = bulk density

$$\text{ie} \quad MC \text{ (\% v/v)} = \frac{\text{Weight of water}}{\text{Weight of dry soil}} \times \frac{\text{Weight of dry soil}}{\text{Total volume of soil}}$$

The PWP can be successfully measured in the laboratory, since variations in moisture content with changes in water potential are less marked than at FC (see Subsection 6.6.8). The usual method is to drain an initially wet sample to equilibrium at -15 bar (pF 4.2) and measure the water content of the soil. This requires pressure plate or membrane apparatus.

Pressure plates tend to give lower values than pressure membranes for the same potential. The membrane method may be more accurate but pressure plates are more practical for large numbers of samples. The plates must be kept free from micro-organisms which tend to block the pores, and excessive drying due to leakage of gases must be prevented.

6.6.6   Available water capacity   Available water capacity (AWC), also sometimes termed available water-holding capacity, is defined as the volume of water retained between FC and PWP. It is a concept that applies to freely drained soils, but is of little value in soils with impeded water movement or with high water-tables.

Values of AWC are most commonly used for determination of the depth and frequency of irrigation required. They are frequently quoted as mm m^{-1} of soil, the calculations being as follows:

a)   Single horizon

    Thickness   30 cm
    Bulk density   1.55 g cm^{-3}
    FC   8.8 g/100g (or 8.8% w/w)
    PWP   3.2 g/100g

    AWC = (8.8 - 3.2) x 1.55 = 8.68 g/100 cm^3 (or 8.68 cm m^{-1}
                          or 8.68% v/v)

    $= \dfrac{8.68 \times 300}{100}$ = 26 mm (per 30 cm of soil) = 87 mm m^{-1}

b)   Multiple horizons

Depth (cm)	Thickness (cm)	AWC (mm)
0 - 30	30	26
30 - 55	25	22
55 - 120	65	42

    Total AWC = 26 + 22 + $\dfrac{(45 \times 42)}{65}$

            = 77 mm in top 1 m of soil

Results may also be quoted for the actual or potential rooting depth of the soil profile.

Wherever possible actual measurements should be used, but if determinations are not available, an estimate of AWC is given by the values shown in Table 6.12. Note, however, that not all the calculated 'available' water is necessarily accessible to plants, since the accessibility depends on:

a)   rooting depth and root concentrations at different depths;

b)   root dimensions and conductive properties;

c)   hydraulic conductivity of the soil as a function of depth and water content.

In turn, (a) and (b) all depend on species, stage of growth, time of year, and soil properties such as the presence or absence of an impenetrable layer or water-table. Other factors which affect the effective AWC include:

d)   imperfect drainage when the soil has more than -0.05 bar potential;

e)   upward movement of water from the GWT (or perched water-table);

f)   salinity of soil water, which influences crop water uptake; the reservoir effect of some stones, mainly limestones and chalk and some types of sandstone, but this will rarely contribute more than 1 or 2% to the AWC (Reinhart, 1961; Coile, 1953; and Hansen et al, 1980);

g)   shattered rock for which the AWC cannot be sampled or even estimated, although roots may be able to penetrate it and extract water;

h)   temperature, although this is usually only of minor importance (see Taylor, 1958; Taylor and Stewart, 1960).

6.6.7   Readily available water capacity   Not all the water held between FC and PWP can be considered as equally available to plants. Milthorpe (1960a) has suggested that at tensions up to 1 bar water is freely available to most crops, whilst Hansen et al (1980) consider that 75% of the AWC can be easily extracted. A rule of thumb is that the total readily available water capacity (TRAWC) value is half to two thirds of the total AWC of a profile.

The proportion of water held at the low tensions represented by the TRAWC may sometimes be more important than the total AWC in determining crop response to soil moisture conditions and the timing of irrigation (Salter and Williams, 1965). It is this readily available water held at low tensions within the larger soil pores which is particularly affected by soil structural conditions.

Estimated AWC values for subsoils of different textures and stone contents   Table 6.12

Texture 1/	Stone-free temperate AWC 2/ ($mm\ \overline{m}^{-1}$) A	Tropical AWC (to nearest 10 $mm\ m^{-1}$) Stone and gravel content (and % reduction of stone-free tropical AWC)					
		Stone-free = A less 10%	Few (10%)	Common (25%)	Many (50%)	Abundant (70%)	Dominant (90%)
cS	80	70	60	50	40	20	10
S	90	80	70	60	40	20	10
fS	110	100	90	70	50	30	10
vfS	200	180	160	140	90	50	20
LcS	80	70	60	50	40	20	10
LS	120	110	100	80	50	30	10
LfS	140	130	110	90	60	40	10
LvfS	200	180	160	140	90	50	20
cSL	120	110	100	80	50	30	10
SL	150	130	120	100	70	40	10
fSL	170	150	140	110	80	50	20
vfSL	200	180	160	130	90	50	20
L	170	160	140	110	80	50	20
ZiL	190	170	150	130	90	50	20
cSCL	140	120	110	90	60	40	10
SCL	150	140	130	100	70	40	10
fSCL	170	150	140	110	80	50	20
CL	150	130	120	100	70	40	10
ZiCL	170	150	140	110	80	50	20
SC	110	100	90	70	50	30	10
fSC	130	120	110	90	60	40	10
ZiC	160	140	130	110	70	40	10
C	140	130	120	90	60	40	10

Notes: 1/ Textures quoted are the nearest USDA equivalents to the textures used in the sources.
2/ Values derived from measurements on UK soils, see Salter and Williams (1967, 1969) and Hall et al (1977); see Annex B for FC and PWP values. Figures for some tropical soils are given by Pidgeon (1972).

The values in this table must be considered indicative only; actual values will depend on soil properties such as OM content, structure, bulk density, clay mineralogy etc.

cf Figures quoted by FAO (1977 b, p 86):

Textural grouping	AWC ($mm\ m^{-1}$)
fine	200
medium	140
coarse	60

Coarse-textured soils hold a comparatively large proportion of their small volumes of available water at low tensions - often more than half the available water is held at tensions below 0.3 bar. As soils become finer textured, the TRAWC increases to a maximum with the loams and silt loams.

The effectiveness of the TRAWC depends upon the depth of crop roots and the moisture extraction pattern. These factors must also be taken into account when using TRAWC values as a basis of determining irrigation schedules. Some indicative figures are shown in Annex F.

6.6.8   Water retention curves 1/   By determining water contents at a number of tensions between 0.05 and 15 bar, a complete water retention curve can be plotted as illustrated in Figure 6.3. Each curve should be constructed from about six experimental measurements at pF values of about 1.7, 2.0, 2.5, 3.0, 3.5 and 4.2 (0.05, 0.1, 0.33, 1, 3 and 15 bars respectively). Oven dry is taken as equivalent to pF 7.0.

The tabulated numbers in Figure 6.3 indicate the different water contents held between the given tensions by three different soils. In general, sands have low residual water contents and AWCs, but high drainable water-levels, whilst clays have low drainable water contents, moderate to low AWCs, and high residual water values. Thus a clay with a water content as high as about 30% by volume can still contain little or no plant-available water.

From the saturated state, clay soils generally show a slow and regular decrease in water content with increasing pF. Sandy soils, in contrast, show only a slight decrease in moisture content in the lower pF range until a point where only a small rise in pF causes a considerable discharge of water, due to emptying of a relatively large number of pores in a particular diameter range. Examples of these trends are illustrated in Stakman (1974). The intersection point of each curve with the water-content axis, at pF = 0, gives the water content of the soils under saturated or near-saturated conditions, which means that this point indicates approximately the total pore-volume percentage, provided no air entrapment has taken place.

Using water retention curves and mechanical analysis, it is possible to prepare profile diagrams illustrating volume relationships, as shown in Figure 6.4. This also shows that texture is not the only limiting factor: note the low aeration and AWC in the compact, in situ, clay soil, combined with the high residual water content; the better-structured alluvial clay shows none of these problems.

Temperature changes affect water retention curves as follows (after Taylor and Ashcroft, 1972):

1/   Also called: characteristic curves; moisture characteristic curves; water release curves; and moisture retention or release curves.

Illustrative soil-water retention curves   Figure 6.3

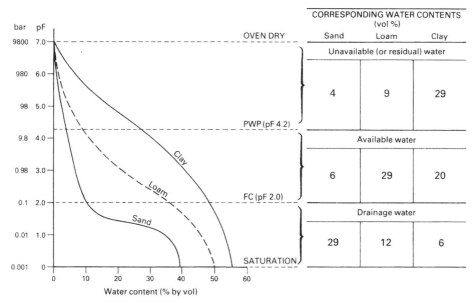

	CORRESPONDING WATER CONTENTS (vol %)		
	Sand	Loam	Clay
Unavailable (or residual) water			
	4	9	29
Available water			
	6	29	20
Drainage water			
	29	12	6

Note: Because of hysteresis effects, the curve for a particular soil will differ slightly according to the previous wetting and drying history and whether the soil is being wetted or dried

Illustrative volume relationships in two clay soils   Figure 6.4

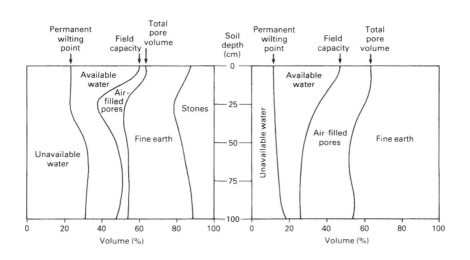

(a)   In situ clay profile   (b)   Alluvial clay profile

a) temperature influence on soil-water tension is small in moist soils at low tensions. The influence increases as the soils become drier and approach the PWP;

b) temperature differences cause more marked changes in water retention curves of fine-textured soils than of coarse ones.

Temperature effects thus vary amongst soils and for different water contents of the same soil; ideally a calibration curve should be constructed for each soil under study, but the effects are very small over most temperature ranges encountered in crop production.

Water retention curves are also subject to hysteresis effects depending upon their previous wetting and drying history; some examples are given in Taylor and Ashcroft (1972, Chapter 7).

6.6.9   Interpretation of available water capacity   Because of the possible variations in AWC noted above, the conditions of measurement of PWP and FC should be clearly stated in all tests. These details are not always reported, and interpretation of results must always be made with care, and should take into account other factors such as the presence of shallow groundwater or subsurface seepage.

The AWC of soil profiles can be used as factors in land suitability classification (see Table 5.7, p 51) and in irrigation planning, provided that the effective rooting depth, percentage root distribution and percentage of readily available water are taken into account. Wherever possible, AWC values for suitability assessment should be derived for the particular crop varieties and cropping patterns at site, but a general grouping for irrigation suitability is shown in Table 6.13.

Broad groupings of AWC values for irrigation planning   Table 6.13

Rating for irrigation suitability	Available water capacity (mm m^{-1})
Low	< 120
Medium	120 – 180
High	> 180

The importance of the AWC depends on the frequency of wetting expected and the duration of dry periods between rainfall and irrigation applications. In many situations crops must rely on water stored in the soil (residual moisture) at the end of a rainy season.

6.6.10   Redox   potential   The   electrical   energy   concept/process   of   oxidation-reduction   or   redox   potential   (E_h)   is   useful   in   describing soils with poor drainage or fluctuating water-tables. As instanced by Patrick and Mahapatra (1968), the most important change in a soil as a result of waterlogging is conversion of the root zone from an aerobic to an anaerobic (or nearly anaerobic) environment where oxygen is absent or limiting.   In the anaerobic state several oxidation-reduction systems that are usually present in oxidised forms are reduced as a result of the activity of anaerobes (micro-organisms).

The   most   prominent   change   occurring   in   soil   as   a   result   of submergence or groundwater percolation is the decrease in redox potential, measured in millivolts.   Aerated soils have redox potentials in the range +700 to +400 mV.   Moderately reduced conditions are indicated by potentials around +300 or +200 mV, whilst waterlogged, highly reduced soils exhibit potentials as low as -200 or -300 mV.

When a soil is being depleted of oxygen by submergence or lateral seepage, the reduction of oxidised inorganic soil components follows a well-authenticated sequence.   Nitrates and manganese dioxide are the first to be reduced, at fairly high redox potentials (about +250 mV).   Hydrated ferric oxide is reduced to ferrous at intermediate potentials of approximately +100 mV. Sulphates, however, lose their oxygen to become sulphides only at extremely low potentials around -150 mV.

Field measurement of redox potential is impracticable and even greenhouse pot methods are poorly replicable.   The usual laboratory test is by contact between a platinum electrode and a soil solution.

6.7   Soil particle density

Soil particle density (ρp) is used together with knowledge of soil bulk density and water content to calculate air space and percentage water saturation.   It is defined as the mass per unit volume of the soil matrix material and is numerically equal to the specific gravity (SG), where:

$$SG = \frac{\text{Mass of soil matrix material}}{\text{Mass of equal volume of water}}$$

and ρp is measured in g cm^{-3} or kg m^{-3}.

Actual values depend upon chemical and mineralogical composition of the particles, as well as upon their degree of hydration.   The density of various soil constituents can be obtained from tables; the generally used value of 2.65 g cm^{-3} represents an average figure for non-ferruginous and non-humose subsoils.   Humus (SG 1.37), ferric oxide (limonite, SG 3.6 to SG 3.7), heavy minerals and volcanic parent materials can change the level from 2.65 g cm^{-3}, but it is usually sufficiently accurate for the majority of soils used for agriculture.

6.8 Soil structure, strength and consistency

6.8.1 Soil structure   The most important structural features of soils are the size, shape and stability of the peds and their penetrability by water, air and roots; Table 6.14 lists the main influences on these.

Although different types and sizes of structure are readily recognisable in soil profiles, and form a part of standard morphological descriptions, plant development in many soils depends critically on microstructures and micropore distributions, which cannot be easily assessed in the field. It is therefore difficult to quantify the effects of different structures on plant growth, in particular without calibration against similar, known conditions.

Some indication of the suitability of a soil as a rooting medium can be gained by observation of the rooting patterns of existing vegetation, but this may often be of limited use - as, say, in very dry desert conditions with little or no vegetative cover, where the soils are scheduled for ploughing and irrigation, or where the natural vegetation and the proposed crop have very different rooting patterns.

For these reasons, direct assessment of soil structure is usually made by field description only, except in the implementation or research phases of a project. The effects of soil structure on characteristics such as water movement or retention may be measured separately, as described above. Soil stability can be measured by wet sieving (Dermott, 1967) and the dispersion ratio method (Middleton, 1930); the different methods are discussed in Pringle (1975) and Williams et al (1966). A comparatively easy-to-do stability classification procedure has been developed based on the response of aggregates dropped into water (Emerson, 1967; Greenland et al, 1975), but the variability within classes tends to be high, and the results may not be of great practical use.

Soil structural limitations can have noticeable effects on plant roots and their actions; these effects include:

a)   restriction in root exploration volume through compaction;

b)   loss of roots through excess water and/or toxic materials generated in the profile;

c)   restricted water uptake due to reduced water-holding capacity;

d)   loss of roots through movement of material in unstable soils. In addition, surface caps reduce water infiltration, and hinder germination and subsequent development of small seeds.

Root restrictions can be reflected in plants' aerial growth, which may often be stunted and show reduced vigour, frequently with pale foliage that recalls the results of nitrogen or magnesium deficiency. The affected areas usually occur in patches, but may extend over whole fields. Additional nitrogen fertiliser is often required to produce normal crops.

Summary of soil structural influences   Table 6.14

Influences promoting good structure		Influences promoting poor structure	
1. Clay content	In general the more clay the more stable the structure, particularly in sandy soils	1. High sand and silt contents	Sands tend to be loose or very friable; silts can form thick, almost impervious caps
2. Organic matter	In general, the higher the OM content of mineral soils the more stable the structure	2. Poor drainage/ wet climates	Wet soils tend to have weak, poorly developed structure; reduction of $Fe(II)$ to $Fe(III)$, dissolving of $CaCO_3$, prevention of shrinkage on drying and effects on biological action can all play a part
3. Cementing agents	Iron oxides often produce very stable structures. Better-developed, more stable structures are often associated with free $CaCO_3$ or gypsum	3. High Na or H levels	High exchangeable Na tends to cause deflocculation of clay particles and hence poor structure. Replacement of exchangeable cations by H (eg with poor reclamation techniques) produces similar effects
4. Good drainage	Good drainage assists structure formation and allows other beneficial influences to act; structures are more mechanically stable when drier	4. Effects of land use	Structural breakdown by fracture (in over-tilled soils); compaction by livestock and/or implements; smearing by implements; deflocculating effects of Na and, to some extent, NH_4 fertilisers on soils low in OM and divalent cations
		5. Drought	Excessive and/or prolonged drying can reduce OM contents, causing loose topsoils in silty or sandy textured soils, or hard surface capping in clayey topsoils

Sources: Batey and Davies (1971) and ILACO (1981).

Some crops are more susceptible to structural limitations in soils than others. In general legumes, maize, groundnuts, melons and barley have low tolerances. Sugarcane does not yield well if the soil is compacted.

6.8.2   Soil strength   Soil strength (cohesion, internal friction) is an important feature of soils in relation to their response to tillage, and their resistance to fracture, compression, smearing, moulding and compaction (or 'poaching', or 'puddling'). However, most measurements are again difficult to interpret without calibration against known conditions, and the variability of soil strength over short distances can be very high, particularly because values are highly dependent on soil-water content and physical disturbance. In addition, many standard soil strength measurements are designed for civil engineering purposes, in which the soil is highly disturbed both in the test and in use, and are not representative of natural properties.

Direct soil strength measurements (see BSI, 1975) are not therefore a normal part of routine survey work, although assessment of consistency (see Subsection 6.8.3) is often used as an indication of characteristics linked to soil strength. Approximate indirect estimates of soil strength used, for example, in compaction studies, can be made with a penetrometer, but such measurements are greatly influenced by the size, shape and composition of the penetrating head and, above all, by soil moisture contents and operator variation. The results should therefore only be used to give rough comparative figures for similar situations, unless very careful standardisation is ensured.

6.8.3   Soil consistency   Soil consistency is commonly used to describe the 'feel' of the soil and includes soil properties such as friability, plasticity, stickiness and resistance to compression and shear, all of which have obvious importance for cultivation operations. Apart from the usual survey descriptions, consistency changes are sometimes described in terms of the various limits summarised in Table 6.15, although these are more often used in soil mechanical studies than in soil surveys; methods are given in BSI (1975). Briefly, the limits are defined as follows:

a)   The Plastic Limit (or Lower Plastic Limit) is the moisture content at which soil consistency changes from friable to plastic, and is taken to represent the minimum moisture content at which a soil can be puddled.

b)   The Liquid Limit (or Upper Plastic Limit) is the water content at which soil cohesion is so reduced that the soil mass will flow when a force is applied, and is similar in value to the Sticky Point which is the water content at which a dry soil, when slowly wetted, begins to adhere to a smooth surface.

c)   The Shrinkage Limit is the water content at which further loss of water causes no further change in soil volume.

In practice these limits are usually determined under artificial conditions in laboratory tests and applications to field situations must be made with care, although empirical relationships can

Indicative variation of soil consistency with water content

Table 6.15

	Moisture status				
	------- Dry ---------	------ Moist ------	------------------ Wet --------------		
Consistency	Hard – loose	Friable – soft	Plastic – sticky	Viscous – sticky	
Cohesion/ adhesion effects on workability	Possible clod formation, or disintegration into loose particles; too dry to work	Optimum soil working conditions	Too wet – compaction, smearing	Too wet – soil flow	
Atterberg and other limits 1/	Shrinkage Limit	Plastic Limit	Liquid Limit		
		----- Friability ----- index	----- Plasticity ----- index		
			Sticky Point		

Note: 1/ The Plastic and Liquid Limits are also known respectively as the Lower and Upper Plastic Limits; collectively they are also called the Atterberg limits or constants.

sometimes be established. In general, if the Plastic Limit is higher than the FC, the soil should normally be easily workable without damage. If the Liquid Limit is equal to or lower than the FC, then the soil will slake and begin to flow under wet conditions, but if the Liquid Limit is significantly above the FC there will be no appreciable slaking (Boekel, 1963). The Sticky Point indicates the moisture content at which a soil will begin to scour during cultivation.

Warkentin (1974) has shown how a relationship between a plasticity index and the Liquid Limit can indicate the nature and amount of surface in a clay sample and therefore the clay minerals present.

6.9   Soil texture and particle size distribution

6.9.1   Introduction   Particle size analysis, also referred to as mechanical analysis, is used to determine the proportion of different-sized particles in a soil and hence its textural class. In practice the measurements are also used as basic indicators of soil physical and chemical properties, and as a standard check on the finger texturing of surveyors in the field. The proportions of individual particle ranges are often needed in correlation studies (eg to relate AWC to silt and/or clay contents) and for size-grading studies to indicate, for example, potential compactability. It is important that a standard method be used because results are greatly affected by pretreatment and by the method itself. Procedures are discussed in Dewis and Freitas (1970).

6.9.2   Pretreatment   In a natural soil, the particles are held together either by electrostatic interaction or by a variety of organic and inorganic substances, the latter including carbonates, gypsum and oxides of iron and aluminium. Several analytical pretreatments have therefore been proposed for removal or reduction of these cohesive bonds in order to ensure a well-dispersed sample, the problem always being to decide what treatment will subsequently allow the most accurate reflection of soil particle behaviour in practice. Most pretreatments are based on oxidation of organic matter (if present as more than 0.5% by weight) with hydrogen peroxide, and dispersion by gentle mechanical shaking (eg end-over-end mixing for 24 h) in fresh Calgon solution [1]. More vigorous treatments, such as reaction with HCl to remove carbonate, are not normally recommended, since they break down cemented aggregates too far.

Dewis and Freitas (1970) suggest that for highly saline soils some modification is necessary to allow for the weight of soluble salts removed during the pretreatment, and soils high in gypsum need washing to remove sufficient gypsum to prevent flocculation of the clay. However, prolonged washing should be avoided, as it will tend to remove soluble salts and lead to unrepresentative deflocculation of the clay. Soils high in colloidal iron or aluminium oxides, or soils from volcanic efflata, may also need

---

[1]   40 g sodium hexametaphosphate with 10 g anhydrous Na_2CO_3 in 5 ℓ of distilled water.

certain special pretreatments involving reagents such as NaOH, $(NH_4)_2CO_3$ with NaOH, Na_2CO_3, NH_4OH and/or various sodium phosphates. Alternatively, ultrasonic dispersion methods may be used.

6.9.3 Sieving   Pretreated samples are wet-sieved to separate the sand fractions down to about 0.05 mm diameter and, after drying at 105°C, subdivisions of the sand can be determined by dry sieving and weighing; sieve mesh numbers and sizes are listed in Annex J, Subsection J.2.4.   It is vital to use a carefully standardised method, since variations in, for example, time and degree of agitation during the test can greatly influence results.

6.9.4 Sedimentation techniques   For the silt and clay size fractions sedimentation techniques are used; these are based on the Stokes' Law relationship between the diameter of suspended particles and their rate of settlement in a liquid at constant temperature:

$$v = x^2 g (\rho_P - \rho_L)/18\eta$$

where   v = rate of settlement
        x = diameter of particle
        g = gravitational constant
        ρ_P = density of particles
        ρ_L = density of liquid
        η = viscosity of liquid

The equation assumes the particles to be rigid, smooth spheres, rather than the plate-like shapes of many soil particles, so that quoted results should always be considered as 'equivalent spherical diameters'.   Note also that the soil-particle density affects the equation; care should be taken in interpreting values for soils with particle densities that differ markedly from the average of 2.65 g cm^{-3}.

The procedure after pretreatment is to allow the dispersed soil to settle slowly in a column of water at a measured, reasonably constant, temperature (variation usually about ± 2°C).   One of the following two sampling techniques is then used to measure the soil still suspended;   by choosing suitable sampling times, the measurements can be used to calculate different particle size fractions in the soil.

a)   The pipette method - This is the more accurate procedure, whereby an aliquot is pipetted from a given depth at given times and the oven-dry soil contents determined by weighing after evaporation of the water.

b)   The hydrometer (or Bouyoucos) method - This is a less accurate but much quicker method, which employs a specially calibrated hydrometer in place of the pipette.   The special calibration is necessary because as the particles settle the hydrometer does not float at a fixed depth;   the size of the hydrometer also means that it does not give a point measurement like the pipette.   For most agricultural purposes however, the Bouyoucos method is sufficiently precise.

6.9.5   <u>Interpretation of particle size analysis</u>   The results of particle size analyses are usually quoted as percentages by weight of the whole soil or of the 'fine earth' fraction of < 2 mm diameter; size limits for the fractions used in three common systems are summarised in Figure 6.5.   The proportions of sand, silt and clay are used with the familiar triangular texture diagram to determine the textural class of the soil;   a version of the diagram, redrawn on 90° axes, is presented in Annex H.   Alternative presentations include the particle size grading curve (% by weight passing through a sieve plotted against sieve mesh diameter on a log scale);   this is very commonly used by engineers to assess the compactability of a soil, or its likelihood of producing siltation problems in drains (see Dieleman and Trafford, 1976, p 101).

The limits of both soil particle sizes and textural classes, and the method of measurement, are to a greater or lesser extent arbitrary, and the practical applicability of the results will depend on the systems used and the soil under study.   In some circumstances, finger texturing can give a better assessment of agricultural properties (Batey and Davies, 1971).   Soil particle sizes and textural classes should therefore only be used as broad indicators of likely soil behaviour, unless detailed local correlations between texture and husbandry are available.

Particle size limits: comparison of systems   Figure 6.5

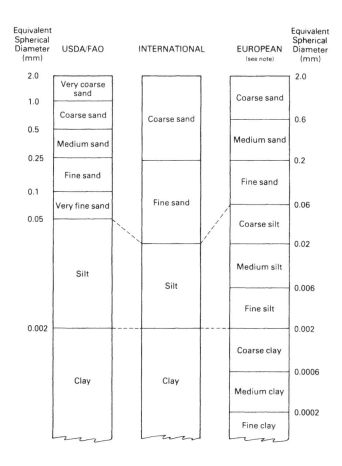

Note: Including: Soil Survey of England and Wales
                   British Standards Institution
                   Massachusetts Institute of Technology.

Source: Kohnke (1968).

Chapter 7

Soil Chemistry

7.1   Introduction

This chapter covers the main soil chemical analyses used in consultancy work, with the exception of soil salinity and sodicity, which are discussed in Chapter 8. The content is restricted to an outline of the methodology and a general interpretation of results for most tropical or subtropical conditions. Crop-specific information is summarised in Annex F, and references to individual measurements are given in the text where appropriate. Further descriptions of laboratory methods are given in standard texts, which include Hesse (1971), Black (1965), Dewis and Freitas (1970) and Jackson (1958). Aspects related to developing countries are discussed in FAO (1980a), and soil laboratory design and equipment are covered in Ferguson (1973), Golden et al (1966) and Cottenie (1980). A general text on soil and plant analysis is provided by Walsh and Beaton (1973), whilst soil field testing kits are compared by McCoy and Donohue (1979). Tomlinson et al (1977) discuss laboratory schedules and simplified procedures for rapid routine testing. Some indicative costs and sample weights are summarised in Annex A, page 198.

Leaf analysis may also be appropriate to serve some projects; a sampling method is described in Benton-Jones et al (1971). Where enough test results are available for correlation with crop yields, a preliminary assessment of deficient and non-deficient soils can be made using the rapid technique of Cate and Nelson (1965).

7.2   Problems in interpretation of soil chemical measurements

The literature of soil science abounds with articles on methods of laboratory analyses, their relevance to particular soils and ways of improving their accuracy. Various qualitative relationships between soil behaviour and the soil characteristics supposedly measured by these standard methods are also widely discussed. However, references to general quantitative interpretations of the results (as opposed to specific interpretations, such as research station reports) are very few, and tend to be extremely elementary, particularly for tropical soils. This arises because there are at least four basic difficulties in making generalisations from soil analytical results:

a) Soil samples may not be representative of the natural soil from which they are taken – not only because of soil spatial variability and often inadequate field-sampling techniques (see Section 3.7), but also because of the effects of pretreatments such as drying (foot of this page). Even for samples from a single site, the chemical results can be greatly influenced by cropping history, management practices and/or time of year when sampled.

b) Soil analytical data are usually derived from standard methods. Whilst these are often chosen or modified to reflect actual soil conditions as far as possible, there is seldom a direct relationship between the conditions of measurement and the environment of the plant. For example, the quantity of an 'available' nutrient measured by a laboratory extraction will not necessarily reflect, let alone accurately measure, the availability of that nutrient at the point of absorption on a root hair (eg Barber et al, 1963).

c) Interpretations of laboratory results are seldom, if ever, universally applicable. Different methods may have to be employed on different soils to measure the same soil characteristic, and other influences (eg clay mineralogy, climate) will affect the interpretation of the same analysis on soils from different sites. Particularly important to interpretations of crop nutrient responses are the interactions between soil chemical constituents (some of which are summarised in Figure 7.1) and the effects of soil conditions on movement of nutrients and their uptake by plants. These conditions include salt concentration in the soil solution, soil nutrient buffering capacity, soil-water potential, soil structure and hydraulic conductivity (Mengel, 1982).

d) Variability of soil analytical results tends to be high (see below).

7.3 Variability of soil analytical results

Apart from problems of field sampling (Section 3.7), analytical results can vary as a result of laboratory procedures. The reasons for these inaccuracies can be summarised under five headings:

a) Pretreatment effects – The most common pretreatment of soils consists of drying, followed by grinding and sieving: the chemical effects, summarised in Table 7.1, depend on the time and temperature of drying, and many result from changes in microbiological conditions. Soils are normally only air dried before chemical analysis, since oven drying produces unacceptably large chemical changes in the soil constituents.

b) Intrinsic errors in methods – The accuracy of laboratory determinations used for routine soil chemical analysis is normally well within the spatial variability of natural soils (see Subsection 3.7.1). Estimated random errors of some standard analyses are summarised in Table 7.2.

Indicative summary of principal plant-nutrient interactions   Figure 7.1

Element acting	Element affected												
	B	Ca	Cu	Fe	K	Mg	Mn	Mo	N	Na	P	S	Zn
B	/	-	-	-	-	-	-	-	-	-	A	-	-
Ca 1/	A	/	-	A	A	A	A	-	-	A	A	-	A
Cu	-	E	/	A	-	-	A	A	-	-	A	-	A
Fe	-	-	A	/	A	-	A	-	-	-	A	-	-
K	A 2/	A	-	E	/	A	E	A	A	A	-	-	-
Mg 1/	-	A	A	A	A	/	A	E	-	A	E	E	A
Mn	-	-	-	A	-	-	/	A	-	-	-	-	-
Mo	-	-	A	A	-	-	A	/	-	-	-	A	-
N	A	E	A	-	A	E	-	E	/	-	E	-	A
Na	-	A	-	-	A	A	-	-	-	/	A	-	A
P	-	A	A	A	A	E	A	E	A	-	/	E	A
S (as SO_4^{2-})	-	-	-	-	-	-	-	A 3/	-	-	E	/	-
Zn	-	-	A	A	-	A	A	-	-	-	A	A	/

A = Antagonises action.
E = Enhances action.
- = Insufficient data or effects too variable to summarise simply.

This is a highly generalised table; interactions vary greatly depending on plant type and soil conditions, the latter including, in particular, pH, temperature, drainage status, relative levels and forms of the elements occurring, and presence or absence of other elements. In particular, where the acting element is present in amounts limiting to crop growth, applications of additional quantities up to a 'sufficient' level may enhance the action of the affected elements which are otherwise antagonised.

An entry in the table means no more than that the potential interaction indicated should be borne in mind during project implementation.

Notes: 1/ Ca or Mg as carbonates may enhance action of other elements in acidic soils by raising pH.
        2/ With high B contents, high K levels can enhance B toxicity, depending on conditions, and may reflect changes in the Ca:B ratio.
        3/ But note that the acidifying effect of gypsum, by decreasing CO_3^{2-} and OH^-, may increase Mo uptake. S as sulphur has acidifying effect and can enhance uptake of other elements, sometimes to toxic levels (notably of Al and Mn).

Sources: Davies (1980), Kumar et al (1981), Olsen (1972), Lucas and Knezek (1972), Russell (1973), Richards (1954), Wright (1976), Phosyn Chemicals Ltd (1980) and Samuels (1982).

Summary of effects of air drying on soil   Table 7.1

Soil constituent or property	Effect of air drying
Carbon	Total C: unaffected
	Organic C: increasing oxidation with time and temperature
Manganese	Exchangeable Mn increases
Nitrogen	Total N: little effect
	Water-soluble N and water-soluble organic matter increase with time and temperature
pH	For sulphur-rich soils, levels can be drastically altered
Phosphorus	Low pH soils: P soluble in H_2O or dilute acid tends to increase
	High pH soils (dry when sampled): P levels tend to decrease
	P fixing capacity of some soils changes with drying (may be linked to Fe and Al changes)
Potassium	Depends on clay minerals present, also on original level of exchangeable K:
	If < 1 me/100 g soil, exchangeable K tends to increase
	If > 1 me/100 g soil, more K tends to become fixed
Sulphur	With some soils more S is releasable to extracting solutions

Source: Hesse (1971).

Soil chemistry

Random analytical errors for some standard chemical analyses   Table 7.2

Soil constituent or property	Analytical method	Approximate random error in quoted result (± %)
Bicarbonate	HCl	2
Carbon (organic)	Walkley–Black	5
Carbonate	Calcimeter	2
CEC	$BaCl_2$/EDTA	10
Chloride	Mercuric nitrate	2
EC	1:5 soil:H_2O extract	10
Exchangeable cations	Sodium acetate extraction	5
Gypsum	Difference in Ca content of saturation extract, and of 1:5 soil:water extract	10
Nitrogen (total)	Kjeldahl semi–micro	5
pH	1:5 soil:H_2O extract	0.2 units
Phosphorus (available)	Olsen	10
Soluble cations	Saturation extract	5
Sulphate	$BaCl_2$ gravimetric	15

Note: The above errors typically occur in a single laboratory: between laboratories the variation is likely to be very much greater, see Section 7.3e.

c) Operator errors — Errors in laboratory data may also be due
    to operator faults, such as incorrect preparation of standard
    solutions, misuse of instruments or incorrect calculations.
    Such errors can normally be detected by an adequate programme
    of sample replication (see Subsection 3.7.8).   Analyses
    commonly showing high variability are listed in Table 7.3.

d) Contamination effects — Sample contamination can arise in
    several ways, perhaps the most common being dust
    contamination: soil and vegetation samples, in particular,
    should be prepared in well-separated rooms, to avoid
    contamination of soils with organic matter and vegetation with
    iron, manganese and other metals.   The exact siting of air
    conditioner intakes, extractor fans etc can also be very
    important.   Note that use of brass sieves whilst engaged in
    copper or zinc analyses should be avoided.   Handling money
    whilst preparing samples can cause copper contamination, and
    the use of talcum powder can cause boron contamination.

e) Variation between laboratories — The results of duplicate
    sample analyses are usually consistent when the work is
    undertaken by a single reputable laboratory, but variations
    between laboratories are notoriously high.   Even for simple
    analyses differences can be over 100%.

Soil analyses commonly showing high variability   Table 7.3
or inaccuracy

Analysis	Reasons for variability or inaccuracy
Phosphate	(Colorimetric method) — Sensitivity of colour development to pH of solution, time allowed for colour complex to form and temperature
Exchangeable bases	Poor cation standards and faulty extraction procedures
pH and EC	Variations in the standing time and poor soil mixing
CEC	Variations in leaching rate; the effects of pre-washing; using sum of individual cations rather than separate measurement
Carbon (organic)	Lack of attention to the size of the reaction vessel and the temperature achieved

7.4 Summary of soil chemical data interpretation

For the reasons outlined above, soil chemical analyses should normally be regarded as providing chiefly 'order-of-magnitude' results. Unless supplemented by detailed correlations from crop trials, such analyses cannot be used for accurate fertiliser or management recommendations, and their main use is usually to indicate potential excess or deficiency problems in soils.

Despite the drawbacks, a large number of soil analyses - at considerable cost - will continue to be required, not least by clients who want to see evidence of 'objective' data in soils reports.

Within the limitations, routine soil chemical data are capable of interpretation in general terms: Table 7.4 provides a general summary of methods and broad quantitative interpretations, which are discussed individually in Sections 7.5 to 7.15. It must be emphasised, however, that local data related to crop production should be used wherever possible in preference to these general guidelines.

7.5 Soil reaction (pH)

7.5.1 Background   Routine soil pH measurements are usually made on extracts from soil suspensions, which vary from saturated soil pastes to soil suspensions at a ratio of 1:5 soil:dilution medium. The dilution medium normally consists of water, or dilute solutions of $CaCl_2$ or KCl.

Dilution ratio has been extensively discussed (Snyder, 1935; Chapman et al, 1941; Peech et al, 1947; Peech, 1965), and most sources indicate that pH values in water increase with the dilution of the suspension; Dewis and Freitas (1970) indicate that pH values for 1:5 suspensions may generally be 0.5 to 1.5 units higher than values for corresponding saturated pastes. However, the pH does not always rise, nor is any increase necessarily proportional to the dilution (Loveday et al, 1972). Measurements in other dilution media, such as $CaCl_2$ or KCl solutions, are sometimes preferred because the concentrations of the test solutions are more representative of the salt concentrations in natural soil solutions, and the values obtained are less dependent on the dilution ratio (see Schofield and Taylor, 1955). Also, by using a standard salt solution, the effects of natural variations in pH for a given soil (eg due to differences in the quantity of soil moisture) are effectively swamped, and a standard figure is obtained.

For temperate soils, and many tropical ones, the use of 0.01 M $CaCl_2$ suspensions is often favoured; values of pH in such $CaCl_2$ suspensions are typically 0.5 to 0.9 units lower than in water, the difference usually being greater for neutral than for acidic soils. The use of KCl solutions may depress the pH by one to two units compared with measurements in an equivalent aqueous suspension.

Brief summary of recommended routine soil chemical analyses and their interpretation   Table 7.4

Analysis	Recommended method(s)	Units	Rating	Range	General interpretation	Section reference
pH	1:2.5 soil:water suspension	–	Very high	> 8.5	Alkaline soils: Ca and Mg liable to be unavailable; may be high Na; possible B toxicity; otherwise as below:	7.5
			High	7.0-8.5	Decreasing availability of P and B to deficiencies at higher values. Above 7.0 increasing liability of deficiency of Co, Cu, Fe, Mn, Zn	Interpretation 7.5.3
			Medium	5.5-7.0	Preferred range for most crops; lower end of range too acidic for some	
			Low	< 5.5	Acid soils: possibly Al toxicity and excess Co, Cu, Fe, Mn, Zn; deficient Ca, K, N, Mg, Mo, P, S (and B below pH 5)	
CEC	a) Unbuffered 1 M KCl at pH of soil	me/100 g soil	Very high	> 40	Normally good agricultural soils – only small quantities of lime and K fertilisers required	7.6
			High	25-40		
	b) Na or NH4 acetate at pH 8.2, 7.0		Medium	15-25	Normally satisfactory for agriculture, given fertilisers	Interpretation 7.6.3
			Low	5-15	Marginal for irrigation (FAO (1979a) quoted low is 8-10 me/100 g soil)	
			Very low	< 5	Few nutrient reserves. Usually unsuitable for irrigation, except rice	
BSP	Calculation: total exchangeable bases/CEC	%	High	> 60	Generally fertile soils	7.6.4
			Medium	20-60	Generally less fertile soils	
			Low	< 20		
			Eutric	> 50		
			Dystric	< 50		
Exchangeable cations						7.7
Ca	As CEC	me/100 g soil	High	> 10	Response to Ca fertiliser expected at levels < 0.2 me/100 g soil. If high Na levels, response occurs with higher Ca levels	7.7.3
			Low	< 4		
Mg	As CEC	me/100 g soil	High	> 4.0	Mg deficiency more likely on coarse, acidic soils. With high Ca, Mg is less plant available	7.7.4
			Low	< 0.5		
K	As CEC	me/100 g soil	High	> 0.6	Response to K fertiliser unlikely. High K effects often similar to high Na, but depends on soil type – especially texture	7.7.5
			Low	< 0.2	Response to K fertiliser likely	
EPP	Calculation: K^+/CEC	%	High	≥ 25%	Very approximate upper limit) (cf ESP ≥ 15%)) 1/	
			Low	≤ 2%	Very approximate lower limit)	
Na	As CEC	me/100 g soil	High	> 1)		
ESP	Calculation: Na^+/CEC	%	High	> 15%)	Alkali or sodic soils 1/	7.7.6
		%	High	> 15%	50% yield reduction for sensitive crops 1/	
				15-25%	50% yield reduction for semi-tolerant crops 1/	
				35%	50% yield reduction for tolerant crops 1/	
Al:CEC	1 M KCl unbuffered	%	High	> 85	Tolerated only by few crops	7.7.8
			Medium	30-85	Generally toxic	
			Low	< 30	Sensitive crops affected	

Notes: See page 114.   cont

Soil chemistry

Table 7.4 cont

Analysis	Recommended method(s)	Units	Rating	Range	General interpretation	Section reference
Exchangeable cation ratios						
Ca:Mg	As CEC	–		⩾ 5:1	Possible Mg and (with high pH) P inhibition	7.7.9
				3–5:1	Normal range	
				< 3:1	Possible P inhibition and Ca deficiency	
Mg:K	As CEC	–		< 1:2	Mg uptake may be affected	
P	See Subsection 7.8.2	ppm		Depends on method	See Table 7.17	7.8 Interpretation 7.8.3
Organic N	Micro Kjeldahl	% by weight	High Medium Low	> 0.5) 0.2–0.5) < 0.2)	Interpretation depends on soil and location	7.4 Interpretaion 7.9.3
Organic C	Walkley–Black	% by weight	High Medium Low	> 10) 4–10) < 4)	Interpretation depends on soil and location	7.10 Interpretation 7.10.3
OM	Organic C x 1.72		–	0.5–1	Usual OM range weak ochric A of xerosols	Table 5.6
				< 0.5	Usual OM range very weak ochric A of yermosols	
C:N	Calculation	–	–	10:1	'Normal' temperate zone value; slightly lower in tropics. Straw residues increase C:N; legume residues decrease C:N	7.10.4
Free carbonate	Acid treatment	% by weight	–	> 40	'Extremely calcareous'. Problems of P and micronutrient availability in calcareous soils [2]	7.11 Interpretation 7.11.3
				> 15	Calcic horizons [2]	
Gypsum	Bower and Huss (1948)	% by weight	–	< 2 2–25 > 25	Favours crop growth No adverse effect if powdery [2] Can cause substantial yield reduction; possibly Ca imbalance [2]	7.12
S	See Subsection 7.13.2	ppm		Depends on method	See Table 7.22	7.13 Interpretation 7.13.3

Notes: [1] Not sharply defined limits.
[2] Physical impedance to root penetration if cemented.

Field pH values also depend on localised redox states, the concentrations of soluble salts in the soil solution and the concentration of CO_2 in the soil air, all of which are constantly changing. There are also likely to be appreciable differences between samples taken at the same time from the same depth within the area of any one field, and seasonal differences may be quite marked.

For all these reasons, pH can never have a precise significance in agriculture (Russell, 1973) and highly accurate measurement is not required. Readings to the nearest 0.2 of a unit are quite acceptable for most purposes.

7.5.2 Recommended method of pH measurement   For most routine survey work pH determination in a soil/water suspension normally provides adequate information, and has the merit of simplicity. The advantages of using thick pastes, which approximate more closely to actual field conditions, are largely offset by the increased time required for sample preparation, particularly of heavy-textured soils.

Measurements of pH should therefore normally be made on a 1:2.5 soil:water suspension, equilibrated for 1 h, and measured with a glass-and-calomel electrode or a combination electrode using KCl gel; details are given in Hesse (1971, p 30), and Dewis and Freitas (1970, p 65).

7.5.3 Interpretation of pH measurements   For the reasons outlined above (Section 7.5.1), pH values do not have precise significance, but some generalisations can nevertheless be made for interpretive purposes. All interpretations must, however, take account of the methods of measurement used and their effects on quoted pH values.

The pH tolerance limits of different plants vary greatly, but for most commercial crops a neutral range is most suitable, with pH values (of a 1:2.5 soil:water suspension) between about 6.3 and 7.5; Annex F gives preferred values for some individual species. General relationships of pH and nutrient availability are outlined in Figure 7.2, and some broad effects of high and low pH values are as follows:

a) Low pH values (< 5.5)

   i) Phosphate – Phosphate ions combine with iron and aluminium to form compounds which are not readily available to plants.

   ii) Micronutrients – All micronutrients except molybdenum become more available with increasing acidity; deficiencies are therefore rare below about pH 7.

   iii) Aluminium – Al ions are released from clay lattices at pH values below about 5.5 and become established on the clay complex. Soils with low pH should therefore be tested for exchangeable Al as a measure of potential Al toxicity (see Subsection 7.7.8).

Effect of pH on availability of common elements in soils   Figure 7.2

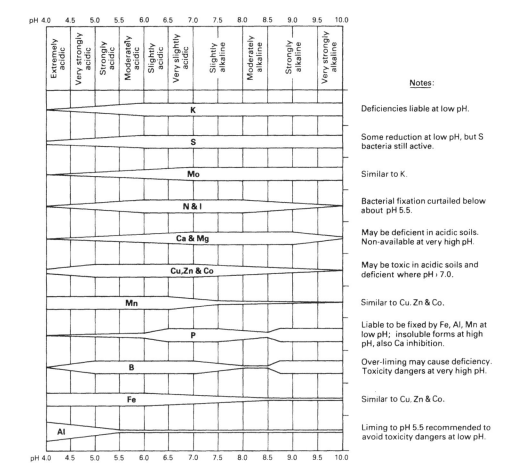

Notes:

Deficiencies liable at low pH.

Some reduction at low pH, but S bacteria still active.

Similar to K.

Bacterial fixation curtailed below about pH 5.5.

May be deficient in acidic soils. Non-available at very high pH.

May be toxic in acidic soils and deficient where pH › 7.0.

Similar to Cu, Zn & Co.

Liable to be fixed by Fe, Al, Mn at low pH; insoluble forms at high pH, also Ca inhibition.

Over-liming may cause deficiency. Toxicity dangers at very high pH.

Similar to Cu, Zn & Co.

Liming to pH 5.5 recommended to avoid toxicity dangers at low pH.

Width of bands give qualitative indication of availability of elements

Source: After Truog (1948).

iv) <u>Nitrification</u> – Below about pH 5.5 bacterial activity is reduced and nitrification of organic matter is significantly retarded.

v) <u>Effect on drainage pipes</u> – If soil pH values are < 4, only supersulphated concrete (as opposed to ordinary concrete) should be used in pipe construction; for pH values between 6.5 and 4.1 high alumina or supersulphated concrete should be used. Above pH 6.5 porous concrete pipes can be used, provided sulphate levels are sufficiently low – see Table 8.11, page 175.

b) <u>High pH values (> 8.0)</u> 1/

i) <u>Phosphate</u> – In the presence of calcium, phosphate tends to be converted to calcium phosphate, and availability of P to plants is reduced. Above pH 8.5 the presence of sodium (see (iii) below) may increase phosphate availability by formation of soluble sodium phosphate.

ii) <u>Boron</u> – Boron toxicity is a common feature of saline and sodic soils (see Chapter 8).

iii) <u>Sodium</u> – Most soil pH values > 8.6 indicate an exchangeable sodium percentage > 15, and the possibility of soil structural and reclamation problems (see Chapter 8).

iv) <u>Nitrification</u> – High pH decreases bacterial activity and hence nitrification of organic matter.

v) <u>Micronutrients</u> – Availability is reduced with increasing pH, except for molybdenum.

Some analytical cross-checks involving pH values are given in Section 8.6.(d).

c) <u>Lime requirement</u> – Soil reaction values are sometimes used as a basis for calculating a general 'lime requirement' of a soil (Peech, 1965). Values thus calculated must be viewed with suspicion if they are based on experience in areas outside the area of study, where the climate, soils, crops and/or economic conditions may be different. In particular, values developed for temperate zones are not always applicable to tropical areas, since the purpose of liming in the two cases is different. Liming is discussed in more detail in Subsection 7.7.7, but as a general rule tropical liming is aimed only at raising the pH to about 5.5, which is usually sufficient to prevent toxicity, particularly of Al (Pearson, 1975).

1/ But note that effects on micronutrients and nitrification may occur at pH values lower than 8.0.

117

d) Effects of different dilution media – The reductions in measured pH using $CaCl_2$ and KCl suspensions have already been discussed in Section 7.5.1. If a higher reading is obtained in salt solution than in water, this almost invariably indicates poor nutrient availability and the likelihood of strong phosphate fixation. The rise in pH is possibly associated with H ion absorption by polymerised aluminium ions (Bache, 1970). Certain rare clay minerals also produce this effect.

7.6 Cation-exchange capacity

7.6.1 Background   Cation-exchange capacity (CEC) measurements are commonly made as part of the overall assessment of the potential fertility of a soil, and possible response to fertiliser application. The CEC results can sometimes also be used as a rough guide to the types of clay minerals present. It should be noted, however, that CEC is an arbitrary concept arbitrarily defined (Russell, 1973), although for many soils the appropriate measurement can give values that, at a given pH, are related to the sum of the cations held by the permanent negative charge on the clay particles and the cations held by the organic matter.

Most CEC estimates are derived from the amount of a particular cation that a soil can hold when leached by a buffered solution containing that cation. Two commonly used leaching solutions are 1 M ammonium acetate buffered at pH 7, and 0.25 M barium chloride with triethanolamine buffered at pH 8.2. For carbonate-rich soils acidification may precede the leaching to remove free (as opposed to exchangeable) calcium and magnesium. An alternative estimate of CEC is given by determining the total amount of exchangeable cations displaced from the soil, by leaching with a neutral salt such as 1 M KCl or 0.5 M $BaCl_2$. These determinations normally underestimate the CEC, since the leaching solutions do not replace all the simple Al ions and probably none of the polymerised Al (Sawhney, 1968).

Measured CEC values often depend critically on pH, as indicated by the very variable ionic adsorptions summarised in Table 7.5. Different concentrations of leaching solution and different solutions used for washing out the excess can also produce very marked effects, some of which are summarised in Table 7.6.

The increase in apparent CEC with increasing ammonium acetate concentration is well known for weathered tropical soils high in iron and aluminium oxides, and the CEC of allophane is particularly susceptible to pH changes, a hysteresis loop being produced by subsequent acid and alkali treatments. Measurements indicate that the CEC of allophane is near zero at pH 3.5 and over 200 me/100 g at pH 10.7, but even at constant pH 7 hysteresis effects can produce variations in CEC from 50 to 150 me/100 g, a spread of 100 me/100 g (see Jackson, 1965, p 594). Corresponding spreads at pH 7 for other materials (in me/100 g) include: zero for kaolinite and quartz; 0.5 for gibbsite; 10 for montmorillonite and 18 for halloysite (Aomine and Jackson, 1959, p 212).

Variation of ion adsorption with pH and concentration   Table 7.5
of leaching solution

Approximate pH	Concentration of NH_4Cl solution							
	1.0 M		0.2 M		0.02 M		0.002 M	
	Adsorption (me/100 g soil)							
	NH_4	Cl	NH_4	Cl	NH_4	Cl	NH_4	Cl
3	11.4	10.7	5.1	1.4	3.5	1.0	3.2	0.3
5	12.6	9.5	5.7	0.8	4.4	0.4	3.6	0.1
6.5	13.4	9.2	6.1	- 0.1	5.0	0.1	4.3	0
7.5	16.5	9.5	7.8	1.1	6.3	- 0.1	4.8	0
8.5	22.8	6.2	9.2	0.6	8.2	- 0.5	5.5	0

Measurements made on red earth from Nachingwea, Tanzania (pH in a
1:1 soil:water extract = 4.78)

Source: Barber and Rowell (1972).

Variation of CEC with washing agent and concentration   Table 7.6
of leaching solution

Washing agent	CEC (me/100 g soil)			
	Concentration of ammonium acetate			
	1.0 M 1/	0.2 M	0.02 M	0.004 M
Absolute alcohol	10.1	-	-	4.5
Industrial methylated spirits 1/	9.7	7.9	6.5	5.4
Water	4.3	-	-	3.9

Measurements made at pH 7.0 on red earth from Nachingwea, Tanzania.

Note: 1/   'Standard' method.

Source: Barber and Rowell (1972).

The conditions of all CEC measurements should always be chosen with
care, therefore, and the results correlated wherever possible with
known soil performance;   soils known to be high in allophane should
have CEC measurements made at the pH of the soil.   In some studies
CEC determinations at two different pH values (or more) may be more
meaningful than a single measurement.   Where this is impractical,
then a pH of about 5.5 may be the appropriate value to take, as it
often represents the value which is to be limed for in the tropics
(Subsections 7.5.3 and 7.7.7).

For some soils a separate CEC measurement may be dispensed with altogether, and instead the sum of the exchangeable cations (including H and Al, and extracted with KCl at the unbuffered pH of the soil) may be used to give a more meaningful value for the CEC, although the accuracy is limited because errors in the individual methods are cumulative.

7.6.2 Recommended methods and units for CEC measurements   Standard methods for CEC determinations are given in Hesse (1971, p 101f); for most projects the following methods are suitable:

a)   for variable-charge soils: 1 M KCl extraction at the unbuffered pH of the soil, but note that some exchangeable H and Al may not be measured;

b)   for neutral soils which are not calcareous or saline: ammonium acetate extraction, adjusted to pH 7;

c)   for other soils, notably saline and saline-sodic soils, and ones containing Ca and Mg carbonates: sodium acetate extraction pH 8.2. The advantage of this is that the extract can also be used for measurement of individual exchangeable cations (see Subsection 7.7.2). The CECs in soils not having high sulphate contents can also be measured using 0.5 M $BaCl_2$/triethanolamine buffered to pH 8.2.

Leaching of soluble salts prior to CEC measurement is not recommended because of hydrolysis effects (even if alcohol is used). For saline soils a separate determination of soluble salts (Chapter 8) should be made, and the value subtracted from the total exchangeable plus soluble ions as determined here. The CEC results are quoted as me/100 g of soil or per 100 g of clay.

7.6.3 Interpretation of CEC measurements   From the discussion in Section 7.6.1 it is apparent that any interpretation of quoted CEC values must start with an evaluation of the method used and its applicability to the soil being considered (see Dewis and Freitas, 1970). In general, topsoil values for the CEC measured on the whole soil (< 2 mm) may be summarised as shown in Table 7.7, bearing in mind that CEC values seem to be higher with Ba, Sr, Ca or Mg as index cations than with NH_4 or Na:

Rating of CEC results for topsoils   Table 7.7

CEC (me/100 g of soil)	Rating
> 40	Very high
25 – 40	High
15 – 25	Medium
5 – 15	Low
< 5	Very low

The FAO (1979a) quote CEC values of 8 to 10 me/100 g of soil as indicative minimum values in the top 30 cm of soil for satisfactory production under irrigation, provided that other factors are favourable. Lower values than these should be highlighted in land suitability classifications. Any CEC values of < 4 me/100 g soil (measured at the pH of the soil) indicate a degree of infertility normally unsuitable for irrigated agriculture, although rice is tolerant of slightly lower values.

It should also be noted that CEC is essentially a property of the colloidal fraction of soil, derived mainly from the clay and organic matter fractions, although the silt-sized particles sometimes contribute significantly. Hence, assessment of CEC values should always take into account the clay content and mineralogy of the sample. The CEC values associated with the main clay minerals and organic matter are summarised in Table 7.8.

Approximate CEC values of clay minerals and organic matter   Table 7.8

Type	Lattice	Nutrient reserves	Approximate CEC at pH 7 (me/100 g of clay)
Kaolinite and halloysite	1:1	Few nutrient reserves	< 10
Illite	2:1	Reserves of potassium	15 – 40
Montmorillonite	2:1	Generally with reserves of Mg, K, Fe etc	80 – 100
Vermiculite	2:1	Generally with reserves of Mg, K, Fe etc	About 100
Organic matter	–	–	About 200

For soils low in organic matter, the CEC can be usefully expressed as a proportion of the clay, thus:

$$\text{CEC (me/100 g clay)} = \frac{\text{CEC (me/100 g soil)}}{\%\ \text{clay}} \times 100$$

This CEC value can be used as a rough guide to clay mineralogy, although it should be used with caution since most soils contain several clay minerals and, in addition, a number of hydrous oxides. Its use is illustrated in the following example of two soils with the same CEC value for the whole soil:

Soil	CEC (me/100 g soil)	Clay content (%)	CEC (me/100 g clay)
A	7.5	68	11.0
B	7.5	20	37.5

Soil A probably has a clay fraction dominated by kaolinite or halloysite minerals with few nutrient reserves, but the clay fraction of soil B has a greater proportion of 2:1 lattice minerals and can be expected to have more nutrient reserves, despite the fact that the sample has the same overall CEC and a substantially lower clay content than soil A.

Because both CEC and moisture retention properties are related to the texture of soil, there generally exists a fair correlation between CEC and saturation percentage particularly in soils with similar parent materials and history of formation.

Finally it should be noted that soils with high allophane and/or Fe and Al oxide contents have CECs which are particularly susceptible to the pH and cation concentration of the saturating solution (see Subsection 7.6.1). Properties of such variable-charge soils are discussed in Uehara and Gillman (1981), and Juo (1981).

7.6.4   Base saturation   The proportion of the CEC accounted for by exchangeable bases (Ca, Mg, K and Na) is frequently used as an indication of soil fertility. However, the base saturation percentage (BSP) does not distinguish between different bases and imbalances in their relative proportions can cause severe plant nutrition problems (see Subsection 7.7.9).

It has often been assumed that for optimum agricultural production a neutral soil is required, with all the acidity neutralised by base saturation of the clay complex exchange sites. However, in acid soils in particular, the limiting factor is often the concentration of aluminium ions in the soil solution, which in turn depends on the concentrations of the other ions involved in the exchange reactions. It has also been shown (Clark, 1966) that some soils are base saturated at pH 5, which may explain why acid-sensitive crops can be grown on some tropical soils with pH values about 5.5, and why liming does not increase their yield (Russell, 1973).

The BSP values are used in soil classification by FAO-Unesco (1974) as indications of soil fertility status; two BSP ranges are used based on ammonium acetate extractions of samples from 20 to 50 cm depth:

> 50%: eutric (or more fertile) soils
< 50%: dystric (or less fertile) soils

A general interpretation of BSP values is as follows: low < 20; medium 20 to 60; high > 60 (Varley, personal communication, 1983).

7.7 Individual exchangeable cations

7.7.1 Background   The levels of exchangeable cations in a soil are usually of more immediate value in advisory work than the CEC, because they not only indicate existing nutrient status, but can also be used to assess balances amongst cations. This is of great importance because many effects, for example on soil structure and on nutrient uptake by crops, are influenced by the relative concentrations of cations as well as by their absolute levels.

7.7.2 Recommended methods and units for exchangeable cation measurements   Because of the strong effect of pH levels on clay content charge – and hence on ion adsorption (Subsection 7.6.1) – it is recommended that exchangeable cation measurements be made using an unbuffered saturating solution such as 1 M KCl at the pH of the soil. A method is described by McLean (1965).

There are however, at least four main drawbacks to this approach: the conditions of measurement will differ from soil to soil; laboratories may not be geared to varying requirements; interpretation and comparison of results may be difficult; and clients may be chary of accepting a non-standard method. On many projects, therefore, the use of more widely accepted methods will be necessary, and the following are recommended (Hesse, 1971, p 101f):

a)   for soil without free carbonate: the ammonium acetate extraction at pH 7;

b)   for soils with free carbonate: the sodium acetate method at pH 8.2.

Both methods tend to dissolve any soluble salts present in the soil. Results are normally quoted as me/100 g of soil.

7.7.3 Interpretation of exchangeable calcium measurements   Indices of plant-available Ca are of little value, since the availability varies enormously from soil to soil, and is highly dependent on a number of other factors. Normally Ca deficiency as a plant nutrient occurs only in soils of low CEC at pH values of 5.5 or less. Large inputs of K fertiliser or high natural K reserves may, however, inhibit plant uptake of Ca in soils having a more neutral reaction.

Calcium may also be effectively deficient at high pH levels when there is an excessive Na content. This can occur when the measured available Ca levels are well in excess of those considered adequate for less extreme soils. However, in these high Na (or sodic) soils additional Ca may not only be required for nutritive purposes, but also to promote flocculation of the clay micelles, and hence assist in the formation of more stable soil structures.

As a rule of thumb, some response to Ca fertiliser, when applied as a plant nutrient, may be expected from most crops when the exchangeable Ca in the soil is < 0.2 me/100 g of soil. However, even when Ca levels are much higher than this, the application of Ca fertiliser can lead to greatly improved yields if it is the pH and not plant nutrient Ca which is the limiting constraint (Subsection 7.6.4). Calcium deficiencies are reported for groundnuts at Ca values < 0.8 me/100 g (Meredith, 1965), and cotton is known to have a high Ca requirement.

When free carbonates (Section 7.11) are present in a soil, measurement of both exchangeable Ca and Mg must be viewed with caution, even when the sodium acetate method has been employed. A proportion of the free carbonate will be taken into solution, and thus raise the measured Ca level, sometimes leading to values exceeding the CEC. In such soils there is no really satisfactory method of obtaining a true estimate of exchangeable Ca and Mg. The subtraction of free carbonate values (Section 7.11) from the apparent exchangeable Ca and Mg values, is likely to result in low results, since methods used for free carbonate measurement generally extract more carbonate than do leaching agents. Similarly, the amounts of soluble Ca and Mg in the saturation extract are generally lower than in ammonium acetate or sodium acetate extracts, so the subtraction of soluble cations from the apparent exchangeable cation values (a method used satisfactorily for determining exchangeable sodium in the presence of soluble sodium) is invalid for Ca and Mg determinations. A crude approximation of the value of exchangeable Ca + Mg can be obtained by subtracting exchangeable Na + K from the CEC when the soil is totally base saturated. High values of exchangeable Ca can be taken as those above about 10 me/100 g soil.

Although it is known that Ca ions have an affinity for phosphate, the effect of the interaction on availability to plants is not well understood. It should be noted, however, that in calcareous soils and soils with high exchangeable Ca, P may be less available to plants. Cation ratios are discussed below and in Subsection 7.7.9.

7.7.4   Interpretation of exchangeable magnesium measurements   The presence of Mg deficiency in a crop may not only be associated with low Mg content in the soil, but also with the presence of large amounts of other cations, particularly Ca and K. With increasing Ca:Mg ratios above about 5:1, the Mg may become progressively less available to plants, although soils can remain fertile over a very wide range of Ca:Mg ratios. When Mg is present in very much larger amounts than Ca, the latter may become somewhat less available, and soil structure become weaker due to increased deflocculation of the clay. In neither event is it possible to give precise Ca:Mg ratios within which fertility is assured and beyond which deficiency symptoms will occur; some rough figures for cation balances are given in Subsection 7.7.9.

Deficiency symptoms resulting from low absolute values of exchangeable Mg, rather than cation imbalance, have been reported in acid, coarse-textured soils having exchangeable Mg levels of < 0.2 me/100 g soil (Heald, 1965), but in the tropics 0.5 me/100 g is often the deficiency threshold; high levels of Mg may be taken as those above 4.0 me/100 g (Varley, personal communication, 1982). Ratings for exchangeable Mg values are given in Table 7.9.

Exchangeable magnesium ratings   Table 7.9

Level (me/100 g)	(ppm)	Rating	Comments
< 0.2	< 30	Low	Quick-acting Mg fertilisers may be required
0.2-0.5	30-60	Medium	Use Mg limestone when lime is needed
> 0.5	> 60	High	Mg usually sufficient in soil

Note: Mg uptake may be affected by high exchangeable K levels (see Subsection 7.7.5).

Source: Adapted from MAFF (1967).

7.7.5   Interpretation of exchangeable potassium measurements
Exchangeable K levels are of only limited value for predicting crop response since they give no direct indication of the capacity of the soil to release currently unavailable K over a period of time (see Mengel and Busch, 1982). Regrettably there is no widely accepted test that either measures exchangeable K plus some index of the rate of K release, or that can be used to estimate the rate of release of K as supplementary data to the exchangeable values (Mengel, 1982). Values for total K might be useful in this respect, but as an example, Varley (personal communication) quotes the contrasting results for a Nepal soil and one from St Helena. The former, containing mica minerals, gave a total K value of 20 000 ppm (2%) but exchangeable K as only 0.1 me/100 g soil. The latter, however, with total K of only 2 000 to 3 000 ppm gave exchangeable K as more than 2.0 me/100 g soil.

It should be remembered that exchangeable K levels usually alter when the soils are dried. As a general rule, samples with large amounts of available K lose some by fixation, and those with low amounts have their exchangeable K augmented from sources that are non-available in the field.

The quantities of exchangeable K, as determined by ammonium acetate extraction, often differ little from the K extracted by dilute acid solutions (Fassbender, 1980) and the results can usually be interpreted in the same general manner using the values given in Table 7.10.

Critical values of available potassium   Table 7.10

Available K (me/100 g soil): Ammonium acetate extraction				
High values	Medium values	Low values	Comments	Source
0.8–0.4	0.4–0.2	0.2–0.03	Based on Malawi soils	Young and Brown (1962)
> 0.5	0.5–0.25	< 0.25	From a general appraisal of USA soils	G W Thomas, personal data (1966)
> 0.8	0.8–0.5	0.5–0.3	New Zealand	Metson (1961)
> 0.6	0.6–0.15	< 0.15	UK soils	MAFF (1967)

In general terms, therefore, a response to K fertiliser is likely when a soil has an exchangeable K value of below about 0.2 me/100 g of soil and unlikely when it is above 0.4 me/100 g of soil. However, these limits should not be considered as definitive, since they are subject to variation dependent both on the nature of the soil, the environment and the crop. Experience in Zimbabwe specifically emphasises the relationship of deficiency levels and soil texture; for three main textural groups of Central African soils, the indices of availability shown in Table 7.11 are quoted.

Boyer (1972) quotes the following absolute and relative minimum quantities of exchangeable K to avoid deficiencies in humid tropical soils, although Jones and Wild (1975) point out that the values are only approximate, and will vary with crops and production levels:

Minimum absolute level: between 0.07 and 0.20 me/100 g

Minimum relative level: at least 2% of the sum of all exchangeable bases

Relationship of exchangeable potassium and soil texture   Table 7.11

Rating (for Central African soils)	Exchangeable K (me/100 g) (Ammonium acetate extraction)		
	Sands	Sandy loams	Typical red–brown clays
Deficient (response to K likely)	< 0.05	< 0.1	< 0.15
Marginal (some response likely)	0.05 – 0.1	0.1 – 0.2	0.15 – 0.3
Adequate (response unlikely, but maintenance of K usually desirable)	0.1 – 0.25	0.2 – 0.3	0.3 – 0.5
Rich (no K required)	> 0.25	> 0.3	> 0.5

Source: As interpreted in the Ministry of Agriculture, Harare, Zimbabwe.

Adverse cation ratios – notably high K:Mg ratios leading to Mg deficiency – may be caused by fertiliser use. For soils with low CEC values Ca and Mg applications may be required in addition to K dressings to maintain the balances at favourable values. With high exchangeable potassium percentages (EPP) – say above 25% – the permeability and structure of the soil may be adversely affected, but not as seriously as with high Na percentage. For glasshouse and fruit crops, Mg uptake may be inhibited if the K:Mg ratio exceeds 2:1 in soils low in Mg (MAFF, 1967).

7.7.6   Interpretation of exchangeable sodium measurements   Although Na may, in particular circumstances, be utilised by some plants as a partial substitute for K, it is not an essential plant nutrient. Its absence, or presence in only very small quantities, is therefore not usually detrimental to plant nutrition. However, when Na is present in the soil in significant quantities, particularly in proportion to the other cations present, it can have an adverse effect, not only on many crops, but also on the physical conditions of the soil.

When reviewing the levels of Na determined by analysis, care must be taken to distinguish between exchangeable Na and soluble Na. The leaching agents involved in determining exchangeable bases take into solution both the exchangeable and any soluble forms of Na that are present. To establish the amount of exchangeable Na in sodic soils, the quantity of soluble Na, determined from the saturation extract (Section 8.5), must be subtracted from the total value obtained for Na in the leaching agent (taking care to use the same units) ie:

Exchangeable Na = Extracted Na - Soluble Na

A widely used measure of the effects of high sodium levels is the exchangeable sodium percentage (ESP), which is defined as:

$$ESP = \frac{Exchangeable\ Na}{CEC} \times 100$$

An ESP value of 15 is often regarded as the boundary between sodic and non-sodic soils, although it has long been realised that this is an arbitrary figure (Richards, 1954), since the properties of soils often exhibit no sharp change as the content of exchangeable Na increases. In some soils an exchangeable Na content of 2 to 3 me/100 g may be a more suitable criterion for distinguishing sodic samples (Richards, 1954), although the particular circumstances under which this limit is best applied have not been defined. In general, soils with exchangeable Na > 1 me/100 g should be regarded as potentially sodic.

In general terms, high ESP values have a greater deleterious effect upon soils with 2:1 lattice clays than on those with 1:1 clays. However, although the onset of adverse physical conditions occurs generally at lower ESP levels in montmorillonitic soils, further comparatively large increases in Na content may not cause much additional deterioration.

In addition to the effects caused by soil physical response to high ESP levels, crops have different tolerances to the presence of exchangeable Na, as indicated by Tables 7.12 and 7.13. The use of ESP values in sodic soil and water investigations, including discussion of the sodium adsorption ratio (SAR), is discussed in Sections 8.3 and 8.4.

7.7.7   Exchange acidity and lime requirement   In acid soils the exchange capacity equals the sum of the exchangeable bases plus the exchange acidity. The exchange acidity is derived from the hydrogen and aluminium ions located on the exchange complex, and is closely related to both the base saturation and the pH of the soil (Subsections 7.5.3(c) and 7.6.4).

The exchange acidity provides a basis from which the lime requirement can be calculated, the formula used depending on the final pH required and the depth of soil involved (see Section 7.5.3). Peech (1965) and Jackson (1958) should be consulted for details of the calculations. Crude estimations of the lime requirement can also be obtained from pH values (Peech, 1965), and from the BSP and CEC (Metson, 1961). The nomograph involved in the latter method is given in Figure 7.3.

In the tropics liming is usually only aimed at preventing Al toxicity, by raising the pH to about 5.5 (Pearson, 1975), although Juo (1981) quotes satisfactory results on soils with pH values raised to only pH 4.7 to 5.2; Sanchez (1976, p 241) quotes high coffee yields from unlimed soils with pH values as low as 3.8. An application of 1.65 t ha^{-1} of $CaCO_3$ equivalent for every 1 me/100 g of soil of exchangeable aluminium is often used as a measure of the lime requirement for very acid tropical soils.

Crop tolerance to ESP   Table 7.12

ESP	Type of crop affected	Growth responses under field conditions	Crop examples
2 - 10	Extremely sensitive	Sodium toxicity symptoms even at low ESP values	Deciduous fruit, nuts, avocado, cassava, citrus
10 - 20	Sensitive	Stunted growth at low values even though the physical condition of the soil may be good	Beans, sugarcane
20 - 40	Moderately tolerant	Stunted growth due to both nutritional factors and adverse soil conditions	Clover, oats, tall fescue, rice, dallis grass
40 - 60	Tolerant	Stunted growth usually due to adverse physical conditions of soil	Wheat, cotton, alfalfa, barley, tomatoes, beets
> 60	Most tolerant	Stunted growth usually due to adverse physical conditions of soil	Crested, fairway and tall wheat grass, rhodes grass

Source: Bower (1959).

Crop yield reduction at different ESP levels

Table 7.13

Sensitive 50% yield reduction at ESP < 15%		Semi-tolerant 50% yield reduction at ESP 15–25%		Tolerant 50% yield reduction at ESP 35%	
Avocado	Persea americana	Dwarf kidney bean	Phaseolus sp	Alfalfa	Medicago sativa
Green bean	Phaseolus vulgaris	Ladino clover	Trifolium repens	Barley	Hordeum vulgare
Corn	Zea mays	Carrot	Daucus carota	Sugarbeet	Beta vulgaris
Tall fescue	Festuca arundinacea	Lemon	Citrus spp	Cotton	Gossypium hirsutum
Peach	Prunus persica	Lettuce	Lactuca sativa	Dallis grass	Paspalum dilatatum
Sweet orange	Citrus spp	Oats	Avena sativa	Onion	Allium cepa
Grapefruit	Citrus paradisi	Rice	Oryza sativa	Bermuda grass	Cynodon dactylon
		Sorghum	Sorghum vulgare		
		Wheat	Triticum vulgare		
		Sugarcane	Saccharum spp		

Note: The relationships tabulated above should only be considered as broad guidelines, since local conditions can markedly affect crop response.

For soils that are to be irrigated, the ESP that will develop in equilibrium with the irrigation water is of major importance, rather than values measured in advance from the soil. Given adequate drainage, the ESP values that are likely to develop can be approximately predicted from the SAR of the irrigation water (see Chapter 8).

Sources: Lunt (1963), FAO–Unesco (1973, p 472) and FAO (1979a, p 22).

Nomogram for estimation of lime requirement from CEC data   Figure 7.3

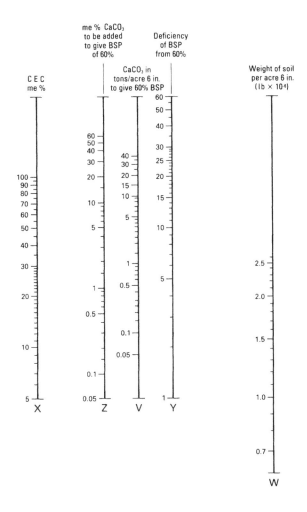

Note: Factors for conversion to metric units are:

  1 lb = 0.45 kg   1 acre 6 in = 4.1 × 10³ m² 15 cm
  1 ton = 1,016 kg   = 4.1 × 10⁻¹ ha 15 cm
              1 ton per acre 6 in = 2,490 kg per ha 15 cm

Example:

            CEC = 40 me/100g (from ammonium acetate method)
            BSP = 35%
            Deficiency of BSP from 60% = 25%
            Soil density = 2.2 × 10⁶ lb per acre 6in (= 2.4 × 10⁶ kg per ha 15 cm)

Align the number 40 of scale X and the number 25 of scale Y and read the CaCO₃
to be added on scale Z (10 me/100 g).   Then align this 10 on scale Z with 2.2 on
scale W, and read the lime requirement on scale V (5 tons per acre 6 in = 12.4 t
per ha 15 cm).

Source: Adapted from Metson (1961).

However, it should be noted that acidic soils with high contents of variable-charge surfaces (eg allophane, iron and aluminium oxides and some humic materials - see Subsection 7.6.1) require larger applications of lime to achieve neutrality than other acidic soils of similar texture, since the Ca is used not only in neutralisation of Al but in development of surface charge (Juo, 1981). Liming to pH values higher than 5.5 often results in no increase in crop yields and may even produce decreases due to cation imbalances; in particular, as pH levels are raised by liming, K deficiency symptoms are sometimes produced, possibly by the additional adsorption of K on to selective sites on the clay complex (Magdoff and Bartlett, 1980). Lime responses are discussed by Pearson and Adams (1967).

7.7.8   Exchangeable aluminium   Aluminium is not present in an available form in soils having a pH above about 5.5 (soil:H_2O 1:2.5), and tests for exchangeable Al should therefore only be made on distinctly acid soils. Although there is some dispute over the merits of unbuffered or buffered extracting solutions (Yuan and Fiskell, 1958; Coleman et al, 1959; Krupsky et al, 1961), probably the most common method of determining exchangeable Al is by the use of unbuffered 1 M KCl (Coleman et al, 1959; McLean, 1965; Hesse, 1971, p 357), although there is some doubt as to the replacement of all Al ions by this method (Sawhney, 1968).

At present there are no generally accepted critical levels for exchangeable Al, but the values given in Table 7.14 may be used as a rough guide. Absolute levels of 2 to 3 me/100 g soil are excessive for some crops (Chapman, 1966).

Exchangeable aluminium levels   Table 7.14

Exchangeable Al percentage = Exchangeable Al/CEC (%)	Effects	Reference
30	Sensitive crops may be affected	–
60	Generally toxic. Only very low Al concentrations expected if no electrolytes present	Nye et al (1961)
	60% tolerated by sugarcane	Evans (1965)
85	May be tolerated by some crops in some conditions (tea, rubber, cassava, pineapple and some tropical grasses and legumes are notably Al tolerant)	Sanchez (1976)

Toxicity of Al may be temporarily alleviated by the addition of large quantities of phosphate fertiliser, but this is likely to be too expensive in normal practice. The Al tends to combine with phosphate to form insoluble compounds, and soils high in exchangeable Al therefore tend to convert phosphate fertilisers to non-available forms. Through time this phosphate becomes preferentially combined with free iron, thus releasing the aluminium. Permanent correction is best achieved by liming the soil to a pH value between about 5.2 and 5.5 (Pearson, 1975, and Subsection 7.7.7).

7.7.9   Exchangeable cation balance   The effects of the major cations in soils on plant growth are often closely interlinked, as described in the preceding subsections. Table 7.15 summarises some of the general relationships.

7.8   Phosphorus

7.8.1   Background   Of all the major plant nutrients, phosphorus possibly has the most complicated chemistry in the soil, at least as far as assessment of P levels and the P fertiliser requirements are concerned. It occurs in soils in both organic and inorganic forms, the latter usually being the more important for crop nutrition. Inorganic P can occur as various compounds of Ca, Fe and Al, in solution, in surface films, in the solid state, or as exchangeable phosphate anions held by the positive charges on the edges of clay plates. Exchange reactions involving adsorbed phosphate ions are very slow compared with the reactions of individual cations, and slow release of such phosphate can occur at least over the length of a cropping period. Diagramatically, the P present in soil can be summarised thus:

'Non-available' $\rightleftharpoons$ Potentially available $\rightleftharpoons$ Available
Sometimes called 'fixed' phosphate

Available P is sometimes referred to as the 'intensity' of the nutrient in the profile, and potentially available P as the 'capacity'. Ideally, any assessment of soil P should take into account the concentration in the soil solution, as a measure of available P, and also the amount of fixed P and the rate at which this can be released. However, this is too complex for routine procedures, and there is no generally accepted interpretation of attempts to produce such an assessment.

Several standard 'available P' methods are in common use, their applicability depending on soil conditions (mainly pH); most of them measure the available phosphate, as defined above, and some of the fixed phosphate. The most important methods are discussed by Fassbender (1980).

7.8.2   Recommended   methods   and   units   for   available   phosphorus determinations   The most widely applicable method of determining soil P is probably by Olsen's method of bicarbonate extraction (Black, 1965). The method is temperature sensitive, however, and is preferred for soils of pH > 7. For acid soils the Bray, Truog or Morgan methods may be used (Dewis and Freitas, 1970; Hesse, 1971).

Exchangeable cation ratios: summary of critical and   Table 7.15
common values

Cation ratio	Approximate value	Effects
Ca:Mg	≥ 5:1	Mg increasingly unavailable with increasing Ca. With high pH also, P availability may be reduced
	3:1 to 4:1	Approximate optimum range for most crops
	< 3:1	P uptake may be inhibited (Yates, 1964)
	1:1	Suggested lowest acceptable limit (Fauck et al, 1969). With lower values, Ca availability slightly reduced
	Ca:Mg ratios commonly decrease with depth, and often with cultivation	
K:Mg	> 2:1	Mg uptake may be inhibited
	< 3:2	Field crops)) Recommended
	< 1:1	Vegetables and sugarbeet) levels) (Doll and
	< 3:5	Fruit and greenhouse crops) Lucas, 1973)
K:CEC (EPP)	2%	Suggested minimum level to avoid K deficiency in humid tropical soils (Boyer, 1972; but cf Jones and Wild, 1975)
	> 25%	K-rich soils (rather rare); similar effects to high Na − see Chapter 8
Na:CEC (ESP)	≥ 15%	Sodic soils (USDA, 1954b; see Chapter 8). Effects usually gradual, no sharp change at ESP = 15
Al:CEC	≥ 30%	Sensitive crops possibly affected
	≤ 60%	Low Al levels expected if no electrolytes present. Tolerated by sugarcane
	85%	May be tolerated by some crops in some conditions

Some additional ratios relating to individual crops are given in Annex F.

The results of the measurements are normally expressed as 'ppm of phosphate' (as P_2O_5); the conversion factors for elemental P and P_2O_5 are as follows:

To convert P to P_2O_5 — multiply by 2.29
To convert P_2O_5 to P — multiply by 0.44

For the reasons outlined in Section 7.8.1 above, it is not recommended that total P be measured in routine analysis.

7.8.3   Interpretation of available phosphorus measurements   Because of the variety of methods used, no one general interpretation of available P can be given. The values for Olsen's method are presented in Table 7.16 and the results for various individual methods are summarised in Table 7.17.

General interpretation of available phosphorus   Table 7.16
determined by Olsen's method

Characteristic crop demand	Examples	Indicative available P values (ppm)		
		Deficient	Questionable	Adequate
Low P	Grass, cereals, soyabeans, maize	< 4	5 – 7	> 8
Moderate P	Lucerne, cotton, sweetcorn, tomatoes	< 7	8 – 13	> 14
High P	Sugarbeet, potatoes, celery, onions	< 11	12 – 20	> 21

Sources: Cooke (1967, p 293) after Bingham (1962).

For the acid fluoride (Bray) method — and, indeed, for most acid extractions — low values certainly indicate deficiencies, but high values cannot be unambiguously interpreted. High laboratory values can result from soils with low or even deficient P levels.

Although it is not recommended that anion—exchange resin methods be employed, values based on such methods may sometimes be encountered. Interpretation guidelines are given below in Table 7.18, but are directly relevant only to Central African soils, and caution must be exercised in extrapolation to other areas.

General guidelines for available phosphorus interpretation   Table 7.17

| Method 1/2/ | Indicative available P values (ppm) | | | Appropriate soils | Comments |
| | High | Medium | Low | | |
	Fertiliser response unlikely	Fertiliser response probable	Fertiliser response most likely		
Olsen 3/ 0.5 $\overline{M}$ NaHCO$_3$	> 15	15 – 5	< 5	All soils, especially where pH is ⩾ 7	Probably most useful general method but temperature control is essential
Bray 4/ Dilute HCl/NH$_4$F	> 50	50 – 15	< 15	Acid soils	High values difficult to interpret
Nelson Dilute HCl/H$_2$SO$_4$	> 30	30 – 10	< 11	Some acid soils	Values based on North Carolina experience
Truog Dilute H$_2$SO$_4$	> 40	40 – 20	< 20	Acid soils	Best suited to temperate areas
Bingham H$_2$O solution	> 2	2 – 1	< 1	All soils	.
Morgan Na acetate/acetic acid	> 15	15 – 5	< 5	Acid soils	USA chiefly
ADAS NH$_4$ acetate/ acetic acid	> 40	40 – 2	< 2	Acid soils	UK experiments. Extracts P in solution and readily soluble Ca phosphate

Notes: 1/   See Olsen and Dean (1965);   additional methods are discussed in Fassbender (1980).
         2/   Note that extractions at higher temperatures usually give higher P values;   the stability of
              the molybdenum blue colour needs checking if laboratory temperature is > 25°C or < 20°C
              (Dewis and Freitas, 1970).
         3/   See also Table 7.16.
         4/   Note that there are several 'Bray methods' and No 2a is referred to here.

Source: Adapted from Olsen and Dean (1965).

Critical levels for available phosphorus using the   Table 7.18
anion-exchange resin method

| Available P by resin extraction (ppm) | ----- Interpretation for Central African soils ----- | |
	Available P status	Yield increases expected with adequate phosphate fertiliser 1/
< 3	Acutely deficient	Very large; up to double the yield
3 - 6.5	Deficient	Large; increases of one to two-thirds
6.5 - 13	Marginal	Small; increases of less than one-third
13 - 22	Adequate	No appreciable response likely with general crops, but maintenance dressing desirable
> 22	Rich	No response likely

Note: 1/   Assuming other nutrients are adequate.

Source: As interpreted in the Ministry of Agriculture, Harare, Zimbabwe.

Where total P levels are known, it may be of use to remember that the average for the top 15 cm throughout the United States has been found to be about 0.06% or 600 ppm (Lipman and Conybeare, 1936), and that total soil P only rarely exceeds 0.2% or 2 000 ppm (Metson, 1961). The following rating is used by J Varley (personal communication, 1982) for total P in tropical soils, as determined by perchloric acid digestion: low < 200 ppm, medium 200 to 1 000 ppm and high > 1 000 ppm.

Interactions of P with Fe, Zn and Cu in soils are discussed in Section 7.14, and it should be noted that the presence of chlorides in soils can reduce plant uptake of P. Uptake of P by sugarcane is also known to be reduced by Al (Evans, 1959), although this may be due to effects on the cane roots.

7.9   Nitrogen

7.9.1   Background   Nitrogen occurs in soils in several forms: organic compounds, nitrate and nitrite anions, and ammonium ions, which can occur as exchangeable cations; nitrates are the main forms of N used by plants. Apart from application of N fertilisers, the main source of N in soils is the breakdown and humification of organic matter; slow decomposition of humus releases NH_4 ions which are subsequently oxidised to nitrite and nitrate. This supply can be greatly influenced by microbial activity. Levels of NO_3 are always high after a long dry spell and are the source of a sudden green flush in tropical plants at the onset of the rain.

137

Different measurements of soil N give divergent results because varying proportions of the different types of N are extracted. Levels of N in a given soil can also vary widely depending on the season and cultivation history (Mengel, 1982). Levels of N are particularly susceptible to change during storage of soil samples: storage time, temperature and moisture content all have effects on oxidation and microbial activity, and hence on N levels (Breimer and Slangen, 1981).

7.9.2   Recommended methods and units for nitrogen measurements   There are many published methods of determining the forms of N present in soils, but they are often too complex in operation, or too specific in their objectives, to be of use in routine analysis. The most common standard method is probably the catalytic oxidation of organic and chemically combined N and subsequent alteration to NH_4, usually by the micro Kjeldahl process (Hesse, 1971, p 197). This method also extracts some of the interstitial NH_4 held in clay lattices, but in most tropical soils the error thus introduced is probably very small. A review of the method has been produced by Bremner (1965). Results are invariably quoted as percentages of N by weight in the soil.

7.9.3   Interpretation of nitrogen measurements   Except in detailed management studies, N measurements are difficult to interpret, since the types of N present and their relevance to crop nutrition are not usually known. Even within specific environments there seems to be no general agreement on ratings of N values measured by the same method. The ratings in Table 7.19 are therefore presented as a very general assessment of total N contents:

Broad ratings of nitrogen measurements   Table 7.19

N content Kjeldahl method (% of soil by weight)	Rating
> 1.0	Very high
0.5 - 1.0	High
0.2 - 0.5	Medium
0.1 - 0.2	Low
< 0.1	Very low

Source: Adapted from Metson (1961).

Note that apart from the interactions summarised in Figure 7.1, N uptake is reduced by the presence of chlorides in a soil. The effects of low pH on N availability are also particularly marked, in that microbiological activity is considerably reduced and available N is consequently very low, whatever the total N. At very low pH values, no breakdown of OM occurs and peats can form.

In view of the uncertainty in interpretation, N measurements are of little value at the prefeasibility and feasibility levels of survey, unless correlatory data are available. Experienced soil surveyors can gauge the approximate N content in a soil, and absolute values are of little importance, since almost all commercial agricultural developments require the application of nitrogenous fertilisers. Guidance for such applications can only be made after detailed field trials, when leaf analysis and crop yield data are likely to be of more use than soil analyses. Soil N measurements have become such an ingrained part of survey analyses, however, that they will undoubtedly continue.

7.10   Organic carbon

7.10.1   Background   Measurements of organic C, like those of soil N, are very widely quoted, but they suffer from similar difficulties of experimental determination and interpretation. They are often made as a measure of the quantity of organic matter in a soil, which in turn is taken as a crude measure of fertility status.

Problems can arise experimentally because of the presence in soils of other forms of C, notably charcoal in regions of human habitation, and undecomposed plant residues, both of which sources may influence the analytical results whilst contributing little to the nutrients immediately available to crops. As with N analyses, seasonal, storage and pretreatment changes in soil organic matter can greatly affect measured values.

7.10.2   Recommended methods and units for organic carbon measurement   Most routine organic C determinations should be made by the Walkley–Black dichromate method (Hesse, 1971, p 245). It should be noted that this method requires a reaction temperature of $130°C ± 5°C$. If the reaction is carried out in too large a vessel then the heat sink is large and this temperature is never reached, resulting in low reported values of organic C.

The results are usually quoted as the percentage by weight of organic C in the soil. For the Walkley–Black method the recovery of organic C is conventionally taken as 75%, giving a conversion factor to total organic C of 1.33, but this figure can vary depending on the type of soil and organic matter under investigation. Sometimes organic C values are multiplied by a further factor to convert them to percent organic matter; one convention is to assume the organic C is 58% of the total organic matter (giving a conversion factor of 1.72) but, again, proportions vary.

7.10.3   Interpretation of organic carbon measurements   Interpretation of organic C data suffers from the same drawbacks as those for N (Section 7.9.3) and guidelines are difficult to find. As a rough approximation, the values shown in Table 7.20 may be used.

Broad ratings of organic carbon measurements   Table 7.20

Organic C content Walkley–Black method (% of soil by weight)	Rating
> 20	Very high
10 – 20	High
4 – 10	Medium
2 – 4	Low
< 2	Very low

Note: Chloride reacts with dichromate, and C values for very saline soils will be correspondingly lowered.   A similar effect is produced by the presence in soil of elemental C, and of ferrous and manganous oxides.

Source: Adapted from Metson (1961).

Organic C levels are also used in a number of regression equations for prediction of water-holding properties of soils (eg Salter and Williams, 1969).   Apart from these, organic C measurements are of little immediate value, except in detailed or management studies. Organic matter levels of the ochric A horizon are used to separate yermosols and xerosols (see Chapter 4, and FAO–Unesco, 1974, p 25).   Soil organic matter, including organic C, N and fertility, is discussed by Schnitzer and Khan (1978).

7.10.4   C:N ratios   The C:N ratios are commonly quoted in soil reports as indications of the type of organic matter present and, in particular, the degree of humification.   Generally applicable interpetations are difficult to make, however, and the relevance of the figures to most consultancy projects is small.

For temperate soils, the equilibrium C:N ratio has often been taken as 10:1 with somewhat lower values in soils with higher temperatures and more microbiological activity.   However, analysis of 172 samples by Leighty and Shorey (1930) showed a range of values from 35:1 to 3:1, and that the ratio was not an accurate indication of the amount of organic matter.

Incorporation of only partially decomposed crop residues can greatly affect the C:N value.   Undecomposed straw residues tend to increase the ratio, whilst legume residues high in nitrogen tend to reduce it.

7.11   Free carbonate

7.11.1   Background   Carbonates in soil profiles may be derived from carbonate-rich rocks (especially from calcite $CaCO_3$ and dolomite $CaCO_3 + MgCO_3$) but are more commonly encountered as a secondary deposition from groundwater.   In calcareous soils the carbonate is often associated with the silt fraction (eg Deb, 1963) and can seriously affect laboratory particle size analysis (see Section 6.9).

When the soil pH is below 7.0 Ca and Mg carbonates are seldom present; when the pH exceeds 7.0 then free carbonates should be determined on a routine basis. The presence of free carbonate normally indicates that the clay complex is dominated by exchangeable Ca, which usually implies favourable soil physical conditions. Excess Ca, however, can lead to deficiencies of minor elements (see Section 7.7), as well as antagonising the action of others (see Figure 7.1).

7.11.2 Recommended methods and units for free carbonate measurements
Both gravimetric and manometric methods can be used (Hesse, 1971, p 52f), although the former is more time-consuming. Free carbonate is usually expressed as calcium carbonate equivalent (percent by weight in soil), although in fact it can be derived from magnesium or other carbonates. The level of exchangeable or soluble cations present will normally indicate which carbonates are present.

7.11.3 Interpretation of free carbonate measurements   There are no precise ratings for levels of free carbonate, but values of over about 40% can be considered as extremely calcareous (Avery, 1964). Values of 70% or more occur in horizons in arid zones. If Mg (as opposed to Ca) carbonate is suspected this should be taken into account in the analysis and interpretation; $MgCO_3$ is less soluble than $CaCO_3$. The Na and K bicarbonates are usually present if the soil has a pH between 8.5 and 9.5, but Na and K carbonates are only present above pH 9.5.

A $CaCO_3$ equivalent of > 15% is used in the FAO definition of a calcic horizon and > 40% for the calcareous material underlying rendzina soils (see Table 4.1 and Annex C, Table C.1).

High levels of carbonate (> 15%) affect the physical, as well as the chemical, properties of a soil. According to Massoud (1973) the available moisture capacity remains low irrespective of the measured clay content, although other experience (eg BAI, 1973) suggests that this is not always so.

The effect of cation imbalance on the availability of nutrients to plants has already been mentioned with respect to Ca and Mg (Subsection 7.7.4). Where a significant amount of free carbonate is present in a soil, other essential nutrients may be less available, particularly if they are in relatively limited supply, and growing crops should be examined for nutrient deficiency symptoms.

Further interpretation and cross-checks for consistency of analytical results involving carbonate values are given in Sections 8.6(c) and (d).

Soil chemistry

7.12   Gypsum

For most purposes the method of Bower and Huss (Hesse, 1971, p 85), is adequate. This is a rapid conductance method, in which gypsum is precipitated from a soil extract with acetone, redissolved in water and the conductivity of the solution measured as an indication of the quantity of gypsum present. Gypsum is widely used as an important remedial amendment for sodic soils (see USDA, 1954b, and Cairns et al, 1977) and also in the reclamation of weakly structured, salt-water affected soils (Pizer, 1966). By reducing CO_3^{2-} and OH^- ions, gypsum can have an acidifying effect on soils and thus affect nutrient availability (see Section 7.5).

Gypsic horizons for the FAO-Unesco (1974) soil legend are defined as follows:

a)   horizon thickness > 15 cm and gypsum content ≥ 5% more than in C horizon; and

b)   horizon thickness (cm) x % gypsum ≥ 150.

At gypsum contents between 14 and 80% cemented and indurated layers can occur; these usually impede root growth and have adverse water retention and transmission properties (van Alphen and Romero, 1971). A general rating for gypsum contents is given in Table 7.21.

General interpretation of gypsum measurements   Table 7.21

Gypsum content (% of soil by weight) 2/	Effects
< 2	Favours crop growth
2 - 25	Little or no adverse effect if in powdery form
> 25	Can cause substantial reductions in crop yield 1/

Notes: 1/ Crop reductions which can occur at high gypsum contents are, in part, attributed to cation imbalance of Ca with Mg or K (see Subsection 7.7.9).

2/ a) Gypsum content: % by weight = me/100 g soil x 0.086.

b) A saturated gypsum solution contains about 240 g ℓ^{-1} of $CaSO_4.2H_2O$ at 25°C; it has an EC of about 2.2 mS cm^{-1}, and S content of about 0.09%.

Source: Adapted from van Alphen and Romero (1971).

7.13   Sulphur

7.13.1 Background   Sulphur is a necessary plant nutrient, but is rarely
determined on a routine basis because of interpretive and
analytical problems, which are due to the complex nature of S
compounds in soils, and the ability of plants to absorb gaseous S
compounds directly from the air, obtaining over 50% of their
requirements in this way under certain conditions when soil S is
limiting (Spedding, 1969).   In humid climates S in aerobic soils
normally occurs in organic forms;   decomposition of these produces
inorganic S (usually sulphate) which is then available to plants.
In anaerobic soils various sulphides and elemental S can occur;
the former may also be found in recent marine deposits, and the
latter in volcanic soils.

In humid regions the inorganic forms of S are usually found in the
subsoil and organic forms in the topsoil;   in arid regions
inorganic S (notably in gypsum) may occur throughout the profile.

7.13.2 Recommended method and units for sulphur measurement   The method
chosen for S analysis depends on the soil type and the forms of S
in the soil to be analysed.   These are discussed at length in
Hesse (1971, p 301), but for most sulphate S determinations a
sodium acetate/acetic acid extraction buffered to pH 4.5 is
recommended;   this appears to provide a good index of sulphur
status in soils with low S contributions from OM (Williams,
1975).   Results are usually expressed as ppm of soil.

7.13.3 Interpretation of sulphur measurements   Although sulphate S is
the form assimilated by plants, measurements of sulphate seldom
give a reliable estimate of S levels in soils, since the ion is
often readily removable by dissolution, and the measurement is
greatly dependent on the conditions of sampling.   Similarly,
measurements of organic S in themselves are not necessarily related
to the rate of release of S to more available forms.

Only very approximate limits can therefore be given, as shown in
Table 7.22.   The S in acid sulphate or thionic profiles is a
special case, described by Pons (1970) and Bloomfield and Coulter
(1973).   Critical levels for S also depend greatly on the crop;
for example, cereals and grains will flourish in soils too
deficient in S for alfalfa and clover;   which are themselves less
demanding than cotton, tomatoes and tobacco (Chapman, 1966).   In
soyabean pot cultures, Singh and Singh (1979) found that S levels
between 80 and 120 ppm antagonised Mo uptake, probably as a result
of the similar ionic radius of both MoO_4^{2-} and SO_4^{2-} (Davies,
1980);   similar inhibiting effects on Se have also been reported
(Hurd-Karrer, 1935;   Shubert et al, 1961).   Other experiments
(Kumar and Singh, 1980b) showed that S enchanced P uptake, except
at very high levels of S of about 80 to 120 ppm.   Other
interactions of S can arise from the acidifying effects of applied
elemental sulphur or gypsum.

Interpretation of sulphur measurements   Table 7.22

S measurement	Approximate S level	Effects	Reference
Total S	< 200 ppm	Deficiency likely	Malkerns Research Station (1959)
Available S (Morgan's reagent) 1/	< 3 ppm	Deficiency likely	Chapman (1966)
Available S (in saturated extract)	> 30 me ℓ^{-1}	Excess	Chapman (1966)
Extractable S (various methods)	6–12 ppm	Upper limit for expected response to S	Reisenauer et al (1973)

Note: 1/   See Table 7.17, and Hesse (1971, p 308).

High levels of Ba in soils can make any free sulphate highly insoluble, and calcium can also cause sulphate precipitation, thus reducing S availability (Hesse, 1958).

Interpretation of sulphate levels is discussed in Section 8.6(c).

7.14   Micronutrients

7.14.1 Background and summary   Only a small minority of the laboratories engaged in routine soil analysis have the facilities to carry out the determination of available micronutrients on a regular basis. In soil or land suitability surveys, it is therefore common to request the determination only of those that a field assessment of the crops, or previous experience with similar soils, has indicated may be critical.   Six trace elements are considered essential for plant growth: these are the micronutrients B, Cu, Fe, Mn, Mo and Zn.   A further four elements, Na, Co, Cl and I are required in small amounts for animal nutrition.

Analytical techniques for the determination of trace elements are described in the standard texts already mentioned (Section 7.1), and notably in Walsh and Beaton (1973).   A review of the effects of sample drying is given in Shuman (1980).   The role of micronutrients in plants and their reactions in soil is extensively covered in Davies (1980), Mortvedt et al (1972) and in FAO Soil Bulletins Nos 17 and 48 (Sillanpää, 1972, 1982), the 1982 volume containing considerable analytical data on a country by country basis.   Table 7.23 summarises reported levels in a variety of soils.

Indicative total trace element levels in soils   Table 7.23

Element	Concentration in soil (ppm) 1/ Approximate mean	Usual range
B	20	2 – 270
Cd	0.35	0.01 – 2
Co	8	0.05 – 65
Cr	70	5 – 1 500
Cu	30	2 – 250
I	5	0.1 – 25
Mn	1 000	20 – 10 000
Mo	1.2	0.1 – 40
Ni	50	2 – 750
Pb	35	2 – 300
S	700	30 – 1 600
Zn	90	1 – 900

Notes: 1/   These figures are derived from a number of different tests on a variety of soils, and are only broadly indicative, partly because most analytical methods fall short of extracting true 'total' amounts of the elements; values in contaminated soils or near mineral deposits can be considerably greater than the higher values indicated here.

          2/   cf Table 7.25 for foliar analysis figures.

Sources: Bowen (1979), Swaine (1969), Aubert and Pinta (1977), Davies (1980) and Fairbridge and Finkl (1979, p 572f).

With the exception of boron, little comment has been made in the past on trace-element levels which may cause toxicity problems; more attention has been paid to the critical levels below which deficiency symptoms may appear. Table 7.24 indicates probable deficiency limits related to the determination used.

Standard methods for measuring trace elements in soils are not widely agreed; the most common methods are designed to quantify water-soluble, cation-exchangeable, readily reducible or complexed forms. The levels estimated by these methods do not correlate well with field calibration studies, and consequently there is a wide range of 'critical' levels reported in the literature. Availability of trace elements to plants is also influenced by many soil and environmental factors, which must all be taken into consideration when interpreting soil test data (Cox and Kamprath, 1972; Lucas and Knezek, 1972). In particular Cu, Fe, Mn and Zn are affected by the redox potential of the soil, and their availability can differ considerably from day to day.

Indicative micronutrient deficiency levels in soils   Table 7.24

Element	Interacting and possibly interacting factors 1/	Extracting agent	Deficiency levels in soils (ppm) 2/
B	Texture, pH, Ca, K	Hot H_2O	0.1 – 0.7 (cf Table 7.28)
Cu	N, Fe, Mg, Mo, P, Zn	Ammonium acetate (pH 4.8)	0.2
		0.5 M EDTA	0.75
		0.43 M HNO_3	3 – 4
		Biological assay	2 – 3
		1 M HCl	100
		0.1 M HCl	0.09 – 1.06
Fe	pH, K, Mn, Ca, Mg, P, Cu, Mo, Zn	Ammonium acetate (pH 4.8)	2
		DTPA + $CaCl_2$ (pH 7.3)	2.5 – 4.5
Mn	pH, OM, K, Mo, P, Fe, Cu, Zn	0.05 M HCl + 0.025 M H_2SO_4	5 – 9
		0.1 M H_3PO_4 and 3 M $NH_4H_2PO_4$	15 – 20
		Hydroquinone + ammonium acetate	25 – 65
		H_2O	2
Mo	pH, Fe, Mn, P, S, Cu	Ammonium oxalate (pH 3.3)	0.04 – 0.2
Zn	pH, Cu, N, P, Ca	0.1 M HCl	1.0 – 7.5
		Dithizone + ammonium acetate	0.3 – 2.3
		EDTA + $(NH_4)_2CO_3$	1.4 – 3.0
		DTPA + $CaCl_2$ (pH 7.3)	0.5 – 1.0

Notes: 1/ Climatic and crop factors, although highly important, are not considered here; see Chapman (1966) for further details.
        2/ Within the ranges quoted the values obtained must be viewed in relation to interacting factors which are summarised in Figure 7.1 and Table 7.26; consult Cox and Kamprath (1972) and Olsen (1972) for details. Figures quoted here are for 'available' measurements.

Sources: Adapted from Cox and Kamprath (1972) and Sillanpää (1972).

These difficulties have led to an increase in the use of foliar, rather than soil, analyses to investigate micronutrient levels. Table 7.25 lists some broad ranges for interpretation of foliar analyses, and Table 7.26 summarises some of the factors which should be considered when interpreting soil test results for individual micronutrients.

Indicative ratings for foliar analyses of micronutrients   Table 7.25

| Micronutrient | Concentration in mature leaves (ppm) | | |
	Deficient	Sufficient	Excessive or toxic
B	< 15	20 – 100	> 200
Cu	< 4	5 – 20	> 20
Fe	< 50	50 – 250	?
Mn	< 20	20 – 500	> 500
Mo	< 0.1	0.5 – ?	?
Zn	< 20	25 – 150	> 400

Sources: Jones (1972); for more specific figures see Chapman (1966) and Jones (1967); cf Tables 7.23 and 7.24 for soil analysis figures.

7.14.2   Boron   This is probably the most common trace element analysed in commercial surveys, usually by hot-water extraction. Its presence in soil is often associated with evaporation of groundwater, and it is therefore commonly encountered at high concentrations in arid zones. Toxic levels are often present in saline-sodic soils (see Chapter 8). Toxic B is also associated with groundwater in volcanically active areas.

Deficiencies in B are commonly encountered when light-textured acid soils are leached drastically by rain or irrigation, or are limed. Liming can, however, decrease toxicity symptoms in plants grown on high B soils, which has led to speculation about a possible chemical interaction between calcium and borate, $B(OH)_4$. It appears that plants only grow normally when a critical balance exists in the intake of $Ca + B$ (Olsen, 1972); this balance is species dependent, and is more likely to be physiological than due to chemical interactions in the soil. (As an example, Lal and Tandon (1955) quote a Ca:B ratio of 20 as the optimum value for sugarcane). Potassium has also been demonstrated to reduce B availability at high pH (Sillanpää, 1972); this may reflect changes in the Ca:B ratio (Davies, 1980).

General levels of B in soils are given in Table 7.23, which indicates that most soils have total B levels of < 3 ppm; soils of the humid temperate regions normally have B contents of between 0.2 and 1.5 ppm, whilst arid and semi-arid soils can have

Soil chemistry

Factors contributing to micronutrient deficiences in plants   Table 7.26

	Element	Deficiency conditions or contributory factors

B	Boron	Soils low in total B:
		    alluvials (fluvents)
		    podzols (spodosols)
		    organics (histosols)
		    regosols (udipsamments)
		    low humic gleys (haplaquepts)
		Moderate to heavy rainfall
		Nearly neutral or alkaline soils
		Dry weather
		High light intensity

Cu	Copper	Low soil Cu: < 6 ppm in mineral soils < 30 ppm in organic soils High soil P High organic matter and N High soil Zn Sandy texture

Mn	Manganese	Alkaline soils
		Naturally poorly drained soils:
		    organics (histosols)
		    ground water podzols (aquods)
		    regosols (udents, udipsamments)
		    low humic gleys (haplaquepts)
		    marly or shelly soils (limnic medisaprists
		      (marly), haplaquepts)
		High soil Fe, Cu or Zn
		Dry weather
		Low light intensity
		Low soil temperature

Mo	Molybdenum	Low soil Mo
		Acid conditions generally:
		    acid regosols (udipsamments)
		    acid podzols (spodosols)
		    acid organics (histosols)
		High free Fe:
		    bog Fe (aquods)

Fe	Iron	Low soil Fe Free $CaCO_3$ High HCO_3 Moisture extremes High amounts of heavy metals High soil P Poor aeration (excess CO_2) Temperature extremes Heavy manuring (alkaline soils) Low organic matter content (acid soils) Excess soil acidity Genetic differences Root damage

Zn	Zinc	Low soil zinc:
		    low humic gleys or humic gleys
		     (haplaquepts, haplaquolls)
		    alluvials (fluvents)
		    regosols (udipsamments)
		    organics (histosols)
		    high rainfall areas
		Calcareous soils
		Low soil organic matter
		Cool temperatures
		High soil P
		Restricted root zones:
		    compacted soils
		    container-grown plants
		Corrals, old orchards, barnyards
		Liberal applications of N

Source: Adapted from Lucas and Knezek (1972).

levels between 10 and 40 ppm (Sillanpää, 1972). A level in the region of 0.5 ppm is quoted as the likely deficiency limit, depending on soil conditions (cf Table 7.24). The range between toxic and deficient levels for plant growth is thus very narrow: Hesse (1971, p 384) quotes a possible range of only 1 to 3 ppm in some soils.

Hot-water extractable B is often well correlated with plant-available B; for very general interpretive purposes, the rating shown in Table 7.27 may be used, but note that different plant species have widely divergent tolerances to B concentration, as shown in Table 7.28; large differences are also found between varieties (Brown et al, 1972). There is also some evidence that B deficiencies occur at lower levels in coarser-textured soils than in finer-textured ones; Stinson (1953) quotes critical levels of hot-water extractable B of 0.3 and 0.5 ppm respectively. Toxicity from B can also be reduced by high OM levels: even semi-tolerant sweet corn can show no apparent symptoms when grown in peats with B levels of 10 ppm (Prasad and Byrne, 1975).

Indicative rating of soil boron levels in the absence of high Ca contents   Table 7.27

B concentration (ppm)		Category
Saturated extract	Hot-water extractable	
–	< 1	Possibly deficient
–	1 – 1.5	Borderline for deficiency
≤ 0.5	1.5 – 3	Satisfactory for most crops
0.5 – 5	3 – 6	Possibly toxic, depending on crop sensitivity
> 10	> 6	Toxic to most crops

Sources: Batey (1971), and Agric Ext Lab (1969); cf Table 7.24.

7.14.3   Zinc   The active form of Zn in the soil is the divalent cation Zn^{2+}. It is generally considered that the concentration of Zn in the soil solution is maintained by sparingly soluble Zn silicates formed by interaction of Zn with amorphous silica (Russell, 1973). Plants vary in their Zn requirements as well as their abilities to extract Zn from the soil. Crops which are sensitive to Zn deficiencies include maize, citrus, legumes and cotton. Sugarcane, sugarbeet and especially tobacco have quite low Zn requirements (Sillanpää, 1972) and deficiency symptoms are rarely encountered for these crops in the field. Possible deficiency levels are summarised in Table 7.24.

Soil chemistry

Plant tolerance types
in descending order of tolerance in each column

Tolerant	Semi-tolerant	Sensitive
Athel	Sunflower	Pecan
(Tamarix aphylla)	(Helianthus annuus)	(Carya illinoensis)
Asparagus	Potato	Walnut
(Asparagus officinalis)	(Solanum tuberosum)	(Juglans regia)
Palm	Acala cotton	Jerusalem artichoke
(Phoenix canariensis)	(Gossypium hirsutum)	(Helianthus tuberosus)
Date palm	Pima cotton	Navy bean
(Phoenix dactylifera)	(Gossypium hirsutum)	(Phaseolus vulgaris)
Sugarbeet	Tomato	American elm
(Beta vulgaris)	(Lycopersicon esculentum)	(Ulmus americana)
Mangel	Sweet pea	Plum
(Beta vulgaris)	(Lathyrus odoratus)	(Prunus domestica)
Alfalfa	Radish	Pear
(Medicago sativa)	(Raphanus sativus)	(Pyrus communis)
Gladiolus	Field pea	Apple
(Gladiolus spp)	(Pisum sativum)	(Malus sylvestris)
Broad bean	Olive	Grape
(Vicia faba)	(Olea europaea)	(Vitis vinifera)
Onion	Barley	Kadota fig
(Allium cepa)	(Hordeum vulgare)	(Ficus carica)
Turnip	Wheat	Persimmon
(Brassica rapa)	(Triticum aestivum)	(Diospyros virginiana)
Cabbage	Corn	Cherry
(Brassica oleracea)	(Zea mays)	(Prunus avium)
Lettuce	Milo	Peach
(Lactuca sativa)	(Sorghum bicolor)	(Prunus persica)
Carrot	Oat	Apricot
(Daucus carota)	(Avena sativa)	(Prunus armeniaca)
	Zinnia	Orange
	(Zinnia elegans)	(Citrus sinensis)
	Pumpkin	Avocado
	(Cucurbita spp)	(Persea americana)
	Bell pepper	Grapefruit
	(Capsicum annuum)	(Citrus paradisi)
	Sweet potato	Lemon
	(Ipomoea batatas)	(Citrus limon)
	Lima bean	
	(Phaseolus lunatus)	

Corresponding B tolerance limits

a)   in saturation extract (ppm) and
b)   in irrigation water (mg ℓ^{-1}) 1/

	Tolerant	Semi-tolerant	Sensitive
a)	2.5–1.5	1.5	0.7
b)	4.0–2.0	2.0–1.0	1.0–0.3

Note: 1/   Indicative levels when B toxicity symptoms show in sand cultures;   levels do not
necessarily indicate yield reductions.

Sources: Adapted from Richards (1954, p 67) after Eaton (1935), and FAO (1979a). Further crops are
given in FAO (1985b, pp 82-83).

Total Zn content in the soil generally falls within the range 10 to 300 ppm, (Sillanpää, 1972; cf Table 7.23), but Zn availability is greatly affected by soil pH. Deficiency symptoms rarely occur in acid soils unless native reserves are very low, but on calcareous soils Zn disorders are common. This is attributed to the formation of very sparingly soluble complexes and carbonates (Lucas and Knezek, 1972). Phosphate can also induce Zn deficiencies, when available P levels in the soil are high or when high levels of P fertiliser are added (Olsen, 1972). When liming reduces Zn availability, Mg can alleviate deficiency symptoms (Olsen, 1972). Zinc also interacts with other micronutrients: Fe uptake and translocation appear to be inhibited by Zn, whilst Cu is reported to lower Zn concentrations in plants. Zinc appears to be mutually antagonistic to both Cu and Fe, since high Zn levels have been reported to induce Cu and Fe deficiencies (Olsen, 1972). Experiments by Kumar and Singh (1980a) indicated that Zn increased S uptake by soyabean at low Zn concentrations (< 5 ppm), whilst Zn levels > 5 ppm decreased S uptake, an effect similar to that in clover noted by Cullen and Arnold (1971). Toxic Zn levels are comparatively rare; Chapman (1966) quotes values of above about 400 ppm of exchangeable Zn as excessive for some crops; total (perchloric acid extractable) Zn levels of above about 150 ppm in soils can be regarded as high, and levels of 40 to 150 as medium.

Several extractants have been used to estimate plant-available Zn in the soil. These include dithizone, dilute HCl, chelating agents (EDTA, DPTA and triethanolamine) and neutral salt solutions (NH_4NO_3, KCl and $MgCl_2$); dithizone and 0.1 M HCl have had wider application than the other extractants. When dithizone is used on acid soils, it is particularly important to remember soil pH when interpreting results, whilst when HCl is used on calcareous soils, free lime content is important (Cox and Kamprath, 1972).

7.14.4 <u>Copper</u>   This exists in soil mainly as the divalent Cu^{2+} ion adsorbed by clay minerals or associated with organic matter. Total Cu in soil falls in the range 2 to 100 ppm. The strong interaction which is generally held to occur between Cu and soil organic matter does not affect Cu availability to plants (Sillanpää, 1972), although it does influence the concentration of Cu in soil solutions (Russell, 1973). The availability is governed primarily by the total amount in the soil (Lucas and Knezek, 1972). Copper availability is also influenced by soil pH, but less so than are other micronutrients. The availability of Cu decreases slowly with increasing pH, but the nature of the interaction is not completely understood. Other plant nutrients influence Cu uptake and utilisation: phosphate reduces the concentration of Cu in roots and leaves of plants, and heavy phosphate fertilisation can induce Cu deficiency. In animals Cu and Mo are antagonistic and there is some evidence for this in plants. High levels of Cu can also induce deficiency symptoms of Fe and Zn; conversely, high levels of Fe and Zn have been found to induce Cu deficiencies.

Most plants are sensitive to Cu deficiency conditions (see Table 7.25). Cereals (oats, wheat, barley, maize) and vegetables are particularly sensitive (Murphy and Walsh, 1972). Toxicity from Cu does not appear to be widespread and is of importance mainly in areas polluted by mining or excessive spraying activity; values above about 150 ppm for total Cu are quoted as toxic for citrus by Chapman (1966). In general, high levels of total Cu (perchloric acid extractable) in soils can be taken as those above about 100 ppm.

Plant-available Cu has been estimated by bioassay and also by chemical extraction procedures. Bioassay has been found to be time-consuming and expensive and has not been used extensively. Chemical extractants have included boiling H_2O, NH_4NO_3, acid ammonium acetate, dilute HNO_3, a citrate/EDTA mixture and dithizone. However, the soil tests developed to date are generally poor indicators of available Cu, although correlations with the need for Cu treatment may be possible in some localities.

7.14.5 <u>Iron</u> This is the fourth most abundant element in the earth's crust, and soils rarely contain < 1%. The major forms of Fe in soils are very sparingly soluble ferric oxides, which occur as coatings of aggregates or as separate constituents of the clay fraction; the very low solubility of these compounds means that Fe concentrations in soil solutions are also very low. Available evidence suggests that plants may take up their Fe in both the divalent ferrous (Fe^{2+}) and trivalent ferric (Fe^{3+}) forms (Russell, 1973). Deficiency symptoms are reported in most groups of plants (Lucas and Knezek, 1972), but Fe toxicity rarely creates problems in the field (Murphy and Walsh, 1972), although Evans (1959) has suggested that it may cause grey node symptoms in sugarcane.

The availability of Fe to plants is influenced mainly by soil pH and the redox equilibria between ferrous and ferric Fe, the solubilities of both falling with increasing pH (Lindsay, 1972). Deficiencies in Fe (see Table 7.25) are most commonly encountered in calcareous soils and in other soils following heavy applications of lime, when the Fe is precipitated as insoluble hydrated ferric oxides (Krauskopf, 1972; Lindsay, 1972). Higher pH also tends to promote the conversion of ferrous compounds to less soluble ferric compounds, in oxidation reactions which are also dependent on the O_2 content of the soil. This can create problems in rice nutrition, when the ferrous Fe is absorbed by the plant roots and is accumulated following oxidation to insoluble forms (Ponnamperuma, 1964).

Uptake of Fe is also inhibited by high phosphate levels, due to the formation of insoluble iron phosphate. In soils with high iron oxide content this reaction limits the availability of phosphate. High levels of Zn, Cu or Mn interfere with the translocation of Fe in plants, and high Fe levels can also interfere with the uptake of these elements. Molybdenum has also been shown to induce Fe deficiencies in tomatoes; amongst the reasons suggested was the formation of insoluble iron molybdates in plant roots.

In conjunction with the CEC, carbonate and other neutralising compounds, the pyrite (FeS_2) content of acid sulphate soils is important in determining potential hazards of development (Pons, 1970); further details are given in Bloomfield and Coulter (1973). A level of > 3 ppm of soluble Fe^{2+} is quoted as a critical limit for potential clogging of subsurface drains, following aeration (ILACO, 1981).

Soil tests for available Fe have not yet become standard, and extractants used include buffered (pH 4.8) 1 M ammonium acetate, H_2SO_4 and various chelating agents (EDTA, EDDHA, DPTA and triethanolamine).

7.14.6 <u>Manganese</u>   In soil Mn originates primarily from the decomposition of ferromagnesian rocks;   total quantities in soil vary from < 100 to several thousand ppm.   It can exist in several oxidation states, the two most important being Mn^{2+} associated with clay minerals and organic matter, and Mn^{4+} present in insoluble oxides.   The chemical behaviour of Mn in the soil is very similar to that of Fe (Krauskopf, 1972).

Plant-available Mn is highly dependent on pH and the redox potential in the soil (Lindsay, 1972).   Plants take up Mn in the divalent form, and the solubility of Mn^{2+} compounds decreases with increasing pH;   above about pH 6.5 deficiency symptoms begin to be seen in some soils (Aubert and Pinta, 1977).   A wide range of crops is sensitive to Mn deficiency, which is common in calcareous soils and soils of high pH, although deficiencies are also frequently induced by liming and poor soil physical conditions, even in soils with pH > 7.   However, since Mn^{2+} is only stable in well-aerated soils which are neutral or acidic, Mn deficiencies also occur in poorly drained soils (even with comparatively low pH values), when conditions encourage the reduction of Mn^{4+} to Mn^{2+} and subsequent leaching.   Some general deficiency levels are summarised in Table 7.25.

Toxic levels of Mn are most common in acidic soils with pH values of about 5.5 or less, but plant response to high levels varies. Relatively tolerant species include sugarbeet, oats, rye and broad beans, whilst swedes and other brassicas are generally highly susceptible (Sillanpää, 1972).   Toxicity of Mn can generally be suspected if more than 1 000 ppm of Mn is present in plant dry matter (Mitchell, 1964).   High levels of total Mn (perchloric acid extractable) in soils can be taken as those above about 2 000 ppm.

The action of Mn is also highly dependent on the presence of other ions: Mn can induce Fe chlorosis, and there is some evidence which suggests that Fe and Zn interfere with Mn uptake;   Mn and P appear to be mutually antagonistic.

Extractants used for estimating plant-available Mn include reducing agents (such as hydroquinone), phosphate solutions (H_3PO_4, $NH_4H_2PO_4$), chelating agents, hot water and ammonium acetate. The easily reducible Mn estimate (hydroquinone extraction) has been most widely used, but the predictive power of all Mn soil tests can be improved if soil pH is considered (Cox and Kamprath, 1972). Crop responses to Mn fertiliser on soils with low extractable Mn are more likely in high pH soils.

7.14.7 <u>Molybdenum</u>   This occurs in very small quantities in the earth's crust and in soil the normal range is between 0.2 and 5 ppm.   The principal reserves in soils are believed to be the oxides MoO_3,

Mo_2O_5 and MoO_2. These oxides are slowly transformed to soluble molybdates ($MoO_4{}^{2-}$), the form in which Mo is taken up by plants. In soil molybdates are strongly adsorbed on the surfaces of ferric oxides and hydrated oxides. Plants require Mo in very small amounts and it can be toxic to grazing animals when the concentration in forage crops is high.

The action of Mo is highly pH-dependent, and it is unique among the micronutrients because its availability is increased by liming, although plant uptake may not always show a corresponding increase, due to the accompanying rise in pH (Fleming, 1980). Deficiencies in Mo (see Table 7.25) are normally associated only with acid soils.

Mo has toxicity effects at levels of about 200 ppm of plant tissue, although contents of more than 10 to 20 ppm can cause nutritional imbalances in ruminants (Kubota et al, 1967). Interpretation of Mo levels in soils is more difficult, because of the numerous interactions which can occur. Grigg (1953) noted in particular the importance of allowing for pH in Mo interpretation; for response to Mo by pasture in New Zealand he found that at pH 6 the critical level was 0.1 ppm, compared with 0.2 ppm at pH 5. Nicholas (1961) quotes deficient levels in tropical soils at about or < 0.01 ppm and adequate levels at 0.1 to 1.0 ppm. However, Cox and Kamprath (1972) indicate that any such interpretations must remain tentative until more is known about Mo interactions in soil.

Molybdenum interacts with both major and minor plant nutrients, and phosphate in particular has been shown to enhance Mo uptake (Olsen, 1972; Barshad, 1951). Sulphate and Mo are mutually antagonistic, possibly because the two anions, being of similar size, compete for adsorption sites on plant roots; gypsum applications have therefore been used to reduce Mo toxicity in soils (Fleming, 1970). Molybdenum also mutually interferes with the uptake of Cu, and even at low levels is antagonistic to Fe uptake. High Mo levels have been shown to reduce P uptake in soyabean (Kumar and Singh, 1980b), but the effect is not seen in all crops.

The most commonly employed extractant for available Mo is 0.275 M ammonium oxalate at pH 3.3, although this is often considered to yield high values which overestimate plant-available Mo (Fleming, 1980). Other extractants which have been used include hot H_2O, ammonium acetate, NaOH, NH_4F and biological assay.

7.15   Trace elements – chromium, cobalt, iodine, nickel, selenium and silicon

7.15.1   Chromium and nickel are often found in high concentrations in association with serpentine rocks, where levels of several thousand ppm are known. Generally, however, levels are of the order of 100 to 300 ppm. High levels of both Cr and Ni can be toxic, but even low levels of Ni may be toxic, especially if associated with high exchangeable Mg:Ca ratios (FAO, 1979a, p 27). Excessive values are listed by Chapman (1966) as being about 5 to 15 ppm for Cr in solution cultures; and about 3 to 70 ppm in soil for Ni (using ammonium acetate extraction).

7.15.2   Cobalt has not been definitely established as an essential plant
          nutrient, but it is required by rhizobia for N-fixation, and
          therefore indirectly by leguminous plants, and is an essential
          element for animal development. The average content of Co in the
          soil ranges from about 1 to 40 ppm. The availability of Co to
          plants is reduced by liming and the element is poorly absorbed by
          plants growing in calcareous soils. When the Co content of forage
          is < 0.1 ppm, deficiencies in grazing animals are likely to
          occur. Chapman (1966) quotes deficiency ranges as lying between
          0.4 and 4 ppm of soil, and 'normal' ranges between 4 and 40 ppm.

7.15.3   Iodine has not been shown to be a directly essential element in
          plant nutrition, although Davies (1980) quotes reports of crop
          responses due to secondary effects which include the stimulation of
          soil-nitrifying bacteria, the dissolution of Mn compounds and the
          alleviation of Cu deficiency (Borst-Pauwels, 1961). Iodine is,
          however, essential for animals, and an I deficency is commonly
          associated with enlarged thyroid glands. Soil contents of I are
          low, from traces to about 80 ppm, most soils containing between 0.1
          and 10 ppm (Davies, 1980). In general I appears to be more
          available in acid soils and its uptake is reduced by liming,
          although some contradicting evidence was found by Whitehead (1975),
          who also reported at least tenfold reductions in plant I levels due
          to farmyard manure applications; other interactions are too varied
          or too little studied to generalise. Analysis of I is normally
          made on plant material rather than on soils.

7.15.4   Selenium is known as an essential plant nutrient only for Astra-
          gallus spp, but it is essential to grazing animals. Most of the
          Se-induced disorders in animals arise from toxicity rather than
          deficiency. In rocks Se is commonly associated with shales, and
          in particular with those containing organic matter; in soils the
          average content of Se is very low, ranging from 0.1 to 2 ppm. The
          Se chemistry in soil has not been extensively studied, although it
          is known that plants take up the element as selenites and selenates
          and plants vary in their roles as Se accumulators. The Se occurs
          most commonly in the form of selanates, SeO_4^{2-}, which are soluble
          at pH values ≥ 7. Singh and Singh (1979) found in pot experiments
          that Se as selenate at levels > 2.5 ppm reduced S, N and P uptake
          by cowpeas.

7.15.5   Silicon is not usually considered an essential element, but plant
          disorders attributable to low Si are reported, and silicate
          applications can dramatically improve yields and quality of, for
          example, sugarcane; reviews of the role of Si in plants appear in
          Comhaire (1966) and Sherman (1969). Plants take up their Si as
          silicic acid and Si deficiencies are likely to occur in those soils
          where soluble Si is low. Silicon is especially important for
          cereals, since it increases the rigidity and strength of their
          cells. It also increases the resistance of some crops to toxic
          levels of Mn (Russell, 1973; Samuels, 1979). Deficiencies in Si
          are normally detected by foliar analyses, and ameliorative
          treatments involve application of soluble silicates or, more
          commonly, basic slag (high in calcium silicates). Critical levels
          for Si are not often reported, but Ross et al (1974) consider a
          value of 77 ppm of acid-soluble Si (modified Truog method) as an
          upper limit for probable economic return to application of silicate

fertilisers on sugarcane, although Wong You Cheong (1970) earlier reported responses at Si levels of up to about 125 ppm. Applications of $CaSiO_3$ can have long-lived effects; cane yields 30% higher than on untreated plots have been reported 6 years after cessation of $CaSiO_3$ applications at a rate of 14 t ha^{-1} on treated plots (Mauritian Sugar Industry Research Institute, 1973).

Chapter 8

Soil and Water Salinity and Sodicity

8.1   Introduction

Salinity and sodicity are considered separately from other chemical
properties because of their common occurrence in arid regions and
the special problems they cause in soil and water management.
Soils with high sodium levels are here referred to as 'sodic'
soils, rather than the older term of 'alkali' soils, to avoid
ambiguity; sodic soils may or may not be strongly alkaline (ie
have a high pH). Saline soils occur where the supply of salts, for
example from rock weathering, capillary rise, rainfall or flooding,
exceeds their removal, for example by leaching or flooding.   Thus
they tend to coincide with areas where evapotranspiration exceeds
precipitation and where there is no lengthy rainy season.
Irrigation is required for crop growth in these areas although it
may itself induce salinisation unless salts are leached regularly
and water-tables are kept low by adequate drainage.   Failure to do
this has contributed to the downfall of many irrigation schemes.

Excessive salts hinder crop growth, not only by toxicity effects,
but by reducing water availability through the action of osmotic
pressure;   nutrient uptake may also become unbalanced.   Standard
references include Richards (1954), FAO-Unesco (1973) and FAO
(1976b, 1976c); a more general description of arid-zone soils is
given by Dregne (1976).

8.2   Electrical conductivity

Electrical conductivity (EC) measurements are used as indications
of total quantities of soluble salts in soils.   The quantities of
salts which pass into solution depend on the relative amounts of
soil and water used, but the relationships are variable.
Determinations of EC should ideally be carried out on undiluted
soil solutions extracted from soils containing water held at the
normal plant-available water potentials (about -0.1 to -15 bar).
However, in practice the extraction procedures are too difficult
and too slow, and routine measurements are made on extracts from
saturated soil pastes or from 1:1 to 1:5 soil:water mixtures.
Methods are given in Richards (1954).   In most consultancy work
Richards EC of the saturation extract (EC_e) is used for base
laboratory work;   field laboratories use 1:2.5 or other extracts
because they are easier to handle.   Values of EC measured by

salinity sensor and pressure vacuum cup and as bulk soil conductivity, are discussed by Yadav et al (1979). Values of EC are invariably quoted in mS cm^{-1} (originally mmho cm^{-1}) at 25°C.

Many interpretations of EC values have been devised, but no universal, precise interpretation is possible because the effects of salinity are modified greatly by other factors such as quality of irrigation waters, soil texture, salt types present, crop varieties and species, soil drainability, stage of crop growth and climate. Some of the more widely used interpretations are given in Tables 8.1 to 8.4; differences in quoted values presumably reflect some of the variability noted above.

General interpretation of EC$_e$ values   Table 8.1

USDA soil class	Designation	EC$_e$ (mS cm^{-1})	Total salt content (%)	Crop reaction
0	Salt free	(0 - 2 ((0 - 2	< 0.15 < 0.15	Salinity effects are mostly negligible Salinity effects are negligible except for the most sensitive plants
1	Slightly saline	4 - 8	0.15 - 0.35	Yields of many crops restricted
2	Moderately saline	8 - 15	0.35 - 0.65	Only tolerant crops yield satisfactorily
3	Strongly saline	> 15	> 0.65	Only very tolerant crops yield satis- factorily

Note: For individual crop responses see Annex F and Tables 8.2 to 8.4.

Sources: Adapted from FAO-Unesco (1973); and Schofield (1942) modified by Richards (1954).

The critical EC levels given in Table 8.1 apply only when measurements are made on the saturation extracts. The values are suitable for fine- and medium-textured soils but may be optimistic for coarse-textured samples. This is because the ratio of the water contents at the saturation percentage to that at the 15 atmosphere percentage (FAP) or -15 bar % (which is in the ratio of about 4:1 for a wide range of medium and fine textures) becomes considerably greater with very coarse textures, since the large pores - filled to capacity when saturated - do not continue to hold

USDA ratings of relative plant tolerances to salt   Table 8.2

Plants are listed within groups in order of decreasing tolerance to salinity. EC_e values
(mS cm^{-1}) correspond to 50% decrease in yield (cf Tables 8.3 and 8.4).

Plant grouping	High salt tolerance	Medium salt tolerance	Low salt tolerance
Fruit crops	Date palm	Pomegranate Fig Olive Grape Cantaloup	Pear Apple Orange Grapefruit Prune Plum Almond Apricot Peach Strawberry Lemon Avocado
Vegetable crops	$EC_e = 12$ Garden beets Kale Asparagus $EC_e = 10$	$EC_e = 10$ Tomato Broccoli Cabbage Bell pepper Cauliflower Lettuce Sweet corn Potatoes (White Rose) Carrot Onion Peas Squash Cucumber $EC_e = 4$	$EC_e = 4$ Radish Celery Green beans $EC_e = 3$
Forage crops	$EC_e = 18$ Alkali sacaton Salt grass Nuttall alkali grass Bermuda grass Rhodes grass Fescue grass Canada wild rye Western wheat grass Barley (hay) Bird's-foot trefoil $EC_e = 12$	$EC_e = 12$ White sweet clover Yellow sweet clover Perennial rye grass Mountain brome Strawberry clover Dallis grass Sudan grass Huban clover Alfalfa (California Common) Tall fescue Rye (hay) Wheat (hay) Oats (hay) Orchard grass Blue grama Meadow fescue Reed canary Big trefoil Smooth brome Tall meadow oat grass Cicer milk-vetch Sour clover Sickle milk-vetch $EC_e = 4$	$EC_e = 4$ White Dutch clover Meadow foxtail Alsike clover Red clover Ladino clover Burnet $EC_e = 2$
Field crops	$EC_e = 16$ Barley (grain) Sugarbeet Rape $EC_e = 10$	$EC_e = 10$ Rye (grain) Wheat (grain) Oats (grain) Rice Sorghum (grain) Sugarcane Corn (field) Sunflower Castor beans $EC_e = 4$	$EC_e = 4$ Field beans Flax $EC_e = 3$

Source: Richards (1954); for an updated table see Appendix II, Table II.3.

Soil and water salinity and sodicity

Table 8.3

Crop tolerance and yield potential of selected crops as influenced by irrigation
water salinity or soil salinity 1/

Crop		Yield potential for EC values shown 2/									
		100%		90%		75%		50%		No yield 3/	
		EC_e	EC_w	EC_e	EC_w	EC_e	EC_w	EC_e	EC_w	EC_e	EC_w
Field crops											
Barley 4/	Hordeum vulgare	8.0	5.3	10.0	6.7	13.0	8.7	18.0	12.0	28.0	19.0
Cotton	Gossypium hirsutum	7.7	5.1	9.6	6.4	13.0	8.4	17.0	12.0	27.0	18.0
Sugarbeet 5/	Beta vulgaris	7.0	4.7	8.7	5.8	11.0	7.5	15.0	10.0	24.0	16.0
Sorghum	Sorghum bicolor	6.8	4.5	7.4	5.0	8.4	5.6	9.9	6.7	13.0	8.7
Wheat 4/6/	Triticum aestivum	6.0	4.0	7.4	4.9	9.5	6.3	13.0	8.7	20.0	13.0
Wheat, durum	Triticum turgidum	5.7	3.8	7.6	5.0	10.0	6.9	15.0	10.0	24.0	16.0
Safflower	Carthamus tinctorius	5.3	3.5	6.2	4.1	7.6	5.0	9.9	6.6	14.5	
Soyabean	Glycine max	5.0	3.3	5.5	3.7	6.3	4.2	7.5	5.0	10.0	6.7
Cowpea	Vigna unguiculata	4.9	3.3	5.7	3.8	7.0	4.7	9.1	6.0	13.0	8.8
Groundnut	Arachis hypogaea	3.2	2.1	3.5	2.4	4.1	2.7	4.9	3.3	6.5	4.4
Rice (paddy)	Oryza sativa	3.0	2.0	3.8	2.6	5.1	3.4	7.2	4.8	11.0	7.6
Sugarcane	Saccharum officinarum	1.7	1.1	3.4	2.3	5.9	4.0	10.0	6.8	19.0	12.0
Corn	Zea mays	1.7	1.1	2.5	1.7	3.8	2.5	5.9	3.9	10.0	6.7
Flax	Linum usitatissimum	1.7	1.1	2.5	1.7	3.8	2.5	5.9	3.9	10.0	6.7
Broadbean	Vicia faba	1.5	1.1	2.6	1.8	4.2	2.0	6.8	4.5	12.0	8.0
Beans	Phaseolus vulgaris	1.0	0.7	1.5	1.0	2.3	1.5	3.6	2.4	6.5	4.2
Fruit crops 7/											
Date palm	Phoenix dactylifera	4.0	2.7	6.8	4.5	11.0	7.3	17.9	12.0	32.0	21.0
Fig	Ficus carica	2.7	1.8	3.8	2.6	5.5	3.7	8.4	5.6	14.0	
Olive	Olea europaea	2.7	1.8	3.8	2.6	5.5	3.7	8.4	5.6	14.0	
Pomegranate	Punica granatum	2.7	1.8	3.8	2.6	5.5	3.7	8.4	5.6	14.0	
Grapefruit 8/	Citrus paradisi	1.8	1.2	2.4	1.6	3.4	2.2	4.9	3.3	8.0	5.4
Orange	Citrus sinensis	1.7	1.1	2.3	1.6	3.3	2.2	4.8	3.2	8.0	5.3
Lemon	Citrus limon	1.7	1.1	2.3	1.6	3.3	2.2	4.8	3.2	8.0	
Apple	Malus sylvestris	1.7	1.0	2.3	1.6	3.3	2.2	4.8	3.2	8.0	
Pear	Pyrus communis	1.7	1.0	2.3	1.6	3.3	2.2	4.8	3.2	8.0	
Walnut	Juglans regia	1.7	1.1	2.3	1.6	3.3	2.2	4.8	3.2	8.0	
Peach	Prunus persica	1.7	1.1	2.2	1.5	2.9	1.9	4.1	2.7	6.5	4.3
Apricot 8/	Prunus armeniaca	1.6	1.1	2.0	1.3	2.6	1.8	3.7	2.5	5.8	3.8
Grape	Vitis spp	1.5	1.0	2.5	1.7	4.1	2.7	6.7	4.5	12.0	7.9
Almond 8/	Prunus dulcis	1.5	1.0	2.0	1.4	2.8	1.9	4.1	2.8	6.8	4.5
Plum 8/	Prunus domestica	1.5	1.0	2.1	1.4	2.9	1.9	4.3	2.9	7.1	4.7
Blackberry	Rubus spp	1.5	1.0	2.0	1.3	2.6	1.8	3.8	2.5	6.0	4.0
Boysenberry	Rubus ursinus	1.5	1.0	2.0	1.3	2.6	1.8	3.8	2.5	6.0	4.0
Avocado	Persea americana	1.3	0.9	1.8	1.2	2.5	1.7	3.7	2.4	6.0	
Raspberry	Rubus idaeus	1.0	0.7	1.4	1.0	2.1	1.4	3.2	2.1	5.5	
Strawberry	Fragaria spp	1.0	0.7	1.3	0.9	1.8	1.2	2.5	1.7	4.0	2.7
Banana	Musa sp.	Generally recommended on soils with $EC_e < 1$ mS cm^{-1}									
Vegetable crops											
Squash, zucchini	Cucurbita pepo melopepo	4.7	3.1	5.8	3.8	7.4	4.9	10.0	6.7	15.0	10.0
Beets 5/	Beta vulgaris	4.0	2.7	5.1	3.4	6.8	4.5	9.6	6.4	15.0	10.0
Squash, scallop	Cucurbita pepo melopepo	3.2	2.1	3.8	2.6	4.8	3.2	6.3	4.2	9.4	6.3
Broccoli	Brassica oleracea botrytis	2.8	1.9	3.9	2.6	5.5	3.7	8.2	5.5	14.0	9.1
Tomato	Lycopersicon esculentum	2.5	1.7	3.5	2.3	5.0	3.4	7.6	5.0	12.5	8.4
Cucumber	Cucumis sativus	2.5	1.7	3.3	2.2	4.4	2.9	6.3	4.2	10.0	6.8
Cantaloup	Cucumis melo	2.2	1.5	3.6	2.4	5.7	3.8	9.1	6.1	16.0	
Spinach	Spinacia oleracea	2.0	1.3	3.3	2.2	5.3	3.5	8.6	5.7	15.0	10.0
Celery	Apium graveolens	1.8	1.2	3.4	2.3	5.8	3.9	9.9	6.6	18.0	12.0
Cabbage	Brassica oleracea capitata	1.8	1.2	2.8	1.9	4.4	2.9	7.0	4.6	12.0	8.1
Potato	Solanum tuberosum	1.7	1.1	2.5	1.7	3.8	2.5	5.9	3.9	10.0	6.7

Notes: See page 161.

cont

Crop		100%		90%		75%		50%		No yield 3/	
		EC_e	EC_w	EC_e	EC_w	EC_e	EC_w	EC_e	EC_w	EC_e	EC_w
Vegetable crops (cont)											
Sweet corn	Zea mays	1.7	1.1	2.5	1.7	3.8	2.5	5.9	3.9	10.0	6.7
Sweet potato	Ipomea batatas	1.5	1.0	2.4	1.6	3.8	2.5	6.0	4.0	11.0	7.1
Pepper	Capsicum annuum	1.5	1.0	2.2	1.5	3.3	2.2	5.1	3.4	8.6	5.8
Lettuce	Lactuca sativa	1.3	0.9	2.1	1.4	3.2	2.1	5.2	3.4	9.0	6.0
Radish	Raphanus sativus	1.2	0.8	2.0	1.3	3.1	2.1	5.0	3.4	8.9	5.9
Onion	Allium cepa	1.2	0.8	1.8	1.2	2.8	1.8	4.3	2.9	7.4	5.0
Carrot	Daucus carota	1.0	0.7	1.7	1.1	2.8	1.9	4.6	3.0	8.1	5.4
Beans	Phaseolus vulgaris	1.0	0.7	1.5	1.0	2.3	1.5	3.6	2.4	6.3	4.2
Turnip	Brassica rapa	0.9	0.6	2.0	1.3	3.7	2.5	6.5	4.3	12.0	8.0
Forage crops											
Tall wheat grass	Agropyron elongatum	7.5	5.0	9.9	6.6	13.0	9.0	19.0	13.0	31.0	21.0
Wheat grass (fairway)	Agropyron cristatum	7.5	5.0	9.0	6.0	11.0	7.4	15.0	9.8	22.0	15.0
Bermuda grass 9/	Cynodon dactylon	6.9	4.6	8.5	5.6	11.0	7.2	15.0	9.8	23.0	15.0
Barley (forage) 4/	Hordeum vulgare	6.0	4.0	7.4	4.9	9.5	6.4	13.0	8.7	20.0	13.0
Perennial rye grass	Lolium perenne	5.6	3.7	6.9	4.6	8.9	5.9	12.0	8.1	19.0	13.0
Trefoil, bird's-foot narrow-leaf 10/	Lotus corniculatus tenuifolius	5.0	3.3	6.0	4.0	7.5	5.0	10.0	6.7	15.0	10.0
Harding grass	Phalaris tuberosa	4.6	3.1	5.9	3.9	7.9	5.3	11.0	7.4	18.0	12.0
Tall fescue	Festuca elatior	3.9	2.6	5.5	3.6	7.8	5.2	12.0	7.8	20.0	13.0
Crested wheat grass	Agropyron sibiricum	3.5	2.3	6.0	4.0	9.8	6.5	16.0	11.0	28.0	19.0
Vetch	Vicia sativa	3.0	2.0	3.9	2.6	5.3	3.5	7.6	5.0	12.0	8.1
Sudan grass	Sorghum sudanense	2.8	1.9	5.1	3.4	8.6	5.7	14.0	9.6	26.0	17.0
Wild rye, beardless	Elymus triticoides	2.7	1.8	4.4	2.9	6.9	4.6	11.0	7.4	19.0	13.0
Cowpea (forage)	Vigna unguiculata	2.5	1.7	3.4	2.3	4.8	3.2	7.1	4.8	12.0	7.8
Trefoil, big	Lotus uliginosus	2.3	1.5	2.8	1.9	3.6	2.4	4.9	3.3	7.6	5.0
Sesbania	Sesbania exaltata	2.3	1.5	3.7	2.5	5.9	3.9	9.4	6.3	17.0	11.0
Sphaerophysa	Sphaerophysa salsula	2.2	1.5	3.6	2.4	5.8	3.8	9.3	6.2	16.0	11.0
Alfalfa	Medicago sativa	2.0	1.3	3.4	2.2	5.4	3.6	8.8	5.9	16.0	10.0
Lovegrass 11/	Eragrostis spp	2.0	1.3	3.2	2.1	5.0	3.3	8.0	5.3	14.0	9.3
Corn (forage)	Zea mays	1.8	1.2	3.2	2.1	5.2	3.5	8.6	5.7	15.0	10.0
Clover, berseem	Trifolium alexandrinum	1.5	1.0	3.2	2.2	5.9	3.9	10.0	6.8	19.0	13.0
Orchard grass	Dactylis glomerata	1.5	1.0	3.1	2.1	5.5	3.7	9.6	6.4	18.0	12.0
Meadow foxtail	Alopecurus pratensis	1.5	1.0	2.5	1.7	4.1	2.7	6.7	4.5	12.0	7.9
Clover, ladino, red, alsike, strawberry	Trifolium spp	1.5	1.0	2.3	1.6	3.6	2.4	5.7	3.8	9.8	6.6

Notes:
1/ These data only serve as a guide to relative tolerances among crops. Absolute tolerances vary depending upon climate, soil conditions and cultural practices. In gypsiferous soils, plants will tolerate about 2 mS cm^{-1} higher EC_e than indicated, but EC_w will remain as here.

2/ EC_e means average root zone salinity in mS cm^{-1} at 25°C. EC_w means electrical conductivity of the irrigation water in mS cm^{-1}. The relationship between soil salinity and water salinity ($EC_e = 1.5\ EC_w$) assumes a 15-20 percent leaching fraction and a 40-30-20-10 percent water use pattern for the upper to lower quarters of the root zone. These assumptions were also used in developing the guidelines in Table 8.9.

3/ The zero yield potential or maximum EC_e indicates the theoretical EC_e at which crop growth ceases.

4/ Barley and wheat are less tolerant during germination and seedling stage; EC_e should not exceed 4-5 mS cm^{-1} in the upper soil during this period.

5/ Beets are more sensitive during germination; EC_e should not exceed 3 mS cm^{-1} in the seeding area for garden beets and sugar beets.

6/ Semi-dwarf, short cultivars may be less tolerant.

7/ These data are applicable when rootstocks are used that do not accumulate Na^+ and Cl^- rapidly or when these ions do not predominate in the soil.

8/ Tolerance evaluation is based on tree growth and not on yield.

9/ Tolerance is an average of several varieties; Suwannee and Coastal Bermuda grass are about 20% more tolerant, while Common and Greenfield Bermuda grass are about 20% less tolerant.

10/ Broadleaf Birdsfoot Trefoil seems less tolerant than Narrowleaf Birdsfoot Trefoil.

11/ Tolerance is an average for Boer, Wilman, Sand and Weeping Lovegrass; Lehman Lovegrass seems about 50% more tolerant.

Sources: Adapted from FAO (1985b), Maas and Hoffman (1977) and Maas (1984).

Relationship of crop yield decrease and EC of soil solution 1/ Table 8.4

Crop		EC_e (mS cm^{-1}) at which yield decreased by		
		10%	25%	50%
Forage crops				
Bermuda grass	Cynodon dactylon 2/	13	16	18
Tall wheat grass	Agropyron elongatum	11	15	18
Crested wheat grass	Agropyron desertorum	6	11	18
Tall fescue	Festuca arundinaceae	7	10.5	14.5
Barley, hay	Hordeum vulgare 3/	8	11	13.5
Perennial rye grass	Lolium perenne	8	10	13
Harding grass	Phalaris tuberosa stenoptera	8	10	13
Bird's-foot trefoil	Lotus corniculatus tenuifolius	6	8	10
Beardless wild rye	Elymus triticoides	4	7	11
Alfalfa	Medicago sativa	3	5	8
Orchard grass	Dactylis glomerata	2.5	4.5	8
Meadow foxtail	Alopecurus glomerata	2	3.5	6.5
Alsike clover	Trifolium hybridum	2	2.5	4
Red clover	Trifolium pratense	2	2.5	4
Field crops				
Barley, grain	Hordeum vulgare 3/	12	16	18
Sugarbeet	Beta vulgaris	10	13	16
Cotton	Gossypium hirsutum	10	12	16
Safflower	Carthamus tinctorius	8	11	12
Wheat	Triticum vulgare 3/	7	10	14
Sorghum	Sorghum vulgare	6	9	12
Soyabean	Glycine max	5.5	7	9
Sesbania	Sesbania macrocarpa 3/	4	5.5	9
Sugarcane	Saccharum officinarum	3	5	8.5
Rice (paddy)	Oryza sativa 3/	5	6	8
Corn	Zea mays	5	6	7
Broad bean	Vicia faba	3.5	4.5	6.5
Flax	Linum usitatissimum	3	4.5	6.5
Field bean	Phaseolus vulgaris	1.5	2	3.5
Vegetable crops				
Beet, garden	Beta vulgaris 4/	8	10	12
Spinach	Spinacia oleracea	5.5	7	8
Tomato	Lycopersicum esculentum	4	6.5	8
Broccoli	Brassica oleracea italica	4	6	8
Cabbage	Brassica oleracea capitata	2.5	4	7
Potato	Solanum tuberosum	2.5	4	6
Sweetcorn	Zea mays	2.5	4	6
Sweet potato	Ipomoea batatas	2.5	3.5	6
Lettuce	Lactuca sativa	2	3	5
Bell pepper	Capsicum frutescens	2	3	5
Onion	Allium cepa	2	3.5	4
Carrot	Daucus carota	1.5	2.5	4
Green bean	Phaseolus vulgaris	1.5	2	3.5

Notes: 1/ In gypsiferous soils EC_e's causing equivalent yield reductions will be about 2 mS cm^{-1} greater than the listed values.
2/ Average for different varieties. For most crops varietal differences are relatively insignificant.
3/ Less tolerant during seedling stage. Salinity (EC_e) at this stage should not exceed 4 or 5 mS cm^{-1}.
4/ Sensitive during germination. Salinity (EC_e) during germination should not exceed 3 mS cm^{-1}.

Source: Bernstein (1964).

water under field conditions. The concentration of salts in the actual soil solution will therefore be greater than is indicated by the normal EC salinity scale, which is based on the assumption of an approximate 4:1 ratio between saturated percentage and FAP. In these circumstances the EC is best measured on a soil:water extract that is related to the actual FAP or to field capacity. This value can then be related to the normal salinity scale; a suitable procedure is discussed by Richards (1954 - method 3b).

When EC is measured on soil:water solutions that are more dilute than the saturation extract (and hence can be tested more easily since suction filtration is not required) the results cannot be interpreted directly from the salinity scale given above. There is no foolproof conversion factor that can be used to compare ECs at different soil:water ratios since even if the moisture characteristics of the soil are known, the solubility of the salts may vary with increasing dilution. However, the following relationships (which are approximate and tentative) may be useful as a rough guide provided the samples do not contain significant amounts of gypsum:

$$EC_e = 2.2 \times EC_{1:1} \text{ (general experience)}$$

$$EC_e = 6.4 \times EC_{1:5} \text{ (Talsma, 1968; Loveday et al, 1972)}$$

(Note that a saturated gypsum solution has an EC of about 2 mS cm^{-1} at 25°C corresponding to a concentration of about 240 g ℓ^{-1} of $CaSO_4.2H_2O$).

The precise conversion factors between ECs at different soil:water ratios are always best checked before use by obtaining the EC_e from a representative number of samples and plotting these values against the EC dilution ratio.

Analytical cross-checks involving EC measurements include the following (Richards, 1954, p 30):

a) EC of a water sample, EC_w, in mS cm^{-1} x 640 ≈ total dissolved solids (TDS) in mg ℓ^{-1} or ppm (valid for EC_w in range 0.1 to 5.0 mS cm^{-1});

b) EC_w in mS cm^{-1} x 10 ≈ total soluble cation concentration in me ℓ^{-1} (valid for EC in range 0.1 to 5.0 mS cm^{-1}).

EC values commonly show very high spatial variability (see Table 6.1, p 60), and individual values must therefore be treated with caution. For detailed work such as mapping isosalinity lines, grid surveying and statistical interpolation are recommended (eg Hajrasuliha et al, 1980).

8.3   Sodium measurements

Exchangeable sodium percentage (ESP) measurements and interpretations have already been presented (Subsection 7.7.6) and their role in defining alkali soils is discussed in Section 8.4 below.

A further important quantity is the sodium adsorption ratio (SAR) which is used in estimations of water quality (see Table 8.8) and for calculating equilibrium ESP values for soils under irrigation. The SAR is defined as follows:

$$SAR = \frac{(Na^+)}{\sqrt{\left(\frac{(Ca^{2+}) + (Mg^{2+})}{2}\right)}}$$

where (Na^+), (Ca^{2+}), (Mg^{2+}) are ionic concentrations in me ℓ^{-1} of solution (Note: units are important – other units, eg m moles ℓ^{-1}, give different values).

At equilibrium, the classical Gapon Equation (Russell 1973, p 94) can be modified to relate concentrations of exchangeable soil cations to ionic concentrations in the water as follows (FAO – Unesco 1973, p 198):

$$\frac{Na_x}{Ca_x + Mg_x} = K \frac{(Na^+)}{\sqrt{\left(\frac{(Ca^{2+}) + (Mg^{2+})}{2}\right)}} = K.SAR$$

where Na_x, Ca_x, Mg_x are the exchangeable cation concentrations in me/100 g soil, and K is a constant depending on soil characteristics, usually with a value between 0.9 and 1.4 when the above units are used.

Richards (1954, p 72) derived the following empirical relationship between ESP and SAR for soils which have reached equilibrium with the applied irrigation water:

$$ESP = \frac{100 (0.01475 \; SAR - 0.0126)}{0.01475 \; SAR + 0.9874}$$

This can be expressed in the form of the nomogram given in Figure 8.1 and, using SAR values, an estimate can be made of the maximum ESP which may develop; interpretation of ESP values is described in Section 7.7.6. The method is generally suitable for solutions with total cation concentrations between about 39 and 110 me ℓ^{-1}; outside this range other regression equations apply (see Doering et al, 1982).

Although good correlation has often been found between measured ESP values and ESP values calculated as above, poor correlation has been reported in some circumstances. At least some of this poor correlation has occurred with high-carbonate waters; for these an adjusted SAR (adj SAR) was recommended (FAO, 1979a; Bower, 1961):

adj SAR = SAR (9.4 - pHc) where pHc is calculated as shown in Table 8.5. However, this may overestimate the problem; current recommended calculations are shown in Appendix II.

Bower (1961) further proposed the following empirical relationship between ESP and SAR for high-carbonate water without residual sodium bicarbonate:

ESP = 2SAR (9.4 - pHc)

Nomogram for determining the SAR value of irrigation water
and for estimating the corresponding soil ESP values

Figure 8.1

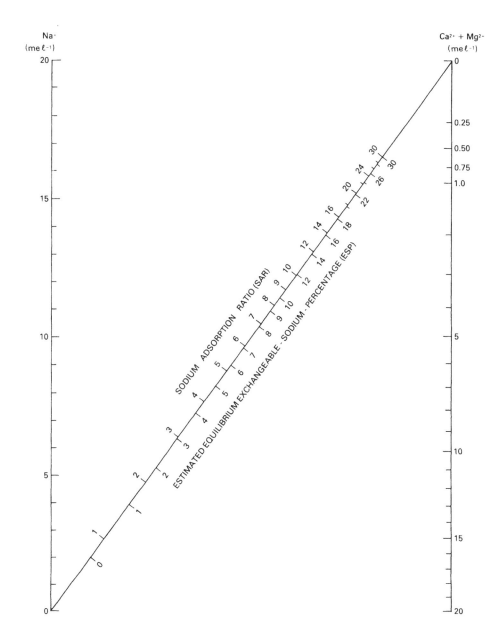

Note: Recent work suggests that these values hold good only for total cation concentrations between 39
         and 110 me ℓ⁻¹; other regression equations apply outside this range (see Doering et al., 1982).

Source: Richards (1954).

Values for calculation of pHc   Table 8.5

The value of pHc is calculated from the pHc equation:

$$pHc = (pK_2' - pK_c') + p\,(Ca + Mg) + p\,(Alk)$$

where the terms on the right-hand side are taken from the following table:

Sum of relevant ionic concentrations 1/ (me ℓ^{-1})	$pK_2' - pKc'$	p (Ca + Mg)	p (Alk)
0.05	2	4.6	4.3
0.10	2	4.3	4
0.15	2	4.1	3.8
0.20	2	4	3.7
0.25	2	3.9	3.6
0.30	2	3.8	3.5
0.40	2	3.7	3.4
0.50	2.1	3.6	3.3
0.75	2.1	3.4	3.1
1.00	2.1	3.3	3
1.25	2.1	3.2	2.9
1.50	2.1	3.1	2.8
2	2.2	3	2.7
2.50	2.2	2.9	2.6
3	2.2	2.8	2.5
4	2.2	2.7	2.4
5	2.2	2.6	2.3
6	2.2	2.5	2.2
8	2.3	2.4	2.1
10	2.3	2.3	2
12.5	2.3	2.2	1.9
15	2.3	2.1	1.8
20	2.4	2	1.7
30	2.4	1.8	1.5
50	2.5	1.6	1.3
80	2.5	1.4	1.1

Note: 1/   From water analysis.   Relevant sums are:

for $(pK_2' - pKc')$: $Ca^{2+} + Mg^{2+} + Na^+$
p (Ca + Mg) : $Ca^{2+} + Mg^{2+}$
p (Alk)   : $CO_3^{2-} + HCO_3^-$

Example of pHc calculation:

Ionic concentrations (me ℓ^{-1}):

Ca^{2+} = 2.32)	3.76	CO_3^{2-} = 0.42
Mg^{2+} = 1.44)		HCO_3^- = 3.66
Na^+ = 7.73		

Sum = 11.49	Sum = 4.08

From Table 8.5 and the equation for pHc:

$pK_2' - pKc'$ = 2.3
p (Ca + Mg) = 2.7
p (Alk)   = 2.4

pHc   = 7.4

Sources: FAO (1979a, p 68) from Wilcox (1966).

Residual sodium carbonate (RSC) was introduced by Eaton (1950) and is defined as:

$$RSC = (CO_3^{2-} + HCO_3^-) - (Ca^{2+} + Mg^{2+})$$

where the items in parenthesis are ionic concentrations in me ℓ^{-1}

The RSC values can be interpreted as follows (Wilcox, 1958):

RSC < 1.25   water probably safe for irrigation

1.25 < RSC < 2.5   water marginally suitable for irrigation

RSC > 2.5   water unsuitable for irrigation

The following interpretation of pHc should be noted (Wilcox, 1966):

pHc	Indication
> 8.4	Tendency to dissolve lime from a soil through which the water moves
< 8.4	Tendency to precipitate lime from the water applied

Further interpretations of SAR values are given in Section 8.4, and in Oster and Rhoades (1975).

8.4   Definitions of saline and sodic soils

Estimates of salinity and sodicity of soils are important because of their effects on crop yields (Tables 8.1 to 8.4). These effects arise when salt concentrations in the soil are high because the resultant salt concentrations in the soil water reduce or even reverse the flow of water into the plants by osmosis. In addition some ions (notably Na^+, Cl^-, SO_4^{2-}) are specifically toxic for some crops. The effects of high sodium are also noticeable in its deleterious effects on soil structure (Subsection 7.7.6).

These effects are often summarised in a simple three-class classification of salt-affected soils, as shown in Table 8.6.

8.5   Salinity and sodicity of irrigation water

Although a detailed discussion of irrigation water is outside the scope of this manual, a basic interpretation is vital for determining irrigability of soils, and determining reclamation practices. Tables 8.7 to 8.10 and Figure 8.2 summarise some criteria for evaluating water quality; note that for water studies the convention is to quote conductivity values in μS cm^{-1}.

USDA classification of salt-affected soils   Table 8.6

Soil	EC_e (mS cm^{-1})	ESP	pH	Description
Saline soils	> 4	< 15	Usually < 8.5	Non-sodic soils containing sufficient soluble salts to interfere with growth of most crops
Saline-sodic soils	> 4	> 15	Usually < 8.5	Soils with sufficient exchangeable sodium to interfere with growth of most plants, and containing appreciable quantities of soluble salts
Sodic soils	< 4	> 15	Usually > 8.5	Soils with sufficient exchangeable sodium to interfere with growth of most plants, but without appreciable quantities of soluble salts

Note: Although fairly widely accepted, the values for EC_e, ESP and pH should be regarded as indicative rather than as fixed critical values. The effects of increasing ESP, for example, gradually worsen rather than rapidly change soil conditions as a value of 15 is reached. Local experience should be compared with measured values wherever this is possible. The presence of gypsum, in particular, in a soil can mitigate the effects of high ESP values.

Source: Richards (1954, p 4).

8.6   Soluble salts: saturation extract

Unless facilities are available that enable extracts to be obtained at field moisture contents, the determination of soluble salts is best carried out on the saturation extract. With greater dilution the quantities of soluble salts become progressively less related to the situation which exists at field moisture levels. Note that soluble salts, although composed of similar ions, are not synonymous with exchangeable ions since they are not held on soil exchange sites.

Commonly determined soluble ions include:

       Cations – Ca, Mg, Na, K
       Anions – CO_3, HCO_3, Cl, SO_4
       Nitrate: where the sum of the cations is significantly greater than the common anions
       Silicate: only in sodic soils with high pH values

USDA classification of irrigation water salinity (sulphate-free waters)   Table 8.7

Salinity class and description	EC range ($\mu S \ cm^{-1}$)	Equivalent salt concentration (approximate)		
		TDS 1/ ($g \ \ell^{-1}$)	TDS 1/ (ppm)	Cl (ppm)
C1 Low salinity water can be used for irrigation with most crops on most soils, with little likelihood that a salinity problem will develop. Some leaching is required, but this occurs under normal irrigation practices, except in soils of extremely low permeability	< 250	< 0.2	< 200	< 60
C2 Medium salinity water can be used if a moderate amount of leaching occurs. Plants with moderate salt tolerance can be grown in most instances without special practices for salinity control	250 – 750	0.2 – 0.5	200 – 500	60–200
C3 High salinity water cannot be used on soil with restricted drainage. Even with adequate drainage, special management for salinity control may be required and plants with good salt tolerance should be selected	750 – 2 250	0.5 – 1.5	500 – 1 500	200–600
C4 Very high salinity water is not suitable for irrigation under ordinary conditions but may be used occasionally under very special circumstances. The soils must be permeable, drainage must be adequate, irrigation water must be applied in excess to provide considerable leaching, and very salt-tolerant crops should be selected	> 2 250	1.5 – 3.0	> 1 500	> 600

Note: 1/   TDS = total dissolved solids.

Source: Adapted from Richards (1954, p 76);   note that further divisions based on SAR are also made;   see Figure 8.2 and Table 8.8.

USDA classification of irrigation water sodicity   Table 8.8
(sulphate-free waters)

Sodium class and description	SAR
S1 <u>Low sodium water</u> can be used for irrigation on almost all soils with little danger of the development of harmful levels of exchangeable sodium. However, sodium-sensitive crops may accumulate injurious concentrations of sodium	< 10
S2 <u>Medium sodium water</u> will present an appreciable sodium hazard in fine-textured soils having high cation-exchange capacity, especially under low leaching conditions, unless gypsum is present in the soil. This water may be used on coarse-textured or organic soils with good permeability	10-18
S3 <u>High sodium water</u> may produce harmful levels of exchangeable sodium in most soils and will require special soil management – good drainage, high leaching and organic matter additions. Gypsiferous soils may not develop harmful levels of exchangeable sodium from such waters. Chemical amendments may be required for replacment of exchangeable sodium except that amendments may not be feasible with waters of very high salinity	18-26
S4 <u>Very high sodium water</u> is generally unsatisfactory for irrigation purposes except at low and perhaps medium salinity, where the solution of calcium from the soil or use of gypsum or other amendments may make the use of these waters feasible	> 26

Note: Irrigation water may sometimes dissolve sufficient calcium from
         calcareous soils to decrease the sodium hazard appreciably, and
         this should be taken into account in the use of C1-S3 and C1-S4
         waters (see Figure 8.2). For calcareous soils with high pH values
         or for non-calcareous soils, the sodium status of waters in
         classes C1-S3, C1-S4, and C2-S4 may be improved by the addition of
         gypsum to the water. Similarly, it may be beneficial to add
         gypsum to the soil periodically when C2-S3 and C3-S2 waters are
         used.

Source: Adapted from Richards (1954); see also Figure 8.2 and Table 8.7.

Guidelines for evaluating irrigation water quality [1]   Table 8.9

Potential irrigation problem	Units	Degree of restriction on use		
		None	Slight to moderate	Severe
Salinity (affects crop water availability)				
Electrical conductivity, EC_w [2]	dS m^{-1} mS cm^{-1}	< 0.7	0.7 – 3.0	> 3.0
(or)				
Total dissolved solids, TDS	mg ℓ$^{-1}$	< 450	450 – 2 000	> 2 000
Infiltration (indicates infiltration rate of water into the soil. Evaluate using EC_w and SAR together) [3]				
Sodium absorption ratio, SAR = 0 – 3 and EC_w =		> 0.7	0.7 – 0.2	< 0.2
= 3 – 6 =		> 1.2	1.2 – 0.3	< 0.3
= 6 – 12 =		> 1.9	1.9 – 0.5	< 0.5
= 12 – 20 =		> 2.9	2.9 – 1.3	< 1.3
= 20 – 40 =		> 5.0	5.0 – 2.9	< 2.9
Specific ion toxicity (affects sensitive crops)				
Sodium (Na) [4]				
surface irrigation	SAR	< 3	3 – 9	> 9
sprinkler irrigation	me ℓ$^{-1}$	< 3	> 3	
Chloride (Cl) [4]				
surface irrigation	me ℓ$^{-1}$	< 4	4 – 10	> 10
sprinkler irrigation	me ℓ$^{-1}$	< 3	> 3	
Boron (B) [5]	mg ℓ$^{-1}$	< 0.7	0.7 – 3.0	> 3.0
Miscellaneous effects on susceptible crops				
Nitrogen (NO$_3$-N) [6]	mg ℓ$^{-1}$	< 5	5 – 30	> 30
Bicarbonate (HCO$_3$) with overhead sprinkling	me ℓ$^{-1}$	< 1.5	1.5 – 8.5	> 8.5
pH (may cause imbalance in nutrient uptake)		Normal range 6.5 – 8.4		

Notes: [1] For normal surface or sprinkler irrigation methods.
        [2] See Note 2 of Table 8.3.
        [3] SAR is sometimes reported by the symbol RNa. See Figure 8.1 for the SAR calculation
            procedure. At a given SAR, infiltration rate increases as water salinity increases.
            Evaluate the potential infiltration problem by SAR as modified by EC_w. Adapted from
            Rhoades 1977, and Oster and Schroer 1979.
        [4] For surface irrigation, most tree crops and woody plants are sensitive to sodium and chloride;
            use the values shown. Most annual crops are not sensitive; use the salinity tolerance tables
            (Table 8.2 and 8.3). With overhead sprinkler irrigation and low humidity (< 30 percent),
            sodium and chloride may be absorbed through the leaves of sensitive crops and can cause damage.
        [5] For boron tolerances, see Table 7.28.
        [6] NO$_3$-N means nitrate nitrogen reported in terms of elemental nitrogen (NH$_4$-N and organic
            nitrogen should be included when wastewater is being tested).

Source: Adapted from California Committee of Consultants (1974), quoted in FAO (1985a).

FAO recommended maximum concentrations of trace elements in   Table 8.10
irrigation waters

Element	Symbol	For water used continuously on all soils (mg ℓ^{-1})	For use up to 20 years on fine-textured soils of pH 6.0–8.5 (mg ℓ^{-1})
Aluminium	Al	5	20
Arsenic	As	0.1	2
Beryllium	Be	0.1	0.5
Boron	B	1/	2
Cadmium	Cd	0.01	0.05
Chromium	Cr	0.1	1
Cobalt	Co	0.05	5
Copper	Cu	0.2	5
Fluoride	F	1	15
Iron	Fe	5	20
Lead	Pb	5	10
Lithium 2/	Li	2.5	2.5
Manganese	Mn	0.2	10
Molybdenum	Mo	0.01	0.05 3/
Nickel	Ni	0.2	2
Selenium	Se	0.02	0.02
Vanadium	V	0.1	1
Zinc	Zn	2	10

These levels will not normally adversely affect plants or soils.

No data were available for mercury (Hg), silver (Ag), tin (Sn), titanium (Ti), tungsten (W).

Notes: 1/   See Table 8.9.
         2/   Recommended maximum concentration for irrigating citrus is
              0.075 mg ℓ^{-1}.
         3/   Only for fine-textured acid soils or acid soils with
              relatively high iron oxide contents.

Sources: Adapted from Environmental Studies Board (1972), quoted in FAO
         (1979a, p 70).

USDA diagram for classification of irrigation water   Figure 8.2

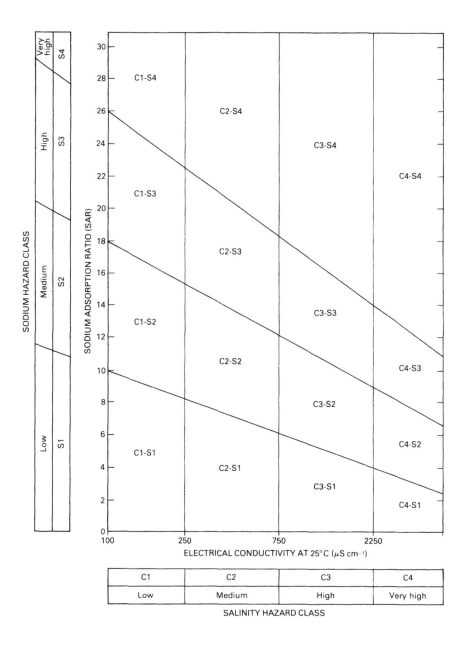

Source: Adapted from Richards (1954, p 80).

Methods are discussed in the standard references (see Section 7.1).

Although excessive quantities of any of the common anions may prove detrimental to plant growth, there do not appear to be any general critical limits at present to reflect crop responses to particular cation or anion levels. Some specific interpretations include:

a) Chloride - Affects similar crops to those affected by Na. Tree crops and woody perennials are sensitive, most annual crops are not - at least at low Cl concentrations; effects may be worse with sprinklers. For highly evaporative climates Nir (1965, quoted in FAO, 1979a) suggests that crop yields are affected when Cl in the saturated extract is > 10 me ℓ^{-1}, and that the Cl will exceed this limit when the following empirical relationship holds:

$$\frac{\text{Cl (ppm in irrigation water) x Saturation \% of soil}}{\text{Annual rainfall (mm)}} > 50$$

With sprinklers, Cl concentrations in the irrigation water of 3 me ℓ^{-1} have caused leaf burn. Critical Cl levels (in % by weight) quoted by ILACO (1981) are as follows: slightly affected 0 to 0.05; moderately affected 0.05 to 0.1; strongly affected > 0.1.

b) Nitrogen - For N (both as NH_3 and NO_3) > 5 ppm, the N-sensitive crops may be adversely affected, but others may benefit (FAO, 1979a). Note that algal growth may be excessive with these N values; in many sprinkler systems mechanical (screens, filters) or chemical ($CuSO_4$) controls may be necessary. Nitrogen-sensitive crops include: sugar (cane and beets), apricots, citrus, cotton.

c) Carbonate and sulphate - ILACO (1981) quote the following ranges (in % by weight):

   i) CO_3^{2-} - Slightly affected 0 to 0.005; moderately affected 0.005 to 0.01; strongly affected > 0.01.

   ii) SO_4^{2-} - Slightly affected 0 to 0.1; moderately affected 0.1 to 0.3; strongly affected > 0.3.

Sulphates will attack concrete pipes and canal linings, producing the most marked effects in porous pipes where the reaction can occur throughout the pipe walls, unless high alumina or supersulphated concrete is used. An example of soil sulphur ratings relating to concrete pipe use is given in Table 8.11.

d) Analytical cross-checks

   i) Total soluble anion concentration and total soluble cation concentration in me ℓ^{-1} are nearly equal.

ADAS soil sulphur suitability ratings for concrete   Table 8.11
pipe use

Water-soluble sulphate in 1:2 soil:water extract (%S)	HCl-soluble sulphate (%S)	Interpretation 1/
≤ 0.06	≤ 0.08	Suitable for porous concrete pipes
0.07 – 0.15	0.09 – 0.20	Suitable for high alumina or supersulphated concrete pipes
> 0.15	> 0.20	Suitable only for high alumina concrete pipes

Note: 1/   See also pH effects, Subsection 7.5.3a.

Sources: Adapted from ADAS (1974) quoted in Dieleman and Trafford (1976).

ii)   pH   and   Ca   and   Mg   concentrations – The
      concentration of Ca and Mg in a saturation extract
      seldom   exceeds   2 me ℓ^{-1} at   pH   9.0   and   above.
      Therefore, Ca plus Mg is low if carbonate ions are
      present in titratable amounts, and Ca plus Mg is
      never high in the presence of a high concentration
      of bicarbonate ions.

iii) Ca and sulphate in a soil-water extract and gypsum
      content of the soil – The solubility of gypsum at
      ordinary   temperatures   is   approximately   30 me ℓ^{-1}
      in distilled water and 50 me ℓ^{-1} or more in highly
      saline solutions.   However, owing to the common ion
      effect, an excess of either calcium or sulphate may
      depress the solubility of gypsum to a value as low
      as 20 me ℓ^{-1}.   Hence, the saturation extract of a
      non-gypsiferous   soil   may   contain   more   than
      30 me ℓ^{-1} of both Ca and sulphate, and that of a
      gypsiferous soil may have a Ca concentration as low
      as 20 me ℓ^{-1}.   As a general rule, any soils with
      saturation extracts that have a Ca concentration of
      more   than   20 me ℓ^{-1} should   be   checked   for   the
      presence of gypsum.

iv)   pH and alkaline-earth carbonates – The pH reading
      of a calcareous soil at the saturation percentage is
      invariably in excess of 7.0 and generally in excess
      of 7.5;   a non-calcareous soil may have a pH reading
      as high as 7.3 or 7.4.

v) pH and gypsum - The pH reading of gypsiferous soils at the saturation percentage is seldom in excess of 8.2 regardless of the ESP.

vi) pH and ESP - A pH reading, at the saturation percentage, in excess of 8.5 almost invariably indicates an ESP of 15 or more, but the converse is not necessarily true.

vii) pH and carbonate and bicarbonate concentrations - If carbonate ions are present in a soil extract in titratable quantites, the pH reading of the extract must exceed 9.0. The bicarbonate concentration seldom exceeds 10 me ℓ^{-1} in the absence of carbonate ions, and at pH readings of 7.0 or less seldom exceeds 4 me ℓ^{-1}.

viii) ESP and SAR - In general, ESP increases with SAR. There are occasional deviations, but generally low SAR values of the saturation extract are associated with low ESP values in the soil, and high SAR values denote high ESP values.

8.7 Leaching requirement

Soil surveyors are sometimes required to estimate the quantities of water necessary to leach soluble salts from a profile in order to produce a soil of specified salinity. Two equations (FAO 1976b, p 37) are commonly used to calculate net leaching requirements under different irrigation conditions:

a) For surface and low-frequency overhead irrigation

$$\text{net LR} = \frac{EC_w}{5EC_e - EC_w}$$

where net LR = the minimum theoretical net leaching requirement (as a fraction of the applied water) needed to control salts in the root zone

EC_w = electrical conductivity of the applied water

EC_e = EC_e corresponding to 90% yield potential (see Table 8.3, p 160) for crop under consideration

b) For high-frequency (near daily) overhead or drip irrigation

$$\text{net LR} = \frac{EC_w}{2 (\text{max } EC_e)}$$

where LR and EC_w are as in (a) above, and

max EC_e = EC_e corresponding to "No yield" potential (Table 8.3, p 160) for crop under consideration

The value obtained for net LR in both (a) and (b) is a theoretical value, and field measurements are needed to confirm that this is appropriate in practice. The value will be lower than that obtained using the now superseded method given by the USDA (Richards, 1954, p 37). Note that calculation of the gross quantity of water to be applied for leaching purposes must also take into account the leaching efficiency, the field application efficiency (allowing for surface and deep percolation losses) and any effective rainfall that occurs. A sample calculation appears in Appendix II.

Table 8.12 gives some indicative leaching requirement figures, although values should always be checked by field experiments. Note that several leachings of the soil with comparatively low water volumes are usually more effective than a single application of the same total volume.

Indicative leaching requirement related to the EC of   Table 8.12
irrigation and drainage waters

Electrical conductivity of irrigation water (μS cm^{-1})	Maximum EC value of drainage water at bottom of root zone (mS cm^{-1})			
	4	8	12	16
	Leaching requirement (%) 1/			
100	2.5	1.2	0.8	0.6
250	6.2	3.1	2.1	1.6
750	18.8	9.4	6.2	4.7
2 250	56.2	28.1	18.8	14.1
5 000		62.5	41.7	31.2

Note: 1/   Fraction of the applied irrigation water that must be leached through the root zone expressed here as a percentage. These figures are only broadly indicative, however, since leaching requirements vary with soil conditions and methods of irrigation.

Source: FAO-Unesco (1973, p 203).

8.8   Soil classification and salinity/sodicity

The classification of saline and sodic soils is based mainly on pH, EC and ESP. Topography and drainage affect the degree of salinisation and ease of reclamation, and clay mineralogy helps control the effect of Na on soil structure and permeability. Broadly speaking, saline soils are solonchaks and sodic soils are solonetz according to the FAO system.   In the USDA Soil Taxonomy salt affected soils form great groups, for example salorthids (see Annex C).   Soils with heavy textures, for example vertisols, or with marked textural changes in the profile, for example planosols, are more susceptible to salinisation and have particular drainage/reclamation problems.

Chapter 9

Soil and Land Suitability Reports and Maps

9.1 Introduction

This chapter is divided into two main parts – the first (Sections 9.2 and 9.3) is a brief guide for those who are not soil specialists on the purpose and scope of soil and land suitability reports and rapid ways of assessing them. The second part (Sections 9.4 and 9.5) discusses the layout and composition of soil and land suitability reports and includes checklists of suggested items essential for soil surveyors to incorporate in these reports.

9.2 The purpose and use of soil and land suitability reports

Soil and land suitability reports often form key documents on the physical background of agricultural development projects. In association with other studies they are essential for project identification and location, and they form a basis for specifying and quantifying such activities as clearance and cultivation practices, design of engineering infrastructures, and economic evaluation of project options.

Unfortunately, soil reports are all too often greatly underused because of the reluctance of any but soil specialists to master the complexities involved, and also because most soil reports make very heavy reading. This is sometimes necessarily so, as the reports may be attempting to serve at least two different purposes, namely to introduce and describe an area, and to act as a reference document during its development. A single format will seldom serve both such purposes well, although much of the tedium can be relieved by the use of summaries, interpretive sections and specialist appendices, each designed for specific, and often different, users.

9.3 Quality of soil and land suitability reports

Assessment of the quality of soil reports depends to a large degree on determining how far the conclusions are based on measurement and how much on interpretation – and the relevance and precision of the measurements with regard to their use in the interpretation. Table 9.1 summarises the main points that should be assessed in any soil or land suitability report when it is to be used as a basis for development activities.

Basic checklist for assessing soil and land suitability reports   Table 9.1

---

1. Read TOR to identify scope of survey, its restrictions and the likely inclusions and omissions in the report text and maps.

2. Estimate how much confidence to place in quoted results.   Read rapidly through section on methods to check:

    a) What properties were studied (see checklist in Annex H, Table H.1); which properties were measured, and which estimated (note especially relative reliance on API and field-work); correlations, if any, with existing crops and agricultural performance.

    b) Accuracy and number of measurements: for instance, relevant details of all significant soil horizons within profiles should be examined, with adequate replication of tests and adequate sampling frequency.   Check whether measurements were made in field or laboratory, and relevance of methods chosen.

    c) Depth, location and intensity of observations (relate to TOR and soil types).   Check that all relevant landscape units were investigated.

    d) Consistency of mapping scales and subjects with observation intensity and intended uses.

3. Check following points on soil classification:

    a) System should be based on soil morphology or observed/measured properties, not merely inferred genesis.

    b) Definitions of all mapping/classification units should be included, if possible with chemical and physical data.   Check that agriculturally significant properties for intended use(s) can be identified.

    c) Specific properties (eg salinity, stoniness) should be adequately described/quantified.

    d) Areas and locations of soil types should be identified, and purity of units should be quantified.

4. Check following points on land suitability classification:

    a) Range of properties and values of class limits used as the basis for the classification should be relevant to the planned development, and not necessarily just standard systems.

    b) Definitions of units should be unambiguous and related to the proposed land use(s).

    c) Specific problems should adequately quantified and based on sufficient measurements.

    d) Areas and locations of classes and subclasses should be given.

    e) There should be adequate interpretation of survey results for practical users; eg management practices for the land units should be specified where appropriate.

9.4 Writing soil and land suitability reports

9.4.1 Introduction   Taking it for granted that the technical content of the report is competently written, the most important sections are the summary and the interpretative parts for individual users. These parts of the report should be particularly well written, informative and easy to find, with good cross-references.

In general each report, and each chapter, should be laid out so that general comments precede detailed discussion, and so that factual material leads on to interpretative conclusions. Careful cross-referencing allows individual readers to pursue specific items to greater depth without wading through irrelevant or highly detailed sections to reach them.

Visual presentation of text and illustrations should be of high quality; general principles are discussed in Reynolds and Simmonds (1982).

9.4.2 Recommended report format   For most reports the format proposed by Smyth (1970) should be used.   Note that even in an integrated study, the soil and land suitability report should be capable of being read as a self-contained report; users may often not have access to other volumes.   In order not to daunt readers, the main soil report should be restricted to about 100 to 120 pages; further information can always be included in a volume of technical annexes.

After the usual introductory pages (title page; lists of contents, tables and figures; key to abbreviations; and acknowledgements) the basic report structure shown below should be followed; this forms a reasonably detailed checklist for most purposes. The list and layout should not, however, be regarded as inflexible: each report should be geared to the particular project's requirements; some possible variations on the general pattern are also indicated in the sections below.

Summary

(Approximately five pages maximum)

a) Location and area of study, with map.

b) Reasons for selection of study area and summarised purpose of study.

c) General description of area.

d) Broad description of soils; classification and mapping units.

e) Tabulated soil areas and percentages.

f) Broad description of land suitability and management.

g) Tabulated land suitability areas and percentages.

The summary should be well written and readable, and should include cross-references to relevant chapters, particularly those on recommendations; the degree of technical depth will depend on the reader aimed at. In a multidisciplinary study a second separate summary may be needed for the main report of the project. Note also that different users are interested in different area figures, which must be carefully specified because confusion is easy; Figure 9.1 shows some of the possible combinations.

Chapter 1   Introduction

(Approximately five pages maximum)

a)   Outline of TOR; brief mention of modifications of TOR.

b)   Aims of survey (in so far as not covered by TOR).

c)   Selection of area(s) of study, where appropriate.

d)   Special features of survey (eg where TOR changed and why).

e)   Outline of report structure (cross-reference to Smyth, 1970), maps provided, and unpublished archival data available on request (and where).

(For Annex: detailed TOR - and names of participants).

Chapter 2   The Environment

(Variable length, depending on complexity)

a)   Location - Latitude and longitude; relation to national or regional geography and administrative areas, main and local towns. Altitude; major landforms and relative relief.

b)   Communications - Roads (and surfaces), tracks (motorable or not), seasonal closures; railways; air and river transport.

c)   Human settlement and present economic activities - Population - numbers, density, distribution; occupations; health and endemic diseases. Industry, agriculture and forestry; main crops, marketing and processing (more details of land use can go into Section 2h).

d)   Infrastructure - Local, regional and national government administrative institutions; agricultural stations; dispensaries and hospitals; schools and colleges; power and water supplies.

e)   Climate - Rainfall quantity, intensity and distribution; wind speeds and directions; maximum, minimum and mean temperatures; bright sunshine hours; solar radiation; evapotranspiration; moisture surpluses/deficits; frost/storm action; seasonal trends; incidence of atypical but crop-damaging weather. Statistical analysis where appropriate; references to standard local works (if any).

Area nomenclature used in development studies

Figure 9.1

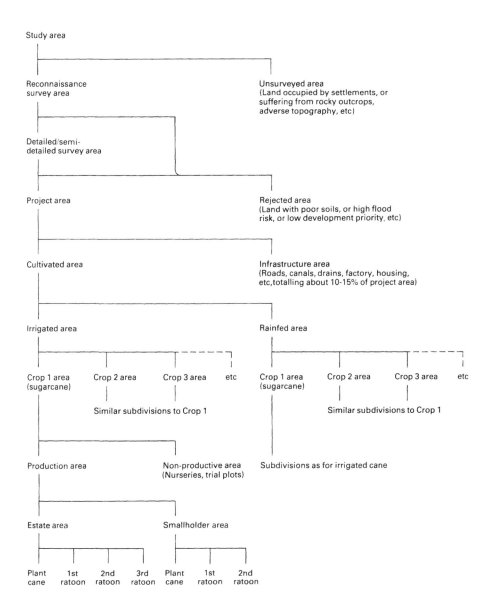

f)   Water resources – River flows and GWT levels (seasonal
     variation); water storage; water quality; regional
     drainage; flood risks, duration and depths.

g)   Geology and geomorphology – Major terrain types, basic
     geology and geomorphology; specific landforms and their
     relationship with soils. References to standard local works,
     if any. Map, if appropriate, with text.

h)   Natural vegetation and present land use – Overall pattern,
     especially as related to landscape features; major tree and
     shrub species and uses. Present land use – major crop and
     weed species, type(s) of cultivation, crop rotations, land
     management, livestock, problems (eg erosion), agricultural
     research. References and, if appropriate, map.

Diagrams: of general location; agro-climatic data; major
geological/geomorphological units (locations and cross-sections);
vegetation/ecology.

(For Annex: detailed list of flora with local, botanical and
common English names).

This is a fairly comprehensive list, usually for a soil and land
suitability survey done in isolation. For integrated studies the
sections on location, climate and water resources can often be
considerably reduced and cross-references to other relevant volumes
inserted instead.

Chapter 3   Methods of investigation

(Approximately six pages maximum)

a)   Level of survey(s) made, grid spacing/site intensity; how/why
     varied; area(s) covered at given intensity.

b)   Numbers of auger and of pit sites.

c)   Numbers of GWT observations.

d)   Depths of all observations; method of description (keep brief
     and cross reference to FAO guidelines).

e)   Processing of data (keep brief and cross-reference to annex on
     details of calculations).

f)   Details of AP cover used – Date(s), scale(s) and area(s)
     covered, quality, limitations.

g)   Map compilation – Scale, base map(s), how boundaries drawn
     (keep brief and cross-reference to chapter on map units),
     number of maps produced and subject(s).

h)   Special measurements – Numbers of sites, replicates and
     brief method descriptions regarding, eg infiltration rate,
     hydraulic conductivity, bulk density etc; also laboratory
     work (numbers of sites, samples, replicates, tests made and

methods used). Cross-reference to annex on detailed methods of investigation, and to chapter and annex containing results.

Diagram: Site intensities. Site locations on main maps - see Section 9.5.

(For Annexes: details of data processing; details of methods of investigation - field and laboratory).

Chapter 4   Soils

(Variable length, depending on soil complexity)

a)   Previous studies and classifications, including level of study, proportions of field-work and API, observations made (depths of profiles, properties recorded).

b)   General description of soils (origin, morphology, chemical, physical) and classification adopted.

c)   Summarised profile descriptions of soil units (in small type - further details in annex).

d)   Correlations with any previous classification(s) - Tabulate grouping; soil names, soil symbols; local, FAO and/or USDA units.

e)   Tabulated physical and chemical data, including means and/or medians, SD/SE, ranges.

Diagrams: Water release curves; infiltration rates; detailed soil/vegetation/geomorphic/topographic relationships.

(For Annexes: see list after Chapter 5 below).

Chapter 5   Soil-mapping units

(Variable length, depending on mapping complexity)

a)   Accuracy of soil boundaries; minimum mapped area.

b)   Discussion of phases, variants, complexes. (Note comments in Section 4.3e on p 38).

c)   Description of individual units (grouped by landscape unit or soil grouping, as appropriate) with tabulations of most important data; include data on slopes, microtopography, erodibility, drainage, potential crop yields etc.

d)   Table of areas and percentages; both of mapping units and individual soils.

e)   Purity of units; major impurities.

Diagrams: of mapping unit occurrence related to landscape; soil associations; general locations of major soil-mapping units. Main soil maps as Section 9.5 below.

Note: The dividing line between Chapters 4 and 5 is somewhat
          arbitrary; they may often be amalgamated.   They are
          best kept separate when the mapping units consist of soil
          complexes (described in Chapter 5) composed of various
          soil series (described in Chapter 4).

(For Annexes:

a)   Detailed soil profile descriptions, with physical and chemical
      data.

b)   Sample data sheets – Soil morphology and soil physical
      tests; auger and pit sheets.

c)   Statistical treatment of results.

d)   Investigations of special sites, eg pilot farm areas).

Chapter 6   Land suitability

(Variable length, depending on complexity)

a)   Objectives of classification and context – Data and
      assumptions in physical, social and economic terms.   Present
      and proposed land usage with respect to land suitability.
      Previous classifications in or near the study area.

b)   Management and improvements envisaged before, at and after the
      time during which the land suitability classification is
      expected to be applicable; agricultural methods; engineering
      installations.

c)   Basis of classification – Modified USBR, FAO principles;
      based on   repayment   capacity   or   downgrading   according   to
      limitations (or however done);   level of technology/management
      assumed;   crop ranges.

d)   Tabulated land class descriptions (see examples in Tables 5.6
      to 5.8):

          i)   general;

          ii)   specific, with quantified limits for subclasses (see
                below).

e)   Criteria chosen for differentiating classes and subclasses.

f)   Description of subclasses, with reference back to table given
      at dii above.

g)   Deficiencies/restrictions;   effects on yields.   Tabulation of
      symbols.

h)   Tabulated subclass areas, percentages, estimated crop yields.

i)   Details of subclasses including impurities (if long, this
      section could be put in small print).

185

j) Social, financial and economic evaluation – assessment and comparison of alternatives. Normally handled by the team economist and/or financial analyst.

k) Source data – Maps, previous reports, local information, either as annex or accompanying documentation.

Diagram: Simplified land class occurrence (and/or on summary map). Main land suitability maps as Section 9.5 below.

Chapter 7   Soil and land management

(Variable length, depending on complexity)

a) Must be well written and explicit; full cross–references to soil and land suitability chapters.

b) Soil fertility – Aims (eg increase Ca, increase all nutrients); recommended applications and types of fertiliser (also those not recommended); specific nutrient problems; field trials; micronutrients; release of nutrients (leaching, split applications).

c) Crop varieties – Normally cross–reference to Sections 2c and 2h.

d) Pests and diseases – Especially soil–related ones, for example nematodes (cross–reference to Section 2h).

e) Erosion – Shelter belts; cover crops and mulching; contour cultivation; prevention of capping.

f) Land preparation – Grading, levelling (especially with respect to problem soils, eg shallow profiles, infertile subsoils); pan prevention, weed control; tillage methods. Cross–reference to Section 2h.

g) Drainage – Infield: type recommended, in general terms. Regional: GWT movement; water quality (eg after leaching). Cross–reference to engineering sections such as 2f and 6b.

h) Irrigation – Discuss choice of method and/or cross–reference to Section 2f to include account of:

    i) deep percolation;

    ii) surface topography/land grading (eg problems of infertile subsoils);

    iii) operation (24 h or not; labour availability);

    iv) installation costs in general terms (eg canal lining).

(For Annex: specifications of designs and design criteria, levelling/grading measurements, details of salinity/sodicity preventative precautions, details of conservation measures etc).

Chapter 8   Summary of technical findings, and specific recommendations

> (Approximately three pages maximum; only required if too long for the initial Summary)
>
> a)   In list form; very brief.
>
> b)   Including need for further work.
>
> c)   Cross-referenced to relevant sections.
>
> In some reports a case can be made for putting Chapters 7 and 8 at the front of the report, since they contain the information that should be acted on.   They can be difficult for a reader to assimilate like this (since he will not have read the soil and land capability chapters) and careful cross-referencing is essential if this approach is adopted.
>
> Supporting photographs – Major land units/vegetation cover. Main soil types.   Major problems (eg erosion).

9.5   Soil and land suitability maps

9.5.1   General choice of map subjects and scales   Soil and land suitability maps are often used without reference to the accompanying report(s).   Choice of map subjects and presentation should therefore aim at being as self-explanatory as possible, and should cover all the major aspects relating to the proposed development.

> a)   Map subjects – These will normally include:
>
>> i)   soils;
>>
>> ii)   land suitability (separate map set for each use envisaged);
>>
>> iii)   specific soil parameters relevant to development, eg salinity, depth (separate set for each development);
>>
>> iv)   summary map of whole area (usually land suitability).
>
> (i) to (iii) normally at same scale; (iv) smaller scale to allow presentation on one sheet.   In surveys of single farms or other small areas, or where soil and land unit boundary configurations are simple, one map sheet may suffice for all the topics.
>
> b)   Map scales – These are conventionally chosen so that the field-sampling intensity corresponds to between 1.0 and 0.25 site cm^{-2} on the final map (FAO, 1979a, p 89).   For many consultancy uses, however, this gives a map that is too small to be useful, and in practice choice of scale depends on:
>
>> i)   scale of available base maps;

      ii)   allowing sufficient space for site numbers to be drawn in;

      iii)   scale that will be of most value to other users (design engineers; agronomists designing cropping patterns; field engineers);

      iv)   production of maps that are small enough to use comfortably in the field.

This usually means for most feasibility studies (medium–intensity surveys) that maps are produced at A1 size (approximately 60 x 85 cm) as a maximum, and at a scale between 1:5 000 and 1:20 000 (1:10 000 most common). The summary maps, often intended for wall display, may be produced up to about A0 size (approximately 85 x 120 cm; paper sizes are summarised in Subsection J.4.5).

9.5.2   <u>Minimum area delineation</u>   The smallest acceptable area delineation is quite arbitrary within certain size limits, but the minimum size chosen usually depends on:

a)   the smallest delineation inside which a simple map unit symbol can be printed legibly; or

b)   the smallest area which can be easily employed by the intended map user.

For the purposes of uniformity, a value of 0.4 cm^2 for the area of the minimum–size delineation is used by Eswaran et al (1977). This area seems to fall within the range of minimum values used by most sources for the smallest areas on soil maps. The USDA (1975a) considers the minimum–size delineation to have an area of 1/16 in.^{2} or 0.403 cm^2. Vink (1963) takes his 'basic mapping unit' as 0.25 cm^2. It should be noted, however, that FAO (1979a, Section 6.2.2) recommend that the minimum area of <u>planning</u> interest should occupy not less than 1 to 2 cm^2 of the final map.

The actual land area which is equivalent to 0.4 cm^2 on a map is a function of the map scale. The values for a series of map scales are given in Table J.4.2.

9.5.3   <u>General map presentation</u>   All maps (soil, land suitability, parameter or theme, summary) should usually include the following features:

a)   title (including client, project, map number and subject, country);

b)   date;

c)   map or sheet code and number;

d)   scale (linear scale, and numerically);

e)   north-point (magnetic/true) and latitude and longitude reference points (or grid system and references);

f)   index diagram to adjoining sheets;

g)   compilation source(s) (eg aerial photographs, topographic maps - authors, reference numbers, scales and dates);

h)   company and associates' names, addresses and logos; where other companies have provided technical inputs such as aerial photography/base maps or cartography they also require mention;

i)   specific subject legend (see Subsections 9.5.4 and 9.5.5);

j)   conventional symbol legend to explain all general symbols (roads, rivers etc);

k)   indication of map reliability/intensity of survey;

l)   cross-reference to accompanying report;

m)   copyright attribution.

The maps should be as self-contained and as easy to interpret as possible.

Note that for some clients a bilingual presentation is preferable or obligatory.

9.5.4   Shading/colour work   General principles include:

a)   darker shades stand out more: reserve for the better quality areas/soils on display maps. On field maps do the reverse to allow data to be written on lighter shaded better areas.

b)   keep the colour range as restricted as possible; make sure that colours 'tone in' by choosing from a manufacturer's co-ordinated colour chart.

Suggested general colour connotations:

Green	Fertile/wet
Blue	Wet
Yellow	Arid/sandy
Brown	Dark - fertile
	Pale - infertile
Grey	Infertile; rock or stones
Red/brown	Iron-rich soils
Black/violet	Urban areas; buildings

Note also that the FAO (1974) presentation uses standard colours for its soil units, and the USBR (1953) has the following colour code for its land classes:

Yellow	Class 1	Brown	Class 4
Green	Class 2	Pink	Class 5
Blue	Class 3	No colour	Class 6

9.5.5   Soil map legends

    a)   Keep as simple as possible.

    b)   General soil properties (texture, depth; fertility; physical properties, major deficiencies).

    c)   Additional data, such as slopes; suitability for different crops; climatic data (or cross-reference).

    d)   FAO symbol/name.

    e)   Composition of complexes (percent of each soil type), and 'reliability' of map (depth and intensity of observations; purity of units).

    f)   Possibly (where easily portrayed) diagrammatic cross-sections showing soil-topography-vegetation associations (or in report).

    g)   Perhaps field investigation sites, but see end of Subsection 9.5.6 below.

9.5.6   Land suitability map legends

    a)   Separate map for each land use envisaged.

    b)   Keep as simple as possible, especially symbols (see Table 9.2 for example).

    c)   Major limitations and/or use restrictions indicated.

    d)   Crop yields indicated, if possible, class by class.

    e)   Management inputs specified.

    f)   Field investigation site locations and numbers - and indication of whether pit or auger; chemical and physical sampling sites (using symbols).

Sites of soil inspections may be shown more easily on land suitability maps because there is usually more room than on soil maps. With small-scale maps a separate set of site location maps may be needed.

Example of crop suitability map legend for multi-use reconnaissance survey

Table 9.2

Land unit — Symbol and name	Description	Gross area (km²)	Max slope (%)	Natural soil fertility	Erosion risk	Crop suitability (rainfed smallholder production) 1/								
						Cassava	Maize	Yam and plantain	Rice	Cocoyam	Cocoa	Oilpalm	Rubber	Forest
Ab Aboa	Basement complex uplands	2 280	40	Mod	**	..	..	...	.	..	..	..	...	...
Ek Ekow	Dissected sedimentary lowlands	200	30	Low	*	..	..	...	.	..	...	..	...	...
El Elele	Undulating coastal plan	4 160	6	Low	*****	.....	..	...	.	...	...	..	...	...
Gu Guitri	Undulating sedimentary lowlands	730	6	Low	***	...	..	...	.	..	...	..	...	...
Ib Ibekwe	Basement complex lowlands	1 340	15	Mod	***	...	..	...	.	...	...	...	...	...
Ik Ikom	Basaltic intrusions – hills and mountains	100	70	Mod	**	...	.	..	.	..	.	.	..	..
Ke Keta	Ridges of beach sand with swampy hollows	310	30	Low	**	..	.	..	.	..	..	..	...	...
Li Likume	Basement complex hills and mountains	1 590	60	Low	**	..	.	..	.	..	...	..	...	...
Ma Manson	Strongly dissected coastal plain	350	30	Low	*	..	.	..	.	..	...	..	...	....
Op Opobo	Marine alluvium–mangrove swamp	1 000	1	Low	*****	.....	.	..	.	...	...	..	...	....
St Stubbs	Dune sand	230	30	Low	***	...	.	..	.	..	...	..	...	...
To Togo	Dissected coastal plain	980	15	Low	***	..	.	..	.....	..	...	..	...	...
Ye Yenagoa	Freshwater alluvium	2 200	2	High	*****	..	..	...	.....	...	...	..	...	...

Key to erosion risk (rainstorm resistance)

*****	Highly resistant
****	Resistant
***	Somewhat erosive
**	Erosive
*	Very erosive

Key to crop suitability

.....	S1	Highly suitable
....	S2	Suitable
...	S3	Marginally suitable or partly unsuitable
..	N1	Largely unsuitable
.	N2	Totally unsuitable

Note: 1/ Data on climate, sociology, management levels etc should be included in text, or summarised for table. These crops and land units are adapted from BAI (1982) and relate to south-eastern Nigeria.

Soil and land suitability reports

Annex A

Survey Planning and Logistics – Tabulated Data

Contents

Summary checklist for items and services to be costed for soil and land suitability surveys   Table A.1

1.	Professional staff services
	(see Tables A.2 and A.3)
		–	including:
			subsistence
			accommodation
			leave/weekends
			field time – pits
			                 augers
			                 mapping
			                 administration

2.	Return air travel	–	including: wives and families where appropriate
3.	Technical and support equipment (see Tables A.5 to A.12)	–	including: UK and local purchases for expendable and non-expendable equipment
4.	Transport of equipment	–	approximately 40 kg per surveyor for long-term project
5.	Local transport (see Tables A.5 and A.10)	–	including: 4WD vehicles tractor bowser bulldozer fuel and maintenance
6.	Local staff (see Table A.3)	–	including: counterparts office staff interpreters drivers labour house staff watchmen/caretakers (office and homes) back-up staff as required
7.	Drafting services (see Tables A.3, A.5, A.8) and A.9)	–	including: rough and final drafts of reports and annexes: diagrams and maps (colour/black and white?)
8.	Final map and report production	–	including: typing printing (colour/black and white?) collating and binding
9.	Office and residential facilities (see Tables A.5, A.11 and A.12)	–	including: soil survey and mapping offices storage, laboratory, soil-drying facilities accommodation (hotel/house/tent?) for team members and, if necessary, support staff all furnishings
10.	Soil laboratory analyses (see Section A.5)	–	including: chemical and physical analyses duplicates, say 5 to 10%, to UK
11.	Report despatch	–	including: packing courier or air freight costs
12.	Physical contingencies	–	10% to allow for illness, overrun etc
13.	Inflation	–	dependent on countries and currencies concerned: in recent years approximately 15% of total annually has often been cited
14.	Overruns due to external factors	–	prorata cost per month of overrun to cover for example: aerial photograph preparation base map production provision of permits/access to area

Note: Period of validity of costs should be stated.   A list of suppliers is given in Table A.16.

Annex A: Planning and logistics

Checklist for timing of soil survey and map preparation 1/ Table A.2

1.	Aerial photography/satellite imagery interpretation	–	Variable. Say 1 to 2 man-weeks at start, depending on size/intensity of survey
2.	Auger borings	–	9 to 20 per day depending on observation depth, soils, vegetation cover and complexity
3.	Pit inspections	–	Up to about 5 per day depending on depth and distance apart
4.	Mapping 2/	–	About 3 days per man-month)) 1 week per month
5.	Administration	–	About 2 days per man-month)
6.	Working month	–	Longer-term projects overseas (over 3 months, say) and UK Inputs: 22 days/month Shorter projects overseas: 26 days/month
7.	Field laboratory tests	–	pH and EC: 100/day
8.	Infiltration tests	–	1 site in triplicate/day
9.	Hydraulic conductivity tests	–	1 site in triplicate/day
10.	Bulk density measurements	–	1 site in triplicate/day
11.	Training survey teams	–	2 days/team/activity
12.	Physical contingencies	–	Usually about 10%

Notes: 1/ Actual survey dates should be adjusted to allow for 'external' influences, such as:
- rainy season(s) and difficulty/impossibility of working;
- public holidays (in particular Ramadan; for dates involved see Freeman-Grenville, 1963);
- special communication problems: frequency of air flights in and out; travelling time for client coordination/laboratory deliveries.

2/ Draughtsman's time additional to this; calculated as follows:

Paper size 3/	Drafting time (man-days) Map complexity		
	Low	Medium	High
A0	3	5	7
A1	2	4	6
A4	0.5	1.5	2

3/ For key to paper sizes see Subsection J.4.5. Interpolate for intermediate paper sizes.

194

Checklist for soil and land suitability staff requirements 1/   Table A.3

Job title	Function/number	Time allocation (for long—term project) UK	Overseas
1. Professional staff			
a) Senior soil scientist	Overall co—ordination and supervision of study	Approximately 1.0 month administration and scrutineering	Usually one supervisory visit of 0.5 month
b) Senior soil surveyor(s)	Direction of field studies and report compilation. Number depends on project size/duration	2.5 to 3.0 months premobilisation and report writing	Whole period of field—work and ad hoc report production
c) Senior soil physicist	Direction of field physical tests, and interpretation of results (including statistics)	About 1.0 month premobilisation and report writing	Whole period of soil physical tests and ad hoc report production
d) Soil surveyor(s)	Field observation and sampling of soil. Number depends on project size/duration	About 0.5 month premobilisation and report collation	Whole period of survey and as needed for ad hoc report
e) Soil physics assistant(s)	Field physical tests on soil. Number depends on project size/duration	About 0.5 month premobilisation and report collation	Whole period of field tests and as needed for ad hoc report
2. Field support staff			
a) Labour	a) For field survey – approximately 5 men per surveyor	Nil	Whole period of soil survey
	b) For physical tests – approximately 3 per physicist	Nil	Whole period of soil physical tests
	c) For trace cutting in thick bush – approximately 5 per survey team	Nil	Until all traces open
b) Drivers	Ideally one for each surveyor and soil physicist and for support vehicles	Nil	Whole period of soil professional staff input
3. Office support staff			
a) Draughtsmen	Drawing diagrams, maps, report covers and arranging for reproduction; sometimes to do planimetry. Normally only 1 or 2 required	Variable – only required if final drawing done in UK, see Table A.2	Variable depending on size and numbers of maps – see Table A.2
b) Typists	General typing duties including report. Usually one sufficient	Whole period of report production; only required if final production done in UK	Whole period of soil technical staff input

Note: 1/   Intermittent inputs will also be required from general project staff, including:

Office manager
Typist/secretary
Cashier/bookkeeper
Messenger/cleaner
Translator

Annex A: Planning and logistics

1. **Office and accommodation**

   a) Size in relation to numbers to be accommodated
   b) Ventilation/air conditioning and lighting conditions
   c) Furniture - basics to be ordered
   d) Noise levels
   e) Dust and dirt (especially in ground-floor offices)

2. **Shipment of baggage**

   a) Best method; name and address of shipper/agent
   b) Arrangements for clearance and collection (customs procedure; documentation)
   c) Costs

3. **AP and map coverage**

   a) Visit to 'surveyor general' to check on map, aerial photo and Landsat data; theme/scale/date/quality/availability/area covered/cost
   b) Order AP and maps

4. **Assessment of field problems**

   a) Objectives of survey with respect to timings and costings
   b) Location and size of gross survey area
   c) Access: methods and timings (vehicle/foot/horseback/boat/helicopter etc); permits (documents needed, and sources); accessibility (trace cutting, topo surveying)
   d) Augers: types and diameters/lengths; local repair facilities
   e) Labour: availability/ability/rates
   f) Preliminary ideas on: soil series/phases and land suitability classes/restrictions: if time, full descriptions of exploratory pits (accurately located)
   g) Special problems: swamps, quicksands, wild animals, minefields ...

5. **Literature**

   a) Local sources/availability
   b) Relevant/related subjects

6. **Transport**

   a) Allocation: one vehicle per soil surveyor
   b) Availability - to be arranged
   c) Spares - to be ordered
   d) Servicing and fuelling - system to be worked out

7. **Local purchase of equipment**

   a) Check availability
   b) Check quality
   c) Check cost (cf(2) above)

8. **Local places to visit**

   Research stations   )
   University departments) For background information; try to assess speed and quality of
   Libraries   ) work, and facilities available to assist survey. Any items of
   Soil laboratory   ) equipment that project could supply.

9. **Survey schedule**

   a) Check with Project Manager and other team members. In particular arrange for clear setting-up period - with, especially, no reports to write. Must allow time to settle in and train teams. Similar clear period at end of project
   b) Determine major deadlines for survey: maps for other disciplines to use; reports on various aspects of work (eg inception report, pilot project report); physical deadlines determined by rainy season(s), manpower availability, Ramadan etc. Deadlines for analyses: AWC measurements for agriculturists
   c) Likely delays: map availability; access after harvest; vehicle repairs; labour. Any extra work likely to be required. Estimate of possible overruns for team members - to be made clear in advance to them
   d) Liaison with other team members - draw up general and individual terms of reference

10. **Extras**

    a) Communications: local/international: short-wave radio, telephone, telex, cable, mail - delays and problems
    b) Recreational facilities: list items to bring
    c) Local supplies: technical and personal: list shortages

Checklist for local support services and equipment   Table A.5

---

1.   Local transport (see also Table A.10)

         4WD (Land Rover, Nissan Patrol etc) 1 per survey team
         1 tractor
         1 x 500 gallon water bowser, or trailer and water drums
         Bulldozer for trace clearance, with support vehicle every 2 weeks for
         maintenance
         Low loader as required
         Fuel, spares and maintenance for above vehicles

2.   Local office equipment

     a)   General

                Stationery
                Telephone/telex
                Electricity
                Copying - photocopier, dyeline

     b)   Specific for soil surveyors

                1 desk and chair per surveyor
                1 large table for maps/AP laydowns (approximately 3 m x 1 m)
                1 light table
                1 draughtsman's table, with stool and cupboard
                Cupboards and shelving (including map cupboard and space
                for AP)
                Notice-boards and wall-map display facility
                Storage space for equipment, including soil physics
                equipment
                Filing cabinet

                Optional   Soil-drying room and shelving
                           Soil laboratory: 2 tables, sink, stools, water supply
                                             laboratory equipment (see Ferguson,
                                             1973)

     c)   Mapping and report production

                Reproduction/collation
                Printing - including covers
                Binding
                Despatch

Annex A: Planning and logistics

Indicative costs and sample weights for common soil analyses

Ref No	Analysis	Method	Approximate minimum air dry weight needed 1/ (g)	UK cost £ per analysis (1988 prices) 2/
A.	Usual routine analysis			
1.	Sample preparation			8
2.	pH			
2.1	1:2.5 suspension	pH meter	20	4
2.2	Saturated paste	pH meter	400 3/	7
3.	EC			
3.1	1:2.5 suspension	EC meter	40	4
3.2	Saturation extract	EC meter	400 3/	7
4.	Cations			
4.1	Exchangeable Ca	)		
4.2	Exchangeable Mg	)		
4.3	Exchangeable K	) 1 M KCl or	)	
4.4	Exchangeable Na	) 1 M ammonium acetate	) 80	20
4.5	Exchangeable Al	)	)	
4.6	Exchangeable H	)		
4.7	CEC	As for exchangeable cations	30	10
4.8	Soluble Ca Mg K Na	Saturation extract	400 3/	14
5.	Available P	Either Olsen or Bray	20	6
6.	Total CO_3	HCl extraction	10	5
7.	Organic C	Walkley-Black	20	10
8.	Total N	Micro Kjeldahl	20	10
9.	Particle size	Pipette 7 fractions	120	30
	Total for routine analysis, non-saline soils 2/: all above analyses except 2.2, 3.2 and 4.8		360	107
10.	Soluble anions in saturation extract			
10.1	CO_3	Titration	)	)
10.2	HCO_3	Titration	)	) 6
10.3	Cl	Ion-selective electrode	) 400 3/	6
10.4	SO_4	Turbidimetry	)	6
10.5	NO_3	Ion-selective electrode	)	6
11.	Trace elements			
11.1	B	Hot-water soluble	38	12
11.2	Cu	Perchloric acid digestion	)	7
11.3	Zn	Perchloric acid digestion	) 2	7
11.4	Mn	Perchloric acid digestion	)	7
	Total for saline soils 2/		800	164
12.	Moisture content at:			
12.1	pF 0	Sand bath	)	12
12.2	pF 2.5	Pressure plate (undisturbed core))		12
12.3	pF 3.0	Pressure plate (undisturbed core)) Depends on core size		12
12.4	pF 3.7	Pressure plate (undisturbed core))		12
12.5	pF 4.2	Pressure plate (repacked core))		12
13.	Moisture content	Oven drying	10	

Notes: 1/ Allowing for one repeat analysis; some extra should be added to allow for wastage during handling.

2/ Indicative commercial laboratory costs for individual analyses including UK VAT at 15%; considerable reductions for various 'packages' are also available.

3/ Only one sample of 400 g is required for all these analyses.

Checklist of technical equipment for soil and land suitability surveys 1/   Table A.7

Description	Estimated cost per article excluding tax (1988 prices) (£ sterling)	No required per surveyor	No of surveyors 1/	Total No of articles required 1/ (including spares)	Estimated total UK cost excluding tax (1988 prices) (£ sterling)
1. Soil survey					
1.1 UK purchases					
1.1.1 Non-expendable					
Desk stereoscope and box (Wild ST4)	709	-	-	1	709
Magnifying eyepiece for ST4	289	-	-	2	578
Pocket stereoscope	20	1	2	2	40
Suunto clinometer 2/	64	1	2	2	128
Prismatic compass (Francis Barker 1216)	73	1	2	2	146
Geological hammer (Estwing 1 kg pick)	28	1	2	2	56
Munsell soil colour chart (+ 3 spare pages)	82	1	2	2	164
Planimeter (Haff 315E)	498	-	-	1 (or 2*)	498
Subtotal	-	-	2	-	2 319
1.1.2 Expendable					
Auger head (assorted types)	38	3*	2	6	228
Trowel	3	1	2	3	9
Magnifying glass (x 10)	7	1	2	2	14
Plastic dropper bottles (20 mℓ)	1	1	2	2	2
Plastic wash bottles (500 mℓ)	1	1	2	2	2
25 ml measuring cylinder	1	-	-	2	2
2 m tape measure	2	1	2	3	6
50 m tape measure 3/	10	-	-	1	10
Auger forms (self-duplicating)	0.3	-	-	500	150
Pit forms (self-duplicating)	0.3	-	-	200	60
Plastic bags (9 in. x 12 in. 500 gauge + ties)	79/1 000 bags	) depending) on total) physical) and	-	1 000	79
Label tags	-	) chemical) samples	-	1 000	10
Marker pens	-	2	-	4	)
Chinagraph or omnichrom pencils	-	4	-	8	)
Clipboards	-	1	-	2	) 30
Set pH papers	-	1	2	2	)
Whistles (Acme Thunderer or similar)	-	2	2	4	)
Subtotal	-	-	2	-	602
1.2 Local purchase (expendable)					
Auger handle and shaft	20	2*	2	4	80
Shovel	10	2	2	4	40
Pickaxe	10	2	2	4	40
100 m polypropylene ropes (for measuring distance) 3/	30	1	2	2	60
Conc HCl 500 mℓ	4	-	-	-	4
Subtotal	-	-	2	-	224

Notes: See page 201.   cont

Annex A: Planning and logistics

Description	Estimated cost per article excluding tax (1988 prices) (£ sterling)	No required per surveyor	No of surveyors 1/	Total No of articles required 1/ (including spares)	Estimated total UK cost excluding tax (1988 prices) (£ sterling)
2. Soil physics					
2.1 UK purchase					
2.1.1 Non-expendable					
Plastic balls (200 mm dia)	20/1 000 balls	-	-	1 000	20
50 kg x 200 g extension spring balance, dial	58	-	-	1	58
Box 24 pF cores (5 cm x 5 cm dia)	234	-	-	say 2 depending on sample Nos	468
Stop-watch	23	-	-	3	69
200 g x 2 g extension spring	8	-	-	1	8
HC bailer (approx 1 m x 6 cm dia))	70	-	-	1	70
HC screen (approx 1 m x 10 cm dia))					
Still well	1	-	-	3	3
HC float	6	-	-	3	18
IR floats and scales	15	-	-	3	45
IR bridges	9	-	-	3	27
Subtotal	-	-	2	-	786
2.1.2 Expendable					
Plastic float	3	-	-	6	18
2 m tape measure	2	-	-	3	6
Soil physics data sheets (self-duplicating)	0.3	depends on No of sample sites	-	100	30
Subtotal	-	-	2	-	54
2.2 Local purchase (expendable)					
Infiltration rings					
i) large (60 cm ID)	8	-	-	3	24
ii) small (30 cm ID)	6	-	-	3	18
Wooden mallets	4	-	-	2	8
Plastic buckets	2	-	-	6	12
Trowels (included above)	-	-	-	-	-
8 mm dia plastic tubing, 10 m coil	1	-	-	1	1
Subtotal	-	-	2	-	63

Notes: See page 201.   cont

Description	Estimated cost per article excluding tax (1988 prices) (£ sterling)	No required per surveyor	No of surveyors 1/	Total No of articles required 1/ (including spares)	Estimated total UK cost excluding tax (1988 prices) (£ sterling)
3. Soil field laboratory (see also Ferguson, 1973)					
3.1 UK purchase					
3.1.1 Non-expendable					
Deioniser and 2 cartridges	386	-	-	1	386
pH meter and electrode	453	-	-	1	453
Spare pH meter electrode	39	-	-	1	39
Conductivity meter	246	-	-	1	246
Subtotal	-	-	2	-	1 124
3.1.2 Expendable					
Plastic jerrycans	6	-	-	3	18
Standard pH solutions/powders	3	-	-	-	3
Plastic beaker and funnel	1	-	-	100	100
Laboratory data sheets	-	-	-	100	4
Box filter papers: 200 Whatman 91, 18.5 cm dia	5	-	-	3	15
Subtotal	-	-	2	-	140
4. Total (items 1 to 3 inclusive)	-	-	-	-	5 312
5. VAT at 15%					797
Subtotal					6 109**
6. Contingencies (10% of total + VAT)	-	-	-	-	611**
7. Grand total	-	-	2	-	6 720**

8. Overall costs	Total cost (£)			Approx total cost (£) per man incl VAT and contingencies 4/
	Without tax	+ 15% VAT	+ 15% VAT + 10% contingencies	
a) Expendable and non-expendable items for whole programme	5 312	6 109	6 720	3 360
b) Expendable items only for whole programme	1 083	1 245	1 370	685
c) Expendable and non-expendable items for whole programme less soil lab	4 052	4 659	5 126	2 563
d) Expendable items only for whole programme less soil lab	943	1 084	1 192	596
e) Expendable items only for whole programme less soil lab and soil physics	826	950	1 045	523

*   For long-term project and depending on survey organisation.
**   Including VAT at 15%.

Notes: 1/   Estimates for medium-intensity survey of about 10 000 ha. General office equipment, aerial photography, satellite imagery, maps, literature and stationery not included; for details of laboratory and office equipment see Ferguson (1973) and Table A.8 respectively.

2/   Alternative Sisteco CM360 clinometer costs £55.

3/   Preferred to the heavier metallic distance measurers: among the best of the latter is the 30 m mainly aluminium land chain by Rabone Chesterman at £154 each.

4/   Calculated by dividing the totals by two; costs for one man only would be higher because he would still require complete sets of equipment.

Checklist of cartographic equipment   Table A.8

Item	Preferred model or specification	Approximate cost (1988 prices) 1/ £ sterling
1. Major equipment		
Ammonia printer (dyeline)	Basic printer plus spares, eg Ozaminor 32FL	1 865
Desk stereoscope	Wild ST4 basic unit	815
	- magnifying eyepiece X 3 or X 8	332
	- cantilever support	392
	- box	263
Drawing board	AO tilting type with stand and parallel motion	154
Grant projector	Standard projector with spare fluorescent tubes	1 490
Light table	AO light box to stand on table, with movable light source	300
Plan file (horizontal)	AO drawers, metal or wood	483
Plan file (vertical)	Elite LEO02W	570
	AO filing strips (per 100)	18
Plan variograph 2/	SGI Rost Mapmaker, Basic with lenses f = 210 mm and f = 150 mm (operating range x 4.1 to 1/4)	3 905
	- darkroom tent	506
	- dust cover	64
Print trimmer/guillotine	1.3 m (51 in.) cut/rail	300
Desk lamp	Adjustable direction	45
2. Drawing materials and associated items		
Cotton wool	1 large packet	2
Drafting rolls	- Polyester film 0.03 in. double matt 1 016 mm (40 in.) x 20 m	24
	- Tracing paper 90 g m-2 smooth 1 016 mm (40 in) x 25 m	13
	- Cartridge paper 155 g m^{-2} 841 mm x 25 m	10
	- Continuous sectional roll 120 g m^{-2} 1 040 mm x 20 m	9
Drawing inks	Non-etching 23 ml bottle (Rotring F) x 2	3
	Etching 30 ml bottle (Rotring K)	4
Lettering stencils	Leroy lettering set 2901	265
	Rotring stencils (various sizes to suit pens)	80
Drawing instruments	Compass, dividers, set square, protractor (360°) eraser etc	45
Mapping pens	Set 8 pens 0.13 to 1.4 mm (Rotring Isograph set)	
	- 2 spare nibs for the 3 smallest sizes, 1 spare for each of the others	110
	- Rotring Foliograph set	140
Adhesive tapes	Invisible masking tape 6 m x 66 mm	3
	Drafting tape 25 m x 50 mm	3
Pencils	- Lead (black) per box of 12	5
	HB for general work	
	2H–4H chisel pointed for lines	
	2H–4H round pointed for lettering	
	- Coloured, per box of 12	4
	- Coloured wax (for use on AP) per box of 12	7
Planimeter	Haff 315E compensating polar	572
Steel ruler	- 1 250 mm graduated in mm	160
	- 300 mm graduated in mm	22
Self-adhesive lettering and shading	Letraset and Letratone or Zip-a-tone sheets (see Table A.9) per sheet:	
	- lettering	6
	- shading	3
T square	1 100 mm (42 in.) mahogany black-edge	34
Scribing film	Roll Keuffel and Esser Stabilene red transparent scribe coat 0.007 in. x 42 in. x 20 yd	256
Scribing equipment	Dual purpose tripod with lens attachment	24
	Scribing point	3

Notes: 1/   Including UK VAT at 15%.

2/   Alternative to Grant projector.

Initial requirements for adhesive type and screens needed for a small drawing office   Table A.9

Typeface	Details of requirements (using Letraset sheets)											
Helvetica medium	Size (pt)	6	8	10	12	14	16	18	20	24		
	Sheet code	1571	1570	1569	1568	729	728	3893	727	726		
	Quantity	2	4	2	2	1	1	1	1	1		
Helvetica light	Size (pt)	6	8	10	12	14	16					
	Sheet code	1575	1574	1573	1572	706	705					
	Quantity	2	4	2	2	1	1					
Helvetica medium italic	Size (pt)	6	8	10	12							
	Sheet code	1783	1782	1781	1780							
	Quantity	1	2	1	1							
Helvetica medium nos	Size (pt)	6/8	10/12									
	Sheet code	2690	2689									
	Quantity	2	2									
Helvetica light nos	Size (pt)	6/8	10/12									
	Sheet code	2692	2691									
	Quantity	2	2									
Letratone	Sheet code	LT29	LT30	LT31	LT32	LT33	LT34	LT35	LT934	LT912	LT68	LT15	LT121
	Quantity	2	2	2	2	2	2	2	1	1	1	1	1
Letraset symbols	Sheet code	553	2450									
	Quantity	2	2									
Mecanorma	Sheet code	531										
	Quantity	1										
Alfac	Sheet code	1231	1232									
	Quantity	1	1									

Annex A: Planning and logistics

For many contracts, the client undertakes to provide 4WD drive vehicles, plus fuel, drivers and so on. In rare cases there may be the opportunity to order new vehicles, either from local distributors or from the country of origin, but the vehicles will still need checking to see that they are suitable and properly equipped. Usually one needs to obtain necessary tools and spares, and fit correct parts and tyres, so this section should be used as an idealistic yardstick.

1. Vehicle type: 4WD, long wheelbase. Examples: long-wheelbase Land Rover or Toyota Station Wagon or pick-up.

2. Fuel: Petrol usually preferable to diesel for ease of do-it-yourself maintenance. Low compression engine since local fuel is usually 'regular' grade.
    Extra fuel capacity is often advisable, especially in remote areas and in reconnaissance surveys over large areas. Toyotas have small tanks. Late Land Rovers have tanks at rear where they are prone to damage (by reversing into things or traversing ditches). A 'centre tank' can be specified to go under passenger seat.
    Extra fuel must usually be carried in jerrycans. These should be of thick-gauge steel and strapped tightly into holders bolted on to vehicle; jerrycans on front of Land Rovers are meant for water (NB fire risk in collisions if these jerries are filled with petrol!); build up of static electricity in plastic jerries is potentially lethal.
    Toyotas need spare fuel filters as they cannot be cleared out.

3. Plugs and points: It is a good idea to check the plug gap, contact breaker gap and ignition timing if the vehicle runs badly or will not start easily. Check especially after a 'service' as garages/workshops frequently mess these settings up.

4. Tyres: Good quality cross-country tyres must be fitted. For desert conditions Michelin Safari tyres or similar should be used on 4WD vehicles. For very soft terrain rear wheels on pick-ups (especially 2WD) should take balloon sand tyres, eg Bridgestone Alligators. On long trips two spare wheels with tyres should be carried, along with a self-vulcanising puncture repair kit and tyre levers. Toyotas have 'split rim' wheels for 'easy' tyre removal, but need special levers and patience.
    Pressures should be checked, especially after someone else has pumped them up or checked them; use a reliable pressure gauge, preferably dial-type. For sandy/soft terrain pressures can be reduced to 15 lb in^{-2}, but for fast driving on a hard surface the pressures should be 25 to 30 lb in^{-2}.

5. Towing hitch: Check that the hitch fits the trailer/caravan/bowser. The 50 mm ball-type is better than the usual pin-type fitted to new Land Rovers, especially for fast towing.

6. Roof: A tropical double-skin roof can be specified for Land Rover station wagons.

7. Essential spares (in each vehicle): Spare engine oil (and gearbox/diff-oil). Spark plugs (to suit comp. ratio and temp). Contact breaker points (to suit model - beware!). Condenser. Fuel filter(s) (for Japanese vehicles). Fan and generator belts. Radiator hoses (top and bottom). Brake/clutch fluid, 1 pint. Battery top-up water (from deioniser in soil laboratory). For long projects, replacement brake shoes, clutch plate and rubber seals for hydraulic systems are advisable.

8. Essential tools (in each vehicle): Starting handle. 'High lift' jack (ensure working) plus handle and jacking block. Wheel spanner (which really fits, and is a good length). Set of 5 or 6 open-jaw spanners (A/F for Land Rovers, metric for Japanese). 2 plain screwdrivers (large and small). 2 crosspoint screwdrivers (large and small). Plug spanner. 10 in. adjustable wrench/spanner. Pliers. Feeler gauge set (inch or metric). 1 lb (500 g) hammer. Foot pump. Tyre pressure gauges (Michelin dial-type best). Nylon tow rope, minimum 10 m. Torch and batteries. Spade for digging out.
    For long trips/remote places advisable to have a set of 3 tyre levers (special type required for split-rim wheels), and a self-vulcanising puncture repair kit and matches. A clamp is needed for this but many locally available clamps break because of inferior castings. Rubber-cement patches are for temporary repairs.

9. Extra equipment: 'Cool seats' in hot climates. A machete or two in areas of dense undergrowth. Padlocks, as required, for tool kit, bonnet and petrol cap. First aid kit. Jump leads. For desert areas a pair of sand ladders. Where there are small stream lines or irrigation ditches etc to cross, a couple of stout long planks. A winch is worth the cost in wet forested areas. Fuel drum(s) and hand pump for base camp.

10. Items to check on receipt of vehicle

    a) Levels of fuel, oil, water (radiator and battery), and brake and clutch fluids. Numbers and condition of tools as above and extra equipment, if needed.

    b) Condition of:
        i)   radiator hoses;   v)   foot and hand brakes;
        ii)   fan and generator belts (and tensions);   vi)   lights;
        iii) tyres and spare tyre(s) and their pressures;   vii) plugs and points.
        iv)   steering;

1. Accommodation: Tents or locally built huts are cheapest and easiest; caravans are expensive and often very hot. Where possible allow for mess and/or office tent where papers, maps, AP etc can be left. Incorporate polythene lining to roof of thatched huts to prevent droppings. For longer-term camps, concrete bases are preferable.
    Site buildings to take full advantage of daytime shade from trees; thatched verandas are also useful for keeping hot sun off walls and windows.
    Mosquito netting should be fitted to doors and windows; groundsheets are advisable.

2. Beds: Lightweight camp beds, preferably metal-framed. If frame beds available, size should not be less than 2.0 x 0.9 m (6 ft 6 in. x 3 ft). Fit with mosquito net, especially where windows and doors are not covered.

3. Cooking: Basic alternatives are:

    Woodfires: tend to be messy; wood may not be available. Very good wood-fired oven can be made from small, clean oil drum buried sideways under earth mound with flat plate as door. Iron plate or mesh needed to form flat tray inside.

    Charcoal: may not be available, but good and reasonably clean.

    Paraffin: pressurised primus-types need to be kept clean; plenty of spare prickers are essential. Paraffin should be stored well away from kitchen and away from food (and flames!).

    Gas: portable cylinder-supplied stoves are probably best, if gas supplies are good.

4. First aid, tools and spares: See Table A.12 and A.14.

5. Food storage/refrigeration: Dry, reasonably cool and fly-proof safe is essential for storing non-perishables. For other food, paraffin refrigerators are excellent but can be temperamental: chimney should be kept very clean - wick should not be turned too high (flame should remain blue) and should be trimmed regularly. Really recalcitrant cases can sometimes be cured by inverting whole fridge (after removal of contents and paraffin tank ...) as this may clear vapour-lock. A few spare lamp glasses are essential.

6. Latrines: Pit latrines should be at least 3 m deep and sited at least 30 m away from any water supply. They should be roofed and provided with a close-fitting pit cover to prevent excess water and flies entering. If there are many people in the camp, provide a separate urinal. Chemical disinfectants should not be used in pit latrines or septic tanks.

7. Laundry: Clothes should be washed often, and dried thoroughly (especially socks and underwear, to avoid athlete's foot etc). In some areas all washing must be ironed to kill eggs or larvae of the bot fly which will otherwise develop under the wearer's skin; clothes dried on the ground are especially vulnerable.

8. Power and lighting: Small petrol generators are noisy and temperamental, and will only provide low-power output such as lights. Diesel generators are more reliable, but small plants cannot run heating or cooling devices such as cookers, refrigerators etc.
    Paraffin-fuelled hurricane lamps are simplest alternative for lighting, although light output is somewhat restricted. Spare wicks and lamp glasses are essential. Paraffin pressure lamps given much brighter light but can be temperamental: plenty of spare prickers, mantles and lamp glasses are essential, and methylated spirits to act as primer.
    Good battery-powered torches and spare batteries are recommended as additional equipment.

9. Refuse disposal: Burn any combustibles; dispose of remainder in pit, covering each addition with layer of soil to discourage scavenging animals and fly breeding.

10. Water supply and disposal

    Drinking water: Should be both boiled and filtered; filter candles should be checked regularly for cracks (and cracked ones replaced) and should be boiled at least weekly, unless of the silver-impregnated type. Note that water-sterilising tablets alone may not be 100% effective (see Table A.13 Item 10).

    Washing water: In areas with schistosomiasis (bilharzia), river water cannot even be used for washing without prior treatment; wells should be sited with care. Note that schistosome larvae are killed if the water is pumped, or stored for two days, or heated to hot bath temperatures. Impurities in rain-water may reduce effectiveness of detergents; grey-looking clothes may not always be the result of lazy laundering.

    Waste water: A suitable surface (ideally a concreted area) with effective drainage should be provided for washing areas to prevent formation of muddy surroundings/mosquito breeding areas. Waste kitchen water should be run through a grease trap before disposal in an earth pit.

Note: 1/ A detailed checklist is given in Table A.12. Further details are discussed in Wood (1978), Ross Institute of Tropical Hygiene (1974), Werner (1980), Cairncross and Feachem (1978) and Hale and Williams (1977).

Annex A: Planning and logistics

Checklist and indicative costs for camping equipment (6-man team for up to about 2 months) 1/   Table A.12

No	Item	Comment/sample type	Packed weight (kg) Per unit	Total	Packed volume (m³) Per unit	Total	Cost (£ sterling) 2/ 1988 prices Per unit	Total
3	2-man tents 3/	For surveyors. Each tent with 3.6 m x 3.1 m floor; 2.57 m high ridge, and 1.1 m wall height.	82.0	246.0	0.31	0.93	952	2 856
1	2-man tents	Dimensions as above; for kitchen/store	82.0	82.0	0.31	0.31	952	952
4	Tent sun linings	)	)					
4	Groundsheets	) One for each of above tents	) 24.0	96.0	0.11	0.44	45	180
4	Mosquito net doors	)	)					
6	Camp beds	213 cm x 90 cm; 61 cm high	11.0	66.0	0.05	0.30	110	660
6	Mosquito net frames	Alloy rods and brackets	0.6	3.6	0.01	0.06	13	78
6	Mosquito nets	Rectangular	0.6	3.6	0.01	0.06	17	102
7	Folding tables	1 each per surveyor; 1 for kitchen	12.7	76.2	0.07	0.49	80	560
7	Folding chairs with arms	1 each per surveyor; 1 for kitchen/spare	4.0	28.0	0.08	0.56	60	420
6	Sleeping bags	eg Rufantuf	2.5	15.0	0.03	0.18	39	234
18	Sleeping bag sheet liners	)					6	108
6	Pillows (washable)	)	-	285.0	-	1.73	5	30
18	Pillowcases	)					2	36
2	Airtight trunks 4/	Lockable - for food/valuables	15.0	30.0	0.16	0.32	52	104
3	Mess kits in chop boxes 4/	Comprehensive 2-man kit, with plates, cutlery, cooking utensils etc	-	90.0	-	0.07	154	462
2	Paraffin stove 4/	2-burner type (unpressurised)	7.5	15.0	0.05	0.10	39	78
18	Metal jerrycans	20 ℓ each; for water, paraffin etc	4.1	73.8	0.03	0.54	11	198
8	Tilley stormlight	1 per surveyor; 1 for kitchen; 1 spare	2.3	18.4	0.10	0.60	32	256
1	Spare set 4/ priming torch vaporiser, mantles and service kit	--------------- negligible -------------					11	11
6	Snake bite kits	1 per surveyor	0.01	0.1	0.01	0.03	5	30
6	First aid kits	1 per surveyor; comprehensive tropical kit in box	2.5	15.0	0.05	0.29	78	468
2	Water filters 4/	eg Puro Filter de Luxe	)				34	68
2	Spare Kieselguhr candles 4/	For water filters	) 0.9	5.6	0.02	0.04	7	14
2	Boxes water-sterilising tablets 4/	100 tablets per box	--------------- negligible --------------				4	8
	Total 2/		-	1 149.5	-	7.06	-	7 913 2/

Notes:
1/ Assuming all in one camp.
2/ Note that costs include UK VAT at 15%, but do not include export packing; otherwise they are FOB delivered UK port. Allowance should also be made for small extras, such as:
    strong string   machetes
    matches   fuel (for stoves and lamps)
    hooks   nylon cord (for clothes-line etc)
    padlocks (for stores, strong boxes)   sledge-hammer
    clothes pegs   polythene bags
    thermos flasks and/or drinking water bottles   cool bags or cool boxes
    can and bottle openers   torches and batteries
3/ Add one extra as office/mess for longer surveys and/or allow for single occupancy. Data are for Safari XXXI tents with steel pegs.
4/ One per surveyor needed if surveyors camping separately.

---

1.   General: Soil surveyors in the tropics are at slightly higher risk than many other workers, since they tend to live and operate in isolated, often unhealthy, areas.   Apart from the usual health-care precautions (mainly concerned with heat, hygiene and insect- and water-borne diseases) which are summarised by, eg, WHO (1982) and the Ross Institute (1974), the following sections outline items of particular importance to soil surveyors.   A good source of do-it-yourself treatments for use in remote situations is Werner (1980), which includes detailed checklists of medical supplies and some 'public health' disease prevention practices;   tropical diseases are covered in more detail in Adams and Maegraith (1980).   A series of publications on specific aspects of tropical diseases and preventive medicine is produced by the Ross Institute;   topics include insecticides, malaria and anti-malarial drugs, flies, excreta disposal, water supplies and primary health care.   Recommended vaccinations on a country-by-country basis are listed in DHSS (1982 and subsequent years).

2.   Rabies: During most surveys there is usually a high risk of staff being bitten by local animals, which should always be considered potential rabies carriers.   Rabies vaccinations are advisable before field-work starts.   If a non-immune person is bitten, the bite should be washed and dressed, and hyperimmune serum should be administered.   Traditional advice is to capture and/or kill the animal, but in practice this is rather academic: if the animal is infected then the immunisation would have to be done anyway;   and even if the tests prove negative, immunisation is often still advisable as an insurance against faulty laboratory diagnosis.

3.   Severe cuts: The first concern is to stop bleeding, the most effective way being to apply pressure, eg with a wad or holding the edges of the wound together;   bleeding can be reduced by getting the patient to lie down.   Once bleeding has stopped, the wound should be cleaned and the edges held together with butterfly clips pending proper attention.

4.   Malaria: Malaria is so common that it is often regarded with little respect, but falciparum malaria can be fatal if not treated.   All unidentified high fevers (where the patient has a temperature of, say, 102°F or more) should be treated as malaria in the first instance if no medical help is available.   Malaria symptoms sometimes produce a recurring pattern of 'well' and 'ill' days in the victim, and even on an 'ill' day high temperatures may occur for only short periods, of less than about an hour;   temperatures should therefore be checked regularly.   Repellents should be used on surveys where mosquito concentrations are high;   applications need repeating every 2 to 3 h.   Appropriate antimalarial prophylaxis should be taken at all times.

5.   Fevers: With all fevers whilst medical help is sought the prime concern must be to try to maintain the patient's fluid balance and keep his temperature down.   Aspirin or paracetamol and bathing with tepid water are the methods usually recommended for the latter;   salt additive may be necessary if fluid intake to prevent dehydration has been very high.

6.   Diarrhoea: An occupational hazard for all surveyors, most, if not all, of whom have their own favourite remedies.   Additional general advice is to cut out solid food and keep up fluid and salt balances.

7.   Soil-associated diseases: Apart from the bot fly (see Table A.11 Item 7) the two commonest problems associated with soils are jiggers and hookworm.   Risks are highest in inhabited areas where faecal contamination is most likely, and care should be taken, for example, not to sit down in shorts and open sandals on the topsoil when describing a profile ....   Away from settlement some risk also exists from contamination by, for example, dog hookworm species, which can cause irritation on the skin and, if ingested, serious cyst infections;   soil textures should therefore never be tested for in the mouth, and sleeping on the ground is not advisable (see also next paragraph).

8.   Living in local accommodation: If it is proposed to spend time in local peasant accommodation, medical advice must be sought.   In South America, for example, Chagas disease can be contracted by sleeping on earth floors in local huts, and hookworm (see above) is a common hazard on the ground in many places.

---

cont

9.   Snake bite: Usually a rare occurrence, the greatest danger often being from sluggish snakes, such as puff adders, which do not get out of the way; special gaiters can be worn in high-risk areas. Main points on treatment for snake bite include:

        a)   try to keep victim calm and still, and wash wound with soap and water; some sources also recommend sucking out venom (with suction apparatus). Apply firm, but not tight, ligature above bite, releasing every 20 minutes or so. Take patient to hospital or medical centre;

        b)   try to catch or at least identify snake: method of attack may help - eg sluggish snake on ground (viper), or aggressive one in tree (cobra);

        c)   if medical treatment is unavailable, watch for signs of envenomation:

             -   viper bites will produce swelling of local tissues and, if very severe, gradual development of haemorrhage from urinary tract, bowels and mucus membranes. If no swelling occurs after about 2 h, envenomation has probably not occurred, although patient should be kept under observation;

             -   cobra bites produce paralysis within 6 h of the bite; first signs appear in the eyes as drooping of the eyelids, and in the palate, causing regurgation if water is drunk;

        d)   if envenomation symptoms develop, then a suitable antivenin should be administered. Note that for field use freeze-dried polyvalent antivenins are probably best - get local advice or consult the Ross Institute in London before the survey starts. Field-workers should also be tested for sensitivity to the antivenin before the survey: some people can react very severely and should not receive antivenin, even if bitten.

10.   Water-borne diseases: Note that standard water-sterilising tablets using chlorine as the active agent do not kill all harmful contaminants in water (cysts, in particular, can be resistant, particularly to low chlorine levels). 'Potable aqua' tablets, containing iodine, kill both cysts and bacteria, but long-term use should be avoided because of possible unpleasant side-effects. Note also that chemical sterilisation is effective only on clean water. For drinking purposes, water should be boiled and filtered, using a Doulton or Berkfield filter - or a one-stage process using a filter containing silver (Sterasyl, Katadyn or Metafilter), which both sterilises and filters.

       Bilharzia is widespread in the tropics, and surveyors on field transects should resist the temptation to splash through pools, rivers or streams without waterproof clothing. Although the snail vectors live in slow-moving or still water, fast-flowing water may still be contaminated where it has passed downstream from an infected source. Water downstream of local villages is also highly suspect. For treatment of drinking and washing water see Table A.11 Item 10.

       Further details on water treatment are given in Cairncross and Feachem (1978).

Checklist of medical and related supplies   Table A.14

Item	Purpose	Comments
Small, medium and large lint dressings) Adhesive plasters (various sizes) and) adhesive plasters roll) Triangular, crepe and pressure) bandages) Netelast support bandages) Butterfly wound closures)	General bandaging and wound dressing	
Gentamicin and/or cetrimide cream	General antiseptic treatments	
Cicatrin powder and/or liquid cetrimide	General antiseptic treatment	Better than creams in hot, humid climates
Penicillin) Septrin) Sulphatriad) Tetracycline)	General antibiotic tablets	Take selection because of possibilities of resistance/ sensitivity to individual types
Imodium tablets	Antidiarrhoea	
Thalazole tablets	Antidiarrhoea antibiotic	For more severe bacterial attacks
Proguanil hydrochloride (Paludrin)) Pyrimethamine and sulfadoxine) (Fansidar)) Pyrimethamine and dapsone (Maloprim)) Chloroquine sulphate (Nivaquine))	Antimalaria tablets – note comments in next column	Take usual daily or weekly tablets for prevention, and different ones for treatment. Types to use are very area-dependent; get local (medical) advice
Magnesium trisilicate tablets	Indigestion treatment	Can be essential if local banqueting is undertaken
Paracetamol tablets	General pain relief	
Oil of cloves	Pain reliever for teeth	
Bonjela gel	Antiseptic and pain-relieving gel for mouth	
Ear drops (Auralgicin)	Treatment of ear infections	
Throat pastilles (eg Bradosol lozenges)	Alleviation of sore throat	Sore throats are common afflic- tions of soil surveyors be- cause of dust etc. Use anti- biotics for severe tonsilitis
Eye-bathing solution (Optrex) and plastic eye bath	Eye bathing	Again, surveyors suffer in dusty areas
Eye drops (Ocusol)	Antibiotic eye drops	
Fungicidal cream (Mycil, Mycota)	Athlete's foot treatment	
Antihistamine cream (Anthisan)	Relief of insect bites/rashes	
Antihistamine tablets (Triludan)	Treatment of severe insect bites/rash reactions	Triludan does not cause drowsiness as other antihistamines can do
Snake bite kit with polyvalent antivenin	Treatment of serious snake bites	Freeze-dried antivenins travel best; check applicability to types of snake. Test field- workers for sensitivity to antivenin before survey starts

Essential extras: Cotton wool; sharp, pointed scissors; thermometer; toothed forceps; safety pins;
insect repellent (cream or liquid, and spray/combustible coil); needles;
water-sterilising tablets; dental floss; mosquito net

Vaccinations: Typhoid and paratyphoid (2 doses 4 to 6 weeks apart; booster every 3 years);
cholera 1/ (2 doses; booster every 6 months); tetanus (3 doses - second dose 6 to
12 weeks after first, followed by third after 12 months; booster every 5 years);
gamma globulin - to combat hepatitis Type A (1 dose; booster after 6 months if
necessary); polio (3 doses at 4 to 8 week intervals; booster recommended after
5 years); yellow fever 1/ (1 dose; booster every 10 years); BCG - against
tuberculosis. For some projects, rabies (2 doses, 3 to 4 weeks apart; booster every
year) and typhus vaccinations may be needed

Note: 1/   International vaccination certificate may be required.

    More detailed checklists are given in Werner (1980).

Annex A: Planning and logistics

1. Documents and official papers

   Address lists (private and professional)
   Airline ticket(s)
   Blank travel/expense claim forms
   Business address cards
   Company brochure(s)
   Diary
   Letters of introduction
   Note of travel insurance policies – numbers of
     policies and company telephones/addresses
   Passport – including visas, entry/residence/work
     permits
   Spare passport photos (for visas, travel passes etc)
   UK and international driving licences
   UK receipts for items to be reimported
   Vaccination certificate(s) – cholera, yellow fever,
     other

2. Finance

   Cheque book and cheque card
   Credit/charge card(s) and telephone/telex numbers
     in case of loss/theft
   Local cash (check if there is local limit for entry
     and exit)
   Travellers cheques
   UK cash (for return)

3. Clothing

   Footwear
     – for work (desert/jungle boots; plimsolls)
     – for office (one smart pair for meetings)
     – flipflops/sandals
     – spare laces
   Formal clothes (only if formal meetings expected)
     – suit
     – tie
   Shirts (cotton-based preferable and drip-dry; at
     least some long-sleeve to protect against sun)
     – for field-work (minimum 3)
     – for office
     – casual
   Socks, handkerchiefs and underwear (cotton-based
     preferable; take several pairs)
   Trousers (drip-dry/water washable; plus belts, if
     worn)
     – for field-work (minimum 2)
     – for office
     – casual
     – shorts (for casual and sports use)
   Miscellaneous
     – sun and/or waterproof hat
     – plastic mac, if needed
     – sweater, if needed
     – swimming costume

4. Toiletries etc

   Antiperspirant
   Cotton (black and white) and needles
   Cream detergent (in tube – for washing clothes)
   Flannel
   Hairbrush and comb
   Scissors
   Shampoo
   Shaving cream
   Shoe polish and brushes
   Small mirror with something to hang by
   Soap and soap dish
   Sponge bag
   Razor and blades (or battery shaver and batteries)
   Talcum powder
   Toothbrush and toothpaste
   Towels
   Wire coat hangers

5. Stationery

   A4 notepads/spiral backs
   Airmail paper, envelopes and stamps
   Duplicate book
   Field notebook(s)
   Pencil sharpener
   Pencils, pens, ball-points, felt-tips

6. Photographic articles

   Cable release
   Camera(s) – Polaroid can be very useful
   Carrying case
   Film (black and white/colour; prints/slides)
   Flash attachment (plus bulbs, if used)
   Telephoto/wide angle lenses
   Tripod plus ball and socket
   UV filter

7. Extras

   Airline eyeshade and slippers (for use in plane)
   Alarm clock
   Books and magazines
   Bottle opener
   Briefcase
   Calculator (and batteries)
   Clothes pegs
   Cool bag or cool box, if needed
   Dark glasses
   Folding umbrella
   Invisible adhesive tape
   Matches
   Padlocks
   Pocket phrasebook and/or guidebook
   Postcards, photos to give as souvenirs
   Selection of polythene bags
   Self-indicating silica gel (for long-term
     visits to keep camera etc fungus-free)
   Small SW radio and/or cassette player
   Spare glasses (if worn)
   Strong string
   Thermos and/or drinking water bottle
   Torch (and batteries)
   Tube Araldite glue

8. Field equipment

   Auger/pit description sheets
   Clipboard or small briefcase
   Collapsible auger
   Compass
   Dictaphone
   Drinking water bottle/flask
   Dropper bottle for HCl
   Field bag
   Geological hammer
   Machete, if needed
   Multipurpose pocket knife
   Munsell colour chart
   pH papers/indicator solution
   Pocket stereoscope
   Sample bags
   Stapler
   Suunto level
   Trowel
   Wash bottle
   Whistles
   X10 hand lens
   2 m tape measure
   8 cm hole puncher
   15 cm transparent plastic ruler

9. Medical items

   See Table A.14.

Note: Hand luggage should contain all important documents, cash and valuables, plus airline eyeshade and
         slippers, torch, change of clothing and sponge bag plus contents.

List of selected suppliers of soil survey and related equipment or services [1]   Table A.16

1. Camping gear

    Safariquip
    13a Waterloo Park
    Upper Brook Street
    Stockport, SK1 3BP
    Phone: 061 429 8700, Fax: 061 429 8837

    The North Face (Scotland) Ltd
    P O Box 16
    Industrial Estate
    Port Glasgow
    Scotland, PA14 5XL
    Phone: 0475 41344, Fax: 0475 44572

2. Cartographic and drawing office supplies

  a) Ammonia printers, drawing boards, map cabinet, strip lights and light tables

    UDO Tottenham Court Road
    32 Gresse Street
    London W1P 1PN
    Phone: 071 631 0222, Fax: 071 436 0150

  b) General drawing materials

    London Graphic Centre
    107-115 Long Acre
    London WC2E 9NT
    Phone: 071 240 0095, Fax: 071 831 1544

  c) Plan variograph

    Survey and General Instrument Co Ltd
    Fircroft Way
    Edenbridge
    Kent TN8 6HA
    Phone: 0732 864111, Fax: 0732 865544

  d) Scribing materials and instruments

    Pacer Graphics Ltd
    Berechurch Road
    Colchester
    Essex CO2 7QH
    Phone: 0206 760760, Fax: 0206 762626

3. Laboratory equipment

  a) General

    Gallenkamp
    Belton Road West
    Loughborough
    Leicestershire LE11 0TR
    Phone: 0509 237371, Telex: 34391 GALKMP G

  b) Chemicals

    Supplier of BDH Chemicals:
    John Bell and Crydon
    54 Wigmore Street
    London W1H 04U
    Phone: 071 935 5555, Fax: 071 935 9605

    Supplier of May and Baker Chemicals:
    Downswood Products
    Station Approach
    Park Lane
    Knebworth
    Hertfordshire SG3 6PJ
    Phone: 0438 813456, Fax: 0438 814093

  c) Water purifying (deioniser)

    Elga Products Ltd
    Lane End
    Buckinghamshire HP14 3JH
    Phone: 0494 881393, Fax: 0494 881007

  d) EC Meter (MC3) and case

    Kent Industrial Measurements Ltd
    Howard Road
    Eton Socon
    St Neots
    Huntingdon
    Cambridgeshire PE19 3EU
    Phone: 0480 75321, Fax: 0480 217948

  e) pH meters, replacement electrodes, buffer and potassium chloride solutions (Type 10)

    Analytical Measurements Ltd
    Spring Corner
    Feltham
    Middlesex TW13 4PB
    Phone: 081 890 5079, Fax: 0784 257938

    ELE International Ltd
    Eastman Way
    Hemel Hempstead
    Hertfordshire HP2 7HB
    Phone: 0442 218355, Fax: 0442 52474

4. Laboratory Analyses

    Tropical Soils Analysis Unit
    Overseas Development Natural Resources
     Institute
    Central Avenue
    Chatham Maritime
    Chatham, Kent ME4 4TB
    Phone: 0634 883433, Fax: 0634 880066/880077

5. Optical equipment

  a) General (also repairs)

    Summit Survey Equipment Supplies
    19 Harmill Industrial Estate
    Grovebury Road
    Leighton Buzzard
    Bedfordshire LU7 8TE
    Phone: 0525 378930, Fax: 0525 851641

  b) Compasses, Abney and Suunto Levels, Altimeters, Chronometers

    Hall & Watts Ltd
    Unit 9 Fosters Business Park
    Old School Road
    Hook
    Hampshire RG27 9NY
    Phone: 0256 763402

  c) Pocket stereoscopes

    Casella London Ltd
    Regent House
    Wolseley Road
    Kempston
    Bedfordshire MK42 7JY
    Phone: 0234 841468, Fax 0234 841490

  d) Mirror Field Stereoscopes (TSPI), Desk Stereoscopes (ST4)

    Wild-Leitz (UK) Ltd
    Davy Avenue
    Knowlhill
    Milton Keynes
    MK5 8LB
    Phone: 0908 666663

---

[1] See note on p 212.   cont

e) Desk stereoscopes, zoom stereosketch

Cartographic Engineering Ltd
Landford Manor
Landford
Salisbury
Wiltshire SP5 2EW
Phone: 0794 390392

f) Compensating polar planimeters
(with zero setting and magnifier)

ELE International Ltd (see 3e above)

Hall & Watts Ltd (see 5b above)

6. Soil field equipment

a) Dutch augers and soil physics equipment

Eijkelkamp B V
Nijverheidsstraat 14
6987 EM Giesbeek
Netherlands
Phone: 010 31 8336 1941,
Telex: 35416 EYKEL NL

UK agents for Eijkelkamp:
Van Walt Ltd
Prestwick Lane
Grayswood
Haslemere
Surrey GU27 2DU
Phone: 0428 61660, Fax: 0428 56808

b) Jarrett augers

Leonard Farnell and Co
Station Road
North Mymms
Hatfield
Hertfordshire AL9 7SR
Phone: 0707 264488, Fax: 0707 268347

c) Munsell colour charts

Tintometer Ltd
Waterloo Road
Salisbury
Wiltshire SP1 2JY
Phone: 0722 27242-4, Fax: 0722 412322

d) Polythene bags

Swains Packaging Ltd
Brook Road
Buckhurst Hill
Essex IG9 5TU
Phone: 081 504 9151, Fax: 081 506 1892

e) Estwing geological hammers

Geo Supplies Ltd
16 Station Road
Chapeltown
Sheffield S30 4XH
Phone: 0742 455746, Fax: 0742 403405

f) Allplas 20mm diameter plastic balls
(for in situ bulk density tests)

Capricorn Chemicals Ltd
1 Sugar House Lane
Stratford
London E15 2QN
Phone: 081 519 4933, Fax: 081 519 2818

g) Neutron probe

ELE International Ltd (see 3c above)

7. Aerial photography and remote sensing

a) Air photography and photogrammetry, world-
wide capability

BKS Surveys Ltd
Ballycairn Road
Coleraine
Co. Londonderry, Ulster
Phone: 0265 52311, Telex: 74385

Clyde Surveys Ltd
Clyde House
Reform Road
Maidenhead
Berkshire SL6 8BU
Phone: 0628 21371, Fax: 0628 782234

b) Satellite imagery

Nigel Press Associates Ltd
1 Fircroft Way
Edenbridge
Kent TN8 6HS
Phone: 0732 865023, Fax: 0732 866521

National Remote Sensing Centre
Space Department R190 Building
Royal Aircraft Establishment
Farnborough
Hampshire GU14 6TD
Phone: 0252 541464, Fax: 0252 375016

Land Resource Planning Department
National College of Agricultural
Engineering
Silsoe
Bedfordshire MK45 4DT
Phone: 0525 60428

Eros Data Centre
Sioux Falls
South Dakota 57198
USA
Phone: 010 1 605 5946511

8. Soil and land resource reference and map
libraries

International Soil Reference and
Information Centre
9 Duivendaal or PO Box 353
Wageningen
Netherlands
Phone: 010 31 8370 19063
(includes the former International Soil
Museum)

Land Resources Department
Overseas Development Natural Resources
Institute
Central Avenue
Chatham Maritime
Chatham
Kent ME4 4TB
Phone: 0634 883091, Fax: 0634 880066/880077

Royal Tropical Institute
Mauritzkade 63
Amsterdam
Netherlands
Phone: 010 31 2092 4949

Note: This is a list of suppliers known to stock the items or services listed in 1990. Since addresses, telephone, telex and facsimile numbers (and company names) can change, an up-to-date directory should be used to check the information. An annual register of British firms, compiled both by company and by goods and services, is produced by the confederation of British Industry and Kompass (CBI/Kompass 1990 et seq).

Annex B

Soil Physics

Contents

Annex B: Soil physics

`

List of Figures

Soil Physics

B.1   <u>Field measurement of infiltration using double-cylinder infiltrometer</u> 1/

B.1.1   <u>Introduction</u>

Infiltration rate (see Section 6.2) refers to the vertical entry of water into a soil surface. It should not be confused with hydraulic conductivity, or permeability, which is a measure of the ability of a soil to transmit water in all directions, horizontally as well as vertically. Two figures are of interest – the initial intake rate (say in the first hour) and the equilibrium or basic infiltration rate which is the constant rate that develops after several hours.

Infiltration rates can be measured by observing the fall of water within two concentric cylinders driven vertically into the soil surface layer. The use of a double cylinder (colloquially double ring), with measurement confined to the inner ring, minimises error due to non-vertical flow at the edge of the cylinder. If two metal cylinders are not available the outer one can be replaced by an earth bund.

Water of the same quality as will be applied in irrigation should preferably be used, or misleading results may ensue. Quirk (1957), for instance, has demonstrated substantial increases in infiltration rates by increasing the electrolyte concentration of the applied water. The soil sites to be tested should be thoroughly pre-wetted: soaking for a few hours using an earth bund at each site is sufficient for most soils, but dry, fine-textured soils may need several days' soaking. The site should be covered with polythene after soaking to prevent surface drying. This pre-wetting allows the rings to be inserted more easily than in dry soils, and helps to reduce variations in rates caused by differences in initial moisture status (see Turner and Sumner, 1978).

The test should normally be run until the steady state is reached (usually from about 3 to 5 h). The amount of water required depends on soil conditions. One 200 ℓ drum may suffice on impermeable clays whereas sandy soils may take four or five drums. The test does not work well on cracked clays as the water disappears too fast and the results are too variable to be reliable. Where necessary, sample(s) should be taken at depths of 25, 75 and 150 cm to determine the moisture content of the soil before the test commences. The initial water content during soil survey operations is likely to be variable, but the test cannot be done on saturated soil. If there is temporary waterlogging of the site, wait until downwards draining recommences. Evaporation rates are usually too low to be significant, but if the infiltration rate is very low and the weather is hot and dry it is necessary to correct for evaporation.

At least three replicates should be run at each site, preferably close to a sampled profile pit so that complete data on the soil are obtained. The test can be made on bare or vegetated soil, but note that the rate under grass is usually substantially higher than on cultivated land. The vegetation must be clipped down so that it does not break the water surface, and loose material which would float should be cleared off.

B.1.2   <u>Equipment (for three replicate tests made concurrently)</u>

Three steel cylinder sets, 40 cm high (old oil drums provide ideal material). Seam is ground smooth on inside. One end should be bevelled from outside to inside. For ease of transport replicate cylinders should be of slightly different diameters to allow concentric stacking when not in use: the inner ones should be about 30 cm and the outer ones about 60 cm in diameter; these dimensions allow an old inner tube to be floated in the outer ring to dissipate the force of water during refills.

One hardwood 15 x 15 cm timber (optionally having 0.6 cm steel plate bolted to one side).

1/   Adapted to standard BAI practice from FAO (1979a).

Annex B: Soil physics

Means of storing and transporting water (water trailer or drums, plus six buckets).

7 kg sledge-hammer, or heavy weight with handle.

Three still wells (20 to 25 cm lengths of 10 cm ID perforated plastic drainpipe) used to reduce movement of water surface caused by refills.

Three floats and scales; three float guides.

Three old inner tubes.

Piece of sacking or plastic sheet.

Auger and shovel.

Knife or shears for cutting vegetation.

Three stop-watches.

Standard observation forms (see Figure B.1) and clipboard (preferably one per replicate).

Graph paper (1 mm squared, Chartwell Ref C14G).

B.1.3   Procedure

a)   Record information on the soil surface at the time of the test, eg the presence of any surface litter, the condition of the soil surface (cultivated, crusted, cracked, etc) and any other conditions which might affect the rate of water intake. Select sites about 10 m apart in areas representative of the soil to be tested, and sample the soil just outside the rings to determine the initial moisture content.

b)   Drive cylinders into the pre-wetted soil to approximately 15 cm depth, placing the driving plate over the cylinder with the heavy timber on top. Rotate the timber every few blows and check that penetration is uniform and vertical (alternatively, in very hard soils, cut soil round rim with knife to insert cylinder, and then seal the edges with bentonite). Firm the soil next to the inside and outside of the cylinders. Place sacking or plastic sheet and inner tubes (or similar) over the soil to dissipate the force of the water and reduce turbidity. Get everything ready for all three replicates before starting the test.

c)   Fill both cylinders to a depth of about 15 to 20 cm at the first site, and record the time and the height of the water in the inner cylinder, using the floating scale in the still well.

d)   Repeat for each replicate. Remeasure the levels after 1, 5, 10, 20, 30, 45, 60, 90, 120 min and each hour for the remainder of the test (more often if the infiltration rate is rapid). Also record levels immediately before and immediately after each refill; if no floating scale is available, levels can be measured from the water surface to the top of the cylinder before and after topping up. The water in the outer cylinder should be kept at approximately the same level as in the inner one to avoid any interaction affecting the inner level. The fall in level should be kept to less than half the maximum depth of water.

e)   Record readings on a standard form (see Figure B.1) and calculate the rates. The curves of infiltration vs time should be plotted on graph paper and the cumulative amount of water infiltrated also plotted as a check (see Figure 6.1). There is ample time to do this in the field between measurements and it should be done at once so that errors can be rectified. If one cylinder gives a widely different rate from the others (perhaps because of a hidden insect burrow) reject the result when making the averages.

f)   After the test period remove the cylinders and dig a cross-section through the centre of the 30 cm cylinder site in order to describe the soil morphology and observe and record the outline of the wetted soil. In some conditions auger borings can be used to delineate the wetted area; in sandy or moist soil the wetting pattern may be too deep or indeterminate to outline. Where field capacity and/or bulk density measurements are to be made, the digging-out process should be left for 48 h until these other measurements are complete.

g)   From the graph the values of the maximum initial infiltration rate and the basic rate can be obtained. Measurements should be made at several sites on the same soil series to obtain a reliable average. The infiltration rates for various soils can then be compared, and the diagram of the wetting pattern is helpful in explaining differences between them (for example, claypan soils may have rapid initial intake which soon decreases to very slow, whereas loamy friable soils may have a lower initial intake rate but a higher final rate).

For corresponding graphs, see Figures 6.1 and B.2

BOOKER AGRICULTURE INTERNATIONAL LTD **BAI**		Project D SS		Site No. 64	
INFILTRATION RATE MEASUREMENT					

Prewetting 44 gal application to 12m² area 72 hr previously	Source/EC of water Well 250m SE of site	Date 27 Jul 83	Author PLS	Sheet No. I	Replicate No. I

Height of zero above soil surface (cm) O		Depth of insertion of ring (cm) 15		Land class 2s	

Surface features Slight capping 0-2cm (see profile field sheet) No vegetation.			Soil DHO I		

Local time	Interval (mins)	Cumulative time (mins)	Depth of water in infiltrometer (cm)	Intake (cm)	Cumulative intake (cm)	Infiltration rate (cm/hr)	
						Immediate	Mean
8.00	0	0	12.0	–	–	–	–
	2	2	9.1 – 10.9	2.9	2.9	87.0	87.0
	3	5	10.1 – 12.1	0.8	3.7	16.0	44.4
	2	7	11.7	0.4	4.1	12.0	35.1
	1	8	11.6	0.1	4.2	6.0	31.5
	4	12	11.1 – 13.5	0.5	4.7	7.5	23.5
	4	16	13.0	0.5	5.2	7.5	19.5
	8	24	12.2	0.8	6.0	6.0	15.0
	8	32	11.5	0.7	6.7	5.25	12.6
	13	45	10.2 – 14.8	1.3	8.0	5.2	10.7
9.00	15	60	13.5	1.3	9.3	5.2	9.3
	15	75	12.2 – 14.5	1.3	10.6	5.2	8.5
	15	90	13.2	1.3	11.9	5.2	7.9
	15	105	12.1	1.1	13.0	4.4	7.4
10.00	15	120	11.0 – 15.1	1.1	14.1	4.4	7.0
	15	135	13.9	1.2	15.3	4.8	6.8
	15	150	12.7	1.2	16.5	4.8	6.6
	15	165	11.6 – 13.3	1.1	17.6	4.4	6.4
11.00	15	180	12.2 – 14.8	1.1	18.7	4.4	6.3
	15	195	13.6	1.2	19.9	4.8	6.1
	15	210	12.5	1.1	21.0	4.4	6.0
	15	225	11.5	1.0	22.0	4.0	5.9
12.00	15	240	10.5	1.0	23.0	4.0	5.8

Diagram/Comments:

Moisture penetration to 80 cm

Replicate locations :   1 ×   × 2   N

Pit 64

× 3

Annex B: Soil physics

h) <u>Notes:</u> 1. During filling, care must be taken to avoid puddling or disturbance of the soil surface by using the sacking or plastic sheet. Care should also be taken to prevent the rings becoming undermined and thus allowing seepage to occur.

2. During the experiment, the water in the outer guard ring and inner ring should be kept at the same level by the addition of water to the outer one after topping up the inner.

3. On dry clay soil it is particularly difficult to hammer the rings into the ground. The hard soil surface tends to shatter and it becomes exceedingly difficult to obtain a good seal. If necessary after pre-wetting, the rim should be sealed with bentonite to make absolutely sure there is no leakage.

B.1.4 Alternative forms of treatment of infiltration data

In some circumstances, soil surveyors may be required to produce or interpret more detailed analyses of infiltration results. Amongst the more common concepts used are the following:

a) The cumulative infiltration - (F) - This is the characteristic which is measured in the field. The measurements are plotted on log-log paper (F against time). Usually a straight line is obtained (see Figure B.2) and, therefore, the accumulated intake can be represented by the equation $F = at^n$, where t is in minutes; n varies from 0.5 to 1.0, and a, related to the order of magnitude of the water intake, is given by a = F after 1 min. The n value for sandy soils (high F value) is usually 0.8 or more; n values < 0.5 indicate the occurrence of soil cracks. In moist soils a values are lower (and n values higher) than the corresponding values in dry soils.

It is preferable to express the time in hours (T) because, in this way, it may be easier to visualise the order-of-magnitude of infiltration. The equation of F expressed in hours can be obtained by the reading of the F value at t = 60 min or by computation:

$$F = 0.24t^{0.70} = 4.2T^{0.70}$$

In homogeneous soils, deviations from a straight line can result from inadvertent delays in the first readings and from the presence of small cracks and holes (see Section 6.1.6). In heterogeneous soils, particularly in soils having a strongly developed crack system and large biogenic holes, very high initial n values (0.9 or more) may occur, which diminish after some time.

b) The instantaneous intake rate - (IR) - This is the volume of water infiltrating through a horizontal unit area of soil surface at any instant. It shows, in general, a rapid decline at the beginning followed by a more stable, very slow decline, after some 3 to 4 h of infiltration (see Figure 6.1).

c) The average infiltration rate - $(\overline{IR})$ - This is the cumulative infiltration F divided by the time since infiltration started.

d) The Philip equation - A commonly used equation for extrapolating infiltration results to the steady state is the Philip equation (Philip, 1954), which refers to the isothermal movement of water into homogeneous soils in one dimension. The equation can be stated as follows:

$$F = at^{0.5} + bt$$

where   F = cumulative depth (cm), t = time (s)

and a and b are constants (soil sorptivity (or storage capacity) and ability to transmit water, respectively) depending on soil properties and initial moisture content.

B.1.5 Worked example of the Philip equation

F = 10.6 cm, t = 75 x 60 = 4 500 s, $t^{0.5}$ = 67.08   )
) from field data
F = 16.5 cm, t = 150 x 60 = 9 000 s, $t^{0.5}$ = 94.87)

therefore   10.6 = 67.08 a + 4 500 b)   (1)
) from the Philip equation (see B.1.4d)
                           16.5 = 94.87 a + 9 000 b)   (2)

multiplying equation (1) x 2 gives:

21.2 = 134.16 a + 9 000 b   (3)

subtracting (2) from (3) gives:

4.7 = 39.29 a

216

Infiltration rate: alternative presentation   Figure B.2

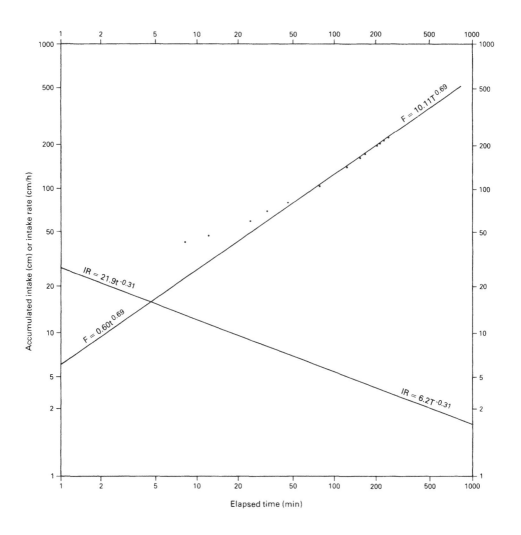

Graph paper required: 3 cycle logarithmic two way scale

A4 size - Chartwell reference no. C5933

For corresponding readings see Figure B.1

217

therefore $\qquad a = 0.12$ $\qquad\qquad\qquad\qquad\qquad\qquad\qquad\qquad$ (4)

substituting (4) in (1) gives:

$$10.6 = (0.12 \times 67.08) + 4\ 500\ b$$

therefore $\qquad b = 0.0006$

and the whole equation becomes

$$F = 0.12\ t^{0.5} + 0.0006\ t \qquad\qquad\qquad\qquad\qquad (5)$$

When $t = 180$ min $= 10\ 800$ sec, $t^{0.5} = 103.9$:

therefore, from (5) $\qquad F = (0.12 \times 103.9) + (0.0006 \times 10\ 800)$

$\qquad\qquad\qquad\qquad\qquad = 18.95$ cm (compared with 18.7 cm as measured)

When $t = 6$ h $= 21\ 600$ sec, $t^{0.5} = 146.97$:

therefore from (5) $\qquad F = (0.12 \times 146.97) + (0.0006 \times 21\ 600)$

$\qquad\qquad\qquad\qquad\qquad = 30.6$ cm

therefore $\qquad \overline{IR}$ (6 h) $= 5.1$ cm h^{-1}

The equation is, however, only approximate, and direct measurement is preferred wherever possible.

B.2   Field measurement of hydraulic conductivity: auger-hole and inverse auger-hole methods [1]

B.2.1   Introduction

Hydraulic conductivity (see Section 6.3) refers to the subsurface movement of water within a soil, both vertically and horizontally.   Measurements are mostly used in connection with the design of drainage systems in wet land and in canal seepage investigations.   The most common measurements made involve the combined horizontal and vertical water movements under saturated conditions, although for special investigations unsaturated flows and/or the vertical and horizontal components may be considered separately.   Similar remarks to those for infiltration rate (Subsection B.1.1) apply to water quality and pre-wetting when making hydraulic conductivity tests by the inverse auger-hole method, but the quantity of water is very much lower;   often only half a 200 ℓ drum will be sufficient.

B.2.2   The auger-hole method

The auger-hole method is based on a hole bored into the soil to a certain depth below the water-table.   When equilibrium is reached with the surrounding groundwater, a volume of water is removed from the hole and the surrounding groundwater allowed to seep in to replace it.   The rate at which the water rises in the hole is measured and then converted by a suitable formula to the hydraulic conductivity (K) for the soil.   The simplest approach is provided in van Beers (1976) and in FAO (1979a, Appendix B.1) which set out in simple and convenient detail the method and calculations to be used for practical purposes;   the experimental set-up is shown in Figure B.3, and photographs of the installation are given in van Beers (1976, p 10).   For tests made by BAI, note that the pointer is set at ground level, not on a stand, which simplifies the calculation slightly.

The auger-hole method gives the average permeability of the soil layers extending from the water-table to a small distance (a few decimetres) below the bottom of the hole.   If there is an impermeable layer at the bottom of the hole, the value of K is governed by the soil layers above this impermeable layer.   The radius of the column of soil of which the permeability is measured is about 30 to 50 cm.

The use of this method is limited to areas where a high GWT occurs (at least during part of the year) and to soil where a boring of known shape can be maintained throughout the test.   Hence in certain sandy soils it is necessary to use a perforated tube as support for the sides (see notes below).   The method is unsuitable for use in very stony or coarse soils because of the difficulty of augering a uniform hole in such materials.   It is also unusable when artesian conditions occur, or in soils containing small sand lenses within less permeable material.   In general, however, the method is suitable for most agricultural soils.

Normally three replicates are run consecutively at each investigation site, since the time involved for each replicate is of the order of half an hour.

B.2.3   Equipment for auger-hole method

8 cm diameter auger

---

[1]   Much of this section is reproduced from ILRI (1974) by permission.

The auger-hole method for hydraulic conductivity measurement   Figure B.3

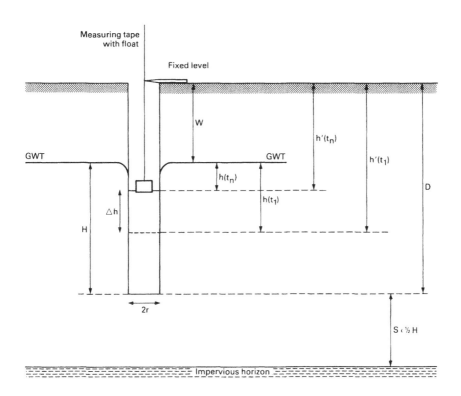

$W =$	depth of GWT below fixed level
$H = D - W =$	depth of the auger hole below water-table
$h'(t_1), h'(t_n) =$	depth of GWT in the hole below fixed level at the time of the first reading (t_1) and after some readings (at time t_n). Usually about five readings are taken
$\triangle h = h'(t_1) - h'(t_n) = h(t_1) - h(t_n) =$	the rise of water-level in the hole during the time of measurements
$\bar{h} = h(t_1) - 0.5 \triangle h =$	average head during time of measurements
$S =$	depth of impervious horizon below the bottom of the hole
$r =$	radius of hole

Source: Adapted from van Beers (1976).

Annex B: Soil physics

Tape measure and holder/pointer

Float (with attachment for tape measure)

Stop-watch

7.5 cm diameter bailer (can be made from a 1 m length of 7.5 cm (3 in.) diameter drainpipe, fitted with a flap valve at the base and rope handle at the other end)

Screen or perforated plastic tube

> The tube material should be about 2 mm thick, with perforations of about 0.5 mm diameter spaced about 4 mm apart, in order to cover about 20% of the tube surface. The screen or tube should be cylindrical, 10 cm in outer diameter for a hole with a diameter of 10 cm, and 1 m in length. For drawing the tube out of the borehole a kind of drawhook is useful. A reinforcement with a riveted ring is fitted to each end of commercial screens, and cams for holding a drawhook can be secured to the top of the screen tube.

Side scratcher to clean sides of hole and remove smearing effects

> Can be made from a tube approximately 7 cm diameter by 9 or 10 cm long, with nails protruding 0.3 or 0.4 cm. Cylinder should be fitted with coupling to allow connection to auger handle.

Special cutter (to give flat bottom to hole)

Bucket and water supply

Standard observation forms and clipboard

Graph paper (semi-log)

For inverse auger-hole method:

> Small piece of sacking (or coarse sand fill) to prevent erosion of bottom of hole.

B.2.4   Procedure for auger-hole method

a) Select sites (and about 3 to 5 stations at each) as for infiltration tests (see Subsection B.1.1), and record site information (as Subsection B.1.2a).

b) Auger and describe a hole of about 8 cm diameter (or more) to a depth below the water-table, and finish off with a special cutting tool to obtain a flat bottom to the hole; take care to eliminate smearing of the sides of the hole (Smitham, 1970). The hydraulic conductivity is calculated as below using the graphs provided in van Beers (1976) or Maasland and Haskew (1957). If the hole penetrates two or more horizons below the water-table, the hydraulic conductivity of each horizon can be determined approximately using the method described by van Beers (1976). In such heterogeneous profiles, however, the piezometer method may be preferable (see eg Johnson et al, 1952; Luthin and Kirkham, 1949; review by Luthin, 1957; ASAE, 1962).

c) Place the tape holder near the hole so that the steel tape, with float attached, hangs exactly vertically in the hole.

d) Lower the float to the groundwater surface and record this level.

e) Lift the float carefully from the hole and turn the tape holder sideways so that it is clear of the edge of the hole.

f) Bail the water from the hole until the level is reduced by about 20 to 40 cm (this may take one or two bailings), rapidly return the pointer to its former position, and lower the float to the surface of the groundwater. The readings should then be started as soon as possible.

g) Take about five readings at regular intervals of about 5 to 30 sec, depending on the hydraulic conductivity; time between readings should correspond to a rise of about 1 cm in water-level. The steel tape or float may tend to stick to the wall of the cavity, and so should be tapped regularly. All readings, including groundwater level and depth of the hole, are taken at the contact of the tape on the pointer.

h) Measure the depth of the hole.

i) The hydraulic conductivity is calculated as below, or using the graphs provided in van Beers (1976) or Maasland and Haskew (1957).

k) Note: In unstable sandy soils it is necessary to use the perforated tube as a support for the sides. The auger is used up to the point where the hole becomes unstable, then the perforated tube is lowered into the hole. By moving the bailer up and down a mixture of sand and water will enter the bailer and the tube can be pushed downwards to the desired depth.

A good rule of thumb is that the rate of rise in mm s^{-1} in an 8 cm diameter hole to a depth of 70 cm below the water-table approximately equals the K value of the soil in m day^{-1}.

Care should be taken to complete the measurements before 25% of the volume of water removed from the hole has been replaced by inflowing groundwater. After that, a considerable funnel-shaped water-table develops around the top of the hole. This increases resistance to the flow around and into the hole. The effect is not accounted for in the formulae or flow charts developed for the auger-hole method and, consequently, it should be checked that $\Delta h < 0.25h(t_1)$ (see Figure B.3).

It often happens that h is relatively large for the first reading, due to water dripping along the walls of the hole directly after bailing. If this occurs, the first measurement should be discarded.

B.2.5   Auger-hole method: calculation for the single-layer situation

Ernst (1950) found that the relation between the hydraulic conductivity of the soil and the flow of water into the auger hole depends on the boundary conditions; this relation, derived numerically, is given as:

$$K = C \frac{\Delta h}{\Delta t}$$

where   K = hydraulic conductivity (m day^{-1})

C = geometry factor = $f (\bar{h}, H, r, S)$ (see Figures B.3, B.4 and B.5)

$\frac{\Delta h}{\Delta t}$ = rate of rise of water-level in the auger hole (cm s^{-1})

In Figure B.5, C is given as a function of $\bar{h}/r$ and H/r when $S > 0.5 H$; in Figure B.4 a similar relation applies when $S = 0$. The use of these figures to calculate K is illustrated in Figure B.6.

After the readings have been taken, the reliability of the measurements should be checked. The Δh value of each measurement is therefore computed to see whether the consecutive readings are reasonably consistent. If the value of Δh decreases gradually, the readings may be averaged up to $\Delta h = 0.25h(t_1)$ or, as in Figure B.6, up to $\Delta h = 7$ to 8 cm. Both conditions are satisfied in the example and therefore K can be computed.

B.2.6   Auger-hole method: calculation for the two-layer situation

If a soil profile consists of two layers of different hydraulic conductivity, K_1 in the upper layer and K_2 in the lower layer, these values for the separate layers can be determined if the water-table is well within the upper layer (see Figure B.7). Two successive measurements are made in a single auger hole, the second measurement being taken after the original hole has been deepened. The hole is bored to at least 40 cm below the water-table, but should not extend further than 20 cm above the lower layer. The deepened borehole must reach at least 50 cm into the lower layer. The measurement in the shallow hole gives K_1, the hydraulic conductivity of the upper layer, in the same way as for a homogeneous profile.

$$K_1 = C_1 \, (\Delta h/\Delta t)_1$$

where $C_1 = f (\bar{h}_1, H_1, r, S_1 > 0.5H_1)$

The rate at which the water in the deep hole rises is thought to be the result of two components (see Figure B.7):

a)   inflow from the upper layer with hydraulic conductivity K_1 only, the lower layer being considered impermeable;

b)   inflow from the lower layer, which is considered to consist of inflow from the whole profile with hydraulic conductivity K_2 minus inflow from the upper layer, also with hydraulic conductivity K_2, the lower layer being considered impermeable.

hence $(\Delta h/\Delta t)_2 = K_1/C_0 + K_2/C_2 - K_2/C_0$

where $C_0 = f (\bar{h}_2, H_0; r, S_1 = 0)$ (see Figure B.4)

$C_2 = f (\bar{h}_2, H_2, r, S_2 > 0.5H_2)$ (see Figure B.5)

Rearranging gives:

$$K_2 = \frac{C_0 \, (\Delta h/\Delta t)_2 - K_1}{C_0/C_2 - 1}$$

Nomograph for determination of C in auger hole method for S = 0   Figure B.4

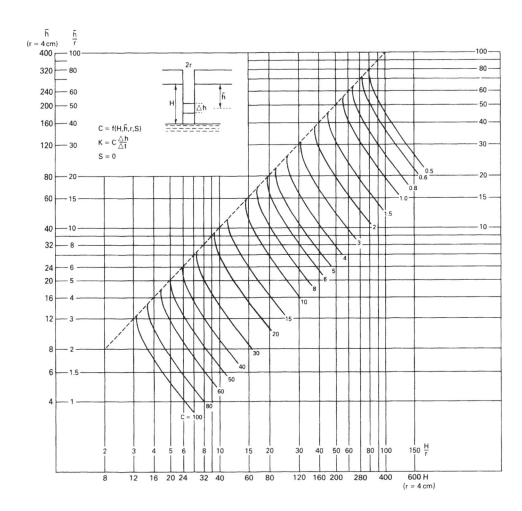

Source: Adapted from Ernst (1950).

Nomograph for determination of C in auger-hole method for S > 0.5H   Figure B.5

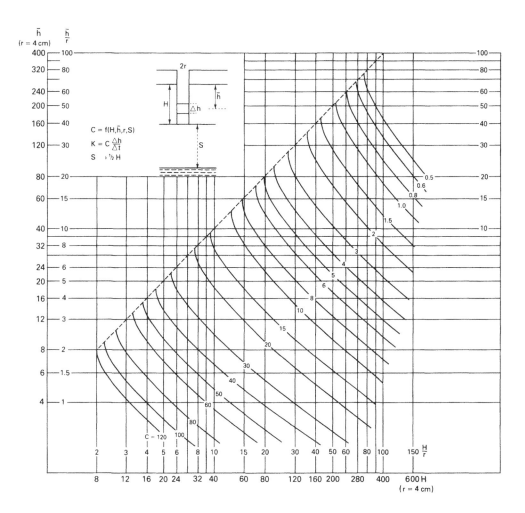

Source: Adapted from Ernst (1950).

223

Auger-hole method: worked example of field data sheet

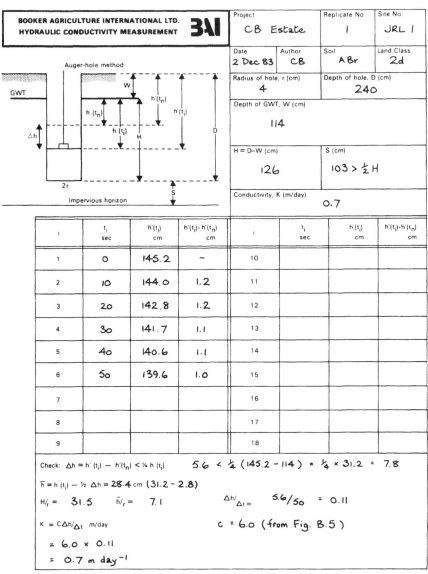

BOOKER AGRICULTURE INTERNATIONAL LTD. HYDRAULIC CONDUCTIVITY MEASUREMENT 3AI				
Project CB Estate		**Replicate No** 1		**Site No.** JRL 1
Date 2 Dec 83	**Author** CB	**Soil** ABr		**Land Class** 2d
Radius of hole, r (cm) 4		**Depth of hole, D (cm)** 240		
Depth of GWT, W (cm) 114				
H = D–W (cm) 126		**S (cm)** 103 > ½ H		
Conductivity, K (m/day) 0.7				

Auger-hole method

i	t_i sec	$h'(t_i)$ cm	$h'(t_i) - h'(t_n)$ cm	i	t_i sec	$h'(t_i)$ cm	$h'(t_i)-h'(t_n)$ cm
1	0	145.2	–	10			
2	10	144.0	1.2	11			
3	20	142.8	1.2	12			
4	30	141.7	1.1	13			
5	40	140.6	1.1	14			
6	50	139.6	1.0	15			
7				16			
8				17			
9				18			

Check: $\Delta h = h'(t_i) - h'(t_n) < \frac{1}{4} h(t_i)$ $5.6 < \frac{1}{4}(145.2 - 114) = \frac{1}{4} \times 31.2 = 7.8$

$\bar{h} = h(t_i) - \frac{1}{2} \Delta h = 28.4$ cm $(31.2 - 2.8)$

$H/_r = 31.5$ $\bar{h}/_r = 7.1$ $\Delta h/_{\Delta t} = 5.6/50 = 0.11$

$K = C\Delta h/_{\Delta t}$ m/day $c = 6.0$ (from Fig. B.5)

  $= 6.0 \times 0.11$

  $= 0.7$ m day^{-1}

Diagram /comments:

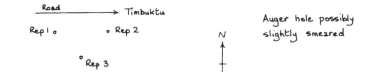

Road ———→ Timbuktu

Rep 1 o   o Rep 2

   o Rep 3

N ↑

Auger hole possibly slightly smeared

Hydraulic conductivity: the auger-hole method for two layers   Figure B.7

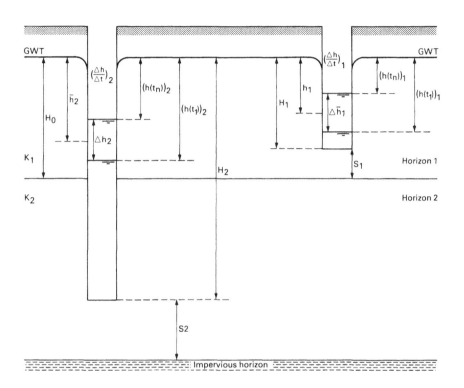

Example

$(\triangle h/\triangle t)_1 = 0.16$

$\left.\begin{array}{l} H_1 = \qquad 70.0\ cm \\ \bar{h}_1 = \qquad 30.0\ cm \\ S_1 > 0.5\ H_1 \end{array}\right\}$ $C_1 = 9.4$ (Figure B.5)

$K_1 = C_1(\triangle h/\triangle t)_1 = 9.4 \times 0.16 = 1.5\ m\ day^{-1}$

$(\triangle h/\triangle t)_2 = 0.26$

$\left.\begin{array}{l} H_2 = \qquad 150.0\ cm \\ \bar{h}_2 = \qquad 40.0\ cm \\ S_2 > 0.5\ H_2 \end{array}\right\}$ $C_2 = 3.9$ (Figure B.5) $\left.\begin{array}{l} H_0 = 100.0\ cm \\ \bar{h}_2 = 40.0\ cm \\ S_1 = 0 \end{array}\right\}$ $C_0 = 6.3$ (Figure B.4)

$K_2 = \dfrac{C_0(\triangle h/\triangle t)_2 - K_1}{(C_0/C_2)-1} = \dfrac{6.3 \times 0.26 - 1.5}{(6.3/3.9)-1} = 0.22\ m\ day^{-1}$

Source: Adapted from van Beers (1976).

225

Annex B: Soil physics

Calculation for auger-hole method using the Ernst formula

Using the auger-hole method, Ernst (1950) derived a formula for the calculations of K which avoided the need to use the nomographs. He showed that the hydraulic conductivity can be computed from a formula which, for a homogeneous soil with an impermeable layer at a depth such that $S \geqslant 0.5$ H, is:

$$K = \frac{4\ 000\ r^2\ \Delta h}{(H + 20r)\ (2 - \frac{\overline{h})}{H)}\ \overline{h}\ \Delta t}$$

In this formula, K is expressed in m day^{-1}. All other quantities are in cm or in seconds.

K = hydraulic conductivity (m day^{-1})

$\overline{h}$ = distance between groundwater level and the average level of the water in the hole for the time interval t (cm)

H = depth of hole below the GWT (cm)

r = radius of auger hole (cm)

S = depth of the impermeable layer below the bottom of the hole or the layer, which has a permeability of about 1/10 or less of the permeability of the layers above (cm)

The advantage of the formula is that it can be used on a computer but, since it is an empirically derived expression, it does not show the exact relationship that should theoretically exist between the different quantities, although the value of K will be sufficiently accurate (maximum error about 20%) if the following conditions are met:

r   > 3 and < 7 cm

H   > 20 and < 200 cm

$\overline{h}$ > 0.2H

D   > H

Δh < $0.25h(t_1)$

When the impermeable layer is at the bottom of the hole (S = 0), the following equation can be used:

$$K' = \frac{3\ 600\ r^2\ \Delta h}{(H + 10r)\ (2 - \frac{\overline{h})}{H)}\ \overline{h}\ \Delta t}$$

The measurements should be completed before $h(t_n) < 0.75h(t_1)$ or $\Delta h > 0.25h(t_1)$.

B.2.8   Hydraulic conductivity measurement above the water-table: the inverse auger-hole method

The inverse auger-hole method (Kessler and Oosterbaan, 1974; FAO, 1979a, Appendix B.1) is an auger-hole test above the water-table, and is described in French literature as the Porchet method.   It consists of augering a hole to a given depth, filling it with water and measuring the rate of fall of the water-level.   The experimental arrangement is very similar to that of the auger-hole test, except that water is poured into the hole, rather than removed, and a fall in level is recorded rather than a rise; an example is illustrated in Figures B.8 and B.9.

Due to the swelling properties of soil, a K value obtained by this method may differ from one obtained if the soil is saturated.   If this change of structure is significant, it has to be taken into consideration when the measured K is evaluated.   Similarly, the test should not be done on a dry soil but only after saturating the test site, eg by conducting the test immediately after an infiltration measurement.

The calculation is as follows:

For the cylindrical auger hole and its flat base

$$A(t_i) = 2\pi r\ h(t_i) + \pi r^2$$

where $A(t_i)$ = area through which water passes
                     into the soil at time t_i (cm^2)

r = radius of the auger hole (cm)

$h(t_i)$ = water-level in hole at time t_i (cm)

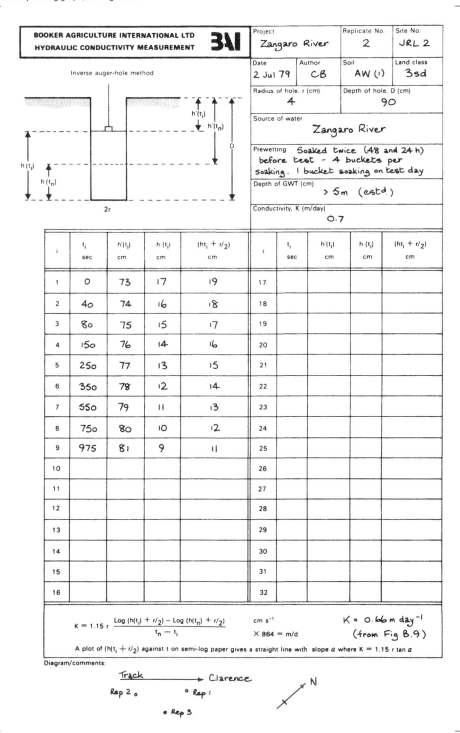

BOOKER AGRICULTURE INTERNATIONAL LTD
HYDRAULIC CONDUCTIVITY MEASUREMENT   3AI

Inverse auger-hole method

Project	Replicate No.	Site No.
Zangaro River	2	JRL 2

Date	Author	Soil	Land class
2 Jul 79	CB	AW (1)	3sd

Radius of hole, r (cm)	Depth of hole, D (cm)
4	90

Source of water   Zangaro River

Prewetting   Soaked twice (48 and 24 h) before test - 4 buckets per soaking. I bucket soaking on test day

Depth of GWT (cm)   > 5m (estd.)

Conductivity, K (m/day)   0.7

i	t_i sec	$h'(t_i)$ cm	$h(t_i)$ cm	$(ht_i + r/2)$ cm	i	t_i sec	$h'(t_i)$ cm	$h(t_i)$ cm	$(ht_i + r/2)$ cm
1	0	73	17	19	17				
2	40	74	16	18	18				
3	80	75	15	17	19				
4	150	76	14	16	20				
5	250	77	13	15	21				
6	350	78	12	14	22				
7	550	79	11	13	23				
8	750	80	10	12	24				
9	975	81	9	11	25				
10					26				
11					27				
12					28				
13					29				
14					30				
15					31				
16					32				

$$K = 1.15\, r\, \frac{\text{Log}(h(t_i) + r/2) - \text{Log}(h(t_n) + r/2)}{t_n - t_i}$$ cm s⁻¹   × 864 = m/d

$K = 0.66\ m\ day^{-1}$ (from Fig B.9)

A plot of $(h(t_i) + r/2)$ against t on semi-log paper gives a straight line with slope a where $K = 1.15\ r\ \tan a$

Diagram/comments:

Track ⟶ Clarence   N

Rep 2 °   ° Rep 1

° Rep 3

Inverse auger-hole method: worked example - graph   Figure B.9

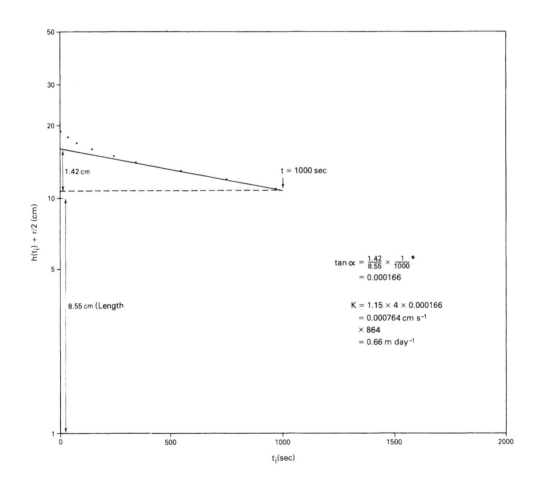

$$\tan \alpha = \frac{1.42}{8.55} \times \frac{1}{1000}$$

$$= 0.000166$$

$$K = 1.15 \times 4 \times 0.000166$$

$$= 0.000764 \text{ cm s}^{-1}$$

$$\times 864$$

$$= 0.66 \text{ m day}^{-1}$$

Graph paper required: 2 cycle logarithmic one way scale (also called semilog paper) 0.1, 0.5 and 1 cm.

A4 size - Chartwell reference No C5521

For corresponding readings see Figure B.8.

• Note: The numbers 1.42 and 8.55 represent length measurements in cm on the original graph paper (reduced in scale for reproduction here)

Supposing that the hydraulic gradient is approximately one, then according to Darcy's Law (Subsection 6.3.2) the volume rate of flow is given by:

$$Q(t_i) = KA(t_i) = 2K\pi r \ (h(t_i) + r/2)$$

where $\quad$ $Q(t_i)$ = volume rate of flow at time t_i (cm^3 s^{-1})

If during the time interval (dt) the water-level falls over a distance (dh), the volume rate of flow into the soil equals:

$$Q(t_i) = -\pi r^2 \frac{dh}{dt}$$

Combining the last two equations gives:

$$2K\pi r \ (h(t_i) + r/2) = -\pi r^2 \frac{dh}{dt}$$

Integration between the limits

$$t_i = t_1, \ h(t_i) = h(t_1) \text{ and}$$
$$t_i = t_n, \ h(t_i) = h(t_n)$$

gives:

$$\frac{2K}{r} (t_n - t_1) = \ln \ (h(t_1) + r/2) - \ln \ (h(t_n) + r/2)$$

Changing to common logarithms and rearranging gives:

$$K = \frac{1.15r \ [\log \ (h(t_1) + r/2) - \log \ (h(t_n) + r/2)]}{t_n - t_1}$$

$$= 1.15r \ \tan \alpha$$

By plotting $(h(t_i) + r/2)$ against t_i on semi-logarithmic paper, a straight line with a slope α is obtained (see Figure B.9). (Note that the slope calculation includes a direct length measurement on the graph paper).

B.2.9   **Inverse auger–hole method: constant level methods**

In some circumstances it is desirable to maintain the level of water in the hole during an inverse auger–hole test so that the water remains, say, within a particular soil horizon. The hydraulic conductivity is then calculated from readings of the volume of water necessary to maintain the constant level in the hole. The experimental set-up, although somewhat more complex than for the falling-head method, is not difficult to use: full details are given in FAO (1979a, pp 173ff). Note the comments on values obtained as given in Subsection 6.3.5.

B.3   **Field measurement of bulk density**

The preferred method is normally the replacement method described in Subsection 6.4.2, but where an undisturbed sample is required, a core-sampling method can be used (see BSI, 1975); this sample can also be used for moisture retention measurements (see Section 6.6). The coring cylinder consists of a thin-walled stainless steel cylinder, chamfered on the outer rim of one end to given a cutting edge. The usual dimensions are of the order of 5 cm length by 5 cm diameter.

The cylinder is driven or pressed vertically into the soil, taking care to prevent lateral movement, until the end of the cylinder is flush with the soil surface inside the core. The cylinder is carefully removed after digging around the core and then the sample is trimmed flush with the ends of the cylinder. If the intact core is not required for other purposes, eg field moisture content or soil moisture release, the sample can be emptied into a polythene bag in situ for subsequent weighing and drying at 105°C for 24 h in the laboratory. When field moisture content is also required, the core should be capped and sealed prior to transport to the laboratory.

Where conditions are favourable (moist stone-free soils of intermediate texture), small cylinders of 2.5 cm diameter can be used. Such small cylinders present less of an obstacle to adequate replication of measurements and provide something approaching a point estimate of density in situations where density varies considerably with depth.

B.4   **Soil and site factors affecting drainage design**

B.4.1   The main soil and site factors are:

a)   the depth to an impermeable barrier, if any;

b)   the layering of the soil above the barrier;

Annex B: Soil physics

c) the hydraulic conductivity of the layers;

d) the site slope.

B.4.2   The depth of the impermeable barrier

In these terms an impermeable barrier is defined as a layer having a hydraulic conductivity of less than 1/10 that of the layer above. The presence of an impermeable barrier interacts with the hydraulic conductivity to affect drain spacing.

Changes in the depth of barrier in relation to the drain are particularly important whilst the two are close together. Once the depth of the barrier below the drain becomes equivalent to 1/3 of the drain spacing, further changes have no effect and the barrier may be considered as being at infinity. In practice if there are no layers in the top 3 m of a soil which might be considered as impermeable barriers, little error will be introduced if the soil is considered as infinitely deep.

B.4.3   Layering of the soil

In principle the steady-state approach allows a top layer situation to be considered. There are more sophisticated approaches including computer methods which allow a large number of layers to be considered but these are outside the scope of this discussion.

B.4.4   Hydraulic conductivity

Hydraulic conductivity is defined as the rate of water movement when the hydraulic gradient is unity. The hydraulic conductivity of the upper horizons is the most important factor in determining the drain spacing. Techniques for measuring the hydraulic conductivity are discussed in Chapter 6, and the auger-hole method (most commonly used in BAI field measurements) is described in Section B.2.

B.4.5   Site slopes

Basically all drainage formulae assume installation on flat land but the results will not be seriously in error for slopes of 1% or 2% whichever way this is with respect to the drains. For the more normal situation where the drains are laid across the slope, Awan (1980) showed that this only affected the highest and the lowest drains in a series of parallel drains even when the slope was 5% to 10%.

B.5   Available water capacity of ADAS field textures

ADAS textural class	Upper limit (field capacity) (% water W/W)	Lower limit (permanent wilting point) (% water W/W)	Available water capacity (mm m^{-1})
Coarse sand	8	4	83
Sand	14	4	150
Fine sand	19	4	200
Very fine sand	20	4	225
Loamy coarse sand	13	7	108
Loamy sand	18	7	158
Loamy fine sand	22	7	217
Loamy very fine sand	25	7	217
Coarse sandy loam	19	9	125
Sandy loam	26	9	175
Fine sandy loam	28	9	192
Very fine sandy loam	28	9	217
Loam	30	13	175
Silty loam	34	10	200
Silt loam	39	16	192
Sandy clay loam	26	15	150
Clay loam	34	18	183
Silty clay loam	43	20	192
Sandy clay	29	19	142
Silty clay	47	25	183
Clay	42	25	175

Source: Salter and Williams (1967, 1969); for ferralitic soils, see also Pidgeon (1972).

Annex C

Soil Classification

Contents

Annex C: Soil classification

List of Figure and Tables

Soil Classification

C.1 Recommended procedure for soil classification by the FAO-Unesco system

The following procedure is a simple method of rapid classification of soils on the FAO system. Tables C.1 to C.3 give a concise summary of the system. A list of the main units and a key to the mapping symbols are presented in Tables 4.1 to 4.3. See Appendix II for the Revised 1988 System.

a) Decide which is the main group (of Table 4.1) that best reflects the soil, either by use of horizon sequence (Table C.2) or using the flow diagram shown in Figure C.1.

b) Decide which adjective (Table 4.2) applies to soil - Most of the FAO soils are associated with only a few descriptive adjectives, and selection of the correct one is usually fairly simple. Occasionally it is easier to start with the adjective and use Table 4.2 to indicate the range of possible soils from which to select the most appropriate.

In some areas the main soil types may not readily fall into any of the defined FAO units. Such soils are likely to occur only rarely, but when they are found it is permissible to use a new unit, provided it is carefully defined. An example is the use of the adjective gypsic, which is strictly only applicable to xerosols and yermosols. Gypsic horizons - or horizons with very high gypsum contents - may occur in several of the soils which can be qualified by the adjective calcic. For such soils the use of gypsic, calci-gypsic or gypsi-calcic may better reflect the soil characteristics.

C.2 Correlation of the FAO system with the USDA Soil Taxonomy

Although Soil Taxonomy (USDA, 1975b) is not recommended for many BAI uses (see Subsection C.4.3), soil classification on this system is often required by clients. Table C.4 indicates the main correlations between this system and the FAO Legend (see Chapter 4) for quick reference. Once the main correspondence has been established from this table the detailed descriptions in the text of Soil Taxonomy should be checked to confirm the final classification. Correlations with other soil classifications (USDA 1938 and 1960 (Seventh Approximation), French, Canadian and USSR) are given in Soil Taxonomy pp 433ff and Tables C.11 to C.13; the system is summarised in the following section.

C.3 Summary of USDA Soil Taxonomy

Soil Taxonomy is a hierarchical system which classifies soils at six levels of detail (categories) from soil series (the most detailed category) to soil orders (the broadest category). Table C.5 illustrates the system for one soil order (entisols), and simplified keys to the orders and nomenclature are presented in Tables C.6 to C.11. Further details are given in Soil Taxonomy (USDA, 1975, pp 83ff and additions, 1982) and the drawbacks for commercial consultants are summarised in Subsection C.4.3. A simplified extract for field use is given in Creutzberg (1982) and a flow-diagram approach in Thomas et al (1980).

C.4 Comparison of international soil classifications

C.4.1 Introduction

Soil classification systems have been developed in many countries, but the majority have restricted international appeal because they are applicable to only a small range of soil types, or are based on theoretical concepts (eg of soil genesis) which are difficult to apply unambiguously - and, particularly, to quantify.

Of the systems designed for international use, including within the tropics and subtropics, only three have been widely employed (although they all have drawbacks, see Young, 1976, Chapter 13): Soil Taxonomy (USDA, 1975 and 1982 - derived from the Seventh Approximation); the CCTA Soil Map of Africa (D'Hoore, 1964) and the FAO-Unesco Soil Map of the World

Simplified flow-diagram representation of FAO-Unesco soil legend   Figure C.1

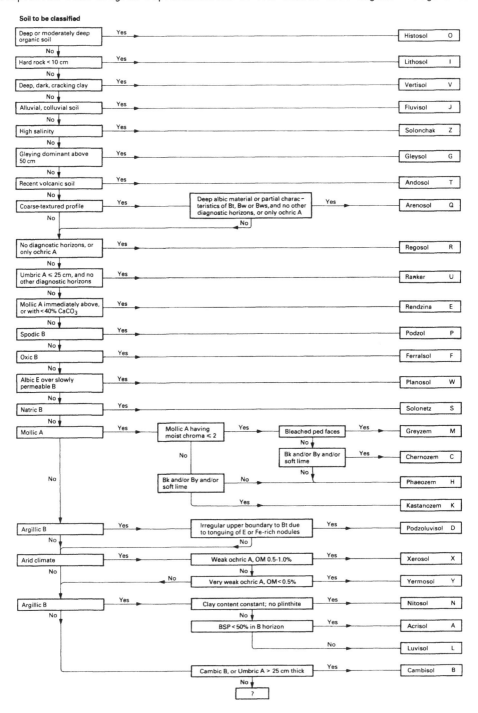

Notes: 1. Soil legend highly abbreviated; does not include, e.g. most depth criteria, nor criteria for buried soils.
Details in FAO-Unesco (1974, p 43f) should always be checked.

2. See Table C.1 for horizon symbols.

3. See Appendix I for Revised 1988 Classification

232

(FAO-Unesco, 1974 revised 1988).   The French and Russian systems, although intended for world-wide application, are based on inferred soil genesis and are of limited value in tropical soil consultancy work.

C.4.2   The CCTA Soil Map of Africa

The CCTA map was the first attempt to map a whole continent from field observation, rather than from genetic supposition.   Although based on the French ORSTOM and Belgian INEAC systems it found wide acceptance, especially in West Africa.   It is a 'natural' system, in which soils are treated idealistically as 'unanalysed entities', but in practice it is quite subjective and a soil profile will often not fit into any grouping or will overlap two or more.   It is hardly used outside Africa.   A summary of the system is given in Young (1976, p 238).

C.4.3   USDA Soil Taxonomy

The USDA Soil Taxonomy was designed primarily to group the 10 000 US soil series into meaningful exclusive groups.   It is an 'artificial' system in that selected, observable soil characteristics (particularly the 'diagnostic horizons') are used to differentiate classes, although these characteristics are largely chosen to reflect agriculturally significant properties.   It attempts to be comprehensive in that any profile can in theory be unambiguously allocated to a class − but in order to do this, the class definitions are necessarily lengthy, complicated and legalistic.   The classes have also been encumbered with a strange and extensive new terminology (see Section C.3), which in itself tends to confuse just as much as it clarifies by avoiding the use of non-explicit terms such as podzol, latosol and laterite.

The system is thus cumbersome, and to most intended users of other disciplines (let alone some soil surveyors!) is almost unintelligible − at least without a lot of additional detailed study.   Such study might be justified if Soil Taxonomy (and its totally analogous ancestor the Seventh Approximation) were the ultimate in soil classification, but it also suffers from a number of serious theoretical defects as discussed, for example, by Butler et al (1961), Avery (1968) and Webster (1968a, 1968b).   The system is defended by Bunting (1969) and Mitchell (1973a).   Two further disadvantages preclude its use as routine on commercial consultancy surveys:

a)   the class definitions depend too heavily on laboratory analyses or long-term soil measurements, making classification in the field at best a provisional exercise only;

b)   the classes identified reflect a regional bias, and make it unsatisfactory for world-wide use − tropical soils and some desert soils are treated inadequately, and hydromorphic soils are not separately recognised at the higher levels of classification.

C.4.4   FAO-Unesco Soil Map of the World

The FAO Soil Map of the World legend probably represents the best system for commercial consultancy use although it, too, is far from ideal.   Whilst it is an artificial system which borrows much from the USDA classification, it is very much simpler, and the parameters chosen for class definitions are designed to reflect natural classes.   It is the only truly international system, incorporating soil units used all over the world, and most soils can be accommodated on the basis of their field descriptions.   It has the great merit for commercial consultancy use of allowing rapid classification into classes which are more or less well known locally and internationally, but it is intended for mapping soils on a continental scale, and the units are inevitably very broad.

Simple correlation tables for the main classification systems, particularly as they apply to tropical conditions, are given in Sanchez (1976), and are reproduced in Tables C.11 to C.14.

C.5   Diagnostic horizons in mineral soils (Soil Taxonomy)

The FAO system has borrowed from Soil Taxonomy (USDA, 1975b) the concept of precisely defined, selected horizons for classifying soil profiles.   Table C.15 is a list of greatly abbreviated definitions from Soil Taxonomy.   Before using these terms in written reports reference should be made to Chapter 3 in Soil Taxonomy or pp 23-27 in the FAO-Unesco Soil Map of the World (Volume 1 - Legend).   Terms such as the anthropic epipedon could not be applied at the scale of the FAO World Map, and information on the distribution of duripans and petrogypsic horizons, etc, is not available in many countries.   Where the information exists they are shown on the map as phases.   Although the specifications are the same, Soil Taxonomy eschews the use of H, A, B and E horizon designations in favour of 'epipedons' and 'subsurface horizons'.   An epipedon is simply the uppermost soil horizon, which may be thinner than the A horizon or include some of the B horizon.   Mixing to 18 cm depth in both systems allows for the effect of ploughing.   Most diagnostic horizons have a minimum thickness of 15 cm and must usually occur within 100 or 125 cm of the surface to apply in the classification.

Annex C: Soil classification

Key to diagnostic horizons and properties of the FAO–Unesco system (1974 Classification)   Table C.1

Diagnostic horizon and symbol 1/		Characteristics	Major diagnostic property 2/	Characteristics
Albic	E	Bleached, usually sandy material	Abrupt textural change	Lower horizon has at least double the clay content of upper horizon (within distance 8 cm if top has < 20% clay)
Argillic	Bt	Textural B with clay skins		
Calcic	Bk	Secondary CaCO₃ accumulation > 15%	Albic material	Bleached material, excluding albic E horizons
Cambic	Bw	In situ weathered B – most B horizons not meeting criteria for argillic, natric, spodic or oxic	Aridic moisture regime	Profile not moist throughout for more than 2 consecutive months, nor moist in some part for more than 3 months
Gypsic	By	CaSO₄ enriched		
Histic	H	High OM (> 20%); peaty	Ferralic properties	(Cambisols and arenosols). B horizons with CEC of < 24 me/100 g clay
Mollic A	Amo 3/	'Fertile earth' topsoil. 1% < OM < 20% and base saturation > 50% (cf umbric A)	Ferric properties	(Luvisols and acrisols). Red mottles or concretions; CEC < 24 me/100 g clay
Natric	Btn	Sodic B. Exchangeable Na > 15%. Columnar structure	High salinity	EC_e > 15 mS cm⁻¹ or, if pH > 8.5, then EC_e > 4 mS cm⁻¹
Ochric A	Aoc 3/	A horizon of dry areas. Pale. OM < 1%	Hydromorphic properties	Saturation by groundwater or histic H horizon, or dominant hues N or bluer than 10Y, or evidence of reduction and segregation of Fe
Oxic	Bws	Highly weathered. Mainly kaolinite and sesquioxides. CEC < 16 me/100 g clay. No clay skins or weatherable minerals		
Spodic	Bh, Bs or Bhs	OM and/or sesquioxide-rich horizon of podzols	Plinthite	Iron–iron clay with quartz, commonly red mottles. Irreversible change to ironstone on drying
Sulphuric	Bsu 3/	Sulphide-rich; pH < 3.5	Sulphidic materials	Waterlogged soil with ≥ 0.75% sulphur and less than three times as much CaCO₃ as sulphur
Umbric A	Aum 3/	'Infertile earth' topsoil. 1% < OM < 20% and base saturation < 50% (cf mollic A)	Takyric features	Heavy soil textures, polygonal cracks, massive or platy crust
			Vertic features	(Cambisols and luvisols). Cracks > 1 cm wide within 50 cm of surface and at least to top of B horizons

Source: FAO–Unesco (1974, pp 20–32), FAO (1977).   See Appendix I for Revised 1988 Classification.

Notes:
1/ Horizon symbols are used as quantitative estimates and do not necessarily imply that criteria for diagnostic horizons are met.   Most diagnostic horizons are at least 15 cm thick and must usually occur within 100 or 125 cm of the surface to apply in the classification.
2/ Full details are given in Section C.5.
3/ No standard FAO symbols exist; these are suggested forms for BAI use, though they depart from the FAO aim of single-letter horizon suffixes.   Alternatives, limited to one letter, might be mollic a, ochric o, sulphuric d and umbric i.

Additional horizon symbol suffixes	Master horizons and layers
b = Buried or bisequal	H Organic horizons, waterlogged
c = Concretions	O Organic horizons, well drained
g = Mottling reflecting oxidation changes	A Mineral surface horizons with OM or
m = Strongly cemented	soil formation
p = Disturbed by tillage	E Eluvial horizons
q = Accumulation of silica	B Mineral subsurface horizons, with
r = Strong reduction due to groundwater	accumulation or pedological alterations
u = (Only used when separating two horizons	C Unconsolidated mineral material
without suffixes; Aul, Au2, etc)	R Continuous indurated rock
x = Fragipan	
z = Accumulation of salts more soluble than gypsum	

Vertical differentiation of similar horizons is denoted by suffixed numbers:
eg: (from surface) Bt1 – Bt2 – Bt3

Lithological discontinuities are denoted by prefixed numbers:
eg: Bt1 – Bt2 – 2Bt3

FAO—Unesco soil classification – horizon sequences (1974 Classification) 1/   Table C.2

This is a highly condensed summary; full definitions should always be checked

Soil symbol and name	Diagnostic horizon(s) and other criteria 2/	Other properties present (or absent) in some units
A Acrisols	Aum or Aoc Bt (BSP < 50%)	Ferric or hydromorphic properties; plinthite; not acidic. Not soils D, N, W. cf soil L
B Cambisols	Aum or Aoc Bw	Bk; By; vertic or ferralic properties. Not acidic. Not soils T, V
C Chernozems	Amo ⩾ 15 cm; moist chroma ⩽ 2 Bk and/or By and/or soft lime < 125 cm	Bt ± hydromorphic properties < 50 cm; tonguing of A into B. Not soils E, V, W, T. cf soils H, M and K
D Podzoluvisols	Aum or Aoc E tonguing into, or Fe nodules overlying, Bt	Hydromorphic properties < 50 cm
E Rendzinas	Amo calcareous C	Not soil V
F Ferralsols	A (all types) Bws	Plinthite < 125 cm
G Gleysols	A or H (all types) hydromorphic properties < 50 cm	Bw; Bk; By; plinthite < 125 cm. Not recent alluvium; if Amo present, no bleached ped coatings. Not soils J, V
H Phaeozems	Amo	Bt ± hydromorphic properties < 50 cm; calcareous > 20 and < 50 cm. If Amo present no bleached ped coatings. Not soils E, V, W, T. cf soils C, M and K
I Lithosols	A < 10 cm R	–
J Fluvisols	Aum or Aoc or H (both types)	Calcareous > 20 cm and < 50 cm; sulphidic material, sulphuric horizon < 125 cm
K Kastanozems	Amo ⩾ 15 cm; moist chroma > 2 Bk and/or By and/or soft lime < 125 cm	As for soil C. Not soils E, V, W, T. cf soils C, H and M
L Luvisols	Aum or Aoc Bt (BSP ⩾ 50%)	Bk and/or soft lime < 125 cm; vertic ferric properties; albic E; plinthite < 125 cm; hydromorphic properties < 50 cm. Not soils W, N, D. cf soil A
M Greyzems	Amo ⩾ 15 cm; moist chroma ⩽ 2 bleached coatings on peds usually Bt	Hydromorphic properties < 50 cm. Not soils E, V, W, T. cf soils C, H and K
N Nitosols	Aum or Aoc (BSP ⩾ 50%) Bt	No Amo, no E. High OM in Bt. No tonguing

Notes: See page 236.   cont

Annex C: Soil classification

Table C.2 cont

This is a highly condensed summary; full definitions should always be checked

Soil symbol and name	Diagnostic horizon(s) and other criteria 2/	Other properties present (or absent) in some units
O Histosols	H	–
P Podzols	A or H (unspecified) Bh	E; thin iron pan; hydromorphic properties < 50 cm
Q Arenosols	Aoc coarse, unconsolidated	Albic material ⩾ 50 cm, or characteristics similar to Bt, Bw or Bws. Not recent alluvials
R Regosols	Aoc unconsolidated	Calcareous > 20 cm and < 50 cm. Not soils V, T, Q
S Solonetz	Aoc or Amo Btn	Hydromorphic properties < 50 cm. Not soil W
T Andosols	Amo or Aum, or Aoc over Bw and BD < 0.85 g cm–3 volcanic material > 60%	Bw. Not soil V
U Rankers	Aum < 25 cm	Not recent alluvials. Not soil T
V Vertisols	Clay ⩾ 30% 0 to 50 cm cracks ⩾ 1 to 50 cm wide 25 to 100 cm one or more of: gilgai, slickenslides, parallelepiped structure	–
W Planosols	A or H (all types) E (at least partly hydromorphic) over slowly permeable horizon (not Bh)	Bt; Btn; fragipan
X Xerosols	Aridic moisture regime weak Aoc one or more of: Bw, Bt, Bk, By	Not soil V. cf soil Y
Y Yermosols	Aridic moisture regime very weak Aoc one or more of: Bw, Bt, Bk, By	Takyric features. Not soil V. cf soil X
Z Solonchaks	Amo or Aoc or H high salinity	Bw; Bk; By; takyric features; hydromorphic properties. Not recent alluvials

Notes: 1/ Permafrost criteria have been omitted; depth criteria indicate distance below soil surface.
2/ For abbreviations see Table C.1. (Note: some symbols used here are not in FAO–Unesco system – see Table C.1).

Source: FAO–Unesco (1974, p 32f). For 1988 Classification see Appendix I

FAO–Unesco soil phases (1974 Classification)   Table C.3

Phase name	Description
Cerrado	(Central Brazil) strongly depleted soils of old land surfaces characterised by tall grass and low contorted trees
Duripan	Duripan (silica cementation) within 100 cm of soil surface
Fragipan	Fragipan (hard, seemingly cemented, loamy) within 100 cm of soil surface. May be 15 to 200 cm thick
Lithic	Continuous, coherent hard rock between 10 and 50 cm of soil surface
Petric	Layer of coarse fragments at least 25 cm thick within 100 cm of soil surface. Includes ironstone, plinthite
Petrocalcic	Petrocalcic horizon (very hard continuously cemented carbonate) within 100 cm of soil surface
Petrogypsic	Petrogypsic horizon (cemented gypsum) within 100 cm of soil surface
Petroferric	Petroferric horizon (iron–rich indurated horizon) within 100 cm of soil surface
Phreatic	GWT between 3 and 5 m from soil surface
Saline	$EC_e > 4$ mS cm^{-1} within 100 cm of soil surface. (Not used for solonchaks)
Sodic	ESP > 6% within 100 cm of soil surface. (Not used for solonetz)
Stony	Gravel, stones, boulders or rocks on surface prevent mechanised agriculture

Abbreviated from FAO-Unesco (1974, pp 5-7). For 1988 Classification see Appendix I.

Annex C: Soil classification

FAO-Unesco (1974)/USDA Soil Taxonomy soil classification correlation   Table C.4

FAO-Unesco soil units	USDA Soil Taxonomy equivalent (and other selected equivalents)
A ACRISOLS	ULTISOLS pp
Ao Orthic acrisols	Hapludults, haplustults (red-yellow podzolic soils pp, Australia)
Af Ferric acrisols	Palexerults, paleustults pp (ferralitic soils pp, CCTA)
Ah Humic acrisols	Humults
Ap Plinthic acrisols	Plinthaquults, plinthudults, plinthustults
Ag Gleyic acrisols	Aquults pp
T ANDOSOLS	ANDEPTS
To Ochric andosols	Dystrandepts pp
Tm Mollic andosols	Eutrandepts pp
Th Humic andosols	Dystrandepts, hydrandepts pp
Tv Vitric andosols	Vitrandepts
Q ARENOSOLS	PSAMMENTS pp
Qc Cambic arenosols	-
Ql Luvic arenosols	Alfic psamments
Qf Ferralic arenosols	Oxic quartzipsamments
Qa Albic arenosols	Spodic udipsamments
B CAMBISOLS	INCEPTISOLS pp
Be Eutric cambisols	Eutrochrepts pp, ustochrepts pp, xerochrepts pp, eutropepts (typische braunerde, Germany)
Bd Dystric cambisols	Dystrochrepts, dystropepts (sols bruns acides, France)
Bh Humic cambisols	Haplumbrepts, humitropepts pp
Bg Gleyic cambisols	Aquic dystrochrepts, aquic eutrochrepts
Bx Gelic cambisols	Pergelic cryochrepts
Bk Calcic cambisols	Eutrochrepts pp, ustochrepts pp, xerochrepts pp
Bc Chromic cambisols	Xerochrepts pp
Bv Vertic cambisols	Vertic tropepts
Bf Ferralic cambisols	Oxic tropepts
C CHERNOZEMS	BOROLLS pp
Ch Haplic chernozems	Haploborolls, vermiborolls (typic chernozems, USSR)
Ck Calcic chernozems	Calciborolls pp
Cl Luvic chernozems	Argiborolls pp
Cg Glossic chernozems	-
F FERRALSOLS	OXISOLS (ferralitic soils CCTA)
Fo Orthic ferralsols	Orthox pp, torrox pp, ustox pp
Fx Xanthic ferralsols	Orthox pp
Fr Rhodic ferralsols	Orthox pp, torrox pp, ustox pp
Fh Humic ferralsols	Humox (humic latosols, Brazil)
Fa Acric ferralsols	Acrox
Fp Plinthic ferralsols	Plinthaquox pp
J FLUVISOLS	FLUVENTS
Je Eutric fluvisols	-
Jc Calcaric fluvisols	-
Jd Dystric fluvisols	-
Jt Thionic fluvisols	Sulfaquepts, sulfic haplaquepts pp (catclays, acid sulphate soils, Indonesia)
G GLEYSOLS	(Aquic suborders pp)
Ge Eutric gleysols	Haplaquents, psammaquents, tropaquents, andaquepts, fragiaquepts, haplaquepts, tropaquepts pp
Gc Calcaric gleysols	-
Gd Dystric gleysols	Haplaquents, psammaquents, tropaquents, andaquepts, fragiaquepts, haplaquepts, tropaquepts pp
Gm Mollic gleysols	Haplaquolls
Gn Humic gleysols	Humaquepts
Gp Plinthic gleysols	Plinthaquepts
Gx Gelic gleysols	Pergelic cryaquepts
M GREYZEMS	BOROLLS pp
Mo Orthic greyzems	Argiborolls pp (grey forest soils, USSR)
Mg Gleyic greyzems	Aquolls pp
O HISTOSOLS	HISTOSOLS (bog soils, USSR)
Oe Eutric histosols	-
Od Dystric histosols	-
Ox Gelic histosols	-
K KASTANOZEMS	USTOLLS (chestnut soils of the dry steppes, USSR)
Kh Haplic kastanozems	Haplustolls, aridic haploborolls
Kk Calcic kastanozems	Calciustolls, aridic calciborolls
Kl Luvic kastanozems	Argiustolls, aridic argiborolls
I LITHOSOLS	LITHIC subgroups (except haplumbrepts)

Note: See page 239.   cont

Table C.4 cont

FAO–Unesco soil units	USDA Soil Taxonomy equivalent (and other selected equivalents)
L LUVISOLS	ALFISOLS pp
Lo Orthic luvisols	Hapludalfs, haploxeralfs pp (sols lessivés modaux, France; parabraunerde, Germany)
Lc Chromic luvisols	Rhodoxeralfs, haploxeralfs pp (terra rossa, Italy)
Lk Calcic luvisols	Haplustalfs pp
Lv Vertic luvisols	Vertic haploxeralfs
Lf Ferric luvisols	(Ferruginous tropical soils, CCTA)
La Albic luvisols	Eutroboralfs
Lp Plinthic luvisols	Plinthustalfs, plinthoxeralfs
Lg Gleyic luvisols	Aqualfs
N NITOSOLS	some ULTISOLS and ALFISOLS
Ne Eutric nitosols	Tropudalfs, paleudalfs, rhodustalfs pp (eutrophic brown soils, CCTA)
Nd Dystric nitosols	Tropudults, rhodudults, rhodustults palexerults pp (krasnozems pp, USSR)
Nn Humic nitosols	Tropohumults, palehumults pp (humic ferrisols CCTA, ferrisols CCTA)
H PHAEOZEMS	UDOLLS pp
Hh Haplic phaeozems	Hapludolls (brunizem, France; degraded chernozem, USSR)
Hc Calcaric phaeozems	Vermudolls pp
Hl Luvic phaeozems	Argiudolls
Hg Gleyic phaeozems	Argiaquolls
W PLANOSOLS	
We Eutric planosols	Albaqualfs, paleargids pp, palexeralfs pp, paleustalfs pp
Wd Dystric planosols	Albaquults
Wm Mollic planosols	Argialbolls, mollic albaqualfs
Wh Humic planosols	–
Ws Solodic planosols	(Solods, USSR; solodized solonetz, CCTA)
Wx Gelic planosols	–
P PODZOLS	SPODOSOLS pp
Po Orthic podzols	Orthods pp
Pl Leptic podzols	–
Pf Ferric podzols	Ferrods
Ph Humic podzols	Humods
Pp Placic podzols	Placorthods, placohumods (thin ironpan podzols, UK)
Pg Gleyic podzols	Aquods
D PODZOLUVISOLS	Glossic great groups of alfisols
De Eutric podzoluvisols	Glossudalfs, glossoboralfs (sols lessivés glossiques, France)
Dd Dystric podzoluvisols	–
Dg Gleyic podzoluvisols	Glossaqualfs, aquic glossudalfs, aquic glossoboralfs
U RANKERS	LITHIC HAPLUMBREPTS
R REGOSOLS	ORTHENTS, PSAMMENTS pp
Re Eutric regosols	–
Rc Calcaric regosols	–
Rd Dystric regosols	–
Rx Gelic regosols	Pergelic cryorthents, pergelic cryopsamments pp
E RENDZINAS	RENDOLLS
Z SOLONCHAKS	(Salic great groups)
Zo Orthic solonchaks	Salorthids
Zm Mollic solonchaks	Salorthic calciustolls, salorthidic haplustolls
Zt Takyric solonchaks	–
Zg Gleyic solonchaks	Haplaquepts pp
S SOLONETZ	(Natric great groups)
So Orthic solonetz	Natrustalfs, natrixeralfs, natrargids, nadurargids
Sm Mollic solonetz	Natralbolls, natriborolls, natrustolls, natrixerolls
Sg Gleyic solonetz	Natraqualfs
V VERTISOLS	VERTISOLS
Vp Pellic vertisols	Pelluderts, pellusterts, pelloxererts
Vc Chromic vertisols	Chromuderts, chromusterts, chromoxererts, torrerts
X XEROSOLS	MOLLIC ARIDISOLS (sirozems, USSR)
Xh Haplic xerosols	Mollic (xerollic or ustollic) camborthids and durorthids
Xx Calcic xerosols	Mollic (xerollic or ustollic) calciorthids pp
Xy Gypsic xerosols	Mollic (xerollic or ustollic) calciorthids pp
Xl Luvic xerosols	Mollic (xerollic or ustollic) haplargids and durargids
Y YERMOSOLS	TYPIC ARIDISOLS
Yh Haplic yermosols	Camborthids, durorthids
Yk Calcic yermosols	Calciorthids pp
Yy Gypsic yermosols	Gypsiorthids pp
Yl Luvic yermosols	Argids
Yt Takyric yermosols	–

Note: pp = part of; CCTA refers to Soil Map of Africa legend (D'Hoore, 1964).

The USDA Soil Taxonomy hierarchical system

Table C.5

Order	Suborder	Great group	Subgroup	Family 1/	Series
Alfisols					
Aridisols					
Entisols	Aquents				
	Arents				
	Fluvents	Cryofluvents	Typic cryofluvents	Coarse loamy, mixed, acid	Susitna
		Torrifluvents	Typic torrifluvents	Fine loamy, mixed (calcareous), mesic	Jocity and Youngston
			Vertic torrifluvents	Clayey over loamy, mixed, (calcareous), hyperthermic	Glamis
	Orthents	Cryorthents	Typic cryorthents	Loamy-skeletal, carbonatic	Swift Creek
			Pergelic cryorthents	Loamy-skeletal, mixed (calcareous)	Durelle

Note: 1/ The complete name of a family consists of the name of the subgroup modified by adjectives to describe properties that are defined in Chapter 18 of Soil Taxonomy.

Source: USDA (1975b, p 84).

Simplified key to USDA Soil Taxonomy Soil Orders   Table C.6

Soil Order 1/ (and great group termination)	Symbol	Brief description 2/ and basis for suborder definition	Nearest FAO–Unesco equivalents 3/		
Alfisols (–alfs)	H	Soils with an argillic or natric horizon Suborders: moisture and temperature regimes	Luvisols Podzoluvisols	and	(Nitosols) (Planosols) (Solonetz)
Aridisols (–ids)	E	a) Soils which are dry for all or nearly all of the year, and have a calcic, petrocalcic, gypsic, petrogypsic, cambic, argillic or natric horizon, or b) which are saturated for > 1 month (most years) within 100 cm and have a salic horizon within 75 cm Suborders: nature of argillic horizon	Solonetz Yermosols Xerosols		
Entisols (–ents)	J	Soils with no diagnostic horizons Suborders: moisture regime, particle sizes, stratification	Arenosols Fluvisols Regosols	and	(Gleysols)
Histosols (–ists)	A	Soils with OM > 30% to at least 40 cm Suborders: moisture regime, nature of OM	Histosols		
Inceptisols (–epts)	I	Soils having no sulphidic material within 50 cm, and with one or more of: a) Umbric, mollic or plaggen epipedon (or histic in volcanic soils) b) Cambic, calcic, placic, duripan, fragipan, sulphuric horizon c) ESP ≤ 15% within 75 cm Suborders: moisture and temperature regimes, mineralogy, epipedon	Andosols Cambisols Rankers	and	(Fluvisols) (Gleysols) (Solonchaks)
Mollisols (–olls)	G	Soils with a mollic epipedon and BSP > 50% Suborders: moisture and temperature regimes, mineralogy, albic horizon	Chernozems Greyzems Kastanozems Rendzinas	and	(Gleysols) (Planosols) (Solonchaks)
Oxisols (–ox)	C	Soils with an oxic horizon (or continuous plinthite < 30 cm if waterlogged and reduced all year) Suborders: moisture regime, amount of OM	Ferralsols		
Spodosols (–ods)	B	Soils with a spodic (sometimes a placic) horizon Suborders: moisture regime; nature of spodic horizon	Podzols		
Ultisols (–ults)	F	Soils with a warm temperature regime and an argillic horizon (or fragipan with clay skins > 1 mm thick) and BSP < 35% Suborders: moisture regime; amount of OM	Acrisols	and	(Nitosols)
Vertisols (–erts)	D	Soils with > 30% clay to 50 cm, cracks ≥ 1 cm wide at 50 cm, gilgai, slicken-slides, and/or wedge-shaped peds Suborders: moisture regime	Vertisols		

Notes: 1/ See also Table C.10 for formation of soil names.
       2/ These descriptions and numerical limits are highly abbreviated, full definitions are given in Soil Taxonomy (USDA, 1975b, pp 91ff).
       3/ These are very broad equivalents because of overlap between the two systems; see Table C.4 for more detail. The full definitions should always be checked before final classification of a soil.

Annex C: Soil classification

USDA Soil Taxonomy nomenclature: formative elements in names of soil orders   Table C.7

Formative element	Derivation 1/		Mnemonicon	Connotation
Alfi		–	Pedalfer	Aluminium and iron
Aridi	L	aridus, dry	Arid	Mainly arid soils
Enti		–	Recent	Young soils
Histo	Gr	histos, tissue	Histology	Organic soils
Incepti	L	inceptum, beginning	Inception	Slightly developed
Molli	L	mollis, soft	Mollify	OM rich
Oxi	F	oxide, oxide	Oxide	Oxic horizon
Spodo	Gr	spodos, wood ash	Podzol	Spodic horizon
Ulti	L	ultimus, last	Ultimate	Weathered
Verti	L	verto, turn	Invert	Self—mulching

Note: 1/   F = French, Gr = Greek, L = Latin.

Adapted from: USDA (1975b, p 87).

Soil Taxonomy nomenclature: formative elements in names of suborders   Table C.8

Formative element	Derivation 1/		Mnemonicon	Connotation
Alb	L	albus, white	Albino	Presence of albic horizon
And	Japanese	ando	–	Ando—like
Aqu	L	aqua, water	Aquarium	Aquic moisture regime
Ar	L	arare, to plough	Arable	Mixed horizons
Arg	L	argilla, white clay	Argillite	Presence of argillic horizon
Bor	Gr	boreas, northern	Boreal	Cool
Ferr	L	ferrum, iron	Ferruginous	Presence of iron
Fibr	L	fibra, fiber	Fibrous	Least decomposed stage
Fluv	L	fluvius, river	Fluvial	Flood plain
Fol	L	folia, leaf	Foliage	Mass of leaves
Hem	Gr	hemi, half	Hemisphere	Partial decomposition
Hum	L	humus, earth	Humus	Presence of organic matter
Ochr	Gr	ochros, pale	Ochre	Presence of ochric epipedon
Orth	Gr	orthos, true	Orthophonic	The common ones
Plagg	German	plaggen, sod	–	Presence of plaggen epipedon
Psamm	Gr	psammos, sand	Psammite	Sand texture
Rend	Polish	rendzina	–	High carbonate content
Sapr	Gr	sapros, rotten	Saprophyte	Most decomposed stage
Torr	L	torridus, hot and dry	Torrid	Torric moisture regime
Trop	Gr	tropikos, solstice	Tropical	Continually warm
Ud	L	udus, humid	Udometer	Udic moisture regime
Umbr	L	umbra, shade	Umbrella	Presence of umbric epipedon
Ust	L	ustus, burnt	Combustion	Ustic moisture regime
Xer	Gr	xeros, dry	Xerophyte	Xeric moisture regime

Note: 1/   Gr = Greek, L = Latin.

Adapted from: USDA (1975b, p 88).

242

Soil Taxonomy nomenclature: formative elements in names of great groups   Table C.9

Formative element	Derivation 1/		Mnemonicon	Connotation
Acr	Gr	akros, at the end	Acrolith	Extreme weathering
Agr	L	ager, field	Agriculture	An agric horizon
Alb	L	albus, white	Albino	An albic horizon
And	Japanese	ando	Ando	Ando-like
Arg	L	argilla, white clay	Argillite	An argillic horizon
Bor	Gr	boreas, northern	Boreal	Cool
Calc	L	calcis, lime	Calcium	A calcic horizon
Camb	L	cambiare, to exchange	Change	A cambic horizon
Chrom	Gr	chroma, colour	Chroma	High chroma
Cry	Gr	kryos, icy cold	Crystal	Cold
Dur	L	durus, hard	Durable	A duripan
Dystr, dys	Gr	dys, ill: trophe, turn	Dyslexia	Low base saturation, dystrophic
Eutr, eu	Gr	eu, well: trophe, turn	Eulogy	High base saturation, eutrophic
Ferr	L	ferrum, iron	Ferric	Presence of iron
Fluv	L	fluvius, river	Fluvial	Flood plain
Frag	L	fragilis, brittle	Fragile	Presence of fragipan
Fragloss	Compound of fra(g) and gloss		–	See the elements frag and gloss
Gibbs	Modified from gibbsite (rock mineral)		Gibbsite	With gibbsite in sheets or nodules
Gloss	Gr	glossa, tongue	Glossary	Tongued
Gyps	L	gypsum, gypsum	Gypsum	Presence of a gypsic horizon
Hal	Gr	halos, salt	Halophyte	Salty
Hapl	Gr	haplos, simple	Haploid	Minimum horizon
Hum	L	humus, earth	Humus	Presence of humus
Hydr	Gr	hydor, water	Hydrophobia	Presence of water
Luv	Gr	louo, to wash	Ablution	Illuvial
Med	L	media, middle	Medium	Of temperate climates
Nadur	Compound of na(tr) and dur		–	See the elements natr and dur
Natr	L	natrium, sodium	–	Presence of natric horizon
Ochr	Gr	ochros, pale	Ochre	Presence of ochric epipedon
Pale	Gr	paleos, old	Paleosol	Excessive development
Pell	Gr	pellos, dusky	–	Low chroma
Plac	Gr	plax, flat stone	–	Presence of a thin pan
Plagg	German	plaggen, sod	–	Presence of plaggen epipedon
Plinth	Gr	plinthos, brick	–	Presence of plinthite
Psamm	Gr	psammos, sand	Psammite	Sand texture
Quartz	German	quarz, quartz	Quartz	High quartz content
Rhod	Gr	rhodon, rose	Rhododendron	Dark-red colour
Sal	L	sal, salt	Saline	Presence of salic horizon
Sider	Gr	sideros, iron	Siderite	Presence of free iron oxides
Sombr	French	sombre, dark	Sombre	A dark horizon
Sphagn	Gr	sphagnos, bog	Sphagnum	Presence of sphagnum
Sulf	L	sulfur, sulphur	Sulphur	Presence of sulphides or their oxidation products
Torr	L	torridus, hot and dry	Torrid	Torric moisture regime
Trop	Gr	tropikos, of the solstice	Tropical	Humid and continually warm
Ud	L	udus, humid	Udometer	Udic moisture regime
Umbr	L	umbra, shade	Umbrella	Presence of umbric epipedon
Ust	L	ustus, burnt	Combustion	Ustic moisture regime
Verm	L	vermes, worm	Vermiform	Wormy, or mixed by animals
Vitr	L	vitrum, glass	Vitreous	Presence of glass
Xer	Gr	xeros, dry	Xerophyte	A xeric moisture regime

Note: Gr = Greek, L = Latin.

Source: After USDA (1975b, p 89).

Table C.10

USDA Soil Taxonomy: great group formative elements arrayed according to suborder (columns) and orders (rows)

Order formative element	alb	and	aqu	arg	bor	fibr	fol	fluv	hem	hum	ochr	orth	psamm	sapr	trop	ud	umbr	ust	xer	Sub-order only 1/
Suborder formative element																				
ert (Vertisols)																Pell Chrom		Pell Chrom	Pell Chrom	Torr
ent (Entisol)			Hydr Cry Trop Hapl Fluv Psamm Sulf					Cryo Torri Usti Xero Udi Tropo			Cry	Cryo Torri Ust Xer Ud Trop	Quartzi Torri Usti Xero Udi Tropo							Ar
ept (Inceptisols)		Cry Dur Hydr Eutr Dystr Ultr Plac	Cry Plac And Trop Fragi Hal Hum Hapl Plinth Sulf								Fragi Dur Cry Ust Xer Eutr Dystr				Us Eu Dys Humi Sombri		Fragi Cry Hapl Xer			Plagg
id (Aridisols)				Nadur Dur Natr Pale Hapl								Dur Sal Pale Calci Camb Gypsi								
od (Spodosols)			Fragi Cry Dur Plac Trop Hapl Sider							Placo Tropo Fragi Cryo Haplo		Plac Fragi Cry Hapl Trop								Ferr
ult (Ultisols)			Alb Plinth Fragi Trop Ochr Umbr Pale							Pale Tropo Haplo Plintho Sombri						Fragi Plinth Pale Rhod Trop Hapl		Plinth Pale Rhod Hapl	Pale Haplo	

Note: See page 245.

cont

Table C.10 cont

Order formative element	alb	and	aqu	arg	bor	fibr	fluv	fol	hum	hem	ochr	orth	psamm	sapr	trop	ud	umbr	ust	xer	Sub-order only 1/
oll (Mollisols)	Natr Argi		Cry Dur Natr Calci Argi Hapl		Pale Cryo Natr Argi Vermi Calci Haplo											Pale Argi Verm Hapl		Dur Calci Pale Argi Verm Hapl	Duri Natri Calci Pale Argi Haplo	Rend
alf (Alfisols)			Natr Trop Fragi Gloss Alb Ochr Umbr Dur Plinth		Pale Fragi Natri Cryo Eutro Glosso											Agr Fragi Natr Trop Ferr Gloss Pale Hapl Rhod Fragloss		Plinth Dur Natr Pale Rhod Hapl	Plintho Duri Natri Rhodo Pale Haplo Fragi	Rend
ox (Oxisols)			Gibbsi Plinth Ochr Umbr Dur Plinth						Sombri Gibbsi Haplo Acro			Gibbsi Acr Eutr Umbri Hapl Sombri						Acr Eutr Hapl Sombri		Torr
ist (Histosols)						Cryo Sphagno Boro Medi Luvi Tropo		Boro Cryo Tropo		Cryo Boro Tropo Medi Luvi Sulfi Sulfo				Cryo Boro Tropo Medi						

Note: 1/   Suborders in this column have no great group recognised at present.

Example:

$$\frac{Fragi}{(c)} - \frac{ochr}{(b)} - \frac{ept}{(a)}$$

a)   Order element   (ept = inceptisol)
b)   Suborder element   (ochr = ochrept, ie inceptisol with ochric epipedon)
c)   Great group element (Fragi = with fragipan)

Adapted from: Buol et al (1980) and USDA (1975b, 1982).

245

Annex C: Soil classification

Correlation of USDA Soil Taxonomy Orders with the previous USDA and other classification systems

Table C.11

Order	Former great soil groups included
Entisols	Azonal soils, some low humic gleys, lithosols, regosols
Vertisols	Grumusols, tropical dark clays, regurs, black cotton soils, dark magnesium clays
Inceptisols	Andosols, hydrol humic latosols, sol brun acide, some brown forest, low humic gley, humic gley
Aridisols	Desert, reddish desert, sirozem, solonchak, some brown and reddish-brown soils, associated solonetz
Mollisols	Chestnut, chernozem, brunizem, rendzina, some brown forest, brown, associated humic gley and solonetz
Spodosols	Podzols, brown podzolic, groundwater podzols
Alfisols	Grey-brown podzolic, grey wooded, non-calcic brown, degraded chernozem, associated planosols and half-bog, some terra rossa estruturada and eutric red-yellow podzolics, some latosols and lateritic soils
Ultisols	Red-yellow podzolic, reddish-brown lateritic, humic latosols, associated planosols, and some half-bogs, latosols, lateritic soils, terra roxa and groundwater laterites
Oxisols	Low humic latosols, humic ferruginous latosols, aluminous ferruginous latosols, some latosols, lateritic soils, terra rossa legitima, groundwater laterites
Histosols	Bog soils, organic soils, peat, muck

Sources: Sanchez (1976), adapted from Soil Survey Staff (1960), Thorp and Smith (1949) and Cline et al (1955); see also Simonson (1982) for other US systems.

Approximate correlation between the FAO-Unesco, USDA Soil Taxonomy and the French soil classification systems

Table C.12

FAO-Unesco legend	Soil Taxonomy	French classification
Fluvisols	Fluvents	Sols minéraux bruts et sols peu évolués d'apport alluvial et colluvial
Regosols	Psamments	Sols minéraux bruts et sols peu évolués d'apport éolien
Arenosols (ferralic)	Oxic quartzipsamments	Sols ferralitiques moyennement ou fortement désaturés, à texture sableuse
Gleysols		
Eutric and dystric	Tropaquepts	Sols hydromorphes humifères à gley
Humic	Humaquepts	Sols humiques à gley
Plinthic	Plinthaquepts	Sols hydromorphes à accumulation de fer en carapace ou cuirasse
Andosols	Andepts	Andosols
Planosols	Paleudalfs and paleustalfs	Sols ferrugineux tropicaux lessivés (pro parte)
Cambisols		
Dystric	Dystropepts	Sols ferralitiques fortement et moyennement désaturés, rajeunis (pro parte)
Eutric	Eutropepts	Sols ferrugineux tropicaux (non lessivés) Sols ferralitiques faiblement désaturés, rajeunis
Humic	Humitropepts	Sols ferralitiques fortement et moyennement désaturés, humifères, rajeunis
Luvisols	Alfisols	Sols ferrugineux tropicaux lessivés
Acrisols	Ultisols	Sols ferralitiques fortement désaturés
Ferralsols	Oxisols	Sols ferralitiques
Lithosols	Lithic subgroups	Lithosols et sols lithiques

Source: Sanchez (1976) from Aubert and Tavernier (1972).

Approximate correlation between the French and USDA soil classification systems   Table C.13

French classification	Soil Taxonomy (orders, suborders, or great groups)
I. Sols Minéraux Bruts (as far as recognised as soil)	Orthents, psamments, fluvents
II. Sols Peu Evolués	
Humifères	Orthents, humitropepts
A allophanes	Andepts, eutrandepts, vitrandepts
Non climatiques	Orthents, fluvents, psamments, tropepts
IV. Andosols	Andepts
Saturés	Eutrandepts
Désaturés	Hydrandepts, dystrandepts
VII. Sols Brunifiés des pays Tropicaux	Eutropepts, tropudalfs
IX. Sols Ferrugineux Tropicaux	
Peu lessivés	Ustropepts
Lessivés	Haplustalfs, paleustalfs, plinthustalfs
Appauvris à pseudogley	Tropaqualfs
X. Sols Ferralitiques	
Faiblement désaturés	
Typiques	Eutrorthox, eutrustox
Appauvris, remaniés	Alfic eutrustox
Rajeunis	Ustropepts, eutropepts
Moyennement désaturés	
Typiques	Haplorthox, haplustox
Humifères	Haplohumox, sombrihumox
Appauvris	Ultic and alfic haplorthox
Remaniés	Oxic subgroups of udults, haplorthox, haplustox
Rajeunis	Typic dystropepts and oxic dystropepts
Fortement désaturés	
Typiques	Haplorthox, acrorthox, oxic psammentic dystropepts
Humifères	Haplohumox, acrohumox, sombrihumox
Appauvris	Ultic subgroups of haplorthox
Remaniés	Haplorthox, acrorthox
Rajeunis	Oxic dystropepts
Lessivés	Paleudults, oxic tropudults, oxic rhodudults
XI. Sols Hydromorphes (with the exception of the Sols Hydromorphes Organiques et Moyennement Organiques)	
Minéraux ou peu humifères	
A gley	Tropaquents, tropaquepts
Lessivés	Tropaqualfs, tropaquults
A pseudogley	Aquic subgroups of tropudalfs and tropudults
A accumulation de fer	Petroferric subgroups of
en carapace ou cuirasse	aquox, aquults and aquepts

<u>Sources</u>: Sanchez (1976) from Aubert and Tavernier (1972).

Annex C: Soil classification

Correlation of the Belgian soil classification system 1/ with the French and USDA systems   Table C.14

Belgian system (INEAC)	French system (ORSTOM)	Soil Taxonomy
Recent tropical soils	Sols peu evolués	Entisols
Non-hydromorphic	D'apport modal	Fluvents
Hydromorphic	D'apport hydromorphique	Aquents
Brown tropical soils	Sols bruns eutrophes tropicaux	Eutropepts
Black tropical clays	Vertisols	Vertisols
Recent textural soils	Sols halomorphes	Natrustalfs
Solonetz	A alcali et lessivés	
Podzols	Podzols alios	Aquods
Kaolisols	Sols ferralitiques et ferrugineux	Oxisols, ultisols
Ferrisols	Sols ferrugineux	Alfisols, entisols
Hydroferrisols	Sols hydromorphes minéraux gley de surface	Aquepts, aquults
Hygroferrisols		
Intergrading to brown and recent tropical soils	Sols faiblement ferralitique	Dystropepts, tropudults
Typic and intergrading to hygroferralsols	Sols ferralitiques typiques	Orthox; oxic subgroups of udults
Hygroxeroferrisols		
Intergrading to brown and recent tropical soils	Sols faiblement ferralitiques	Ustropepts, tropustults
Typic and intergrading to ferralsols	Sols ferralitiques typiques rouges et jaunes	Ustox; oxic subgroup of ustults
Humiferous ferrisols	Sols ferralitiques humifères d'altitude	Humox, humults
Xeroferrisols	Sols ferrugineux tropicaux lessivés	Ustalfs
Ferralsols	Sols ferralitiques	Oxisols, ultisols
Hygroferralsols	Sols ferralitiques lessivés	Orthox, tropudults
Typic, with plinthite	En argile	
Hygroxeroferralsols	Sols ferralitiques lessivés modal	Ustox
Typic, with plinthite	En argile	Tropustults
Arenoferralsols	Sols ferralitiques lessivés podzoliques	Oxic quartzipsamments

Note: 1/   As used in Zaire.

Sources: Sanchez (1976) adapted from Jurion and Henry (1969), Sys et al (1961), Aubert (1968); see also Duchaufour (1963).

248

USDA Soil Taxonomy diagnostic horizons (terms in brackets are the 1974 FAO-Unesco equivalent)   Table C.15

AGRIC HORIZON	This horizon is found directly under the plough layer and has clay and humus accumulated as thick, dark lamellae to the extent that they occupy at least 15% of the soil volume.
ALBIC HORIZON (Albic E horizon)	A pale, light-coloured eluvial horizon, depleted of clay humus and/or iron.
ANTHROPIC EPIPEDON	A surface horizon like the mollic epipedon but contains > 250 ppm of citric acid soluble P_2O_5.
ARGILLIC HORIZON (Argillic B horizon)	In general, this is a B horizon that has at least 1.2 times as much clay as the eluvial horizons above, or 3% more clay content if the eluvial layer has < 15% clay, or 8% more clay if eluvial layer has > 40% clay. It is formed by illuviation of clay, and illuviation argillans are usually observable unless there is evidence of stress cutans. It should be > 1/10 as thick as all overlying horizons or more than 15 cm, whichever is thinner.
CALCIC HORIZON (Calcic horizon)	A layer of secondary accumulation of carbonates, usually Ca or Mg, in excess of 15% calcium carbonate equivalent and contains at least 5% more carbonate than an underlying layer.
CAMBIC HORIZON (Cambic B horizon)	An altered subsoil horizon showing structural development but no significant illuviation nor extreme weathering.
DURIPAN	A subsurface horizon at least half-cemented by silica. Air-dry peds do not slake in water.
FRAGIPAN (Fragipan)	A subsurface horizon of high bulk density, brittle when moist, and very hard when dry. It does not soften on wetting, but can be broken in the hands. Air-dry peds slake in water.
GYPSIC HORIZON (Gypsic horizon)	A layer of calcium sulphate accumulation. Contains 5% or more $CaSO_4$ than underlying material.
HISTIC EPIPEDON (Histic H horizon)	A surface horizon containing > 20 to 30% organic matter, depending on clay content, and is water saturated for 30 days at some season of the year unless artificially drained. It is thinner than 30 cm if drained or 45 cm thick if not artificially drained.
IRONSTONE (Ironstone)	Hardened plinthite.
MOLLIC EPIPEDON (Mollic A horizon)	A surface horizon that, when mixed to a depth of 18 cm, contains 1% organic matter, with colour values darker than 5.5 dry and 3.5 moist. The structure cannot be massive and hard. Base saturation > 50%.
NATRIC HORIZON (Natric B horizon)	This horizon meets the requirements of an argillic horizon but also has prismatic or columnar structure and over 15% of the CEC is saturated with Na (ie ESP > 15%).
OCHRIC EPIPEDON (Ochric A horizon)	A surface horizon that is light in colour, with colour values > 5.5 dry and > 3.5 moist and contains less than 1% organic matter.
OXIC HORIZON (Oxic B horizon)	This horizon is at least 30 cm thick. It is highly weathered, with a high content of low charge 1:1 clays and sesquioxides retaining < 10 me of NH_4^+ from 1 N NH_4Cl or < 10 me KCl-extractable bases per 100 g of clay. Only a trace of the clay is water dispersible.
PETROCALCIC HORIZON	An indurated calcic horizon. Hardness 3 or more (Moh's scale) and 50% breaks down in acid, but does not break down in water.
PETROGYPSIC HORIZON	A gypsic horizon is so strongly cemented with gypsum that dry fragments do not slake in water, and roots cannot enter. Usually > 60% gypsum content.
PLACIC HORIZON	Thin ironpan, generally 2 to 10 mm black to dark-reddish pan cemented by Fe and Mn.
PLAGGEN EPIPEDON	A man-made surface horizon > 50 cm thick, created by years of manure additions (mainly in Western Europe).
PLINTHITE (Plinthite)	Not necessarily a horizon, but material rich in iron, poor in humus, which hardens irreversibly to ironstone with repeated wetting and drying. The red, indurating portions are usually mottled with yellowish, greyish, or white bodies.
SALIC HORIZON	A horizon of secondary soluble salt enrichment (> 2 to 3% salt, depending on thickness).
SOMBRIC HORIZON	A dark subsurface horizon, generally in high-altitude tropical soils, containing illuvial humus neither associated with Al nor dispersed by Na. Does not underlie an albic horizon. Base saturation < 50%.
SPODIC HORIZON (Spodic B horizon)	This horizon has an illuvial accumulation of free sesquioxides and organic matter. There are many specific limitations dealing with Al, Fe, organic matter and clay ratios, depending on whether the overlying horizon is virgin or cultivated.
SULPHURIC HORIZON (Sulphuric horizon)	Results from artificial drainage of sulphide-rich soils. Characterised by pH < 3.5 and bright yellow jarosite mottles.
UMBRIC EPIPEDON (Umbric A horizon)	A surface horizon like the mollic epipedon but < 50% base saturated.

Annex C: Soil classification

Annex D

Selected Land Classification Systems

Contents

Annex D: Land capability classification

List of Tables

Selected Land Classification Systems [1]

USDA land capability classification

The system of Klingebiel and Montgomery (1961) is as follows:

Class 1 - Soils in Class 1 have few limitations that restrict their use

These soils are suited to a wide range of plants and may be used safely for cultivated crops, pasture, range, woodlands and wildlife. They are nearly level or only gently sloping and the erosion hazard from wind or water is low. The soils are deep, generally well drained, and easily worked. They hold water well and are either fairly well supplied with plant nutrients or highly responsive to inputs of fertiliser.

The soils in Class 1 are not subject to damaging overflow. They are productive and suited to intensive cropping. The local climate must be favourable for growing many of the common field crops.

In irrigated areas, soils may be placed in Class 1 if the limitation of the arid climate has been removed by relatively permanent irrigation works. Such irrigated soils (or soils potentially useful under irrigation) are nearly level, have deep rooting zones, favourable permeability and water-holding capacity, and are easily maintained in good tilth. Some of the soils may require initial conditioning, including levelling to the desired grade, leaching of a slight accumulation of soluble salts or lowering of the seasonal water-table. Where limitations due to salts, water-table, overflow or erosion are likely to recur, the soils are regarded as subject to permanent natural limitations and are not included in Class 1.

Soils that are wet and have slowly permeable subsoils are not placed in Class 1, although some Class 1 soils may be drained as an improvement measure for increased production and ease of operation.

Soils in Class 1 that are used for crops need ordinary management practices to maintain productivity - both soil fertility and soil structure. Such practices may include the use of fertilisers and lime, cover and green-manure crops, conservation of crop residues and animal manures, and sequences of adapted crops.

Class 2 - Soils in Class 2 have some limitations that reduce the choice of plants or require moderate conservation practices

These soils require careful soil management, including conservation practices, to prevent deterioration or to improve air and water relations when the soils are cultivated. The limitations are few and the practices are easy to apply. The soils may be used for cultivated crops, pasture, range, woodland, or wildlife food and cover.

Limitations of soils in this class may include (singly or in combination) the effects of gentle slopes, moderate susceptibility to wind or water erosion or moderate adverse effects of past erosion, less than ideal soil depth, somewhat unfavourable soil structure and workability, slight to moderate salinity or sodium easily corrected but likely to recur, occasional damaging overflow, wetness correctable by drainage but existing permanently as a moderate limitation, and slight climatic limitations on soil use and management.

These soils provide the farm operators less latitude in the choice of either crops or management practices than those in Class 1. They may also require special soil conserving cropping systems, soil conservation practices, water control devices, or tillage methods when used for cultivated crops. For example, deep soils of this class with gentle slopes subject to moderate erosion when cultivated may need terracing, strip cropping, contour tillage, crop

---

[1]   Sources: Sections D.1 to D.3 - Olson (1974), Section D.4 - USDI (1953).

rotations that include grasses and legumes, vegetated water disposal areas, cover or green-manure crops, stubble mulching, fertilisers, manure and lime. The exact combinations of practices vary from place to place, depending on the characteristics of the soil, the local climate and the farming systems.

Class 3 - Soils in Class 3 have severe limitations that reduce the choice of plants or require special conservation practices, or both

Soils in Class 3 have more restrictions than those in Class 2 and when used for cultivated crops the conservation practices are usually more difficult to apply and to maintain. They may be used for cultivated crops, pasture, woodland, range, or wildlife food and cover.

Limitations of these soils restrict the amount of clean cultivation; the timing of planting, tillage and harvesting, and the choice of crops; or some combination of these limitations. The limitations may result from the effects of one or more of the following: moderately steep slopes; high susceptibility to water or wind erosion or severe adverse effects of past erosion; frequent overflow accompanied by some crop damage; very slow permeability of the subsoil; wetness of some continuing waterlogging after drainage; shallow depths to bedrock, hardpan, fragipan or claypan that limit the rooting zone and the water storage; low moisture-holding capacity; low fertility not easily corrected; moderate salinity or sodium; or moderate climatic limitations.

When cultivated, many of the wet, slowly permeable, but nearly level soils in this class require drainage and a cropping system that maintains or improves the structure and tilth of the soil. To prevent puddling and to improve permeability, it is commonly necessary to supply organic material to such soils and to avoid working them when they are wet. In some irrigated areas, part of these soils have limited use because of a high water-table, slow permeability and the hazard of salt or sodic accumulation. Each distinctive kind of soil in Class 3 has one or more alternative combinations of use and practices required for safe use, but the number of practical alternatives for average farmers is less that that for soils in Class 2.

Class 4 - Soils in Class 4 have very severe limitations that restrict the choice of plants, require very careful management, or both

The restrictions in use for soils in Class 4 are greater than for those in Class 3 and the choice of plants is more limited. When these soils are cultivated, more careful management is required and conservation practices are more difficult to apply and maintain. Soils in Class 4 may be used for crops, pasture, woodland, range, or wildlife food and cover.

Soils in this class may be well suited to only two or three of the common crops, or the harvest produced may be low in relation to inputs over a long period of time. Use for cultivated crops is limited as a result of the effects of one or more permanent features such as: steep slopes, severe susceptibility to water or wind erosion, severe effects of past erosion, shallow soils, low moisture-holding capacity, frequent overflows accompanied by severe crop damage, excessive wetness with continuing hazard of waterlogging after drainage, severe salinity of sodium, or moderately adverse climate.

Many sloping Class 4 soils in humid areas are suited to occasional but not regular cultivation. Some of the poorly drained, nearly level soils put in Class 4 are not subject to erosion but are poorly suited to intertillable crops because a long time is required for drying out in the spring and productivity for cultivated crops is low. Some of these soils are well suited to one or more special crops, such as fruits and ornamental trees and shrubs, but this suitability itself is not sufficient to place a soil in Class 4.

In subhumid and semi-arid areas, soils in Class 4 may produce good yields of adapted cultivated crops during years of above average rainfall: low yields during years of average rainfall; and failure during years of below average rainfall. During the low rainfall years soil must be protected, even though there can be little or no expectancy of a marketable crop. Special treatments and practices to prevent soil blowing, conserve moisture and maintain soil productivity are required. Sometimes crops must be planted or emergency tillage used for the primary purpose of maintaining the soil during years of low rainfall. These treatments must be applied more frequently or more intensively than on soils in Class 3.

Class 5 - Soils in Class 5 have little or no erosion hazards but have other limitations, impractical to remove, that limit their use largely to pasture, range, woodland, or wildlife food and cover

Class 5 soils have limitations that restrict the kind of plants that can be grown and that prevent normal tillage of cultivated crops. Though nearly level, some are wet, are frequently overflowed by streams, are stony, have climatic limitations, or some combination of these limitations. Examples of Class 5 soils are: those on bottom lands subject to frequent overflow that prevents the normal production of cultivated crops; nearly level soils with a growing season that prevents the informal production of cultivated crops; level or nearly level, stony or rocky soils; and ponded areas where drainage for cultivated crops is not feasible but where soils are suitable for grasses or trees. Because of those limitations, cultivation of the common crops is not feasible, but pastures can be improved and benefits from proper management can be expected.

Class 6 - Soils in Class 6 have severe limitations that make them generally unsuited to cultivation and limit their use largely to pasture or range, woodland, or wildlife food and cover

Physical conditions of soils placed in this class may require application of range or pasture improvements such as seeding, liming, fertilising, and water control with contour furrows, drainage ditches, diversions or water spreaders. They may have continuing limitations that cannot be corrected, such as steep slope, severe erosion hazard, effects of past erosion, stoniness, shallow rooting zone, excessive wetness of overflow, low moisture capacity, salinity or sodium, or severe climate. One or more of these limitations may render soils generally unsuitable for cultivated crops. But they may be used for pasture, range, woodlands or wildlife cover, or some combination of these.

Some soils in Class 6 can be safely used for the common crops, provided intensive management is used. Some are also adapted to special crops such as sodded orchards, blueberries or the like, requiring soil conditions unlike those demanded by the common crops. Depending upon soil features and local climate, they may be well or poorly suited to woodlands.

Class 7 - Soils in Class 7 have very severe limitations that make them unsuited to cultivation and that restrict their use largely to grazing, woodland or wildlife

The physical condition of Class 7 soils make it impractical to apply such pasture or range improvements as seeding, liming, fertilising, and water control with contour furrows, ditches, diversions or water spreaders. Soil restrictions are more severe than those in Class 6 because of one or more continuing limitations that cannot be corrected, such as very steep slope, erosion, shallow soil, stones, wet soil, salts or sodium, unfavourable climate, or other limitations that make them unsuited to common cultivated crops. They can be used safely for grazing, woodland, wildlife food and cover, or for a combination of these under proper management.

Depending upon the soil characteristics and local climate, soils in this class may be well or poorly suited to woodland. They are not suited to any of the common cultivated crops; in rare instances, some of these soils may be used for special crops under unusual management practices. Some areas of Class 7 may need seeding or planting to protect the soil and to prevent damage to adjoining areas.

Class 8 - Soils and landforms in Class 8 have limitations that preclude their use for commercial plant production and restrict their use to recreation, wildlife, water supply or to aesthetic purposes

Soils and landforms in Class 8 cannot be expected to return significant on-site benefits from management for crops, grasses or trees, although benefits from wildlife use, watershed protection or recreation may be possible.

Limitations that cannot be corrected may result from the effects of erosion or erosion hazard, severe climate, wet soil, stones, low moisture capacity and salinity or sodium.

Badlands, rock outcrop, sandy beaches, river wash, mine tailings and other nearly barren lands are included in Class 8. It may be necessary to give protection and management for plant growth to soils and landforms in this class in order to protect other more valuable soils, to control water or for wildlife or aesthetic reasons.

The assumptions on which land capability classification is based are defined by Klingebiel and Montgomery (1961) as follows:

a) A taxonomic (or natural) soil classification is based directly on soil characteristics. The capability classification (unit, subclass and class) is an interpretation classification based on the effects of a combination of climate and permanent soil characteristics on risk or soil damage, limitations in use, productive capacity, and soil management requirements. Slope, soil texture, soil depth, effects of past erosion, permeability, water-holding capacity, type of clay minerals and the many other similar features are considered permanent soil qualities and characteristics. Shrubs, trees or stumps are not considered permanent characteristics.

b) The soils within a capability class are similar only with respect to degree of limitation in soil use for agricultural purposes or hazard to the soil when it is so used. Each class includes many different kinds of soil, and many of the soils within any one class require unlike management and treatment. Valid generalisations about suitable kinds of crops or other management needs cannot be made at the class level.

c) A favourable ratio of output to input is one of several criteria used for placing any soil in a class suitable for cultivated crops, grazing or woodland use, but no further relation is assumed or implied between classes and output-input ratios. The capability classification is not a productivity rating for specific crops. Yield estimates are developed for specific kinds of soils and are included in soil handbooks and soil survey

Annex D: Land capability classification

reports. A favourable output–input ratio is based on long–time economic trends for average farms and farmers using moderately high level management. The ratio may not apply to specific farms and farmers but will apply to broad areas.

d) A moderately high level of management is assumed – one that is practical and within the ability of a majority of the farmers and ranchers. The level of management is that commonly used by the 'reasonable' men of the community. The capability classification is not, however, a grouping of soils according to the most profitable use to be made of the land. For example, many soils in Class 3 or 4, defined as suitable for several uses including cultivation, may be more profitably used for grasses or trees than for cultivated crops.

e) Capability Classes 1 to 4 are distinguished from each other by a summation of the degree of limitations or risks of soil damage that affect their management requirements for long–time, sustained use for cultivated crops. Nevertheless, differences in kinds of management of yields of perennial vegetation may be greater between some pairs of soils within one class than between some pairs of soils from different classes. The capability class is not determined by the kinds of practices recommended. For example, Classes 2, 3 or 4 may or may not require the same kinds of practices when used for cultivated crops, and Classes 1 to 7 may or may not require the same kind of pasture, range or woodland practices.

f) Surface water or excess water in the soil, lack of water for adequate crop production, stones, soluble salts and/or exchangeable sodium, or hazard of overflow are not considered permanent limitations where their removal is feasible. Improvements are considered to be feasible where the characteristics and qualities of the soil allow removal of the limitation and over broad areas it is now economically possible to remove the limitation.

g) Soils considered feasible for improvement by drainage, irrigation, stone removal, salts or exchangeable sodium removal, or protection from overflow are classified according to their continuing limitations in use or the risks of soil damage, or both, after the improvements have been installed. Differences in initial costs of the systems installed on individual tracts of land do not influence the classification. The fact that certain wet soils are in Classes 2, 3 and 4 does not imply that they should be drained. But it does indicate the degree of their continuing limitation in use or risk of soil damage, or both, even though adequately drained. Where it is considered not feasible to make improvements by one or more of these means, the soils are classified according to present limitations in use.

h) Soils already drained or irrigated are grouped according to the continuing soil and climatic limitations and the risk that affect their use under the present systems or feasible improvements.

i) The capability classification of the soils in an area may be changed when major reclamation projects are installed that permanently change the limitations for use or reduce the hazards or risks of soil or crop damage for long periods of time. Examples include establishing major drainage facilities, building levees or flood–retarding structures, providing water for irrigation, removing stones or large–scale grading of gullied land. Minor dams, terraces or field conservation measures subject to change in their effectiveness in a short time are not included in this assumption.

j) Capability groupings are subject to change as new information about the behaviour and responses of the soils becomes available.

k) Distance to market, kinds of roads, size and slope of the soil area, locations within fields, skill or resources of individual operators, and other characteristics of land ownership patterns are not criteria for capability groupings.

l) Soils with such physical limitations that common field crops can be cultivated and harvested only by hand are not placed in Classes 1, 2, 3 and 4. Some of these soils need drainage and/or stone removal before certain machinery can be used. This does not mean that mechanical equipment cannot be used on soils in capability Classes 5, 6 and 7.

m) Soils suited to cultivation are also suited to other uses such as pasture, range, forest and wildlife. Some not suited to cultivation are suited to all the rest of the named uses; others are suited only to pasture, range or wildlife; others only to forest and wildlife; and a few suited only to wildlife, recreation, and water–yielding uses. Groupings of soils for pastures, range, wildlife or woodland may include soils from more than one capability class. Thus, to interpret soils for these uses, a grouping different from the capability classification is often necessary.

n) Research data, recorded observation and experience are used as the bases for placing soils in capability units, subclasses and classes. In areas where data on response of soils to management are lacking, soils are placed in capability groups by interpretation of soil characteristics and qualifies in accord with the general principles about use and management developed for similar soils elsewhere.

D.2   Land-use capability classification of the Soil Survey of England and Wales

The classes defined by Bibby and Mackney (1969) are:

Class 1 - Land with very minor or no physical limitations to use

Soils are generally well drained, deep loams, sandy loams or silt loams, related humic variants, or peat, with good reserves of moisture or with suitable access for roots to moisture; they are either well supplied with plant nutrients or responsive to fertilisers. Sites are level or gently sloping and the climate is favourable. A wide range of crops can be grown, and yields are good with moderate amounts of fertiliser.

Class 2 - Land with minor limitations that reduce the choice of crops and interfere with cultivations

Limitations may include, singly or in combination, the effects of moderate or imperfect drainage, less than ideal rooting depth, slightly unfavourable soil structure and texture, moderate slopes, slight erosion and slightly unfavourable climate. A wide range of crops can be grown, though some root crops and winter-harvested crops may not be ideal choices because of difficulties in harvesting.

Class 3 - Land with moderate limitations that restrict the choice of crops or demand careful management, or both

Limitations may result from the effects of one or more of the following: imperfect or poor drainage, restrictions in rooting depth, unfavourable structure and texture, strongly sloping ground, slight erosion and moderately unfavourable to moderately severe climate.   The limitations affect the timing of cultivations and range of crops which are restricted mainly to grass, cereal and forage crops. Whilst good yields are possible, limitations are more difficult to overcome.

Class 4 - Land with moderately severe limitations that restrict the choice of crops or require very careful management practices, or both

Limitations are caused by one or more of the following: poor drainage difficult to remedy, occasional damaging floods, shallow or very stony soils, moderately steep gradients, slight erosion and moderately severe climate.   Climatic disadvantages combine with other limitations to restrict the choice and yield of crops and increase risks.   The main crop is grass, with cereals and forage crops as possible alternatives where the increased hazards can be accepted.

Class 5 - Land with severe limitations that restrict its use to pasture, forestry and recreation

Limitations are due to one or more of the following defects which cannot be corrected: poor or very poor drainage, frequent damaging floods, steep slopes, severe risk of erosion and severe climate.   High rainfall, exposure and a restricted growing season prohibit arable cropping although mechanised pasture improvements are feasible.   The land has a wide range of capability for forestry and recreation.

Class 6 - Land with very severe limitations that restrict use to rough grazing, forestry and recreation

Of the following limitations one or more cannot be corrected: very poor drainage, liability to frequent damaging floods, shallow soil, stones or boulders, very steep slopes, severe erosion and very severe climate. This land has limitations that are sufficiently severe to prevent the use of machinery for pasture improvement. Very steep ground which has some sustained grazing value is included. On level of gently sloping upland sites, wetness is closely correlated with peat or peaty or humus flush soils.

Class 7 - Land with extremely severe limitations that cannot be rectified

Limitations result from one or more of the following defects: very poorly drained boggy soils, extremely stony, rocky or boulder-strewn soils, bare rock, scree, or beach sand and gravel, untreated waste tips, very steep gradients, severe erosion and extremely severe climate.   Exposed situations, protracted snow cover and a short growing season preclude forestry although a poor type of rough grazing may be available for a few months.

D.3   Canadian land capability classification for forestry

The land capability classes and limitations, as defined for use in Canada (Canada Department of Forestry, 1965) are as follows; other forestry systems are covered in Laban (1981).

Annex D: Land capability classification

Class 1 - These soils have no significant limitations in use for crops

Class 2 - These soils have moderate limitations that restrict the range of crops or require moderate conservation practices

Limitations of soils in this class may be any one of the following: adverse regional climate, moderate effects of accumulative undesirable characteristics, moderate effects of erosion, poor soil structure or slow permeability, low fertility correctable with consistent moderate applications of fertilisers and lime, gentle to moderate slopes, occasional damaging overflow, or wetness correctable by drainage but continuing as a moderate limitation.

Class 3 - Soils in this class have moderately severe limitations that restrict the range of crops or require special conservation practices

Limitations of soils in this class are a combination of two of those described under Class 2 or one of the following: moderate climatic limitations including frost pockets, moderately severe effects of erosion, intractable soil mass or very slow permeability, low fertility correctable with consistent heavy applications of fertilisers and lime, moderate to strong slopes, frequent overflow accompanied by crop damage, poor drainage resulting in crop failure in some years, low water-holding capacity or slowness in release of water to plants, stoniness severe enough to handicap cultivation seriously and necessitate some clearing, restricted rooting zone or moderate salinity.

Class 4 - Soils in the class have severe limitations that restrict the range of crops or require special conservation practices or both

Limitations of soils in this class include the adverse effects of a combination of two or more of those described in Classes 2 and 3 or one of the following: moderately severe climate, very low water-holding capacity, low fertility difficult or unfeasible to correct, strong slopes, severe past erosion, very intractable mass of soil or extremely slow permeability, frequent overflow with severe effects on crops, salinity severe enough to cause some crop failure, extreme stoniness requiring a lot of clearing for annual cultivation, or very restricted rooting zone, but more than one foot of soil over bedrock or an impermeable layer.

Class 5 - Soils in this class have very severe limitations that restrict their capability to producing perennial forage crops, but improvement practices are feasible

Limitations of soils in this class include the adverse effects of one or more of the following: severe climate, low water-holding capacity, severe past erosion, steep slopes, very poor drainage, very frequent overflow, severe salinity permitting only salt-tolerant forage crops to grow, or stoniness or shallowness to bedrock that make annual cultivation impractical.

Class 6 - Soils in this class are capable of producing only perennial forage crops, and improvement practices are not feasible

Limitations of soils in this class include the adverse effects of one or more of the following: very severe climate, very low water-holding capacity, very steep slopes, very severely eroded land with gullies too numerous and too deep for working with machinery, severely saline land producing only salt-tolerant native plants, very frequent overflow allowing less than 10 weeks' effective growing, water on the surface of the soil for most of the year, or stoniness or shallowness to bedrock that makes any cultivation impractical.

Class 7 - Soils in this class have no capability for arable culture or permanent pasture

Soils in this class have limitations so severe that they are not capable of use for farming or pasture. These soils may or may not be able to support trees, native fruits, wildlife and recreation.

D.4   USBR land classification system

D.4.1   USBR land classes

a) Basic classes - The land classes are based on the economics of production and land development within ecologic areas. Hence, the production and repayment potentials will differ significantly between such areas. Although all classes will be found in any given ecologic area, they will not necessarily be found in a given project area. Four basic classes are used in the Bureau system to identify the arable lands according to their suitability for irrigation agriculture, one provisional class, and one class to identify the non-arable lands. The first three classes represent lands with progressively less ability to repay project construction costs. The excessive deficiency and restricted utility subclasses of Class 4 may have repayment ability ranging from less than that of Class 3 to more than that of Class 1 depending upon the particular utility involved. The number of classes mapped in a particular investigation depends upon the diversity of the land conditions encountered and other requirements as dictated by the objectives of the particular investigation.

b)   Class 1 - Arable - Lands that are highly suitable for irrigation farming, being capable of producing sustained and relatively high yields of a wide range of climatically adapted crops at reasonable cost. They are smooth lying with gentle slopes. The soils are deep and of medium to fairly fine texture with mellow, open structure allowing easy penetration of roots, air and water and having free drainage yet good available moisture capacity. These soils are free from harmful accumulations of soluble salts or can be readily reclaimed. Both soil and topographic conditions are such that no specific farm draining requirements are anticipated, minimum erosion will result from irrigation and land development can be accomplished at relatively low cost. These lands potentially have a relatively high payment capacity.

c)   Class 2 - Arable - This class comprises lands of moderate suitability for irrigation farming, being measurably lower than Class 1 in productive capacity, adapted to somewhat narrower range of crops, more expensive to prepare for irrigation or more costly to farm. They are not so desirable nor of such high value as lands of Class 1 because of certain correctable or non-correctable limitations. They may have a lower available moisture capacity, as indicated by coarse texture or limited soil depth; they may be only slowly permeable to water because of clay layers or compaction in the subsoil; or they also may be moderately saline which may limit productivity or involve moderate costs for leaching. Topographic limitations include uneven surface requiring moderate costs for levelling, short slopes requiring shorter length of runs, or steeper slopes necessitating special care and greater costs to irrigate and prevent erosion. Farm drainage may be required at a moderate cost or loose rock or woody vegetation may have to be removed from the surface. Any one of the limitations may be sufficient to reduce the lands from Class 1 to 2 but frequently a combination of two or more of them is operating. The Class 2 lands have intermediate payment capacity.

d)   Class 3 - Arable - Lands that are suitable for irrigation development but are approaching marginality for irrigation and are of distinctly restricted suitability because of more extreme deficiencies in the soil, topographic or drainage characteristics than described for Class 2 lands. They may have good topography, but because of inferior soils have restricted crop adaptability, require larger amounts of irrigation water or special irrigation practices, and demand greater fertilisation or more intensive soil improvement practices. They may have uneven topography, moderate to high concentration of salines or restricted drainage, susceptible of correction but only at relative high costs. Generally, greater risk may be involved in farming Class 3 lands than the better classes of land, but under proper management they are expected to have adequate payment capacity.

e)   Class 4 - Limited arable or special uses - Lands are included in this class only after special economic and engineering studies have shown them to be arable. They may have an excessive, specific deficiency or deficiencies susceptible of correction at high cost, but are suitable for irrigation because of existing or contemplated intensive cropping such as for market gardens (truck) and fruits; or they may have one or more excessive, non-correctable deficiencies thereby limiting their utility to meadow, pasture, orchard or other relatively permanent crops, but are capable of supporting a farm family and meeting water charges if operated in units of adequate size or in association with better lands. The deficiency may be inadequate drainage, excessive salt content requiring extensive leaching, unfavourable position allowing periodic flooding or making water distribution and removal very difficult, rough topography, excessive quantities of loose rock on the surface or in the plough zone, or cover such as timber. The magnitude of the correctible deficiency is sufficient to require outlays of capital for land development in excess of those permissible for Class 3 but in amounts shown to be feasible because of the specific utility anticipated. Subclasses other than those devoted to special crop use may be included in this class such as those for subirrigation, and sprinkler irrigation which meet general arability requirements. Also recognised in Class 4 are suburban lands which do not meet general arability requirements. Such lands can pay water charges as a result of income derived either from the suburban land and other sources or from other sources alone. The Class 4 lands may have a range in payment capacity greater than that for the associated arable lands.

f)   Class 5 - Non-arable - Lands in this class are non-arable under existing conditions, but have potential value sufficient to warrant tentative segregation for special study prior to completion of the classification, or they are lands in existing projects whose arability is dependent upon additional scheduled project construction or land improvements. They may have a specific soil deficiency such as excessive salinity, very uneven topography, inadequate drainage or excessive rock or tree cover. In the first instance, the deficiency or deficiencies of the land are of such nature and magnitude that special agronomic, economic or engineering studies are required to provide adequate information, such as extent and location of farm and project drains, or probable payment capacity under the anticipated land use, in order to complete the classification of the lands. The designation of Class 5 is tentative and must be changed to the proper arable class or Class 6 prior to completion of the land classification. In the second instance, the effect of the deficiency or the outlay necessary for improvement is known, but the lands are suspended from an arable class until the scheduled date of completion of project facilities and land development such as project and farm drains. In all instances, Class 5 lands are segregated only when

257

Annex D: Land capability classification

the conditions existing in the area require consideration of such lands for competent appraisal of the project possibilities, such as when an abundant supply of water or shortage of better lands exists, or when problems related to land development, rehabilitation and resettlement are involved.

g) Class 6 - Non-arable - Lands in this class include those considered non-arable under the existing project or the project plan because of failure to meet the minimum requirements for the other classes of land, arable areas definitely not susceptible to delivery of irrigation water or to provision of project drainage, and Classes 4 and 5 land when the extent of such lands or the detail of the particular investigation do not warrant their segregation. Class 6 irrigated land with water rights encountered in the classification will be delineated and designated Class 6W. Generally, Class 6 comprises: steep, rough, broken or badly eroded lands; lands with soils of very coarse or fine texture, or shallow soil over gravel, shale, sandstone or hardpan, and lands that have inadequate drainage and high concentrations of soluble salts or sodium. Excluding the position subclasses, the Class 6 lands do not have sufficient payment capacity to warrant consideration for irrigation.

D.4.2   USBR types of land classification

a) Standard types - Three types of land classification, each representing a standard scale of operation, are recognised. These are differentiated primarily on the amount of detail included and the accuracy of the results, and are designated reconnaissance, semi-detailed and detailed. The detail and accuracy necessary or desired in a particular investigation are determined by the objectives of the survey as outlined below. The diversity of land conditions, or anticipated use of the results in more detailed studies, however, may require or warrant a more detailed type of classification. The choice of type of land classification in all cases must be consistent with the purpose of the investigation and governed by the primary applications intended. Generally, only two of the three types of surveys will be utilised in the planning of projects: either a reconnaissance or a semi-detailed classification to provide preliminary information; and a detailed classification to provide information for project authorisation, construction and operation.

b) Reconnaissance - Reconnaissance land classification involves a general outline of land features of conspicuous importance in preliminary planning of irrigation development in a particular region. These surveys are accomplished on maps normally having a scale of 1:24 000 (2 000 ft to the inch) or on contact prints of aerial negatives. Generally, Classes 1, 2 and 3 and the pertinent subclasses are delineated with other lands designated as Class 6. Classes 4 and 5 and their subclasses may be delineated if the specific project conditions warrant. The survey is applicable and must be restricted to these general conditions:

   i)   for use on large areas where only general information on the extent of the arable land is required;

   ii) to determine the extent, location and quality of arable areas with the object of obtaining sufficient information from which to determine the justification of making detailed investigations.

c) Semi-detailed - Semi-detailed land classification involves careful examination of land features at about one-half mile intervals on potentially irrigable areas, whilst non-arable areas are covered in a more general manner. The separations between arable and non-arable lands are identified with considerable accuracy, but boundaries between classes and subclasses are delineated in less detail. This type of classification will be delineated on maps normally having a scale of 1:12 000 (1 000 ft to the inch), preferably aerial photographs adjusted to this scale. Generally, Classes 1, 2, 3 and 6 are mapped. Special subclasses under Classes 4 and 5 are differentiated when conditions warrant. Semi-detailed classifications will be made:

   i)   when the complexity of a particular project area precludes obtaining satisfactory results from a reconnaissance survey;

   ii) when the preliminary analyses of the engineering phases of the project are on a more detailed basis than the ordinary reconnaissance investigations;

   iii) when an unfeasible project is indicated, but more detailed information than contained in a reconnaissance survey is desired to support an unfavourable report;

   iv) when arable areas are involved which are a part of an ultimate project plan, but are not contemplated for irrigation under the initial development.

d) Detailed - Detailed land classification involves the examination of land features in sufficient detail to provide information as to the extent and character of the various lands in each 40 ac tract. Basic data with respect to various soil and subsoil conditions, topography and drainage are therefore obtained in detail for the purpose of determining proper land use, size of farm units, payment capacity or assessments,

258

irrigable area, irrigation requirements, land appraisal, irrigation and drainage systems, land development, costs and benefits. The delineation of this type of classification will be accomplished on maps normally having a scale of 1:4 800 (400 ft to the inch). A smaller scale, but not less than 1:12 000, may be used on fully developed areas or on highly uniform new land areas where no specific problems are associated with soils, topography or drainage and none are anticipated. Under such conditions the degree of accuracy resulting from a detailed survey is obtainable by meeting the general requirements of a semi-detailed survey provided necessary modifications are made, such as in intensity of sampling. Basic topographic maps are prerequisite to a detailed survey, except upon fully developed areas, but coverage may be limited to representative subareas if the information may be extended satisfactorily to the entire area. All classes and subclasses are considered and mapped as necessary to meet the objectives of survey. Detailed land classification will be used in:

i) feasibility investigations for project authorisation, development of the final plan for projects going into construction and Secretarial certification for construction as required by the 1953 Appropriation Act, except for development of suburban areas, supplemental water projects where the lands are fully developed, and highly uniform land areas, provided, in each instance, that no specific problems are associated with soils, topography or drainage, or are anticipated. Under these conditions a modified detailed classification, adequate existing land classification, land capability survey, detailed soil survey or other land inventory whose accuracy, reliability and applicability has been ascertained will be accepted in lieu of a detailed classification upon presentation of a justification by the Regional Director and its approval by the Commissioner. Such an existing land classification, soil survey or other inventory must be transformed in the field into the basic Bureau land classes with an accuracy at least equivalent to that of a semi-detailed survey and be supplemented by the necessary additional field-work, such as appraisal of topography, drainage and soil-water relationships and correlations with payment capacity. Where 100 000 or more acres are involved, the detailed land classification may be restricted initially to representative subareas and the results extended therefrom to the project area. As a minimum, a reconnaissance classification of the entire area will be a prerequisite to the selection of such subareas. This procedure may also be applied where the area involved is < 100 000 ac, provided prior approval is obtained from the Commissioner;

ii) reappraisal of Bureau of Reclamation operating projects except upon presentation by the Regional Director and approval by the Commissioner of a justification showing that:

- the production experience record of the project, supplemented by adequate land and water data, and an acceptable soil or land classification survey are available, or

- the nature of the reappraisal does not require land classification.

e) Minimum requirements – The minimum requirements by types of classification are summarised below in Table D.1 and general specifications for each land class are outlined in Table D.2. The minimum areas for segregation of classes in the field survey generally are subject to some adjustment in the irrigability analysis. In addition to size, the tract must be of such shape and so located as to permit its being farmed as a field. The smaller bodies of Class 6 lying within large arable bodies will be initially segregated in accordance with minimum requirements, although subsequent irrigability studies on an overall tract basis may warrant a change to an appropriate pay class. The depths indicated for the borings or pits and deep holes are minimum unless impervious materials are encountered at shallower depths. The number of examinations and analyses should be increased as necessary to meet specific objectives or with the complexity of the area. Full use should be made of road cuts, stream banks and similar exposures. Borings and pits are required for field appraisals, and samples of each significant layer of soil will be taken of at least the minimum number of borings for field laboratory analyses, including total soluble sales (salinity), pH (alkalinity) and reaction to dilute HCl (total carbonates). The deep holes and pits are required for full profile study of the soil, subsoil and substrata in representative areas and samples will be taken of each significant layer of at least the minimum number of borings for such analyses as: mechanical analysis; available moisture capacity; vertical-and-horizontal core, undisturbed water conductivity; disturbed water conductivity; apparent density; air porosity; exchangeable sodium; total soluble salts, pH; gypsum; and reaction to dilute HCl.

Annex D: Land capability classification

USBR minimum survey requirements

Table D.1

Criteria	Reconnaissance survey	Semi-detailed survey	Detailed survey	
			New lands	Fully developed or highly uniform new land areas
Land classes recognised	1-2-3-6	1-2-3-6	1-2-3-4-5-6	1-2-3-4-5-6
Scale of base maps	1:24 000	1:12 000	1:4 800	1:12 000
Maximum distances between traverses (miles)	1	0.5	0.25	0.5
Accuracy – (%)	75	90	97	97
Field progress per day (one land classifier and crew) – (miles2)	3-5	1-3	0.25-1	1-3
Minimum area of Class 6 to be segregated form larger arable areas – (ac)	4	0.5	0.2	0.2
Minimum area for change to lower class of arable land – (ac)	40	10	2	10
Minimum area for change to higher class of arable land – (ac)	40	20	10	20
Minimum soil and substrata examination Borings or pits (5 ft deep) per square mile Deep holes (10 ft or more) per township	1 1	4 2	16 4	4 2

Source: USBR (1953).

USBR land class general specifications   Table D.2

Land characteristics	Class 1 – arable	Class 2 – arable	Class 3 – arable

Soils

Land characteristics	Class 1 – arable	Class 2 – arable	Class 3 – arable
Texture	Sand loam to friable clay loam	Loamy sand to very permeable clay	Loamy sand to permeable clay
Depth (measurements in cm):			
To sand, gravel or cobble	90 plus – good free working soil of fine sandy loam or finer; or 105 of sandy loam	60 plus – good free working soil of fine sandy loam or finer; or 75–90 of sandy loam to loamy sand	45 plus – good free working soil of fine sandy loam or finer; or 60 to 75 of coarser-textured soil
To shale, raw soil from shale or similar material (15 less in each to rock and similar material	150 plus; or 135 with minimum of 15 of gravel overlying impervious material or sandy loam throughout	120 plus; or 105 with minimum of 15 of gravel overlying impervious material or loamy sand throughout	105 plus; or 90 with minimum of 15 of gravel overlying impervious material or loamy sand throughout
To penetrable lime zone	45 with 150 penetrable	35 with 120 penetrable	25 with 90 penetrable
Alkalinity	pH 9.0 or less, unless soil is calcareous, total salts are low and evidence of black alkali is absent	pH 9.0 or less, unless soil is calcareous, total salts are low and evidence of black alkali is absent	pH 9.0 or less, unless soil is calcareous, total salts are low and evidence of black alkali is absent
Salinity	Total salts not to exceed 0.2%. May be higher in open permeable soils and under good drainage conditions	Total salts not to exceed 0.5%. May be higher in open permeable soils and under good drainage conditions	Total salts not to exceed 0.5%. May be higher in open permeable soils and under drainage conditions

Topography

Land characteristics	Class 1 – arable	Class 2 – arable	Class 3 – arable
Slopes	Smooth slopes up to 4% in general gradient in reasonably large-size bodies sloping in the same plane	Smooth slopes up to 8% in general gradient in reasonably large-size bodies sloping in the same plane; or rougher slopes which are < 4% in general gradient	Smooth slopes up to 12% in general gradient in reasonably large-size bodies sloping in the same plane; or rougher slopes which are < 8% in general gradient
Surface	Even enough to require only small amount of levelling and no heavy grading	Moderate grading required but in amounts found feasible at reasonable cost in comparable irrigated area	Heavy and expensive grading required in spots but in amounts found feasible in comparable irrigated areas
Cover (loose rocks and vegetation)	Insufficient to modify productivity or cultural practices, or clearing cost small	Sufficient to reduce productivity and interfere with cultural practices. Clearing required but at moderate cost	Present in sufficient amounts to require expensive but feasible clearing

Drainage

Land characteristics	Class 1 – arable	Class 2 – arable	Class 3 – arable
Soil and topography	Soil and topographic conditions such that no specific farm drainage requirement is anticipated	Soil and topographic conditions such that some farm drainage will probably be required but with reclamation by artificial means appearing feasible at reasonable cost	Soil and topographic conditions such that significant farm drainage will probably be required but with reclamation by artificial means appearing expensive but feasible

Class 4 – limited arable

Include lands having excessive deficiencies and restricted utility but which special economic and engineering studies have shown to be irrigable

Class 5 – non-arable

Includes lands which will require additional economic and engineering studies to determine their irrigability and lands classified as temporarily non-productive pending construction of corrective works and reclamation

Class 6 – non-arable

Includes lands which do not meet the minimum requirements of the next higher class mapped in a particular survey and small areas of arable land lying within larger bodies of non-arable land

Source: USBR (1953); cf Table 5.7.

Annex D: Land capability classification

D.4.3   USBR map symbols

Standard mapping symbols and components for irrigated land classification surveys of the USBR are as follows:

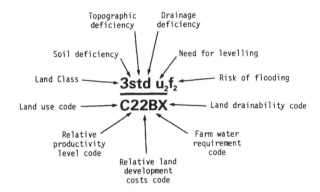

Basic land classes (see D.4.1) and subclasses:

Arable Class 1: 1
Arable Class 2: 2s, 2t, 2d, 2st, 2sd, 2td, 2std
Arable Class 3: 3s, 3t, 3d, 3st, 3sd, 3td, 3std
Limited Arable Class 4: pasture – 4Ps, 4Pt, 4Pd, 4Pst, 4Psd, 4Ptd, 4Pstd
       Similar subclasses for fruit 4F, rice 4R, vegetables in market gardens (truck) 4V,
       suburban 4H, sprinkler 4S and subirrigation 4U

Tentatively non–arable Class 5:
       Pending investigation: 5s, 5t, 5d, 5st, 5sd, 5td, 5std
       Pending reclamation: 5(1), 5(2s), 5(2t), etc
            Project drainage: 5d(1), 5d(2), 5d(2t), etc
            Similar subclasses for flooding: 5f
       Pending investigation or reclamation
            Isolated: 5i(1), 5i(2s), 5i(2t), etc
            Similar subclasses for high 5h and low 5l

Non–arable Class 6: 6s, 6t, 6d, 6st, 6sd, 6td, 6std
       Isolated: 6i(1), 6i(2s), 6i(2t), etc
       Similar subclasses for high 6h, low 6l and water right 6W
       (6W denotes water rights encountered in the classification)

Subclass designation:
       s – soils
       t – topography
       d – farm drainage

Soil appraisals 1/:
       k – shallow depth to coarse sand, gravel or cobbles
       b – shallow depth to relatively impervious substrata
       z – shallow depth to concentrated zone of lime
       v – very coarse texture (sands, loamy sands)
       l – moderately coarse texture (sandy loams, loams)
       m – moderately fine texture (silt loams, clay loams)
       h – very fine texture (clays)
       e – structure
       n – consistence
       q – available moisture capacity
       i – infiltration
       p – hydraulic conductivity
       r – stoniness
       y – soil fertility
       a – salinity and alkalinity

1/   Appraisals are defined further by use of subscript numerals, such as k_1, k_2 and k_3, indicating ranges of depths.

Topographic appraisals:
g – slope
u – surface
j – irrigation pattern
c – brush or tree cover
r – rock cover

Drainage appraisals:
f – surface drainage – flooding
w – subsurface drainage – water-table
o – drainage outlet

Land use:
C – irrigated cultivated
L – non-irrigated cultivated
P – irrigated permanent grassland
G – non-irrigated permanent grassland
B – brush or timber
H – suburban or homestead
W – waste or miscellaneous
ROW – right of way

Productivity and land development:
1, 2, 3, 4 or 6 denote land class level of factor, such as:
  Class 2 productivity, Class 2 development cost – '22'

Farm water requirement:
A – low
B – medium
C – high

Land drainability:
X – good
Y – restricted
Z – poor or negligible

263

Annex D: Land capability classification

Annex E

Proposal Preparation

Contents

Proposal Preparation

E.1   Introduction

This Annex discusses in general terms the principles behind proposal preparation. Checklists covering items that must be discussed and/or timed and/or costed in a proposal have already been given in Annex A.

E.2   Scope, purpose and impact of proposals

A project proposal may often be the principal factor in winning a contract.   It is therefore essential that proposals should be carefully prepared and presented, and wherever possible tailored to suit the client's exact requirements.   These latter may vary considerably, and may not always tally with the issued terms of reference, so wherever possible each project should be discussed in advance with the client, preferably on site.

The contents of proposal documents are often transferred with few changes into the final contract, and should therefore be composed so that only practical suggestions and schedules are submitted, with realistic costings.   There is a tendency, for the sake of completeness, to overspecify a survey – which should be resisted.   A proposal should be as detailed as needed, and as easy as possible to read and understand.   Section E.3 below discusses aspects of the 'polish' required in a submission;   this will depend on the client, and should not be over or underdone.

Notwithstanding the above, another major factor in winning some contracts is the quality of the team proposed, and particularly that of the Team Leader.   The World Bank (1981) normally weights the main components of a proposal as follows:

    40–60%   –   Suitability of key personnel – of which Western (1978) estimates about half is attributed to the Project Manager or Team Leader

    25–40%   –   The work plan

    10–20%   –   The firm's general experience

Western (1978) also states that 'this considered emphasis on the apparent experience and skill of the proposed staff is because major investigators such as the World Bank only invite proposals from a short list of consultant organisations of comparable competence and experience.   The most important difference between these organisations is the staff they can provide for a given contract, and especially the senior staff'.

E.3   Contents and presentation

E.3.1   General requirements

A proposal should indicate a thorough understanding of the scope and requirements of a project, and state in detail the topics, the methodology and the timings that are proposed. These 'technical' aspects should be supported by evidence of competence and experience, both of the consultant company and the individual team members, and by a clear statement of all costs (local and foreign exchange) with schedules of payments.   The client's and the consultant's responsiblities should be carefully distinguished, and clauses inserted to protect the company from the effects of delays or incompetence in inputs not under their control.

In general, the work proposed should not exceed the requirements of the terms of reference, and must be achievable within the specified time limits.   If unrealistic terms of reference have been issued, alternative approaches can be specified (and costed) separately.

Annex E: Proposal preparation

The expanded general list of contents given in Subsection E.3.2 is recommended as a guide for most proposals; actual contents will depend on the project and the place of the soil and land suitability studies within any broader investigations.   The division of material between the main text and the annexes will likewise vary.

For ease of production, the presentation of a proposal should, wherever possible, follow the standard company format.   Careful consideration should be given to the cover design and the quality of paper to be used; an impression of close attention to detail for each individual proposal should be conveyed, rather than an 'off the shelf' approach.   Western (1978) includes the following additional points to be watched in the presentation:

a)   the proposal must succeed on the worth and clarity of its contents.   Be careful with photographs;   they may be interpreted as window-dressing unless they are really germane to the proposal;

b)   printing on both sides of the paper makes for neatness and may have a psychological advantage in reducing the bulk of material confronting busy readers.   There should be an adequate margin for the client reader's notes;

c)   use a standard set of covers, bindings and page sizes, as far as possible;

d)   the contents of the proposal as a whole and of each individual section should follow a logical sequence;

e)   the relative lengths of different sections and subsections should have some regard to their relative importance;

f)   paragraph arrangement and numbering must be uniform throughout;

g)   avoid descriptive material of what the client already knows and avoid detailed reiteration of tasks already described in the terms of reference;

h)   stress any special competence the consultant may have in what are known, and stated by the client, to be follow-up operations and implementation phases in which the consultants would be eligible to participate or for which they could tender;

i)   a map of the project area should be included with an inset map to indicate its location with the country or region.   The map should show and name principal places and geographical features, including all those mentioned in the proposal.

E.3.2   Specific guidelines for proposal contents

a)   Covering letter
   - response to invitation to submit proposal
   - introduce the consultant organisation and very briefly indicate competence
   - mention specific contacts/visits connected with the proposal
   - mention important dates (especially start)
   - mention/introduce Team Leader and/or Project Director
   - indicate willingness to discuss proposal and amend contents

b)   Summary
   - (usually for longer proposals only)
   - outline of work to be done by consultants
   - outline of client's responsibilities
   - summary of any deviations proposed from terms of reference
   - summary of costs
   - view of project in relation to regional/national development

c)   Introduction/background and scope
   - brief summary of and cross-reference to terms of reference
   - outline of proposed study and justification of approach

d)   Work specification
   - work to be undertaken, methodology to be used
   - reference to standards to be adhered to (eg FAO, USDA)
   - reports and maps: presentation and contents, number of copies and (if appropriate) distribution

e)   Personnel and time schedules
   - organisational overview
   - team required and management structure
   - bar chart of activities/individuals 1/
   - liaison with client and other responsible bodies

1/   Note that timings should be referred to date of signing of contract or date of mobilisation, which in turn should be referred to physical constraints such as wet season access or public holidays (see Section A.3).

    – potted CVs (one paragraph each) of key team members
    – use of counterparts
    – mobilisation: cross-reference to payments schedule
    – reporting: schedule and contents; cross-reference to payments

f) <u>Client's responsibilities</u>
    – provision/assistance with entry visas/work permits etc
    – provision of counterparts
    – provision of back-up services (eg accommodation, offices, laboratories, equipment)
    – provision of vehicles
    – provision of background reports, research data, maps, AP, space imagery

g) <u>Project costs</u>
    – equipment, air fares, accommodation
    – personnel (fees)
    – client's cost responsibilities
    – local and foreign exchange items
    – contingency and inflation allowances
    – period for which costs valid
    – schedule of payments: mobilisation

h) <u>Annexes</u>
    – more detailed information on any or all of the following:
    – details of methodology
    – staff CVs in standardised format (an obligatory annex in nearly every proposal)
    – logistical needs (accommodation; vehicle needs; local equipment etc)
    – cost breakdowns
    – any legal necessities including: tax situation; confidentiality; right of access; termination; security of personnel and force majeure; arbitration
    – company description, brochure and project profiles (unless previously sent when prequalifying)

Annex E: Proposal preparation

Annex F

Soil Suitability Guidelines for Major Tropical Crops

Contents

Annex F: Soil suitability for crops

List of Tables

Soil Suitability Guidelines for Major Tropical Crops

F.1   Introduction

The aim of this Annex is to assist the soil surveyor in the field with land evaluation for specific crops.   Only basic information is given here but references to more specific data are included.   The format consists of brief notes outlining the specific soil and climatic requirements for individual crops and, wherever possible, the range of tolerances for each crop.

Compilation of a simple set of notes such as these is hampered for several reasons: first, the authors of the various publications quoted are usually agronomists, and references to soil conditions are often limited and unquantified.   An agronomist and a soil surveyor/land use planner may see the soil in very different ways, with the result that some of the statements used in the notes on crop requirements are ambiguous.   Second, data from different sources are often contradictory, and many quoted figures cannot be used, or have had to be amended.   Third, much of the published information is specific to individual regions and may not apply to other localities;   wherever possible this has been pointed out.   Fourth, the laboratory methods used to measure amounts of major or minor nutrients are seldom stated.   As there can be considerable variation between the results of different analytical methods (see Chapter 7), all figures quoted should be regarded as guidelines only.

F.2   General soil and yield data

It is noteworthy that the optimal soil requirements of the majority of fairly common tropical and subtropical crops, with the exceptions of rice, jute, sago and cocoyams, are very similar;   these general conditions of soil fertility are shown in Table F.1.   In practice ideal conditions are rarely attained, so it is the extent to which a crop will grow and yield economically under various suboptimal conditions that will largely decide its cultivated area.   In addition, on any given project different economic and social constraints will operate;   these are too diverse to be considered here, but are often very important factors in determining the final choice of crops.

Typical attainable yields and general soil limitations of tropical crops are shown in Tables F.2 to F.4.   'Best possible' yields are tabulated in FAO (1979b, p 5), as are general crop growth requirements which have been reproduced here as Tables F.4 to F.8.   Sections F.3 to F.15 following these tables describe the requirements for specific crops;   for some of those not covered in these sections, the following references are recommended:

Alfalfa or lucerne	–	Bolton (1962), Bogdan (1977)
Cassava	–	Jones (1959)
Coconuts	–	Child (1973), Jenkin and Foale (1968)
Cotton	–	Prentice (1972)
Dates	–	Dowson and Atan (1962)
Pasture and fodder	–	Bogdan (1977)
Pulses	–	Smartt (1976)
Root crops	–	Kay (1973)
Sisal	–	Lock (1969)
Soyabean	–	Norman (1963, 1978)
Tobacco	–	Akehurst (1981)
Tropical fruits	–	Samson (1980)
Yams	–	Coursey (1967)

More general crop information is contained in ILACO (1981);   de Geus (1973);   Williams (1975);   Doorenbos and Kassam (1979);   and for East Africa, Acland (1971).   Other crop-related references to note are: Webster and Wilson (1980) and Wrigley (1982) for general tropical agriculture;   Purseglove (1975) for crop botany, and Holm et al (1977) and Kasasian (1971) for major weed species and their control.

Annex F: Soil suitability for crops

General conditions affecting soil fertility

Soil property	Conditions favouring high soil fertility	Conditions unfavourable to high soil fertility
Depth to limiting horizon	> 150 cm	< 100 cm
Texture	Loam, sandy clay loam, sandy clay; clay (if structure and consistence favourable)	Sand, loamy sand; heavy clay
Structure and consistence	Moderate or strong, fine or medium structure; friable consistence	Massive, or coarse structure, with very firm consistence
Moisture conditions	Free drainage with good moisture retention	Substantial drainage impedance; low moisture retention and rapid permeability
Plant nutrients	High levels	Low levels
Cation-exchange capacity	Medium to high levels (> 20 me/100 g in topsoil, > 10 me/100 g in lower horizons)	Low levels
Weatherable minerals	Present within 200 cm	Absent above 200 cm
Reaction	Generally pH 5.0 to 8.0, but varies with crops	See previous column
Salinity	Soluble salts and exchangeable sodium low	High soluble salts or exchangeable sodium
Organic matter	Adequate in relation to levels under natural vegetation	Low levels

Source: Young (1976).

Typical yields of selected tropical crops

Crop	Mean yields for all less developed countries	
	1961-65 $(kg\ ha^{-1})$	1970-72
Banana	12 000	13 000
Beans, field	430	450
Cassava	8 690	9 490
Cocoa	280	320
Cocoyam	4 930	4 960
Coffee	440	500
Cotton	670	780
Groundnut	810	790
Maize	1 130	1 280
Millet, bulrush	520	590
Rice	1 630	1 830
Sisal	780	740
Sorghum	640	730
Soyabean	750	1 120
Sugarcane	46 900	50 700
Sunflower	690	790
Sweet potato	6 480	6 280
Tea	870	970
Tobacco	810	850
Yam	9 130	9 280

Owing to the predominance of unimproved methods, the mean yields for less developed countries are fairly representative of traditional farming.

Source: Mainly after Young (1976, p 310).

Indicative yields of selected tropical and subtropical crops

Table F.3

Common name	Botanical name	Indicative annual yield (t ha^{-1}) [1]	Conditions applying to yield estimate	Plant density	Comments	Source
Alfalfa or lucerne	Medicago sativa	2-2.5 (t per cut)	Hay with 10% moisture content. Good experimental station yield in subtropics; cutting interval 25-30 days; Yield in tropics about 40% lower	-	2-12 cuts per growing season; yields depend on climate, highest in second year. With mild winters, crop grown for 3-4 years. Moisture content of fresh material is ≈ 80%	[4]
		30-40	Fresh material with 20% dry matter: rainfed [2]	-	-	[5]
		80-100	Fresh material with 20% dry matter: irrigated [2]	-	-	[5]
Banana	Musa spp	40-60	Good commercial yield	Between 2 x 2 and 5 x 5 m, or 400-2 500 plants ha^{-1}, respectively	Under poor management, plant crop yields best; under good management, first ratoon is best	[4]
		15-25	Average farmer: rainfed [2]	1 500 trees ha^{-1}		[5]
		35-50	Average farmer: irrigated [2]	1 500 trees ha^{-1}		[5]
Bean	Phaseolus vulgaris	6-8	Fresh	Plant and row spacing vary respectively from 5-10 cm and 50-75 cm	Dry conditions produce more fibrous, lighter leaf	[4]
		1.5-2	Dry. Good commercial yields in favourable conditions under irrigation			
		0.5-1.0	Rainfed: average farmer			[5]
		1.0-1.5	Irrigated: average farmer			[5]
Cabbage	Brassica oleracea var capitata	25-35	Fresh heads: commercial yield under normal rainfed conditions, good management	Depends on head size: spacing 30-50 cm for 1.0-1.5 kg heads; 50-90 cm for heads up up to 3 kg. 30 000-40 000 plants ha^{-1}	Small, poor quality heads produced under limited water supply	[4]
		80	Fresh heads: approx upper limit to commercial yields under ideal climate, with good irrigation and management			
		10-20	Average farmer: rainfed			[5]
		20-40	Average farmer: irrigated			[5]

Notes: See page 279.

cont

271

Table F.3 cont

Common name	Botanical name	Indicative annual yield (t ha^{-1}) [1]	Conditions applying to yield estimate	Plant density	Comments	Source
Cassava	Manihot esculenta	5-15	Average rainfed smallholder yield, moderate to good soil	80-100 cm x 80-140 cm; or ridges 75-120 cm apart	Yields vary depending on soil, climate, cultivar, age at harvest	
		30-40	Normal commercial plantation yield	90 x 90 cm planting gives 10 000 plants ha^{-1}		[10]
		> 50	Commercial plantation yields, using high-yielding cultivars			
		15-20	Average farmers' yields: rainfed			
		25-35	Average farmers' yields: irrigated			[5]
Citrus	Citrus spp	Good commercial yield (and fruits per tree):		4 x 4 m to 8 x 8 m, or 200-800 trees ha^{-1}	Fresh citrus approximately 85% water (except limes, 70%)	[4]
orange		25-40 (400-500)				
grapefruit		40-60 (300-400)				
lemon		30-45				
mandarin		20-30				
		Average farmer yields [2]:				
orange		10-20 (rainfed) 20-30 (irrigated)		250 trees ha^{-1}		[5]
grapefruit		8-15 (rainfed) 15-25 (irrigated)		150 trees ha^{-1}		
mandarin		8-15 (rainfed) 15-25 (irrigated)		200 trees ha^{-1}		
Cocoa	Theobroma cacao	0.6-2.0	Dried beans: approx rainfed yield, good climate and soil plus improved management	1 000-1 500 trees ha^{-1} reduced on thinning to about 650 trees ha^{-1}. Shade trees at about 15 x 15 m	Main feeding roots concentrated 0-15 cm	[14][15]
		0.8-1.5	Dried beans: approx range for smallholders to estates: rainfed			[5]
Coconut	Cocos nucifera	0.026 (t copra per tree)	Minimum average yield of husked nuts for tree selection (see comments at right)	Indicative figure is 148 trees ha^{-1}	Yield figure is equivalent to 33 nuts (or 26 kg copra) per tree yr^{-1}. At 148 trees ha^{-1} this is about 3.7 t ha^{-1} of copra. Good oil extraction: 65% of copra	[11]
		1.5-2.5	Approx range of rainfed yields, smallholder - estate		Copra is 60-65% oil	[5]

Notes: See page 279.

cont

Table F.3 cont

Common name	Botanical name	Indicative annual yield (t ha^{-1}) [1]	Conditions applying to yield estimate	Plant density	Comments	Source
Cocoyam ("old cocoyam")	Colocasia esculenta		Indicative rainfed smallholder yields (but note that yields vary greatly with location):	Under sole crop conditions, high density is about 60 x 60 cm (but wider is recommended in high rainfall areas with heavy cloud) or 27 225 plants ha^{-1}. When intercropped, population as low as 5 000 plants ha^{-1}	High shade tolerance. Yields depend on cultivar, crop duration and cultural conditions. Colocasia is the cocoyam of SE Asian origin; also called taro. High K and Mg needed for good yields	[10]
		≤ 5	Low			
		10	Medium			
		> 20	High			
Cocoyam ("new cocoyam")	Xanthosoma spp	5-7	Low yields under tropical peasant agriculture	0.6 x 0.6 m to 1.8 x 1.8 m: at 0.9 x 0.9 m spacing, plant material is 2.5-5.0 t ha^{-1}	High shade tolerance. Xanthosoma is the cocoyam of tropical American origin; also called tan(n)ia. Use is increasing (especially in W Africa) since it is more blight-resistant than Colocasia. High K and Mg needed for good yields	[10]
		12.5-20	Good average yield under tropical peasant agriculture			
Coffee	Coffea arabica	0.5-1.2	Clean green coffee (hulled beans); rainfed: Indicative smallholder yield	Approx 2.75 x 2.75 m	More susceptible to disease than robusta; so better response to intensive management on estates	
		1.0-2.0	Indicative estate yield			
	Coffea canephora (robusta coffee)	0.5-1.2	Indicative smallholder yield	Approx 3.0 x 3.0 m	Hardier than arabica, requires less inputs	[8]
		0.8-1.5	Indicative estate yield			
Cotton	Gossypium hirsutum	4-5	Good commercial seed yield, irrigated	30 x 50 cm to 50 x 100 cm	35% of yield figure is lint	[4]
		1-1.5	Average farmer: rainfed)			
		2-3	Average farmer: irrigated)		34-38% is lint	[5]
Grape	Vitis spp	15-30	Good average commercial yield in subtropics	1.5 x 3.5 m to 4.5 x 5.5 m	Fresh fruit contains ≈ 80% water. Yields vary greatly with year and individual vine	[4]
		3-5	Average farmer: rainfed [2]	2 000 vines ha^{-1}		
		5-10	Average farmer: irrigated [2]			[5]

Notes: See page 279.

cont

Table F.3 cont

Common name	Botanical name	Indicative annual yield (t ha⁻¹) 1/	Conditions applying to yield estimate	Plant density	Comments	Source
Groundnut	Arachis hypogaea	2-3	Unshelled nut: good commercial yield: rainfed, good management	Row spacing 0.75-1.0 m. 120 000 plants ha⁻¹ (small pod varieties) to 70 000 plants ha⁻¹ (large pod varieties)	Approx 40% of unshelled weight is shell weight	4/
		3.5-4.5	Unshelled nut: good commercial yield: irrigated, good management			4/
		0.5	Typical smallholder yield: rainfed, no improvement		50% oil	6/
		1.0-2.0	Average farmer: rainfed			5/
		1.5-2.0	Average farmer: irrigated			5/
Maize	Zea mays	6-9	Good commercial grain yield: irrigated	Row spacing 0.5-1.0 m. Usually grown in double rows with 5-10 cm spacing, in beds 30-50 cm apart	10-13% of fresh weight is moisture	4/
		> 3	Good, rainfed smallholder yield, with improved seeds and fertiliser			
		0.5-1.5	Typical rainfed smallholder yield with no improvements			6/
		1.5-3.0	Average farmer: rainfed			5/
		4.0-5.0	Average farmer: irrigated			5/
Millet	Various	1.0-2.0	Average farmers' yields: rainfed			
Oil palm	Elaeis guineensis	3-4	Palm oil yield: good estate management. In addition to 0.5-1.0 t ha⁻¹ yr⁻¹ kernels	Varies with climate, rate of palm development etc. Optimum for Africa about 143 palms ha⁻¹ (9 m triangular spacing); for Malaysia 158 palms ha⁻¹	Yields depend greatly on variety; planting intensity; age of trees; time of harvest, etc. Oil contents 50% of both kernels and fresh fruit pulp. Measurements indicate < 1% difference in cumulative yield over 25-30 years caused by 10% variation in planting density	5/13/

Notes: See page 279.

cont

274

Table F.3 cont

Common name	Botanical name	Indicative annual yield (t ha⁻¹) 1/	Conditions applying to yield estimate	Plant density	Comments	Source
Onion	Allium cepa	35-45	Good commercial yield under irrigation	Row spacing 0.5-1.0 m. Usually grown in double rows with 5-10 cm spacing in beds 30-50 cm apart	10-13% of fresh weight is water	4/
		5-10	Average farmer: rainfed			
		10-20	Average farmer: irrigated			5/
Pepper	Capsicum spp	10-15	Fresh fruit: normal commercial yield	90 x 40-60 cm	Yields vary greatly with climate, especially growing period, which affects number of pickings	4/
		20-25	Fresh fruit: approx upper limit to commercial, irrigated yields under ideal conditions			
Pineapple	Ananas comosus	25-35	Fresh fruit: good commercial yield for subtropical production	Usually grown in double rows with 60-30 cm spacings in beds 75-90 cm apart		4/
		15-25	Fresh fruit: equivalent yield in tropics			
		25	Average farmer: rainfed	40 000 plants ha⁻¹		
		40	Average farmer: irrigated			5/
Rice (paddy)	Oryza sativa	6-8	Unhusked grain: good commercial yield under fully controlled irrigation with high inputs	15 x 15 cm to 30 x 30 cm; 10^6 to 15 x 10^6 plants ha⁻¹ after transplanting	Milling percentage of rice is about 65%	6/
		3-4	Unhusked grain: good yields under food irrigation	80-100 kg ha⁻¹ (direct sowing), or 20-40 kg ha⁻¹ (nursery). Nursery area: field area ratio 1:20-1:30		
		0.5-1.5	Unhusked grain: typical unimproved smallholder yields		Lower yield for upland rice; high yield for swamp rice	
		1.5-2.5	Unhusked grain: average farmers' yield: rainfed			
		4.0-5.0	Unhusked grain: average farmers' yield: irrigated			5/

Notes: See page 279.

cont

275

Table F.3 cont

Common name	Botanical name	Indicative annual yield (t ha⁻¹) 1/	Conditions applying to yield estimate	Plant density	Comments	Source
Rubber	Hevea brasiliensis	1-2	Dry rubber: approx average yield under good conditions, improved management (see comments at right)	Approx 385 trees ha⁻¹ at start of tapping	With stimulant tapping (from tapping year 11 onwards) average over 25 year life of tree is about 2 t ha⁻¹ dry rubber. Peak yields from unstimulated tapping in years 11-16	9/
Safflower	Carthamus tinctorius	1.0-2.5	Good rainfed yield			
		2.0-4.0	Good irrigated yield			4/
		0.8-1.3	Average farmer: rainfed		40% oil	
		1.5-2.0	Average farmer: irrigated		40% oil	5/
Sisal	Agave sisalana	2	Average plantation yield of fibre	35 plants m⁻¹ in row: rows 50-80 cm apart	First cutting at 2 years old, then 8 annual cuttings	12/
		1.5-2.0	Dry processed fibre: average farmers' yield: rainfed			5/
Sorghum	Sorghum bicolor	3.5-5.0	Good commercial yield of grain under irrigation	100 000-150 000 plants ha⁻¹	Grain has 12-15% moisture. Spate irrigation yield figure for area with growing period 90 days, ET_m 3/ = 425 mm day⁻¹ and net depth applied water = 300 mm	4/
		0.8-1.3	Indicative yield under spate irrigation, traditional methods			
		2-3	Indicative smallholder yields with irrigation, fertiliser and improved seeds			
		0.2-0.8	Indicative rainfed smallholder yields, no improvements			6/
		1.3-2.0	Average farmer: rainfed			
		4.0-5.0	Average farmer: irrigated			5/

Notes: See page 279.

cont

276

Table F.3 cont

Common name	Botanical name	Indicative annual yield (t ha^{-1}) 1/	Conditions applying to yield estimate	Plant density	Comments	Source
Soyabean	Glycine max	1.5-2.5	Good seed yield: rainfed	30-40 seeds per row: 40-60 cm between rows	Yields highly dependent on water availability, fertilisers and row spacing. Seed contains 6-10% moisture	4/
		2.5-3.5	Good seed yield: irrigated			
		0.8-1.3	Average farmer: rainfed		12-20% oil	
		1.5-2.0	Average farmer: irrigated		12-20% oil	5/
Sugarbeet	Beta vulgaris	40-60	Good fresh beet yield	3-6 beets m^{-1} of row in single or double rows, with spacings respectively 50-70 cm and 50-100 cm	15% sugar	4/
		20-30	Average farmer: rainfed			
		40-45	Average farmer: irrigated			5/
Sugarcane	Saccharum officinarum	70-100	Good yield rainfed estate cane in humid tropics	Rows 1.0-1.4 m apart: 20 000-35 000 sets ha^{-1}	Plant and first ratoon yields are higher than for later ratoons. Normally plant and 3 ratoons for estates; smallholder practices highly variable. Sugar content about 10% of harvested cane, depending on temperature and moisture regime	4/7/
		110-150	Good yield irrigated estate cane in tropics and subtropics			
Sunflower	Helianthus annuus	0.8-1.5	Normal seed yield: rainfed	Row spacing about 90 cm: 60 000 plants ha^{-1}		
		2.5-3.5	Common seed yield: irrigated			4/
		1.0-1.5	Average farmer: rainfed		40% oil	
		1.5-2.0	Average farmer: irrigated		40% oil	5/

Notes: See page 279.

cont

Annex F: Soil suitability for crops

Common name	Botanical name	Indicative annual yield (t ha^{-1}) 1/	Conditions applying to yield estimate	Plant density	Comments	Source
Sweet potato	Ipomoea batatas	2.5-50	Normal range of attainable yields	22.5-30 cm x 60-70 cm	Yields vary greatly depending on cultivar, climate, and cultural conditions	10/
		17.5-20	'Satisfactory' yields			
		5-10	Average farmer: rainfed			
		12-18	Average farmer: irrigated			5/
Tea	Camellia sinensis		General average of made tea:		Yield varies greatly with age and spacing of bushes	
		0.2	For bushes 18-30 months old			
		1.7	For bushes at maturity (say ≥ 10 years)			16/
		1.5-2.5	Fermented leaf: yield range smallholder to estates			5/
Tobacco	Nicotiana tabacum	2.0-2.5	Good commercial yield fresh leaf, under adequate moisture regime	After transplanting: 0.9-1.2 m x 0.6-0.9 m	Leaf quality affected by fertiliser and irrigation practice	4/
		0.5-1.0	Flue-cured leaf } average farmers' yield: rainfed			
		1.0-1.5	Air-cured leaf }			5/
		1.0-2.0	Flue-cured leaf } average farmers' yield: irrigated			
		1.5-3.0	Air-cured leaf }			
Tomato	Lycopersicon esculentum	45-65	Fresh fruit: good commercial yield: irrigated	30-60 cm x 60-100 cm: 40 000 plants ha^{-1}	Frequent irrigation improves quality, reduces dry matter content and acidity	4/
		10-20	Average farmer: rainfed			
		20-40	Average farmer: irrigated			5/

Notes: See page 279.

cont

Table F.3 cont

Common name	Botanical name	Indicative annual yield (t ha⁻¹) [1]	Conditions applying to yield estimate	Plant density	Comments	Source
Watermelon	Citrullus vulgaris	25-35	Good commercial yield: irrigated	60-90 cm x 180-240 cm	Quality very dependent on water regime	[4]
		10-20	Average farmer: rainfed			
		20-40	Average farmer: rainfed			[5]
Wheat	Triticum spp	4-6	Grain: good commercial yield: irrigated	For irrigation; 100-120 kg ha⁻¹ (drilled) or 100-140 kg ha⁻¹ (broadcast)		[4]
		0.7	Grain: indicative smallholder yield – unimproved, rainfed			[6]
		1.3-2.0	Average farmer: rainfed			
		3.0-5.0	Average farmer: irrigated			[5]
Yam	Dioscorea spp	7.5-17.5	Indicative gross smallholder yields (rainfed): W Africa	Common spacings: 1.2 x 1.2 m; 1.2 x 0.9 m; 1.8 x 0.6 m. In general, wider spacings lead to lower yields ha⁻¹	Yields vary greatly, depending on cultivar, climate, soil etc. Note that net yields are about 2.5 t ha⁻¹ less than gross yields, to allow for propagation material	[10]
		12.5-25	SE Asia			
		20-30	W Indies			
		12.0-25.0	Average farmer: rainfed			[5]

Notes:
[1] Examples of actual yields are given, for example, in Ruthenberg (1980); estimates for other crops are summarised in ILACO (1981, p 472f).
[2] Note that at least 50% differences in yields can occur at a result of management and soil differences in a given area.
[3] For perennial crops: lower indicative annual yield figure corresponds to smallholder production; higher to estate. ET_m = maximum evapotranspiration rate.

Sources:
[4] Doorenbos and Kassam (1979)
[5] ILACO (1981, p 519)
[6] Ruthenberg (1980)
[7] R A Yates (personal communication)
[8] H W Mitchell (personal communication)
[9] J S O Suttie (personal communication)
[10] Kay (1973)
[11] Child (1964)
[12] Lock (1969)
[13] Hartley (1977)
[14] Young (1976)
[15] Wood (1975)
[16] Eden (1976)

279

Table F.4

Soil requirements and limitations for selected tropical and subtropical crops 1/

Crop	Texture: Fine	Texture: Medium	Texture: Coarse	Texture: Very coarse tolerated	Drainage: Free essential or desirable	Drainage: Imperfect well tolerated	Drainage: Poor tolerated or needed	Drainage: Tolerance to short periods of waterlogging 2/	Drainage: Minimum groundwater depth (cm) 3/	Minimum rooting depth class 4/	Moisture: Drought resistance 2/	Moisture: High AWC important	Moisture: Low AWC well tolerated	Erosion hazard 2/5/	Reaction: Optimum pH	Reaction: Range of pH tolerance for satisfactory yield 5/	Nutrient needs: General level of requirements 2/	Nutrient needs: Specific requirements	Salinity tolerance 2/
Cereals																			
Barley		+	+		+				60	M	L		+	H	6.5-7.8	5.5-8.0	H		H
Maize	+	+	+		+			M	75	M	H	+	+	H	5.5-7.0	5.0-8.0	H	High N	L
Millet (bulrush)		+	+	+						M	M			M			M		
Millet (finger)	+	+	+						60	S	H	+		M	5.0-6.5	5.0-6.0	M		L
Millet (panicum)		+	+							S	L			L			L		M
Rice (paddy)	+	+					+			S	L			H		4.0-8.0	M		M
Rice (upland)	+	+	+					M/H	50	M	M	+	+	M	5.5-6.5	4.5-7.5	H	High N	
Rice (hungry)	+	+						M	60	M	M			H	6.0-7.0	4.5-7.5	L		
Sorghum	+	+	+					L		M	H		+	H		5.0-8.5	H	High N	M
Wheat	+	+						M		M	M			H			M		H
Fibre crops																			
Cotton	+	+	+		+			M/L	100	M	M			H	5.2-6.0	4.8-7.5	H		H
Hemp		+	+					M/L	75	D						6.0-7.0	H		L
Jute		+	+					H/M	10-75	M						6.0-7.5	M		L
Kenaf		+	+					M/L	75	M						6.0-7.5	M		L
Rosella		+	+					M/L	60	M						6.5-8.0	H		L
Sisal		+		+				L	150	D				L			M		L
Fruit crops																			
Banana		+	+		+	+		M	100	D	L	+			6.0-7.5	4.0-8.0	M/H	High N, K	L
Cashew		+	+						100	D						5.5-7.0	M		M
Citrus		+	(+)					L	130	D	M				5.5-6.5	5.0-8.0	M		L
Date palm		+	+					H	100	D	H		+			6.5-8.0	M		H
Grape		+	+	+	+	+				D		+		H		6.0-7.0	M		L
Mango		+	+						60	D					7.0	5.5-7.5	H		L
Olive	+	+	+							D	H				6.0-6.5	5.5-7.5	H		M
Papaya		+			+			M		D							M	High N, K	L
Pineapple	+	+	+					L	60	S	M			H	6.0-6.5	5.0-6.5	M		L

Notes: See page 282.

cont

Table F.4 cont

Crop	Texture – Fine	Texture – Medium	Texture – Coarse	Texture – Very coarse tolerated	Drainage – Free essential or desirable	Drainage – Imperfect well tolerated	Drainage – Poor tolerated or needed	Drainage – Tolerance to short periods of waterlogging 2/	Drainage – Minimum groundwater depth (cm) 3/	Minimum rooting depth CLASS 4/	Moisture – Drought resistance 2/	Moisture – High AWC important	Moisture – Low AWC well tolerated	Erosion hazard 2/5/	Reaction – Optimum pH	Reaction – Range of pH tolerance for satisfactory yield 5/	Nutrient needs – General level of requirements 2/	Nutrient needs – Specific requirements	Salinity tolerance 2/
Oil crops																			
Coconut		+	+	+		+		H	50	D	L		+	L	6.0–7.5	5.0–8.0	L	Very high K	M
Groundnut		+	+	+	+			L	50	M	M		+	M	5.3–6.6	5.0–7.0	H		L
Oil palm		+	+		+			M/L	100	D	L	+		M	5.5–6.0	4.0–8.0	H		L
Safflower		+	+		+			M/L	75	M				M		5.5–6.5	H		L
Sesame		+	+			+		H	100	M				M		5.0–7.0	H		M
Soyabean		+	+		+			M/L	75	M				M	6.0–7.0	4.5–7.5	M		M
Sunflower	+	+	+		+			M	75	M	M		+	M		6.0–7.5	M		M
Pulses																			
Bean		+	+		+			M/L	30–50	M	L			M	6.0–7.0 ⎱	5.5–7.5 ⎱	M		L
Cowpea		+	+					M/L	40	M		+		M	⎰	⎰			L
Gram	+	+						M/L	30–50	M				M			M		M
Root crops																			
Cassava		+	+	+	+			L	50	D	H		+	M	5.0–5.8	5.5–6.5	M	(Low tolerated)	L
Cocoyam	+	+					+	H		M	L			L	5.8–6.0		H	High K	
Potato		+	+		+			L	30	M	L	+		M		4.5–7.0	M	High K	M
Sweet potato		+	+		+			L	50		M			M			H	High K	
Yams			+					L		D				M		5.5–6.5		High K, Mg	L
Sugar																			
Sugarbeet	+	+						M/H	45	M	L			H	6.0–7.5	4.5–8.5	H		H
Sugarcane	+	+				+		M/H	40	D		+					H	High N	M/L

Notes: See page 282.

cont

Annex F: Soil suitability for crops

Crop	TEXTURE Fine	Medium	Coarse	Very coarse tolerated	DRAINAGE Free essential or desirable	Imperfect well tolerated	Poor tolerated or needed	Tolerance to short periods of waterlogging 2/	Minimum groundwater depth (cm) 3/	MINIMUM ROOTING DEPTH CLASS 4/	MOISTURE Drought resistance 2/	High AWC important	Low AWC well tolerated	EROSION HAZARD 2/5/	REACTION Optimum pH	Range of pH tolerance for satisfactory yield 5/	NUTRIENT NEEDS General level of requirements 2/	Specific requirements	SALINITY TOLERANCE 2/
Tree and shrub perennials																			
Cocoa		+	+		+			L	150	(D)	L	+		M	6.0-7.0	4.5-8.0	M/H		L
Coffee (arabica)		+	+		+			L	100	D	L	+		M	5.0-6.0	4.5-7.0	M/H		L
Rubber		+	+		+			H	75	D	H			H	4.0-6.5	3.5-8.0	L		L
Tea		+	+		+			L	100	D	L	+		H	4.0-5.5	4.0-6.5	M	N when young	L
Coconut	see under oil crops above																	High N	
Oil palm																			
Banana	see under fruit crops above																		
Citrus																			
Vegetables																			
Cabbage		+	+		+			L	50	S					6.0-7.0		H		M
Cucumber		+	+		+			L	50	M					6.5-7.5		H		L
Onion		+	+		+			L	50	S					6.0-7.5		H		L
Tomato		+	+		+			L	50	M					5.0-7.0				L
Others																			
Alfalfa		+	+		+	+		L	50	D	M	+		L	6.5-7.5	6.0-8.0	M/H	Low N, high Ca,S	M
Tobacco	+	+	+		+			L	100	M	L			M	5.5-6.0	5.0-7.5	M		L
air-cured					+													High N, K	
fire-cured					+													High N, K	
flue-cured					+													Low N, high K	

Notes: + = Desirable condition or attribute.

1/ See also Table F.5. Note that CEC and OM criteria omitted because of lack of data; in general the higher the better – the latter up to at least 5%.

2/ L = low; M = medium or moderate; H = high.

3/ Minimum depth during growing period; this level produces about 25% reduction in optimum yield.

4/ D = deep (>90 cm); M = medium (60 to 90 cm); S = shallow (30 to 60 cm).

5/ For conditions before full canopy development and/or without cover crop; note that at maturity sugarcane, tea and rubber have low erosion hazards.

Sources: ILACO (1981, p 569); Young (1976, p 308), including figures from Jacob and Uexküll (1963) and Richards (1954). Sugarcane figures from R A Yates (personal communication).

Table F.5

Indicative soil requirements and tolerances of selected crops

Crop	Requirement for					Tolerance of				
	Water	Clayey texture	Good structure	Calcium	Acid conditions	Water-logging	Drought	Clayey texture	Acid conditions	Salinity
Apple	M/H	M	H	M	L	L	L/M	L	L	L
Barley	L/M	L	L	L	L	L	M/H	M	M	H
Beans	M	M*	M*	M*	L	L/M*	L*	M/H	L	L*
Cherry	M/L	L	M	L	L		M	L/M	M	L
Citrus	M	M	H	H	L	L	M	M*	M	L/M*
Cocoa	M	M	H	M	L	L	M	L	L	L
Coffee	M	L	H	M	L	L	M	L	L	L
Date palm	M	L	H	H	L	L	M	L	L	H
Flax	M	M	H	M	L	L	L	L	L	L
Maize	L/M	L	M	L	L	L	M/H	L	L	L/M
Mangolds	H	L	M	L	L	H	L	H	H	M
Oats	M	L	L	L	M	H	L	H	M	L
Oil palm	H	M	H	L	L	M	L	M	L	L
Pear	H	L/M	H	M	L	M	M	M	L	L*
Peas	M	M	H	M	M	L	L	L	L	L/M
Potatoes	M/H	L	L	L	H	M/H	M	H	H	L
Rice	H	M	L*	L	L	H	L	H*	M	L/M
Rubber	H	H	H	L	M	H	L	H	H	L
Rye	L	L	L	L	L	L	L/M	H	H	M
Sisal	M	M	L*	M	L	M*	M	H*	M	M/H
Sugarbeet	H	M	H	M	L	M	M	H	M	M/H
Tapioca	M	M	M	L	H	L	M	H	M	M/H
Tea	H	L	H	L	H	M	L	L	H	L
Tobacco	M	L	H	M	L	L	M	L	H	L
Wheat	L/M	H	H	H	L	L	M	M/H	L	M

Notes: 1. L = low, M = medium, H = high, * = depending on variety.
        2. See also Table F.4.

Sources: McRae and Burnham (1981) after Vink (1975); see also ILACO (1981, p 569).

Table F.6

Indicative climatic and soil requirements for selected crops

Crop	Total growing period (days)	Mean daily temperature for growth (°C) optimum (and range)	Day length requirements for flowering	Specific climatic constraints/requirements 1/	Soil requirements 2/	Sensitivity to salinity 3/
Alfalfa (Medicago sativa)	100-365	24-26 (10-30)	Day neutral	Sensitive to frost; cutting interval related to temperature; requires low RH in warm climates	Deep, medium-textured, well drained: pH 6.5-7.5	Moderately sensitive
Banana (Musa spp)	300-365	25-30 (15-35)	Day neutral	Sensitive to frost; temperature < 8°C for longer periods causes serious damage; requires high RH, wind < 4 m s⁻¹	Deep, well-drained loam without stagnant water: pH 5-7	Very sensitive
Bean (Phaseolus vulgaris)	Fresh: 60-90 Dry: 90-120	15-20 (10-27)	Short day/ day neutral	Sensitive to frost, excessive rain, hot weather	Deep, friable soil, well drained and aerated: optimum pH 5.5-6	Sensitive
Cabbage (Brassica oleracea)	100-150+	15-20 (10-24)	Long day	Short periods of sharp frost (-10°C) are not harmful: optimum RH 60-90%	Well drained: optimum pH 6-6.5	Moderately sensitive
Citrus (Citrus spp)	240-365	23-30 (13-35)	Day neutral	Sensitive to frost (dormant trees less), strong wind, high humidity; cool winter or short dry period preferred	Deep, well aerated, light- to medium-textured soils, free from stagnant water: pH 5-8	Sensitive
Cotton (Gossypium hirsutum)	150-180	20-30 (16-35)	Short day/ day neutral	Sensitive to frost, strong or cold winds; temperature required for boll development; 27-32°C (20-38°C range); dry ripening period required	Deep, medium- to heavy-textured soils: pH 5.5-8 with optimum pH 7-8	Tolerant
Grape (Vitis vinifera)	180-270	20-25 (15-30)	Day neutral	Resistant to frost during dormancy (down to -18°C) but sensitive during growth; long, warm to hot, dry summer and cool winter preferred/required	Well-drained, light soils are preferred	Moderately sensitive
Groundnut (Arachis hypogaea)	90-140	22-28 (18-33)	Day neutral	Sensitive to frost; for germination temperature > 20 C	Well-drained, friable, medium-textured soil with loose topsoil: pH 5.5-7	Moderately sensitive
Maize (Zea mays)	100-140+	24-30 (15-35)	Day neutral/ short day	Sensitive to frost; for germination temperature > 10°C; cool temperature causes problem for ripening	Well-drained and aerated soils with deep water-table and without waterlogging: optimum pH 5-7	Moderately sensitive
Oil palm (Elaeis guineensis)	365	27 (24-30)		Sensitive to frost; requires high RH, > 1 500 mm well-distributed rainfall and > 1 300 h sunshine	Well-drained, aerated soils with good water-holding capacity and unrestricted rooting medium	Moderately sensitive

Notes: See page 286.

cont

Table F.6 cont

Crop	Total growing period (days)	Mean daily temperature for growth (°C) optimum (and range)	Day length requirements for flowering	Specific climatic constraints/requirements 1/	Soil requirements 2/	Sensitivity to salinity 3/
Olive (Olea europaea)	210-300	20-25 (15-35)		Sensitive to frost (dormant trees less); low winter temperature required (< 10°C) for flower bud initiation	Deep, well-drained soils free from waterlogging	Moderately tolerant
Onion (Allium cepa)	100-140 (+30-35 in nursery)	15-20 (10-25)	Long day/day neutral	Tolerant to frost; low temperature (< 14-16°C) required for flower initiation; no extreme temperature or excessive rain	Medium-textured soil: pH 6-7	Sensitive
Pea (Pisum sativum)	Fresh: 65-100 Dry: 85-120	15-18 (10-23)	Day neutral	Slight frost tolerance when young	Well-drained and aerated soils: pH 5.5-6.5	Sensitive
Pepper (Capsicum spp)	120-150	18-23 (15-27)	Short day/day neutral	Sensitive to frost	Light- to medium-textured soils: pH 5.5-7	Moderately sensitive
Pineapple (Ananas comosus)	365	22-26 (18-30)	Short day	Sensitive to frost; requires high RH; quality affected by temperature	Sandy loam with low lime content: pH 4.5-6.5	Sensitive
Potato (Solanum tuberosum)	100-150	15-20 (10-25)	Long day/day neutral	Sensitive to frost; night temperature < 15°C required for good tuber initiation	Well-drained, aerated and porous soils: pH 4.5-6	Moderately sensitive
Rice (paddy) (Oryza sativa)	90-150	22-30 (18-35)	Short day/day neutral	Sensitive to frost; cool temperature causes head sterility; small difference in day and night temperature is preferred	Heavy soils preferred for percolation losses, high tolerance to O_2 deficit: pH 5.5-6	Moderately sensitive
Rubber (Hevea brasiliensis)	365	28 (26-30)		Sensitive to frost; wide range in temperature unfavourable, strong winds harmful. Pronounced dry season reduces yield	Deep, well aerated, permeable, acid soils. Shallow and peaty soils to be avoided	Very sensitive
Safflower (Carthamus tinctorius)	Spring: 120-160 Autumn: 200-230	Early growth: 15-20 Later growth: 20-30 (10-35)		Tolerance to frost; cool temperature required for good establishment and early growth	Fairly deep, well-drained soils, preferably medium textured: pH 6-8	Moderately tolerant
Sorghum (Sorghum bicolor)	100-140+	24-30 (15-35)	Short day/day neutral	Sensitive to frost; for germination, temperature > 10°C; cool temperature causes head sterility	Light to medium/heavy soils relatively tolerant to periodic waterlogging: pH 6-8	Moderately tolerant

cont

Notes: See page 286.

Table F.6 cont

Crop	Total growing period (days)	Mean daily temperature for growth (°C) optimum (and range)	Day length requirements for flowering	Specific climatic constraints/requirements 1/	Soil requirements 2/	Sensitivity to salinity 3/
Soyabean (Glycine max)	100-130	20-25 (18-30)	Short day/day neutral	Sensitive to frost; for some varieties temperature > 24°C required for flowering	Wide range of soil except drought susceptible and poorly drained: pH 6-6.5	Moderately tolerant
Sugarbeet (Beta vulgaris)	160-200	18-22 (10-30)	Long day	Tolerant to night frost; towards harvest mean daily temperature < 10°C for high sugar yield	Medium- to slightly heavy-textured soils, friable and well drained: pH 6-7	Tolerant
Sugarcane (Saccharum officinarum)	270-1 200	22-30 (15-35)	Short day	Tolerant of only very light frost; during the harvest period cool (10-20°C), dry, sunny weather is beneficial	Deep, well aerated with ground water deeper than 1.5-2 m but relatively tolerant to periodic high water-tables and/or flooding and O_2 deficit: pH 4.5-8.5; optimum pH 6.5	Moderately sensitive
Sunflower (Helianthus annuus)	90-130	18-25 (15-30)	Short day/day neutral	Sensitive to frost	Fairly deep soils: pH 6-7.5	Moderately tolerant
Tobacco (Nicotiana tabacum)	90-120 (+40-60 in nursery)	20-30 (15-35)	Short day/day neutral	Sensitive to frost	Quality of leaf depends on soil texture, and is impaired by salinity: pH 5-6.5	Moderately sensitive
Tomato (Lycopersicon esculentum)	90-120 (> 25-35 in nursery)	18-25 (15-28)	Day neutral	Sensitive to frost, high RH and strong wind; optimum night temperature 10-20°C	Light loam, well drained without waterlogging: pH 5-7	Sensitive
Watermelon (Citrullus vulgaris)	80-110	22-30 (18-35)	Day neutral	Sensitive to frost	Sandy loam is preferred: pH 5.8-7.2	Moderately sensitive
Wheat (Triticum spp)	Spring: 100-130 Winter: 180-250	15-20 (10-25)	Day neutral/long day	Spring wheat: sensitive to frost; winter wheat: resistant to frost during dormancy (> 15°C), sensitive during post-dormancy period; requires a cold period for flowering during early growth. For both, dry period required for ripening	Medium texture is preferred; relatively tolerant to high water-table: pH 6-7	Moderately sensitive

Notes: 1/ Temperatures quoted are optimal, with ranges in parentheses.
2/ Indicative rooting depths and soil-water tension are given in Table F.8.
3/ See also Tables 7.12, 7.13 and 8.2 to 8.4.

Sources: Adapted from Doorenbos and Kassam (1979); see also ILACO (1981, pp 562ff) and Tables 7.12 and 7.13, 8.2 to 8.4. Sugarcane figures amended according to R A Yates (personal communication).

Indicative nutrient and water requirements for selected crops   Table F.7

Crop	Nutrient requirements 1/ N : P : K (kg ha^{-1}/growing period)	Ideal water requirements 2/ (mm/growing period)	Sensitivity to water supply (and ky value) 3/	Water utilisation efficiency for harvested yield (Ey) 4/ kg m^{-3} (and % moisture of product)
Alfalfa (Medicago sativa)	0–40: 55–65: 75–100	800–1 600	Low to medium–high (0.7–1.1)	1.5–2.0 Hay (10–15%)
Banana (Musa spp)	200–400: 45–60: 240–480	1 200–2 200	High (1.2–1.35)	Plant crop: 2.5–4 Ratoon: 3.5–5.6 Fruit (70%)
Bean (Phaseolus vulgaris)	20–40: 40–60: 50–120	300–500	Medium–high (1.15)	Fresh: 1.5–2.0 (80–90%) Dry: 0.3–0.6 (10%)
Cabbage (Brassica oleracea)	100–150: 50–65: 100–130	380–500	Medium–low (0.95)	12–20 Head (90–95%)
Citrus (Citrus spp)	100–200: 35–45: 50–160	900–1 200	Low to medium–high (0.8–1.1)	2–5 Fruit (85%, lime: 70%)
Cotton (Gossypium hirsutum)	100–180: 20–60: 50–80	700–1 300	Medium–low (0.85)	0.4–0.6 Seed cotton (10%)
Grape (Vitis vinifera)	100–160: 40–60: 160–230	500–1 200	Medium–low (0.85)	2–4 Fresh fruit (80%)
Groundnut (Arachis hypogaea)	10–20: 15–40: 25–40	500–700	Low (0.7)	0.6–0.8 Unshelled dry nut (15%)
Maize (Zea mays)	100–200: 50–80: 60–100	500–800	High (1.25)	0.8–1.6 Grain (10–13%)
Olive (Olea europaea)	200–250: 55–70: 160–210	600–800 (per year)	Low	1.5–2.0 Fresh fruit (30%)
Onion (Allium cepa)	60–100: 25–45: 45–80	350–550	Medium–high (1.1)	8–10 Bulb (85–90%)
Pea (Pisum sativum)	20–40: 40–60: 80–160	350–500	Medium–high (1.15)	Fresh: 0.5–0.7 Shelled (70–80%) Dry: 0.15–0.2 (12%)
Pepper (Capsicum spp)	100–170: 25–50: 50–100	600–900	Medium–high (1.1)	1.5–3.0 Fresh fruit (90%)
Pineapple (Ananas comosus)	230–300: 45–65: 110–220	700–100	Low	Plant crop: 5–10 Ratoon: 8–12 Fruit (85%)
Potato (Solanum tuberosum)	80–120: 50–80: 125–160	500–700	Medium–high (1.1)	4–7 Fresh tuber (70–75%)

Notes: See page 288.   cont

Annex F: Soil suitability for crops

Crop	Nutrient requirements 1/ N : P : K (kg ha^{-1}/growing period)	Ideal water requirements 2/ (mm/growing period)	Sensitivity to water supply (and ky value) 3/	Water utilisation efficiency for harvested yield (Ey) 4/ kg m^{-3} (and % moisture of product)
Rice (paddy) (Oryza sativa)	100–150: 20–40: 80–120	350–700	High	0.7–1.1 Paddy (15–20%)
Safflower (Carthamus tinctorius)	60–110: 15–30: 25–40	600–1 200	Low (0.8)	0.2–0.5 Seed (8–10%)
Sorghum (Sorghum bicolor)	100–180: 20–45: 35–80	450–650	Medium–low (0.9)	0.6–1.0 Grain (12–15%)
Soyabean (Glycine max)	10–20: 15–30: 25–60	450–700	Medium–low (0.85)	0.4–0.7 Grain (6–10%)
Sugarbeet (Beta vulgaris)	150: 50–70: 100–160	550–750	Low to medium–low (0.7–1.1)	Beet: 6–9 (80–85%) Sugar: 0.9–1.4 (0%)
Sugarcane (Saccharum officinarum)	100–200: 20–90: 125–160	1 500–2 500 (per year)	High (1.2)	Cane: 5–10 (80%) Sugar: 0.6–1.2 (0%)
Sunflower (Helianthus annuus)	50–100: 20–45: 60–125	600–1 000	Medium–low (0.95)	0.3–0.5 Seed (6–10%)
Tobacco (Nicotiana tabacum)	40–80: 30–90: 50–110	400–600	Medium–low (0.9)	0.4–0.6 Cured leaves (5–10%)
Tomato (Lycopersicon esculentum)	100–150: 65–110: 160–240	400–600	Medium–high (1.05)	10–12 Fresh fruit (80–90%)
Watermelon (Citrullus vulgaris)	80–100: 25–60: 35–80	400–600	Medium–high (1.1)	5–8 Fruit (90%)
Wheat (Triticum spp)	100–150: 35–45: 25–50	450–650	Medium–high (Spring: 1.15 winter: 1.0)	0.8–1.0 Grain (12–15%)

Notes: 1/   Rough figures under irrigation;   actual values will obviously depend on soil, climate, cultivar etc;   also note that:
   1 kg P = 2.4 kg P$_2$O$_5$
   and 1 kg K = 1.2 kg K$_2$0

   2/   Indicative rooting depths and soil-water tensions are given in Table F.8.
   3/   ky = yield response factor = ratio of relative yield decrease
   (1 - actual yield/maximum yield) to relative evapotranspiration deficit
   (1 - actual ET/maximum ET)
   ie ky = (1 - Ya/Ym):(1 - ETa/ETm)
   ky of the total growing period: low: ky < 0.85
                                     medium: ky 0.85–1.0
                                     medium–high: ky 1.0–1.15
                                     high: ky > 1.15
   4/   Ey = water utilisation efficiency = kg of produce m^{-3} of water supplied.

Source: Adapted from Doorenbos and Kassam (1979);   sugarcane figures amended according to R A Yates (personal communication).

Table F.8

Indicative rooting depths and available water extraction for selected crops

Common (and botanical) name	Maximum rooting depths (cm) 1/ All roots	Main nutrient/ water uptake roots	Rooting pattern	Readily available soil water 2/ (% of total AWC)	Maximum tension of readily available water (bar)
Alfalfa (Medicago sativa)	> 300 (developed after first year)	100–200	Deep extensive system capable of deep water extraction	55	1.5
Banana (Musa spp)	≤ 90	50–75	Sparse, shallow system; 60% of water extracted in top 30 cm	35	0.3–1.5
Barley 3/ (Hordeum vulgare)	100–150	< 100	Dense	55	0.4–0.5 (growth) 0.8–1.2 (ripening)
Bean 3/ (Phaseolus vulgaris)	100–150 (tap-root)	50–70	Extensive lateral system concentrated in top 30 cm	45	0.75–2.0
Cabbage (Brassica oleracea)	60	40–50	Extensive shallow system	45	0.6–0.7
Carrot (Daucus carota)	50–100	< 60	Dense	35	0.55–0.65
Citrus (Citrus spp)	100–> 200 (tap-root)	120–160 (> 200 in drought)	Shallow, horizontal lateral roots: 60% < 50 cm 30% 50–100 cm 10% > 100 cm	40 (flowering) 60–70 (fruiting)	0.2–1.0
Cocoa (Theobroma cacao)	≥ 100	≤ 15	Shallow, lateral roots	20	No data
Cotton (Gossypium hirsutum)	≥ 180 (tap-root)	100–170	70–80% concentrated in top 90 cm	65	No data
Cucumber (Cucumis sativus)	> 120	70–120	Moderately dense	50	No data
Date palm (Phoenix dactylifera)	> 300	150–250	Sparse	50	No data
Grape (Vitis vinifera)	> 300	100–200	Most roots concentrated between 50 and 150 cm	40	0.4–0.5 (early) 0.9–1.3 (mature)
Groundnut (Arachis hypogaea)	≥ 180 (tap-root)	50–100	Main rooting system in top 50–60 cm	50	No data
Lettuce (Lactuca sativa)	> 50	30–50	Dense	30	0.4–0.6

Notes: See page 291.

cont

Table F.8 cont

Common (and botanical) name	Maximum rooting depths (cm) 1/ All roots	Main nutrient/ water uptake roots	Rooting pattern	Readily available soil water 2/ (% of total AWC)	Maximum tension of readily available water (bar)
Maize 3/ (Zea mays)	200	80–100	80% of water extracted in top 100 cm	55	0.5-1.0 (growth) 0.8-1.2 (ripening)
Melon (Cucumis melo)	> 150	100–150	Fairly sparse	35	0.35-0.4
Oil palm (Elaeis guineensis)	> 200	< 45	Extensive system but majority of roots near base of palm	40	No data
Olive (Olea europaea)	> 180	120–170	Very extensive lateral roots, > 12 m long	65	No data
Onion (Allium cepa)	50	30–50	Shallow system concentrated in top 30 cm	25	0.45-0.65
Pea (Pisum sativum)	100–150 (tap-root)	60–100	Dense	40	0.3-0.5
Pepper (Capsicum spp)	100	50–100	Roots concentrated in top 30 cm	25	No data
Pineapple (Ananas comosus)	100	30–60	Shallow, sparse system	50	No data
Potato (Solanum tuberosum)	60	40–60	Shallow system; 70% water uptake from top 30 cm	25	0.3-0.5
Rice (Oryza sativa)	100	< 50	Dense surface mat	20	No data
Rubber (Hevea brasiliensis)	300–400	< 200	Deep, extensive system with lateral roots extending 20 m or more	55	No data
Safflower 3/ (Carthamus tinctorius)	100–200	100–200	Deep, extensive system	60	No data
Sisal (Agave sisalana)	> 300	50–100	Sparse	80	No data
Sorghum 3/4/ (Sorghum bicolor)	100–200	100–200	Extensive system; 60–90% of water uptake in top 100 cm	55 (growing period) 80 (ripening period)	0.4-0.5 (growth) 0.8-1.2 (ripening)

Notes: See page 291.

cont

Table F.8 cont

Common (and botanical) name	Maximum rooting depths (cm) 1/ All roots	Main nutrient/ water uptake roots	Rooting pattern	Readily available soil water 2/ (% of total AWC)	Maximum tension of readily available water (bar)
Soyabean (Glycine max)	> 180 (tap-root)	60-130	Roots concentrated in top 30-60 cm	55 (growing period) < 85 and > 50 (germination)	No data
Sugarbeet 3/ (Beta vulgaris)	> 150 (tap-root)	70-120	Dense	50	0.4-0.6
Sugarcane 3/5/ (Saccharum officinarum)	> 300	120-200	Deep extensive root system	40-70	0.1-1.0
Sunflower (Helianthus annuus)	200-300	80-150	Moderately dense	45	No data
Sweet potato (Ipomoea batatas)	> 180	100-150	Fairly sparse	65	No data
Tobacco (Nicotiana tabacum)	> 100 (tap-root)	50-100	Extensive horizontal roots; 75% of water uptake in top 30 cm	35 (early tobacco) 65 (late tobacco)	0.3-0.8
Tomato (Lycopersicon esculentum)	150	70-150	Deep rooting system, but > 80% of water uptake in top 50-70 cm	40	0.8-1.5
Vegetables	> 60	30-60	Shallow	20	0.3-0.7
Watermelon (Citrullus vulgaris)	150-200	100-150	Fairly sparse	45	No data
Wheat (Triticum spp)	120-150 (spring wheat) 150-200 (winter wheat)	100-150 (active depth for spring wheat 90 cm)	Lateral spread 15-25 cm; water uptake figures: 50-60% ≤ 30 cm 20-25% 30-60 cm 10-15% 60-90 cm < 10% > 90 cm	50 (growth period) 90 (ripening)	0.4-0.5 (growth) 0.8-1.2 (ripening)

Notes: 1/ Indicative maximum depths to which roots can penetrate in soils with no impedance; actual root penetration depends on cultivar, water regime (intensity and frequency of rainfall/irrigation) and soil conditions (density, presence of pans, salinity, waterlogging etc).
2/ The portion of the total non-saline AWC extractable before crop evapotranspiration is affected, ie the maximum depletion at which water should be applied. The figures are approximate guides for ETcrop = 4 to 7 mm day⁻¹; for ETcrop ≤ 3 mm day⁻¹ add 30% of values shown, and for ETcrop ≥ 8 mm day⁻¹ subtract 30%. For these values in mm m⁻¹, see Table 6.12. Note: very rough estimates are given; in practice values depend on factors such as climate, cultivars, soil-water characteristic curves, salinity and hydraulic conductivity.
3/ Higher depletion values than those indicated apply during ripening.
4/ But note ability to cease growth during severe water stress without affecting subsequent recovery.
5/ But note ability to produce compensating regrowth following water stress.

Sources: Doorenbos and Kassam (1979); Doorenbos and Pruitt (1977); Hartley (1977); ILRI (1972); Taylor and Ashcroft (1972, pp 434/5); sugarcane figures amended after Thompson (1976).

Annex F: Soil suitability for crops

F.3   Bananas (Musa spp)

F.3.1   Climate and altitude

Optimal mean monthly temperature 25° to 28°C (77° to 83°F), a temperature of 21°C (70°F) or less results in reduced growth. The minimum temperature for growth lies between 15° and 16°C (59° and 61°F). Osborne (personal communication) states that plantation chilling occurs when the temperature in winter months falls below 10°C (50°F), and that 12 h exposure at any temperature below 7°C (45°F) generally causes chilling injury. High temperatures result in leaf and fruit scorch. It has been shown in Martinique that even small changes in altitude may have a significant effect on the period to shooting:

Altitude (m)	up to 137	183-365	397-641
Months to shooting	6-7	9-10	11-13
Approximate temperature drop (°C)	-	1°	3°

An average rainfall of 1 500 to 2 500 mm (60 to 100 in.) well distributed is best, but bananas can be grown in areas with a pronounced dry season. Drought-resistant varieties are available, but they show considerable variation in quality and are mostly grown for local consumption. A dry period can adversely affect the development of the fruit. Relative humidity of at least 60% is preferred.

Wind damage represents a major source of loss in the production of bananas, especially in areas with infestation of the nematode Radopholus similis, which weakens the roots. A wind above 50 km h^{-1} (30 mph) causes considerable damage, and above 100 km h^{-1} (60 mph) results in total loss.

Large amounts of solar radiation are necessary for good banana production.

F.3.2   Soil

Bananas thrive best on a free-draining, well-aerated, deep, fertile loam, but are produced on a wide range of soil types which respond satisfactorily to good drainage practices and adequate fertiliser applications; banana roots will not tolerate any waterlogging. A high AWC is important as the plant is very susceptible to damage from moisture stress. Soils that are freely drained can have a wide range of textures. The delicate root system cannot penetrate compact clays which do not have adequate pore space. Also very fine sands and silts are generally unsuitable due to inadequate drainage and the difficulties of producing economic improvements. Mulching, however, can lead to reasonable production on sandy soils.

The pH can range from 5.5 to 7.5, though the optimum is 6.5. Purseglove (1975) notes that bananas require high nitrogen and potash levels, and quotes levels of adequacy in the soil at 85 to 100 ppm for phosphate and 300 to 350 ppm for potash; critically low levels are quoted as being about 20 ppm for phosphate and 300 ppm for potash. Simmonds (1966) quotes 'lower limits of adequacy' of 1 200 ppm for N, 50 ppm for P_2O_5, and 150 ppm for K_2O.

Dabin and Leneuf (1960) consider $CaO/MgO/K_2O$ = 10:5:0.5 a well-balanced ratio for banana soils in the Ivory Coast. The calcium value ought to be at least twice as high as the magnesium value, since similar calcium and magnesium levels generally result in one or more of a wide range of nutritional troubles or soil problems. To avoid magnesium deficiency the MgO to K_2O ratio should not be lower than 4; if > 25 it will cause potassium deficiency. Magnesium deficiency can be suspected for levels < 1.0 me Mg per 100 g soil, according to figures from the Windward Islands. Other work in the Windward Islands has established a critical level in 'normal' soils of about 0.4 to 0.5 me/100 g of exchangeable potassium. This critical value is misleading in cases where soils contain abundant primary minerals or where the soil contains very high calcium and/or magnesium levels.

Also in the Windward Islands a critical low level of 7 to 9 ppm of Truog acid-soluble phosphate has been established. Zinc deficiency has been shown to be associated with high levels of Truog acid-soluble phosphate. The soil factors responsible are likely to be alkalinity, sandy or gravelly texture, and a high phosphate content. Numerous investigators have observed that zinc deficiency usually occurs on soils of pH 6.0 or higher. Some investigators have reported reduced zinc availability when phosphate fertilisers are added to the soil.

In soils with high levels of organic matter, C:N ratios of 20 or over imply that little nitrogen will be available to bananas. It has been established in Jamaica by Twyford (1973) that the plant cannot tolerate more than 500 ppm salts in the soil (ie sodium content ≤ 1 me/100 g ads).

A crop of 25 t ha^{-1} removes approximately 17 to 28 kg N, 6 to 7 kg P_2O_5, 56 to 78 kg K_2O. The banana plant does not provide complete vegetative protection against erosion, therefore conservation works are important on sloping land.

F.3.3   Mineral deficiency symptoms

Calcium	Yellow bands along leaf margins
Iron	Interveinal chlorosis on young leaves
Magnesium	Interveinal chlorosis with wavy leaf margins
Manganese	Spots on leaves and fruit. Caused by high soil pH
Nitrogen	Slow growth, stunting, and yellowish-green colour of leaves
Phosphorus	Slow growth caused by marginal scorching of leaves
Potassium	Marginal yellowing of leaves. If acute, the leaves turn bright yellow. Nematode attack may accentuate K deficiency
Zinc	Produces bunchy top effect

F.3.4   Miscellaneous

The banana plant has a narrow range of tolerance and requires soils of high fertility. Healthy banana can be considered an indicator plant for the presence of good soil. Further data (yields, rooting depth, plant spacings, etc) are summarised in Tables F.3 to F.8.

F.3.5   References

Arens P L (1978).   Edaphic criteria in land evaluation, pp 24-31.   In: 'Land Evaluation Standards for Rainfed Agriculture'.   World Soil Res Rep 49.   FAO, Rome.

Dabin B and Leneuf N (1960).   Fruits, 15(3), 117-127.

De Geus J G (1973).   Fertiliser guide for the tropics and sub-tropics, pp 617-628.   Centre d'Etude de l'Azote, Zurich.

Doorenbos J and Kassam A H (1979).   Yield response to water.   Irrigation and drainage. Paper 33.   FAO, Rome.

Jardine C G (1962).   Metal deficiencies in bananas.   Nature 194, 1160-1163.

Murray D B (1959).   Deficiency symptoms of the major elements in the banana.   Trop Agric, 36, 100-107.

Osborne R E.   Personal communication.

Purseglove J M (1975).   Tropical crops, monocotyledons: Musaceae, pp 345-377.   Longman, London.

Simmonds N W (1966).   'Bananas'.   Second Edition.   Longmans, London.

Tai E A (1977).   Banana, pp 441-460.   In: 'Ecophysiology of Tropical and Sub-Tropical Crops', Alvim P T and Kozlowski T T (Eds).   Academic Press, New York.

Twyford I T (1973).   Interpretation of soil analysis results.   Banana Res Note No 9. Windward Islands Banana Research Scheme.

Windward Islands Banana Research Scheme (1970).   Annual Report.

F.4   Citrus (Citrus spp)

F.4.1   Climate and location

Mostly grown in subtropical countries between sea-level and 600 m (2 000 ft), between 45°N and 35°S.   On the equator citrus does not do well above 1 830 m (6 000 ft).   Average rainfall should be at least 875 mm (35 in.) unless irrigated.   Citrus is intolerant of high humidity, but mandarins can tolerate wetter conditions than other citrus spp.   High winds can cause much damage and consideration should be given to the provision of windbreaks. Grapefruit can withstand long hot periods better than other citrus fruits.

F.4.2   Soil

Citrus roots have a high oxygen requirement, so the soil should be well aerated, well drained and not too heavy.   Light sandy to medium loam soils are considered best.   Citrus can be grown on poor, sandy soils which are extremely low in natural fertility;   in some cases as many as 12 essential elements (N, P, K, Mg, Ca, S, Mn, Cu, Zn, B, Fe, Mo) have to be applied for normal growth and development.   An excess of phosphorus can cause micronutrient deficiency and impair nitrogen use.   Citrus is also susceptible to magnesium deficiency caused by excesses of calcium or potassium.   As magnesium is a rather weak competitor against calcium or potassium, dolomitic limestone should be used for liming acid soils.   pH should be between 5 and 8.   Citrus spp are sensitive to high B levels (> 0.7 ppm in the saturated extract).

F.4.3   Salinity

Citrus is sensitive to the presence of salt at EC_e values between 1.5 and 2 mS cm^{-1}.

Annex F: Soil suitability for crops

Mineral deficiency symptoms

Boron	Watery spots, dull, brownish-green leaves which tend to curl; small, hard, lumpy fruits often with gummy discolorations in pith
Copper	Causes die-back with blisters on young branches, leaves and fruit
Iron	Interveinal chlorosis, the veins remaining green
Phosphate	Reduced growth rate with thin dull-green foliage
Potassium	Fading and curling of leaves, leaf drop, small fruit
Magnesium	Leaves turn yellow. Cured by Epsom salts ($MgSO_4$)
Nitrogen	Pale green-yellowish leaves, stunted shoots
Zinc	Mottling and reduction in size of leaves

F.4.5   Miscellaneous

Degree of hardiness, beginning with the tenderest: citron, lime, lemon, grapefruit, sweet orange, sour orange, mandarin, kumquat, trifoliate orange. Further details (yields, rooting depth, plant spacing, etc) are summarised in Tables F.3 to F.8.

F.4.6   References

De Geus J G (1973). Fertiliser guide for the tropics and sub-tropics, pp 590-616. Centre d'Etude de l'Azote, Zurich.

Maas E V and Hoffman G J (1977). Crop salt tolerance - current assessment, June 1977. J Irrig Drainage Div. Am Soc Civ Engs.

Purseglove J W (1975). Tropical crops, dicotyledons: Rutaceae, pp 496-522. Longman, London.

F.5   Cocoa (Theobroma cacao)

F.5.1   Climate and altitude

Mean temperature should not fall below 15 to 21°C (60 to 70°F) nor rise above 28 to 30°C (82 to 86°F). Diurnal range not > 9°C (16°F).

Rainfall can vary from 1 250 to 2 500 mm (50 to 100 in.); below 1 250 cm (50 in.) irrigation will probably be necessary. Should be well distributed with no marked dry season, ie when monthly rainfall is below 63 mm (2.5 in.).

High winds can cause extensive damage.

Latitude limits are 15 to 20°N and 15 to 20°S, though most citrus is grown between 10°N and 10°S, below 300 m (1 000 ft); it is grown at up to 1 200 m (4 000 ft) in Venezuela and 900 m (3 000 ft) in Colombia.

F.5.2   Soil

The 'ideal' cocoa soil consists of aggregated sand, silt and clay having a total pore space of about 66% of the soil volume. For satisfactory soil aeration non-capillary pore space must be > 10%. Adequate soil aeration is absolutely essential to satisfactory growth and where annual rainfall is high, over 3 750 mm (150 in.) per annum, only soils having high permeability are capable of supporting thrifty cocoa. The ratio of monovalent to divalent bases (K + Na):(Ca + Mg) is important and should be at or near 1 : 50, otherwise growth and development of the plant will be hindered. The nutrients removed by a crop of cocoa of 560 kg ha^{-1} of dry beans are 25 kg N, 4.5 kg P, 36 kg K; the main feeding roots are concentrated in the top 10 to 15 cm of soil. Some examples of nutrient status in cocoa soils, and a fertility rating system are given in Tables F.9 and F.10.

Consideration should be given to the provision of shade trees, particularly for young cocoa. Shade protection is needed most when soil conditions are least satisfactory and should only be dispensed with when soil conditions are entirely satisfactory or there is danger of a nutrient imbalance developing within the plant due to excessive photosynthesis.

The total rooting depth of cocoa is 2 m, but 80% of the roots occur in the top 15 cm. As the topsoil is the main feeding zone of the plant, topsoils should ideally have an organic matter content > 3%.

The tolerated pH range is 5.5 to 7.5, and the ideal is 6.5 with no layer having pH above 8.0 or below 4.0 within 1 m of the surface. High pH causes malformation of the leaves due to lack of available Fe, Zn, Cu. Preferred values for exchangeable cations are as follows (Wood, 1975): Ca ≥ 8 me/100 g, Mg ≥ 2 me/100 g, K ≥ 0.24 me/100 g. Unless the soil CEC is exceptionally high, the BSP should be ≥ 35% in the subsurface layers; CEC in the surface should be ≥ 12 me/100 g, and ≥ 5 me/100 g in subsurface layers. Ideally OM in the top 15 cm should be ≥ 3% (or 1.75% organic C).

Nutrient status of selected cocoa soils

Table F.9

Locality	pH value	Total nitrogen (%)	C/N ratio	'Available' nutrients (ppm)		Exchangeable bases (me/100 g)			Ratios		Sum of bases (me/100 g)
				P_2O_5	K_2O	CaO	MgO	K_2O	$\frac{CaO}{MgO}$	$\frac{CaO + MgO}{K_2O}$	
Good soils 1/											
West Indies											
Trinidad	6.6	0.28	9.0	42	220	20.5	5.8	0.44	3.5	60	26.7
Tobago	6.7	0.24	11.1	63	105	–	–	0.21	–	–	–
Grenada	6.9	0.21	10.7	75	350	17.7	8.8	0.70	2.0	38	27.2
West Africa											
Ghana	6.4	0.14	11.1	25	205	10.5	2.3	0.41	4.8	31	13.1a
Nigeria	6.5	0.13	10.7	23	156	–	–	0.31	–	–	–
Poor soils											
West Indies											
Trinidad	5.6	0.25	7.7	25	160	6.3	5.5	0.32	1.1	37	12.1
Tobago	6.6	0.21	9.1	24	70	–	–	0.14	–	–	–
Grenada	6.6	0.23	9.4	45	25	6.3	7.9	0.50	0.8	28	14.7
West Africa											
Ghana	5.8	0.09	9.7	18	15	4.7	0.9	0.29	5.2	19	5.9
Nigeria	4.8	0.07	13.4	14	8	–	–	0.15	–	–	–

Note: 1/ Cocoa soils are classified as 'good' when they yield 450 kg ha^{-1} or more of dry, cured cocoa.

Source: de Geus (1973).

Rating of fertility levels for cocoa in top 15 cm of soil

Table F.10

Fertility rating	pH value	Total nitrogen (%)	C/N ratio	'Available' nutrients (ppm)		Exchangeable bases 3/ (me/100 g)			Ratios		Sum of bases (me/100 g)
				P_2O_5 1/	K_2O 2/	CaO	MgO	K_2O	$\frac{CaO}{MgO}$	$\frac{CaO + MgO}{K_2O}$	
High	7.5	0.35	11.5	120	275	24.0	6.0	0.55	4.0	54	50
Medium	6.5	0.20	9.5	60	175	12.0	3.0	0.35	4.0	43	15
Low	5.0	0.05	7.5	20	100	4.0	1.0	0.20	4.0	25	5

Notes: 1/ Truog's method (extracting agent 0.001 M sulphuric acid buffered at pH 3.0 with ammonium sulphate).
   2/ Exchangeable potash.
   3/ Assuming exchange capacity around 24 me/100 g.

Source: Adapted from de Geus (1973).

Annex F: Soil suitability for crops

Soils should not have less than 0.2 ppm of hot-water-soluble boron.

F.5.3   Mineral deficiency symptoms

Aluminium	On very acid soils. Pale interveinal region near tip of older leaves
Iron	Induced by pH of 7 or over and free $CaCO_3$. Dark green veins with pale green between
Magnesium	On acid soils. Old leaves turn pale green
Manganese	On very acid soils. Younger leaves with irregular pale areas
Nitrogen	Small leaves, pale yellowish colour
Phosphorus	Mature leaves paler towards tip and margin
Potash	Pale yellow between veins on older leaves
Zinc	Induced by high pH, or soil compaction causing lack of aeration. Leaves narrow and malformed

Young cocoa leaves are naturally a pale reddish brown and should not be confused with signs of mineral toxicities or deficiencies.

F.5.4   Miscellaneous

Further details (yields, rooting depth, plant spacing, etc) are summarised in Tables F.3 to F.8.

F.5.6   References

Cadbury Limited (1971). Cocoa Growers Bulletin, No 7, 27-30. Birmingham.

De Geus J G (1973). Fertiliser guide for the tropics and sub-tropics, pp 418-439. Centre d'Etude de l'Azote, Zurich.

FAO (1966b). Selection of soils for cocoa. Soils Bull No 5. FAO, Rome.

Purseglove J M (1975). Tropical crops, dicotyledons: Sterculiaceae, pp 571-598. Longman, London.

Smyth A J (1966). Cocoa Growers Bulletin, 6, 7. Cadbury Limited, Birmingham.

Wood C A R (1975). Cocoa. Third Edition. Longman, London.

F.6   Coffee (Coffea spp)

F.6.1   Arabica coffee (Coffea arabica)

a)   General — Arabica is the most widely grown species, forming a high proportion of the crop in Central and South America, and in the African highlands. It produces the highest quality coffee, the best being produced by the pulped and washed method of processing ('mild coffee'). In some countries, notably Brazil, the berries are sun-dried before being hulled to produce 'hard arabica coffee'.

b)   Climate and altitude — Minimum temperature 10°C (50°F) though absolute minimum is the limit of frost, maximum 30°C (85°F), best 15 to 25°C (60 to 75°F). High and low temperatures can be mitigated by mulching and shading.

Rainfall range from 750 to 2 500 mm (30 to 100 in.), best 1 750 to 2 000 mm (70 to 80 in.), but distribution is more important than total. Well distributed with a drier period of 2 to 3 months for the initiation of the flower buds is best. Grown at sea level in subtropics, and at altitude in tropics, eg Tanzania 1 100 to 1 600 m (3 600 to 5 200 ft), Kenya 1 400 to 2 000 m (4 700 to 6 600 ft), Mexico 1 000 m (3 300 ft). Periods of mist and low cloud are beneficial, strong or cold winds are detrimental.

c)   Soil — Deep (ideally over 2 m in dry conditions), slightly acid (pH 6.0 to 6.5, though may be more acid if other conditions are favourable: see 'mineral toxicity symptoms'), friable, permeable, well drained, eg fertile loams of lateritic or volcanic origin with 'reasonable' humus content. Coffee roots have a high oxygen requirement, therefore ill drained and heavy clay soils are unsuitable. Subsoil moisture must be present at all times, but the topsoil must be drier for some part of the year to initiate flower buds through moisture stress. Sandy soils are satisfactory if underlain by a subsoil with a higher clay content. In East and Central Africa adverse soil structure, low moisture-holding capacity, and low base status are often offset by the use of a grass mulch.

A hectare of fast-growing, high-yielding coffee will take up at least 135 kg N, 35 kg P_2O_5 and 145 kg K_2O. Nitrogen requirements are lower for shaded coffee. Most important inorganic constituents are nitrogen and potash, secondly phosphorus. Calcium requirements are fairly high. Magnesium deficiency lowers quality.

d) Mineral deficiency symptoms

Boron	Young leaves malformed, narrow leathery leaves, short weak branches fan branching
Calcium	Convex cupping of leaves and corky growths on veins
Iron	White colour between green veins on leaf. Remedy is to acidify soils with ammonium sulphate
Magnesium	'Herring-bone' pattern on older leaves. High concentrations of K will often cause Mg deficiencies, particularly when heavily mulched
Manganese	Olive-green terminal leaves, dark-green leaves below, mature leaves with yellow mottle
Nitrogen	General yellowing of whole leaf
Phosphorus	Blue-green colour of older leaves and reddish 'autumn tints'
Potash	Brown scorching of entire leaf margin
Sulphur	Causes excessive nitrogen intake (data from Papua New Guinea)
Zinc	Shortened internodes, production of small deformed narrow leaves

e) Mineral toxicity symptoms

Manganese	Causes 'Cafe Macho' in Puerto Rico, excessive Mn uptake under the influence of low soil pH. Remedy by applying lime to increase pH

F.6.2 Robusta coffee (Coffea canephora)

Robusta is grown largely in the wet lowlands of Africa and is a larger, more vigorous and hardier species than arabica, producing beans of lower quality with a higher caffeine content. It is used for blending with arabica and for the production of instant coffee.

Not as specific in its requirements as arabica and shows a wider range of adaptability. Rainfall 1 000 to 2 500 mm (40 to 100 in.), optimum 1 750 mm (70 in.). Temperature 18°C to 32°C (65°F to 90°F). Like arabica, requires a drier period for initiation of flower buds. Can be grown on shallow soils in high rainfall areas and will stand temporary waterlogging.

F.6.3 Liberica coffee (Coffea liberica)

Liberica forms a much larger tree than the other coffee species, with large beans of poor quality and a very bitter taste. It is a tree of hot, wet lowland forests and requires a heavy rainfall and high temperatures. It can be grown on a variety of soils from peat to clays and can be grown on poorer soils and withstand more neglect than robusta or arabica.

F.6.4 Miscellaneous

Further data (yields, rooting depth, plant spacing, etc) are summarised in Tables F.3 to F.8.

F.6.5 References

Acland J (1971). East African crops. Longman, London.

De Geus J G (1973). Fertiliser guide for the tropics and sub-tropics, pp 440-473. Centre d'Etude de l'Azote, Zurich.

Mitchell H W (1988). Cultivation and harvesting of the arabica coffee tree, pp 43-90. In: 'Coffee. Vol 4: agronomy', Clarke R J and McRae R (Eds). Elsevier, Barking.

Purseglove J W (1975). Tropical crops, dicotyledons: Rubiaceae, pp 458-492. Longman, London.

Snoeck J (1988). Cultivation and harvesting of the robusta coffee tree, pp 91-128. In: 'Coffee. Vol 4: agronomy'. Clarke R J and McRae R (Eds). Elsevier, Barking.

Wrigley G (1988). Coffee. Longman, London.

F.7 Maize (Zea mays)

F.7.1 Climate

Maize has a wide range of tolerance to environmental conditions, but the growing season must be frost-free, short-season varieties being able to develop with as little as 300 mm (12 in.) of rainfall in the growing season, as against 750 mm (30 in.) or more for longer-season varieties. The length of the rainfall season determines the type of maize which should be grown. The maturity length of the type chosen should not be much longer than the duration of the rains, otherwise the crop may be left short of moisture at the critical cob-formation stage. Hail can do great damage to the crop.

Annex F: Soil suitability for crops

F.7.2 Soils

Maize can be grown on a wide variety of soils. The best are well-drained, well-aerated, deep, loams and silt loams with adequate organic matter. The pH can range from 5.0 to 8.0, but 6.0 to 7.0 is optimal. A high AWC can have the effect of increasing the length of the growing season through a period of dry weather. For soils with a low moisture potential, or in areas of low rainfall, a low plant density should be used to avoid competition for water and nutrients. Yield increases with planting density on irrigated plots, but the reverse may obtain on unirrigated plots. Nitrogen is the most important nutrient. Young maize has difficulty in taking up phosphorus from the less available phosphate forms in the soil. Potassium removal is very high in maize harvested for silage, at 200 to 300 kg K_2O ha^{-1}. Replacement of the nutrients removed from the soil by a crop of cobs at 6 270 kg ha^{-1} (a high yield) and in the stover, requires approximately 165 kg N, 55 kg P_2O_5 and 135 kg K_2O. Medium tolerance of boron.

F.7.3 Salinity

Medium salinity tolerance. EC_e values at initial yield decline thresholds are 1.8 mS cm^{-1} for forage corn, 1.7 mS cm^{-1} for sweet and grain corn. Yield decrease per unit increase in salinity beyond the threshold, forage corn 7.4%, sweet and grain corn 12%. 50% yield reduction when EC_e = 10 mS cm^{-1}.

F.7.4 Mineral deficiency symptoms

Magnesium	Whitish or yellowish striping between the leaf veins
Nitrogen	Reduced vigour, a pale-green or yellowish colour
Phosphorus	Stunted growth, delayed ripening and purplish colour
Potassium	Small whitish-yellow spots on leaves
Zinc	Chlorotic fading of the leaves with broad whitish areas

F.7.5 Miscellaneous

Table F.11 summarises type of maize, length of rainy season, and planting density at altitudes of about 1 800 m (6 000 ft) in Kenya:

Cropping data for maize   Table F.11

	70-90		90-100		over 100	
Length of reliable rainy season in days	70-90		90-100		over 100	
Maturity length of maize	Early		Medium		Late	
Recommended types	Katumani Composites		Embu Hybrid 511		Kitale Hybrids	
Level of farming and amount of rainfall	Average	Good	Average	Good	Average	Good
Population recommended:						
per acre	14 520	14 520	14 520	21 780	14 520	20 000
per ha	35 800	35 880	35 880	53 820	35 880	49 420

Source: de Geus (1973).

F.7.6 References

De Geus J G (1973). Fertiliser guide for the tropics and sub-tropics, pp 90-108. Centre d'Etude de l'Azote, Zurich.

Harrison M N (1978). Maize in the tropics. Longman, London.

Maas E V and Hoffman G J (1977). Crop salt tolerance - current assessment, June 1977. J Irrig Drainage Div, Am Soc Civ Engs.

Purseglove J W (1975). Tropical crops, monocotyledons: Gramineae, pp 300-334. Longman, London.

Sprague G F (1977). Corn and corn improvement. Agron No 18. Am Soc Agron Inc, Madison, Wisconsin.

F.8   Millet

F.8.1   Introduction

Though there are many forms of millet, all are characterised by an ability to grow and mature under conditions of low soil fertility and low rainfall, with little or no attention. Therefore, in many tropical areas millet forms the staple food, and is also useful for making beer, for cattle fodder, for feeding poultry, and for bedding, thatching, fencing and fuel. Millet is widespread due to its hardiness, not its food value. All forms except finger millet suffer from bird damage. If growing conditions are less than severe sorghum or maize are more attractive crops.

F.8.2   Bulrush millet (Pennisetum typhoides)

The staple food of the drier parts of tropical Africa and India. It can be grown on poor sandy soil in low rainfall areas, and it stores well. Bourke (1963) divides the West African bulrush millets into:

a)   early millets, which occupy the land for 60 to 95 days;

b)   late-season millets, which take 130 to 150 days to maturity.

Deshaprabhu (1966) gives the variation in duration for the rainfed crop in India as:

a)   short-duration types maturing in about 80 days;

b)   medium-duration types maturing in 100 days or more;

c)   long-duration types of 180 days or more.

F.8.3   Climate

The shorter-maturing cultivars require less rainfall than sorghum. Northern limit of cultivation in West Africa is the 250 mm isohyet. Even distribution of rainfall throughout the growing season is more important than total precipitation. Millet does not possess the facility of remaining dormant during periods of drought as does sorghum. The optimum temperature is 30°C to 35°C (86°F to 95°F).

F.8.4   Soil

Cannot tolerate waterlogging. Will give economic yields, albeit low, in soil conditions too poor or exhausted to support other cereals, with the possible exception of hungry rice and some of the small millets.

F.8.5   Miscellaneous

Seldom manured, although fertilisers can dramatically increase yields. Susceptible to bird damage. Further data are summarised in Tables F.3 to F.8.

F.8.6   Finger millet (Eleusine coracana)

Short-duration (95 to 100 days), medium-duration (105 to 110 days) and long-duration (> 120 days) varieties. Can be stored for long periods of up to 10 years without deterioration or weevil damage and is therefore an important famine food. Suffers little from bird damage.

F.8.7   Climate

Dry weather is required for drying the grain at harvest. In drier areas with unreliable rainfall, sorghum and bulrush millet are better suited than finger millet. It will not tolerate such heavy rainfall as rice or maize. In India finger millet is the staple crop in areas with an annual rainfall of 800 to 900 mm. Thomas (1970) says that the crop grows best in Uganda where the average maximum temperature exceeds 27°C (81°F) and the average minimum does not fall below 18°C (64°C). It is the staple crop in Uganda in the drier areas at altitudes from 1 000 to 1 500 m with an annual rainfall of 900 to 1 250 mm.

F.8.8   Soils

The crop is grown on a wide variety of soils, but reasonably fertile, free-draining sandy loams are preferred. It cannot tolerate waterlogging. Thrives best in India on red lateritic loams. On the infertile soil of north-eastern Zambia it is grown in chitemene ash culture (see Purseglove, 1975, p 154).

F.8.9   Common millet (Panicum miliaceum)

Has one of the lowest water requirements of any cereal. It matures in 60 to 90 days and can be grown when the climate is too hot, the rainy season too short and the soil too poor for most other cereals with the possible exception of little millet. Can be grown at high

Annex F: Soil suitability for crops

altitude. The northern limit of cultivation is a June isotherm of 17°C and a July isotherm of 20°C. In India yields of 450 to 650 kg ha^{-1} are obtained, with 1 000 to 2 000 kg ha^{-1} when irrigated.

F.8.10   Little millet (Panicum sumatrense)

Grown throughout India to a limited extent up to altitudes of 2 100 m. Little millet will thrive under such adverse conditions that it can be grown on soils which otherwise produce little or nothing, and it will mature into a crop, albeit a small one, even in famine years. The soft straw is palatable to cattle and the green plant has potential as a quick-growing fodder. The crop, which usually receives little attention, matures in 75 to 150 days and yields 250 to 600 kg ha^{-1}, which in a good season may be increased to 900 kg ha^{-1}.

F.8.11   References

Bourke D O'D (1963). The West African millet crop and its improvement. Sols Afr 8, 121-132.

De Geus J G (1973). Fertiliser guide for the tropics and sub-tropics, pp 109-124. Centre d'Etude de l'Azote, Zurich.

Deshaprabhu S B ed (1966). The wealth of India: raw materials. Publication and Information Directorate, CSIR, New Delhi.

Purseglove J W (1975). Tropical crops, monocotyledons, pp 147-156, 199-220, 204-214. Longman, London.

Thomas D G (1970). Finger millet, pp 145-153. In: 'Agriculture in Uganda', Jameson J D. Oxford University Press.

F.9   Oil palm (Elaeis guineensis)

F.9.1   Climate and location

Oil palm is a typical tree crop of the tropical rain forest region and adjacent moist savanna woodland. In equatorial areas the oil palm is not grown at altitudes above about 600 m (2 000 ft).

Optimum mean daily temperature is 24 to 30°C (75 to 85°F) with a mean minimum of at least 18°C (64°C). Annual sunshine should be at least 1 300 h for high yields of fruit. Ideally commercial oil palm requires at least 1 500 mm (60 in.) annually of well-distributed rainfall with dry periods of no longer than 3 consecutive months.

Excessive rainfall decreases both pollen density and oil content of the mesocarp.

F.9.2   Soil

Oil palm requires deep, permeable soils. The soil should be well structured, and have a moisture-holding capacity of > 150 mm m^{-1} (> 1.8 in. ft^{-1}) plant-available water and an unrestricted rooting medium (< 2% by volume stoniness preferred). Effective soil depth should be > 100 cm.

Waterlogging is harmful: groundwater levels of < 90 cm below the soil surface lasting > 14 days should be avoided. If land is deeply (> about 80 cm) flooded for more than a week in 10 years, it is not suitable.

Terrain should have a slope of < 8° with an upper limit of about 10° unless platforming or terracing already exists, in which case slopes as steep as 16 or 17° can on occasion be utilised without soil erosion. Except in very few localities, virtually all Asian, it is too expensive and labour-intensive to construct new benches/terraces. The most suitable soil textures are in the range sandy loam to light clay. The best soils are rich alluvial deposits on which, with proper use of fertiliser, annual yields of up to 28 t ha^{-1} of fruit or 6 t ha^{-1} palm oil are possible. Unfavourable soils to be avoided are poorly drained, lateritic soils, very sandy coastal soils and deep (> 50 cm or 20 in.) peats.

Soils rich in available and potential plant nutrients are preferable as oil palm has a high nutrient requirement. After 4 years growth the soil fertility is best checked by leaf analysis on frond 17, see eg Hartley (1977) p 483. Nutrients removed by a crop of 15 t ha^{-1} of fresh fruit bunches are: 90 kg N, 20 kg P_2O_5, 135 kg K_2O and 40 kg CaO.

Potassium is the element required in largest amounts. Nitrogen is needed for rapid growth of young palms in the field. Phosphate is best applied with K on certain soils and the need for Mg is exhibited through its deficiency symptoms, often induced by K manuring, although vascular wilt desease (a common ailment of oil palms caused by the fungus Fusarium oxysporum) can produce similar symptoms, since it restricts nutrient supply to the leaves.

Soils should possess high levels of available P and exchangeable potassium should exceed 0.5% meq. Oil palm is cultivable in soils with pH 4.0 to 8.0 but most suited to pH 5.0 to 6.0. Salinity should be < 2 mS cm^{-1} but up to 8 mS cm^{-1} can be tolerated.

F.9.3 Mineral deficiency symptoms

Boron	'Little leaf' occurs in young oil palms and in older palms 'hock leaf' is often prominent where a distinctive hook appears at the apex of one or more pinnae on an affected frond. 'White stripe' may be an early or mild symptom
Magnesium	Deep orange tints, known as Orange Frond, specific and well established. Symptoms only occur in unshaded parts of the pinnae, but note similar effects produced by <u>Fusarium oxysporum</u> (see F.9.2)
Nitrogen	Fronds become pale green, changing to pale or bright yellow as the chlorosis becomes more severe
Phosphorus	No known field symptoms. The need for phosphorus usually detected through foliar analysis
Potassium	Orange spotting and mid-crown yellowing of leaflet fronds leading to terminal and marginal necrosis

F.9.4 Miscellaneous

Further data (yields, rooting depth, plant spacing etc) are summarised in Tables F.3 to F.8.

F.9.5 References

De Geus J G (1973). Fertiliser guide for the tropics and subtropics. Centre d'Etude de l'Azote, Zurich.

Hartley C W S (1977). The oil palm. Second Edition. Longman, London.

ILACO, BV (1981). Agricultural compendium for rural development in the tropics and sub-tropics, pp 486-487. Elsevier Scientific Publ Co, Amsterdam.

Purseglove J W (1972). Tropical crops, monocotyledons, pp 479-510. Longman, London.

Turner P D (1981). Oil palm diseases and disorders, pp 242-273. Oxford University Press.

Turner P D and Gillbanks R A (1982). Oil palm cultivation and management. Incorporated Society of Planters, Kuala Lumpur, Malaysia.

Williams C N and Chew W Y (1979). Tree and field crops of the wetter regions of the tropics, p 200. Longman, London.

F.10 Onion (Allium cepa)

F.10.1 Climate

Onions can be grown under a wide range of climate conditions, but they succeed best in a mild climate without excessive rainfall or great extremes of heat and cold. They are not suited to regions with heavy rainfall in the lowland humid tropics. Cool conditions with an adequate moisture supply are most suitable for early growth, followed by warm, drier conditions for maturation, harvesting and curing. Bulbing takes place more quickly at warm than at cool temperatures provided the minimum photoperiod for the cultivar has been reached. Long-day cultivars developed in temperate countries will not form bulbs in the shorter days of the tropics. Bolting with the production of inflorescences is induced by low temperatures.

F.10.2 Soil

Onions can be grown on a variety of soils, but the soil should be retentive of water, non-packing and friable; a good fertile loam usually gives the best results. They may be grown successfully on some peat soils. In Nigeria river levees are favourite sites, with the crop benefiting when young from a high water-table, which recedes with maturity, permitting bulbing and easy harvest. The optimum soil pH is about 6.0 to 7.0. Excess nitrogen slows down the bulbing process.

F.10.3 Salinity

Tolerant of high B levels, and moderately tolerant of soluble salts. EC_e = 1.2 mS cm^{-1} at initial yield decline threshold. The yield decreases by 16% for each unit increase in salinity beyond the threshold. 50% yield reduction at EC_e = 10 mS cm^{-1}.

F.10.4 Miscellaneous

Further data (yields, rooting depth, plant spacing etc) are summarised in Tables F.3 to F.8.

Annex F: Soil suitability for crops

F.10.5   References

Maas E V and Hoffman G J (1977).   Crop salt tolerance – current assessment, June 1977.
J Irrig Drainage Div, Am Soc Civ Engs.

Purseglove J W (1975).   Tropical crops, monocotyledons: Alliaceae, pp 38–50.   Longman,
London.

F.11   Rice (Oryza sativa)

F.11.1   Wetland rice

a)   Climate – Rice tolerates a very wide range of climatic conditions and can be grown in
temperate or hot tropical climates from sea-level to high altitude.   The average
temperature should lie between 20 and 38°C (68 to 100°F) during the growing season.

Long periods of sunshine are essential for high yields.

Availability of irrigation water is probably more important than rainfall though ideally
both should be in good supply.   Good management is more important than an ideal soil or
climate.

b)   Soil – Rice will grow on a wide range of soils, there being no optimum soil.   The
optimum pH is 5.5 to 6.5 when dry, though this may rise to 7.0 to 7.2 when flooded.
Cultivation is possible on alkaline soils, eg black soils in India have a pH of 8
to 9.   If pH falls as low as 2.0 to 3.4, as is possible in a reclaimed mangrove swamp,
rice cannot be grown.   Medium tolerance of soluble salts (50% yield reduction with
$EC_e = 10$ mS cm^{-1}). EC_e at initial yield decline threshold 3.0 mS cm^{-1}.

Heavy alluvial soils of river valleys and deltas are usually better suited to rice than
lighter soils, though rice can be grown on many soils from sandy loams and shallow
lateritic soils to heavy clays.   It should be possible to puddle the soil to maintain a
high water-table during growth, and to drain the soil for ripening and harvesting.

Due to anaerobic conditions decomposition of organic matter is slowed and nitrogen
fixation takes place by Azotobacter and blue-green algae.

c)   Nutrients – A rice crop producing about 3 360 kg of grain, and an equal amount of
straw, per hectare, removes approximately 54 kg N, 26 kg P and 46 kg K.   The most usual
deficiencies are nitrogen and phosphorus, with potassium and sulphur in limited areas,
and sometimes silica on peaty soils.   The transformation and availability of N and P in
waterlogged soils is covered by Patrick and Mahapatra (1968).

F.11.2   Upland rice

During its shorter growing season upland rice responds to the same soil conditions as
irrigated or 'swamp' rice.   In Asia this is generally appreciated but in Africa much upland
rice is grown on soils that are far too sandy, shallow or pervious, hence the very poor yield
commonly obtained.   Where slopes are rather steep and the rains are marginal or subject to
great seasonal variability, as in West Africa between Senegal and Cameroon, conditions are
even more severe for upland rice.

F.11.3   Miscellaneous

Further details (yields, rooting depth, plant spacing etc) are summarised in Tables F.3
to F.8.

F.11.4   References

De Datta S K (1981).   Principles and practice of rice production, p 618.   Wiley, Chichester.

De Geus J G (1973).   Fertiliser guide for the tropics and sub-tropics, pp 35–66.   Centre
d'Etude de l'Azote, Zurich.

Grist D H (1983).   Rice.   Fifth Edition.   Longman, London.

IRRI (1975).   Major research in upland rice.   International Rice Research Institute,
Los Baños, Philippines.

Maas E V and Hoffman G J (1977).   Crop salt tolerence – current assessment, June 1977.
J Irrig Drainage Div, Am Soc Civ Engs.

Patrick W H and Mahapatra (1968).   Transformation and availability to rice of nitrogen and
phosphorus in waterlogged soils.   Adv Agron 20, 323–360.

Purseglove J W (1975).   Tropical crops, monocotyledons: Gramineae, pp 161–198.   Longman, London.

Young A (1976).   Tropical soils and soil survey, p 229.   Cambridge University Press.

F.12   Rubber (Hevea brasiliensis)

F.12.1   Climate and location

Mainly cultivated between latitudes 10°N and 10°S in the tropical rain forest zone.   Growth is most rapid at altitudes below 200 m (650 ft) and where monthly mean temperatures are 27 or 28°C (80 to 83°F).   Mean annual rainfall should be between 1 500 and 4 000 mm (60 and 160 in.) well distributed through the year with no monthly mean being less than 100 mm (4 in.).   Both excessive amounts of rain and marked dry seasons reduce the yields.

F.12.2   Soil

Rubber has an extensive root system with a tap-root that can go 3 to 4 m deep and shallow (0–30 cm) lateral roots that can extend 20 m or more.   Accordingly the ideal soils are deep (> 100 cm), well aerated, and with a good physical structure and water-holding capacity. Shallow soils permit only surface rooting and are likely to cause water stress.

Rubber requires well-drained soils with a good supply of water throughout the year.   Under poor drainage conditions, the roots atrophy.   Ideal conditions are permeable medium-textured soils, with a GWT permanently at 4 to 6 m.   Sandy clay loams and clay loams are amongst the most suitable textural classes.

Hardpans, stoniness (no gravel within 100 cm soil depth is optimal) and high GWT can greatly restrict root growth.   Soils subject to erosion can be used with suitable conservation techniques.   Gently sloping or rolling terrain with minimal soil erosion (< 8° slope) is desirable.

Rubber will grow adequately in impoverished soils, but is responsive to a good nutrient supply and a balanced manuring programme, based on the nutrient status (actual and potential) of the soil.

Rubber will grow on soils of (surface) pH 3.6 to 8.0 but the most suitable soils are strongly to moderately acid (pH 4.4 to 5.2).   With pH over 6.5 growth is retarded and liming can be deleterious.   Optimal values for CEC are > 15 me but limitations are not experienced above about 4 me.   Ideally there should be > 1% organic carbon in the upper horizon.

Young rubber requires the major nutrients N, P, K and Mg for optimum growth until tapping. Large responses are obtained from applications of these elements but in the case of N an excess should be avoided as this increases vegetative growth and tree height with a greater risk of trunk breakage.

In mature rubber, there is often only a very limited and slow yield response to added fertilisers but a balanced nutrient supply is essential for maintaining vigorous and healthy trees with a high production potential.

An annual yield of 1 500 kg ha^{-1} of latex contains approximately 40 kg N, 10 kg P_2O_5 and 25 kg K_2O.   Soils should possess no deficiency or toxicity levels of trace elements.

F.12.3   a)   Mineral toxicity symptoms

Boron	Faint interveinal mottled yellowing on the leaves followed by severe marginal and tip scorch
Manganese	Dull greenish-brown leaves and occasionally interveinal scorch

b)   Mineral deficiency symptoms

Copper	Can possibly occur, leading to defoliation and death of apical growing points
Magnesium	Chlorosis in interveinal leaf areas spreading inwards from leaf margins
Manganese	Paling of leaf with bands of green tissue outlining midrib and veins (< 50 ppm in leaf)
Molybdenum	Development of a very pale brown scorch around the leaf margins, particularly in the region of the leaf tip. Especially in the most acid soils
Potassium	Marginal and tip chlorosis followed by marginal necrosis
Zinc	Lamina becomes reduced in breadth and is often twisted and there is general chlorosis of the leaf with midrib and main veins remaining dark green in colour

Annex F: Soil suitability for crops

F.12.4   <u>Miscellaneous</u>

Further data (yields, rooting depth, plant spacing, etc) are summarised in Tables F.3 to F.8.

F.12.5   <u>References</u>

Chan H Y (1977). A soil suitability technical grouping system for hevea. Planters
    Bull 152, pp 135-146. Kuala Lumpur.

Coulter J K (1972). Soils of Malaysia. A review of investigations on their fertility and
    management. Soils Fert 35, 475-498.

De Geus J G (1973). Fertiliser guide for the tropics and subtropics. Centre d'Etude de
    l'Azote, Zurich.

ILACO, BV (1981). Agricultural compendium - for rural development in the tropics and
    sub-tropics, pp 508-509. Elsevier, Amsterdam.

Ng S K and Law W M (1971). Pedogenesis and soil fertility in West Malaysia. Nat Resour
    Res Unesco 11, 129-131.

Polhamus L G (1962). Rubber - botany, production and utilisation, pp 179-180. Leonard
    Hill, London.

Purseglove J W (1968). Tropical crops, dicotyledons, pp 146-171. Longman, London.

Pushparajah E (1983). Use of soil survey and fertility evaluations in hevea cultivation.
    Paper Presented at Fourth International Forum on Soil Taxonomy and Agrotechnology
    Transfer, Bangkok.

Pushparajah E and Amin L L (1977). Soils under hevea and their management in Peninsular
    Malaysia, 188 pp. Rubber Research Institute of Malaysia, Kuala Lumpur.

Van Barneveld G W (1977). FAO-UNDP soil surveys and land evaluations for the second
    development programme of the Cameroon Development Corporation, pp 94-102. Tech Rep
    No 7. Soil Science Department, Ekona.

Williams C N and Joseph K T (1970). Climate, soil and crop production in the humid tropics.
    Oxford University Press, Singapore.

F.13   <u>Sorghum (Sorghum bicolor)</u>

F.13.1   <u>Climate</u>

Sorghum is essentially a plant of hot and warm countries. It can tolerate hotter and drier
conditions than maize. The optimum temperature for growth is 30°C (86°F) and it is killed
by frost. Sorghum can also be grown in areas of high rainfall in which waterlogging may
occur, though under these conditions it may be subject to attack by fungus.

Its great merit is its drought resistance which makes it suitable for areas of erratic
rainfall, and its xerophytic characters permit it to survive physiological drought produced
by waterlogging when root functions are temporarily impaired. The plant remains dormant
during drought and resumes growth when conditions become more favourable. Maize subjected
to wilting for a week or more suffers permanent damage to the stomata; sorghum stomata are
scarcely affected by wilting for 14 days, and quickly resume the diurnal rhythm of change in
stomata opening. There may therefore be considerable risk of famine or serious food
shortages in areas which are marginal for maize if this crop is substituted for sorghum.

American high-yielding varieties are unacceptable for human consumption. They also mature
very quickly and are therefore not suited to the length of growing season in Africa where
local varieties are generally well adapted to climatic conditions. Certain new varieties
may alleviate this problem.

F.13.2   <u>Soil</u>

Sorghum can tolerate a wide range of soil conditions. It will grow well on heavy soils,
especially the deep-cracking valley-bottom soils of the tropics. In Sukumaland in Tanzania
it is the principal crop of the clay mbuga soils in the valley bottoms, bulrush millet or
maize being grown on the lighter soil of the hillsides. Sorghum is also tolerant of light
sandy soils. It can be grown over a wide range of pH from 5.0 to 8.5 and tolerates salinity
better than maize; EC_e at initial yield decline threshold, 4 mS cm^{-1}, 50% yield
reduction at 11 mS cm^{-1}.

The root and stubble residues of sorghum leave excess amounts of sugar in the soil causing a
vigorous multiplication of the soil micro-organisms. This microbial activity locks up the
available nitrogen for some time, inducing N deficiency in a succeeding crop.

F.13.3   Miscellaneous

The time to maturity depends on the cultivar and can vary from 100 days to 5 to 7 months.

Sorghum is susceptible to bird damage and storage pests, such as rice weevil, flour beetle and grain moth, which cause considerable losses.   White corneous grains are usually preferred for human consumption, but they are more subject to bird damage as they contain no bitter principle.

The aerial shoots of sorghum contain the cyanogenic glycoside dhurrin, which by enzyme action hydrolyses to give hydrocyanic acid, HCN.   As little as 0.5 g HCN is sufficient to kill a cow, and more than 750 ppm is regarded as dangerous to stock.   The quantity of HCN varies with the cultivar and the growth conditions;   it usually diminishes with age.   Small plants and young tillers have a high dhurrin content, most of which occurs in the leaves. Droughted and freshly ratooned plants are a dangerous source of poisoning, and it is most likely to occur when growth has been resumed after a check.   Nitrogenous manuring increases the HCN content.   The poison is destroyed when the fodder is made into hay or silage.

Further data (yields, rooting depth, plant spacing etc) are summarised in Tables F.3 to F.8.

F.13.4   References

Ayers R S and Westcot D W (1976).   Water quality for agriculture.   Irrigation and drainage, Paper No 29, p 26.   FAO, Rome.

De Geus J G (1973).   Fertiliser guide for the tropics and sub-tropics, pp 109-120.   Centre d'Etude de l'Azote, Zurich.

Doggett H (1988).   Sorghum.   2nd Edition.   Longman, London.

Purseglove J W (1975).   Tropical crops, monocotyledons: Gramineae, pp 261-287.   Longman, London.

F.14   Sugarcane (Saccharum officinarum)

F.14.1   Climate and location

Although often considered only a tropical crop, sugar is successfully grown both in the tropics and subtropics from latitudes of 0 to about 33°.   Ideally, the climate during the growing season should be warm (mean day temperatures around 30°C) with high incident radiation and fully adequate soil water.   During ripening and harvesting high radiation is also needed, but the weather should be drier, with cool but frost-free temperatures (mean day temperatures 10 to 20°C), although very light frosts can be tolerated.   Table F.12 summarises the main effects of climatic factors on cane.

F.14.2   Soil and topography

Ideally (for ease of field layout and harvesting) cane land should have long, smooth slopes of up to 1 to 3°, the higher values referring to heavier soils.   Completely flat land produces problems of low surface runoff and, under surface irrigation, of water distribution, unless artificial slopes are constructed.   With appropriate management techniques (contour ploughing, grass strips etc) slopes of up to about 10° are suitable, but above this an increasing economic penalty is paid in the greater costs of land preparation and the restrictions on layout, and husbandry and harvesting methods.

Ideal soils are well-drained, well-structured loams to clay loams > 1 m deep, with a pore space of at least 50% which at field capacity is only half-filled with water.   GWT levels should be > 1.5 to 2 m, and AWC values $\geqslant$ 150 mm m^{-1}.   Cane can, however, be grown successfully in conditions far from ideal, the main constraints being:

a)   Coarse-textured soils: Build-up of harmful nematode populations, aggravated because cane is perennial.

Limited AWC, which may limit growth (and hence cane yield);   if irrigated, may need frequent water applications and limit range of equipment.

Fertiliser losses by leaching;   split applications help, but increase costs.

b)   Fine-textured soils: Possibly long-lasting drainage problems (but cane is tolerant of short spells of waterlogging or even flooding).

Capping (especially on silty soils) reduces water penetration and aeration.

Compaction: high bulk densities hinder root development.

Annex F: Soil suitability for crops

Summary of climatic and climate-related influences on sugarcane

Parameter	Comments
1. Temperature	Root temperatures more important than air temperatures, an important consideration for irrigated cane. Minimum mean air temperature for active growth: 20°C, but variable (18 to 22°C) depending on variety and cultural factors (especially whether irrigated or not, because of effects on root temperatures). Cane growth is effectively curtailed in air temperatures above about 35°C
2. Radiation	In general, the greater the radiation intensity the higher the yield; day length effects are not well defined in a generalised form
3. Rainfall/irrigation	Main effects are best understood in terms of ET and soil water (see below). Note susceptibility of cane in high humidity to fungal diseases in leaves, sheathes and roots - suitable varietal selection is the only remedy. Rain and floods transmit a variety of fungal, viral and bacterial diseases
4. Soil water	Cane is vegetative material the growth of which is directly proportional to water transpired; maximum yields are therefore produced when water is freely available during growth periods. Note, however, ability of cane to produce compensatory regrowth after periods of water stress. Cane can tolerate short periods (say 1 to 2 weeks) of waterlogging or even flooding
5. Evapotranspiration	See item 4 above. With good management on soils with no limiting characteristics (such as temperatures, pans, salinity, waterlogging etc) approximately 1 t of cane is produced for every 1 cm of water transpired over the growing period

Note that cane is capable of very deep root development which may be advantageous in drought-prone areas; if sufficient water is always available roots will develop in upper layers only - eg Baran et al (1974) quote figures of 45% of all roots in the top 15 cm, 68% in the top 35 cm and 85% in the top 95 cm under a weekly irrigation cycle, compared with, respectively, 32, 50 and 69% under a 5-week cycle.

F.14.3   Salinity and soil reaction

High-fibre and drought-resistant varieties tend to be more resistant than others, but general response is as follows:

EC_e (mS cm^{-1})	Effect on cane
2-3	Noticeable yield reduction
5-9	Severe yield reduction
11-12	Total growth failure

When $EC_e \geqslant 2$, trials should be made to test resistant varieties.

Recoverable sugar is reduced by about 0.2 to 0.3% for each 1 mS cm^{-1} increase in EC_e. Valdivia (1981) notes the following for the cultivar H32-8560 on saline soils:

a)   effect of GWT level:

   for GWT > 2 m, soil EC $\leqslant$ 2.0 mS cm^{-1} tolerated
   for GWT 0.8 to 1.1 m, soil EC $\leqslant$ 8.2 mS cm^{-1} tolerated

b)   low N rates ($\leqslant$ 180 kg ha^{-1}) are recommended for saline soils with EC between 2 and 8 mS cm^{-1};

c)   cane and sugar yields are inversely proportional to soil salinity.

The optimum soil pH for sugarcane is about 6.5, but a wide range is tolerated, from about 4.5 to 8.5.

F.14.4   Mineral deficiency symptoms

Boron	Small, elongated, watery spots which develop striping parallel to the vascular bundles
Calcium	Small chlorotic spots, developing to dark reddish-brown with dead centres and which may eventually coalesce producing rusty appearance to the leaf
Copper	Causes 'droopy top' (failure of the spindle to uncurl, with drooping leaves in heavier canes) plus chlorosis and poor growth. Symptoms resemble those of mosaic disease
Iron	Pale, striped interveinal areas in young leaves; leaves almost white in severe cases
Magnesium	Rare. Young leaves pale green and older leaves yellowish green with small chlorotic (later dark-brown) spots
Manganese	Pale yellowish-green to white interveinal stripes of uniform width, over mid-portion and tips of leaves. Early stages may look like Fe deficiency; acute cases may show complete chlorosis of leaf with reddish-brown necrotic spots, which develop into continuous lengthwise stripes that cause leaf splitting
Nitrogen	Stunted growth, pale leaves, general lack of vigour
Phosphorus	Susceptibility to root rot; stunting; pale green to yellow leaves, with drying of tips and margins
Potassium	Leaf die-back from tips and margins; upper surfaces of midribs become reddish
Zinc	Rare. Pale green along major veins; increasingly pale interveinal areas with increasing deficiency. In severe cases, young leaves are completely chlorotic, with necrosis spreading down from leaf tips

F.14.5   Miscellaneous

Once the full canopy is established (after about 3 to 4 months, depending on climate etc) cane is a good protector against soil erosion, especially because of the long period between harvests, and the retention of ratoons which allows an extensive rooting system to develop.

Further data (yields, soil pH, plant spacing etc) are summarised in Tables F.3 to F.8 and in Table F.12; data on various nutrient levels are given in Evans (1959).

F.14.6   References

Baran R, Bassereau D and Gillet N (1974). Measurement of available water and root development of an irrigated sugarcane crop in the Ivory Coast. Pro Int Soc Sugarcane Technol XV, 2, 726-735.

Barnes A C (1974). The sugarcane. The World Crop Series. Leonard Hill, Aylesbury.

Evans H (1959). Elements other than nitrogen, potassium and phosphorus in the mineral nutrition of sugarcane. Proc Int Soc Sugarcane Technol 10th Congr, Hawaii, pp 473-508.

Thompson G D (1976). Water use by sugarcane. SA Sugar J 61, 593-600 and 627-635.

Yates R A (1978). The environment for sugarcane. In: 'Land Evaluation Standards for Rainfed Agriculture'. World Soil Res Rep No 49. FAO, Rome.

Yates R A (1979). Sugarcane as an irrigated crop, pp 103-113. In: 'Land Evaluation Criteria for Irrigation'. World Soil Res Rep No 50. FAO, Rome.

F.15   Tea (Camellia sinensis)

F.15.1   Climate

Mean minimum temperature not below 13°C (55°F), mean maximum not above about 29°C (85°F).

Rainfall not below 1 125 mm (45 in.) annually, preferably above 1 500 mm (60 in.) and well distributed; should not fall below 50 mm (2 in.) per month.

Hail can cause much damage.

F.15.2   Soil

Best are deep, permeable, well-drained, acid soils (often tropical red earths). pH between 4.0 and 6.0, optimum pH 5.3 to 5.6. Where pH is > 5.8 it can be reduced by applying 150 g of sulphur per 0.5 unit of pH per planting hole. On established tea use aluminium sulphate.

Annex F: Soil suitability for crops

Tea is intolerant of high levels of calcium, but can suffer from calcium deficiency, which is curable by adding gypsum. Magnesium deficiency is common in Sri Lanka, but is correctable by dolomite applications.

Tea is an aluminium accumulator and this element should be available in the soil.

The most important nutrient is nitrogen, secondly phosphorus and potassium. A tea crop of 1 100 kg ha^{-1} removes approximately 62 kg of nitrogen, 34 kg of potash and 11 kg of phosphoric acid.

F.15.3 Mineral deficiency symptoms

Copper	Leaf will not ferment properly
Magnesium	Main veins of leaf show dark green
Phosphate and potash	Excessive leaf loss and thinning of foliage. Eventually leaves become very dark green and turn rough and thick with age
Sulphur	Known as 'tea yellows'. Mottling of leaf, leaf fall may occur. Controlled by applications of ammonium sulphate
Zinc	Short internodes and dwarfed leaves

F.15.4 Miscellaneous

Albizzia spp, Dissotis spp and tropical highland bracken (Pteridium aquilinum) are sometimes used as indicator plants in selecting soils for tea.

Further details (yields, spacing, etc) are summarised in Tables F.3 to F.8.

F.15.5 References

Chenery E M (1966). Factors limiting crop production: Tea. Span 9, 45-48.

Child R (1953). The selection of soils suitable for tea. Pamphlet 5. Tea Res Inst of East Africa, Kericho.

De Geus J G (1973). Fertiliser guide for the tropics and sub-tropics, pp 474-494. Centre d'Etude de l'Azote, Zurich.

Eden T (1976). Tea. Third Edition. Longman, London.

Harler C R (1964). Tea. Third Edition. Oxford University Press.

Purseglove J W (1975). Tropical crops, dicotyledons: Theaceae, pp 599-612. Longman, London.

Tea Res Inst of East Africa (1969). Tea Growers Handbook. Kericho, Kenya.

Annex G

Erosion and the Universal Soil Loss Equation

Contents

Annex G: Universal soil loss equation

List of Figures and Tables

.

Erosion and the Universal Soil Loss Equation

G.1   Background

The Universal Soil Loss Equation (USLE) is a quantitative method of predicting soil erosion losses from rainfall, soil and other factors. Also, once the effects of the various factors are known and an acceptable limit has been fixed for soil losses, the various options of possible cropping systems can be ascertained for given levels of soil protection measures.

The equation was developed in the US in the 1950s by Wischmeier and Smith from trials conducted from the late 1920s. It has been modified for use in various parts of the world, for instance as the Soil Loss Estimation Model for Southern Africa (SLEMSA). However, the usefulness of the equation is that it can be applied anywhere, although the method of quantifying the various factors must be tested and usually modified to suit a particular region.

Discussion of the applications and limitations of the USLE are given in Goldsmith (1977), Hudson (1971), Soil Conservation Society of America (1977) and by Wischmeier and Smith (1978) in USDA Handbook No 537, which is now the standard reference. A recent review of SLEMSA is contained in Elwell and Stocking (1982), and details are given by Elwell (1981); Zachar (1982) presents a more general review of erosion, and economic aspects of soil conservation are discussed by Wiggins (1981).

G.2   The USLE equation

$$A = R \, K \, L \, S \, C \, P$$

where   A = soil loss in short tons (2 000 lb) per acre per year
        R = rainfall erosivity factor
        K = soil erodibility factor
        L = slope length factor
        S = slope gradient factor
        C = crop management factor
        P = conservation practice factor

For a particular region A is determined empirically for a range of representative soils on experimental plots by standardising factors L, S, C and P. L = 1 for a slope 72.5 ft long and S = 1 at 9% gradient (experimental plots used in USA were 1/100 ac and 6 ft wide, with median slope 9%). C = 1 for bare cultivated fallow. P = 1 when ploughed up and down the steepest slope.

Then $A = R \, K$ (the 'Basic Equation')

Since these standards were set using imperial (US) units it only adds to the confusion by trying to convert to metric units. L, S, C, and P are dimensionless as they are merely ratios. The units of R depend on the means of deriving the index, and since the basic equation is empirical they are not important. Similarly K is merely a coefficient to balance A and R, and its units derive from these.

G.3   Annual soil loss (A)

The equation can be used to calculate the amount of soil lost annually by erosion, but it is more useful in selecting crop management practices. For this purpose an acceptable limit has to be defined, which, according to principles of conservation, must not exceed the rate at which the soil can regenerate.

Under natural vegetation it may take hundreds of years for 1 in. of soil to form, but under cultivation the time can be reduced to one-tenth by artificially churning, aerating and leaching. 1 in. in 30 years gives a figure of 5 t^{-1} ac $year^{-1}$. Accepted limits of erosion range from 1 to 5 short t^{-1} ac $year^{-1}$ (2 to 12 t^{-1} ha $year^{-1}$). The higher

309

Annex G: Universal soil loss equation

limits apply to humid tropical regions, where soil formation is rapid, or to hilly areas such as in Sumatra or Swaziland where lower limits would be impossible to achieve.

G.4   Rainfall erosivity factor (R)

Wischmeier found the best correlation of soil erosion with rainfall data when using EI_{30}, the product of the kinetic energy of a storm and its intensity. The kinetic energy (E), in foot-pounds per acre, reflects the greater effect of larger drops, which impact with a higher terminal velocity ($E = 0.5 \, mv^2$, where m = mass; v = velocity). In practice this has to be calculated from an empirically determined relationship with rainfall intensity. The maximum 30 min intensity (I_{30}) is the intensity in inches per hour for the 30 min period of greatest intensity as obtained from recording rain gauges.

Outside the USA this index has been found to give poor correlation, particularly in the tropics. In Southern Africa, Hudson (1971, pp 66 to 67) found that a simpler index, $E \approx 1$, calculated by summing the kinetic energy of rain falling at an intensity > 1 in. h^{-1} for a storm, gave a better correlation.

An index found useful in the West Indies and West Africa (Lal, 1977a) is AI_m, the product of the total rainfall amount (A) and peak storm intensity (I_m).

The main difficulty in formulating an effective erosivity index is the paucity of rainfall intensity records, let alone data on drop size/velocity and hence kinetic energy. However Roose (1977) found that a good correlation held for West Africa with an erosivity index derived only from rainfall amounts, and a recent technique for estimating R from daily rainfall amounts in the USA is given in Richardson et al (1983).

Clearly, for any region the index used must depend on the rainfall data available, and where autographic records exist the best combination of parameters should be tested empirically. Some typical values are given in Table G.1.

Some typical rainfall erosivity indices (R) for different regions of the world   Table G.1

Region	Source	R (mean annual values)
Tunisia	Roose (1977 p 180)	60 – 300
Morocco	Roose (1977 p 180)	50 – 300
Central France	Roose (1977 p 180)	60 – 340
Upper Volta	Roose (1977 p 180)	200 – 600
Ivory Coast	Roose (1977 p 180)	500 – 1 400
North-east USA	Hudson (1971 p 182)	62 – 220
North-central USA	Hudson (1971 p 182)	64 – 260
South-east USA	Hudson (1971 p 182)	142 – 779
Lesotho	Goldsmith (1977)	127

(Note that in Lesotho R varied from 4 in 1965 to 792 in 1975)

It is usually convenient to present the annual rainfall erosivity as a cumulative distribution curve (eg as in Figure G.1), so that the year can be divided into stages during which different cropping (and conservation) practices are carried out. It might be added that the index R can apply to a particular year, or even a storm, or be an average annual figure.

G.5   Soil erodibility factor (K)

This term is considered to be an inherent property of the soil, usually applied to a soil series, and is independent of the effects of management, which are covered by factors P, C and L. Originally it was determined by standardising LS, P and C as described, so that K = A/R. Rainfall was either accurately measured or simulated, and the eroded soil was collected at the base of the slope for weighing. More recently a nomograph (Figure G.2) has been produced by Wischmeier et al (1971) which purports to give a K value from known values of:

a)   percent silt and very fine sand (0.002 to 0.1 mm);

b)   percent sand (0.1 to 2.0 mm);

c)   organic matter content;

310

Specimen curve of cumulative erosivity (R) for maize (see note 1)

Figure G.1

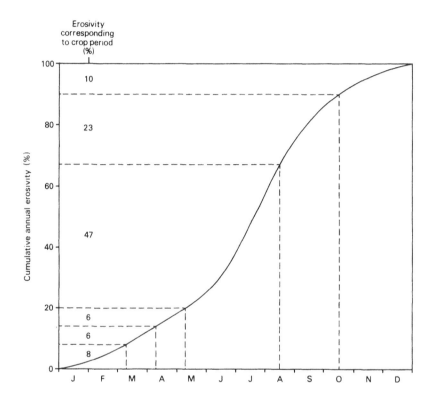

Crop period[2]	Fallow	1	2	3	4	Fallow
C value for maize[3]	–	28	19	12	18	10

Notes

(1) Curve relates to parts of North and South Carolina and Virginia.

(2) Crop periods
1 Seeding
2 Establishment
3 Growing and maturing
4 Residue and stubble

(3) From tabulated data in Wischmeier et al. (1965).

Source : *Adapted from Hudson (1971, p187).*

311

Figure G.2

Nomograph for estimation of soil erodibility (K)

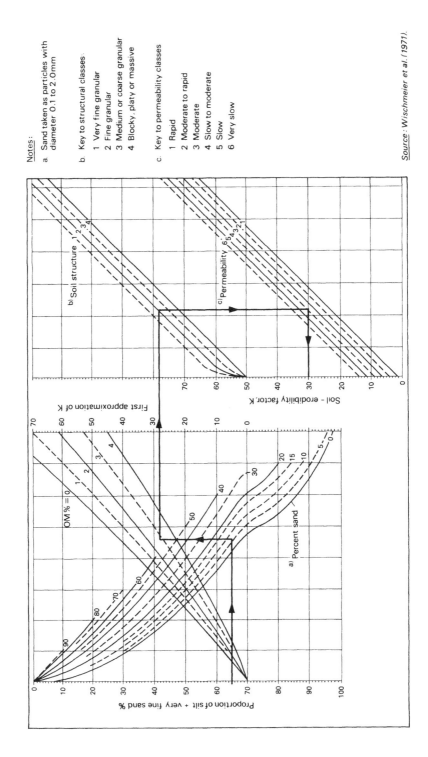

Notes:

a   Sand taken as particles with
    diameter 0.1 to 2.0mm

b   Key to structural classes:
    1 Very fine granular
    2 Fine granular
    3 Medium or coarse granular
    4 Blocky, platy or massive

c   Key to permeability classes
    1 Rapid
    2 Moderate to rapid
    3 Moderate
    4 Slow to moderate
    5 Slow
    6 Very slow

Source: Wischmeier et al. (1971).

d)   topsoil structure grade;

e)   permeability grade.

The procedure is illustrated by the heavy, arrowed line in Figure G.2 for a soil having the following properties:

```
Proportion silt + very fine sand = 65%
Proportion of sand               =  5%
Amount of organic matter         =  2.8%
Structural class                 =  2
Permeability class               =  4
```

From the nomograph the resulting erodibility (K) is 0.30.

Using the nomograph does not necessarily mean that the calculated value of K is accurate or meaningful; it is unwise to use or extrapolate from a nomograph derived in one part of the world to another.   Some typical values for K are given in Table G.2.

Some typical values for soil erodibility (K)   Table G.2

Country	Soil	K
USA	Albia gravelly loam, New Jersey	0.03
	Dunkirk silt loam, New York	0.69
	(Source: Hudson, 1971, p 192)	
Sri Lanka	Unidentified tea soil	0.31
	(Source: Greenland and Lal, 1977, p 205)	
West Africa	Ferralitic soils on tertiary sand	0.05 – 0.10
	Ferralitic soils on granite	0.10 – 0.15
	Ferralitic soils on schist	0.15 – 0.18
	Ferruginous tropical soils on granite	0.20 – 0.30
	(Source: Roose, 1977, p 182)	
Arbitrary relative grades	Erosion risk	
	low	0.09 or less
	low to medium	0.10 – 0.19
	medium	0.20 – 0.29
	medium to high	0.30 – 0.39
	high	0.40 – 0.59
	very high	0.60 or more
	(Source: Goldsmith, 1977)	

Note: The sources mainly quote reported figures from original workers.

G.6   Slope factor (LS)

When land has mechanical protection works, the effective slope length is the distance between channel terraces.   Length (L) and gradient (S) are therefore interconnected in the design and can conveniently be combined.   An approximate LS factor can be rapidly estimated from slope gradient and length data, as shown in Figure G.3, although more sophisticated tables and formulae are also available (Wischmeier et al, 1978).   It should be added that the effect of slope gradient is considered by some to be more pronounced in the tropics (eg Hudson, 1971, p 184) and that such nomographs and formulae should not be used indiscriminately.

G.7   Crop management factor (C)

This is defined as the ratio of soil loss from land cropped under specified conditions to the corresponding loss from bare tilled fallows.   It reflects the protective influence of the vegetation and ground cover, and is the most complicated factor in the equation.

It can be derived experimentally from trial plots, but Wischmeier and Smith (1965) devised a procedure for determining C for a given cropping system and locality.   This is explained more fully in Goldsmith (1977) and summarised by Hudson (1971, pp 185 to 188).

The cropping year is usually divided into five periods: period F (rough fallow), period 1 (seedling), period 2 (establishment of crop), period 3 (growing and maturing crop) and period 4 (residue or stubble).   For each period a value, c, is derived and multiplied by the

Relationship of slope factor (LS) to gradient and length   Figure G.3
of slope

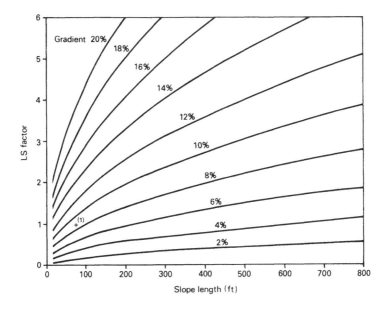

Note: (1) Standard condition : LS = 1 when slope = 9% and length = 72.5 ft.

Source : Hudson (1971, p185).

relevant proportion of the annual rainfall erosivity, r, from a cumulative distribution curve (see Hudson, 1971, p 187).

Thus $\quad C \times R = c_f r_f + c_1 r_1 + c_2 r_2$ etc

This method is more sensitive than simply using annual C and R values since both vary in an interrelated way throughout the year. For instance a season of high erosivity may produce little erosion if it coincides with a mature crop with good ground cover.

Typical values for C are given in Table G.3 for illustrative purposes.

Some typical crop management factor values (C)   Table G.3

Crop management	Level of production	C values [1] at each crop stage period				
		F	1	2	3	4
Continuous dry beans, low residue, ploughed	High	70	78	54	27	62
Continuous dry beans, low residue, disced	High	–	78	54	27	62
Continuous dry beans, low residue, ploughed	Medium	75	80	70	35	75
Maize after maize, grains, or cotton, residue left, ploughed	High	36	52	41	20	30
Maize after maize, grains, or cotton, residue left, ploughed	Low	51	60	45	33	47
Continuous grain sorghum, residue left, ploughed	High	36	63	50	26	30
Continuous grain sorghum, residue left, ploughed	Low	55	70	58	32	50
Continuous small grain, residue left, ploughed	High	36	60	40	10	10
Continuous small grain, residue left, ploughed	Medium	55	70	45	10	10
Grass legume ley	Medium	–	–	–	–	0.6
Alfalfa	Medium	–	–	–	–	2.0

Note: 1/   C values are the ratios in percent compared with loss from continuous bare fallow.

Sources: Goldsmith (1977); Wischmeier and Smith (1965 and 1978).

G.8   Conservation practice factor (P)

This is the ratio by which erosion is reduced by conservation measures from the worst possible case (P = 1.0), in which soil is ploughed up and down the steepest slope.

The effect of contour cultivation is not independent of the slope factor, so that different values of P are given in Table G.4 for different slope gradients, S. The table is highly simplified. See also Tables 13 to 15 in Wischmeier and Smith (1978).

Some typical conservation practice values (P)   Table G.4

Land slope %	Contour ploughing	Contour strip cropping
0 – 7	0.50	0.25
8 – 12	0.60	0.30
13 – 18	0.80	0.40
19 – 24	0.90	0.45

Terraces and contour ridges effectively change the land slope characteristics. The best procedure then is to use the same P value as for contour cultivation and then use an LS value appropriate to the spacing between the terraces.

Annex G: Universal soil loss equation

Annex H

Soil Description and Data Presentation

Contents

Annex H: Soil description

List of Figures and Tables

Soil Description and Data Presentation

H.1   Field descriptions

Use of the FAO (1977a) system (based on USDA, 1951) is recommended, and data collected should include all the items listed in Table H.1.   Examples of soil record sheets for pit sites and for auger sites are presented in Figures H.1 and H.2 respectively.   Surveyors should make a point of periodically rereading the definitions of the words used in the FAO descriptions; standard words are very often used incorrectly in reports and descriptions.

For some surveys the FAO data need modification; some suggestions are:

a)   Slope   – always record actual slope (% and/or degrees) and direction
                    – FAO Class 1 (Flat) needs subdividing for surface irrigation projects; if possible 0 to 0.5°, 0.5 to 1°, 1 to 2°
                    – estimate slope lengths and complexity for surface irrigation projects as a guide to land levelling requirements

b)   Drainage   – this almost always has to be estimated from soil morphology; the system shown in Figure H.3 can be used as a basis, but will often need local modification, as there is no simple relationship between soil–water regime and morphological expression of gleying (Hodgson, 1978, p 36)

c)   Stoniness   – note that FAO terminology (given below) differs from the USDA and international system nomenclature.

                    Gravel: fragments 0.2 – 7.5 cm dia
                    Stones: fragments 7.5 – 25 cm dia
                    Boulders: fragments > 25 cm dia

                    See also boulder, cobble, gravel and stone in the Glossary, Annex L.

d)   Carbonates   – note that texture and structure can affect the vigour of the reaction with HCl; samples should be gently crushed before testing.   Carbonate contents can be estimated using Table H.2.

H.2   Data presentation

Specimen layouts for recording and presenting data are given in Tables H.3 and H.4.   They are designed primarily to be used in reports on semi–detailed or reconnaissance surveys for specific crop or project developments.

H.3   Textural classes

The standard triangular diagram sometimes proves confusing and/or time–consuming when used to classify particle size data; Figure H.4 shows the same information on normal 90° axes.

Annex H: Soil description

---

1. Site details

   Location and reference: Site/station names/numbers; date; location (grid reference, aerial photograph/map sheet number; distance and direction from nearby landmarks); author of information

   Landform/hydrology/geology: At site and surrounding area; landform units; relief/elevation; slope types, lengths and angles; position of site on landform unit; microtopography; erosion hazard; local and regional drainage networks; frequency/depth of flooding; depths to GWT and seasonal variations; quality of surface and groundwaters; river/stream flow rates; main rock formations/parent materials and depth below soil cover

   Vegetation/land use: General description, dominant species, and size/density of natural vegetation at site/in area.   Soil/plant relationships; crop(s) at site/in area - main species, methods of cultivation, cropping history, health and vigour; soil/land use relationships

2. General soil/land information: At site and in area: parent material; surface stoniness/rock outcrops; evidence of erosion; presence of salt/alkali; soil moisture conditions and profile drainage; soil patterns and variability; soil and land evaluation classification; methods of identification of mapped boundaries

3. Soil morphology: Horizon depths and thicknesses
                                  Moisture status
                                  Colour and mottling (Munsell notation): wet/moist/dry
                                  Texture
                                  Structure and structural stability
                                  Consistence: wet/moist/dry
                                  Cutans, pressure faces etc
                                  Cementation/pan formation
                                  Pore size, type and distribution
                                  Rock/stone/nodule content: abundance/size/type
                                  Carbonate/soluble salt accumulation
                                  Artefacts
                                  Biological features
                                  Root size and distribution
                                  Horizon boundaries

4. Field tests: pH; electrical conductivity; saturated hydraulic conductivity; infiltration rate; field capacity; bulk density; carbonate content

5. Laboratory tests: Standard and special problem physical and chemical tests - see Chapters 6 to 8

6. Additional and specialised
   information: Climatic data 2/ (location of weather stations; precipitation amount/intensity/distribution; evaporation data; wind speed and direction; maximum, minimum and mean temperatures; radiation levels; humidity values; frost/hail data; start and finish dates of growing season); clay mineralogy; toxic materials (eg heavy metals); shear stress; bearing ratio

---

Notes: 1/   See sample profile sheets and profile description, Table H.3.
         2/   Preferably long-term daily records, ≥ 30 year period.

Sources: Dent and Young (1981); McRae and Burnham (1981); FAO (1977a, 1979a, 1979c); USDA (1951). For additional data (eg on topography, hydrology and economics) see ILACO, 1981, pp 174ff.

Example of soil record sheet (pit site)   Figure H.1

BOOKER AGRICULTURE INTERNATIONAL LTD 3AI SOIL RECORD SHEET (Pit site)		Project		Date	Author	Site No.

Location		Parent material		Maximum slope		Soil (1) (2)	Land class (1) (2)

Position		Landform			Soil drainage	GWT (cm)

Microtopography/Surface features	Estimated permeability

Vegetation/ land use	Horizons sampled

Depth (cm)	1	2	3	4	5	6
Horizon thickness (cm)						
Moisture status						
Colour moist dry						
Mottles ab/s/ct/col						
Texture						
Structure dev/s/type						
Consistency dry moist wet						
Porosity ab/s/distrib/type						
Concs/Gravels ab/s/consist/shape						
Fauna						
Roots fine medium coarse						
HCl						
Lower boundary						

Comments/Diagram:

(1) Preliminary classification   (2) Final classification

319

Example of soil record sheet (auger site)   Figure H.2

BOOKER AGRICULTURE INTERNATIONAL LTD **3AI** SOIL RECORD SHEET (Auger site)	Project		Date	Author	Site No.

Location	Parent material	Maximum slope	Soil (1) (2)	Land class (1) (2)

Position	Landform	Soil drainage	GWT (cm)

Microtopography/Surface features	Structure

Vegetation/ land use	Roots
	Estimated permeability

Depth (cm)	Colour (field)	Mottles ab/s/ct/col	Texture	Consistency	Moisture status	HCl	Other characteristics

Comment on profile

Diagram	Between sites

(1) Preliminary classification   (2) Final classification

320

Drainage class and water regime related to profile morphology   Figure H.3

DRAINAGE CLASS	Well drained	Moderately well drained	Imperfectly drained	Poorly drained	Very poorly drained
	months	months	months	months	months
WATER REGIME INDICATIVE AVERAGE ANNUAL PERIOD OF WATERLOGGING	At 30 cm: 0 At 60 cm: 0 At 90 cm: ‹1	0 ‹1 ›1	‹1 1-6 -	3-6 ›6 -	›6 - -

INDICATIVE COLOURS (cm scale 0, 30, 60, 90)

Well drained: Occasional ochreous mottles

Moderately well drained: Possibly dull colours; Faint greyish and ochreous mottles

Imperfectly drained: Ochreous mottles; Distinct grey and ochreous mottles prominent at depth

Poorly drained: Rusty root mottles; Prominent grey and ochreous mottles often a grey horizon

Very poorly drained: Dark peaty; Pre-dominant grey with ochreous streaks or tubes along root channels

Grey colours of Munsell chroma ‹3 in values 4 or more or Munsell chroma 3 in values 6 or more

Munsell chroma 1 or less or hues bluer than or greener than 10Y dominant

SOIL DRAINAGE CLASS	KEY TO FIELD IDENTIFICATION
Well drained (excessive)	Coarse-textured soils with small available water capacity and only saturated during and just after heavy rain. Surplus water is removed very rapidly. Any water-table is well below the solum.
Well drained	Soil is rarely saturated in any horizon within 90 cm. Mottling is usually absent throughout the profile.
Moderately well drained	Some part of the soil in the upper 90 cm is saturated for short periods in winter or after heavy rain but no horizon within 50 cm remains saturated for more than one month in the year. Colours typical of well drained soils on similar materials are usually dominant but may be slightly lower in chroma, especially on ped faces and faint to distinct ochreous or grey mottling may occur below 50 cm.
Imperfectly drained	Some part of the soil in the upper 50 cm is saturated for several months but not for most of the year. Subsurface horizon colours are commonly lower in chroma and/or yellower in hue than those of well drained soils on similar materials. Greyish or ochreous mottling is usually distinct by 50 cm and may be prominent below this depth. There is rarely any gleying in the upper 25 cm.
Poorly drained	The soil is saturated for at least half the year in the upper 50 cm but the upper 25 cm is unsaturated during most of the growing season. The profiles normally show strong gleying. A horizons are usually darker and/or greyer than those of well drained soils on similar materials and contain rusty mottles. Grey colours are prominent on ped faces in fissured clayey soils or in the matrix of weakly structured soils.
Very poorly drained	Some part of the soil is saturated at less than 25 cm for at least half the year. Some part of the soil within the upper 60 cm is permanently saturated. The profiles usually have peaty or humose surface horizons and the subsurface horizon colours have low (near neutral) chroma and yellowish to bluish hues.

Source: Adapted from Corbett and Tatler (1970)

Note: This system provides a rough guide for use only when lack of data precludes more precise systems (e.g. the 'wetness classes' of Jarvis and Mackney, 1973). Interpretation of colour/mottling should allow for effects of parent materials, high or low Fe contents, relict features, current flooding, climate etc.

Annex H: Soil description

Modified USDA soil texture triangular diagram   Figure H.4

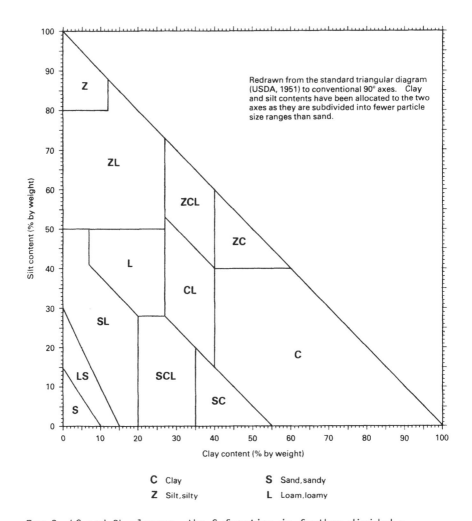

Redrawn from the standard triangular diagram
(USDA, 1951) to conventional 90° axes.   Clay
and silt contents have been allocated to the two
axes as they are subdivided into fewer particle
size ranges than sand.

C Clay	**S** Sand, sandy
Z Silt, silty	**L** Loam, loamy

For S, LS and SL classes, the S fraction is further divided :-

fS ⩾ 50% fS, or < 25% vcS,   S ⩾ 25% vcS and cS, and
    cS and S and < 50% vfS   < 50% any other single
                                           S grade.
vfS ⩾ 50% vfS
                                       cS ⩾ 25% vcS, cS and S, and
                                           < 50% fS or vfS

<u>Source</u>: Adapted from USDA (1951)

322

Field estimation of carbonates 1/   Table H.2

Field description and limits of groups	Estimated $CaCO_3$ range (%)	Audible effects (close to ear) on applying HCl 2/	Visible effects on applying HCl 2/
Non-calcareous	< 0.5	None	None
Barely calcareous	0.5-1.0	Faintly increasing to slightly audible	None
Slightly calcareous	1.0-2.0	Faintly increasing to moderately audible	Slight effervescence confined to individual grains, just visible
Moderately calcareous	2.0-5.0	Moderately to distinctly audible; heard away from ear	Slightly more general effervescence visible on close inspection
Calcareous	5.0-10.0	Easily audible	Moderate effervescence; obvious bubbles up to 3 mm diameter
Very calcareous	> 10.0	Easily audible	General strong effervescence; ubiquitous bubbles up to 7 mm diameter; easily seen

Notes: 1/ This applies principally to $CaCO_3$; estimates of less reactive $MgCO_3$ are less reliable

2/ HCl at 10% strength

Source: Hodgson (1974).

Annex H: Soil description

ZANGARO SERIES - SITE ZA 113

| Soil classification | FAO Chromic Luvisol | USDA Ustic Haplustalf |

Land capability   Modified USBR   2s Irrigable

(1) SITE CHARACTERISTICS

Location/date/author	3 km north of Vindu, 50 m E of main road. Soil Map Sheet 74 27 July 1983/JRL
Landform	Flat plain bordering Zangaro River
Parent material	Alluvial/aeolian deposits
Microtopography	Very flat; maximum slope 0.5°
Surface features	Weak sandy capping; low ridge and furrow; common termite activity
Drainage/GWT	Well drained; no GWT within 3 m
Vegetation/land use	Fallow sorghum/millet plot; 'farmed parkland'
Climate	W African Sub-Saharan. Annual rainfall 900 mm. Temperature ranges: mean maximum 38.4 to 28.9°C; mean minimum 24.4 to 12.3°C

(2) SOIL PROFILE CHARACTERISTICS

Depth (cm)	Field description
0-13	Very slightly moist fine sand; strong brown (7.5YR 5/6) when moist, and reddish yellow (7.5YR 7/6) when dry; very weak very fine platy; slightly hard when dry, friable when moist, non-sticky non-plastic when wet; very few very fine random tubular pores; no reaction with HCl; common fine roots; clear wavy boundary with:
13-56	Very slightly moist fine sand; yellowish red (5YR 4/6 to 5/6) when moist, and light reddish brown (5YR 6/4) when dry; single grain; slightly hard when dry, friable when moist, non-sticky non-plastic when wet; common fine random tubular pores; no reaction with HCl; few fine termite holes; very few fine roots; gradual smooth boundary with:
56-250 (augered below 200)	Very slightly moist fine sandy clay loam; reddish brown (5YR 4/4 to 4/6) when moist, and yellowish red (5YR 5/8) when dry; prismatic; hard when dry, friable to firm when moist, slightly sticky and plastic when wet; continuous thin cutans, possibly of clay with iron oxides, on horizontal and vertical ped faces; common fine, few medium random tubular pores; no reaction with HCl; few fine termite holes; extremely few fine roots; boundary not seen.
250-300	Moist fine sandy clay loam; yellowish red (5YR 5/8) when moist, and reddish yellow (5YR 6/8) when dry; common fine distinct pale yellow and red mottles; firm when moist, slightly sticky and plastic when wet; few hard black manganese nodules; no reaction with HCl; no roots.
REMARKS	A typical Zangaro series profile, characterised by 5YR hues and sandy topsoil; textures increasingly fine with depth.

(3) SOIL ANALYTICAL DATA

Depth (cm)	Texture (lab)	cS	fS	vfS	cZ	Z	C	pH (1:2.5) H2O	KCl	EC (1:2.5) (mS cm^{-1})	C (%)	N (%)	C/N
3-10	fSL	7	46	18	17	4	7	5.8	4.1	0.01	0.60	0.05	12
25-35	fSL	6	38	10	23	8	15	5.4	4.0	0.02	-	-	-
70-80	CL	5	29	10	18	10	28	5.8	4.2	0.01	-	-	-
140-150	CL	6	28	9	18	7	32	5.8	4.2	0.01	-	-	-

Depth (cm)	Exch bases (me/100 g ads) Ca	Mg	Na	K	CEC	BSP	ESP	K2O (ppm) Total	Avail	P2O5 (ppm) Total	Avail
3-10	1.20	0.50	0.15	0.11	2.80	70	5.3	250	48	70	4
25-35	2.50	0.80	0.20	0.16	6.25	59	3.2	-	-	-	-
70-80	5.30	1.10	0.20	0.21	9.75	68	2.1	-	-	-	-
140-150	7.15	1.05	0.20	0.26	12.70	67	1.6	-	-	-	-

cont

(4) SOIL PHYSICAL MEASUREMENTS

(a) Laboratory moisture release determinations

Depth (cm)	Lab texture	Moisture contents (% by weight) at ψ (mb) -10^{-3}	-0.1	1.0	15	Bulk density (g cm^{-3})
5-10	fSL	36.0	19.0	4.4	3.6	1.63
35-40	fSL	37.2	18.6	10.5	9.6	1.54
65-70	CL	37.0	20.5	14.2	12.6	1.55

(b) Field AWC values and related measurements

Depth (cm)	Field texture	Moisture contents (% by volume) at: FC 1/	PWP 2/	Bulk density 1/ (g cm^{-1})	AWC (=FC-PWP) 1/ (% by vol)
5-10	fS	24.9	5.1	1.42	19.8
35-40	fS	23.8	13.0	1.35	10.7
60-65	fSCL	25.8	18.3	1.45	7.5

AWC in top metre of profile: 105 mm

Note: 1/ Average from 3 field replicates, taken 48h after saturation
       2/ At matric potential - 15 bar

(c) Infiltration rates

Replicate position No	Accumulated intake (cm) after 1 h	2 h	3 h	4 h	Final instantaneous rate after 4 h (cm h^{-1})	Basic infiltration rate after 4 h (cm h^{-1})
1	23.1	37.5	51.2	64.2	13.6	13.6
2	22.5	31.5	44.6	56.4	12.4	12.0
3	15.3	27.6	38.7	50.0	12.0	11.2

(d) Hydraulic conductivity measurements

Site data	Depth of test (cm) 0-100			100-200		
Field texture	fS/fSCL			fSCL		
Replicate position No	1	2	3	1	2	3
K (mm h^{-1})	26	-	34	10	11	15

Table H.4

Specimen field sample register layout for soil surveys

Date collected	Collected by	Site and field sample Nos	Depth (cm)	Lab sample No	Analyses requested	Date of delivery to lab	Date results received from lab	Invoice No
19 Oct 83	JRL	ZA 113						
"	"	Position 1a	5-10	A 4031	Moisture content	20 Oct 83	7 Nov 83	03871
"	"	" b	25-30	2	"	"	"	"
"	"	" c	60-65	3	"	"	"	"
"	"	" 2a	5-10	4	"	"	"	"
"	"	" b	25-30	5	"	"	"	"
"	"	" c	60-65	6	"	"	"	"
"	"	Main Pit 3a	5-10	A 4097	Full chem analysis	27 Oct 83	2 Dec 83	03942
"	"	" b	25-30	8	"	"	"	"
"	"	" c	60-65	9	"	"	"	"
22 Oct 83	DRM	BA 319						
"	"	Position 1a	5-10	A 4117	Moisture content	24 Oct 83	10 Nov 83	03953
"	"	" 1b	30-35	8	"	"	"	"
"	"	" 1c	70-75	9	"	"	"	"

Annex J

Tabulated Weights and Measures Data

Contents

Annex J: Weights and measures

CONTENTS (cont)

J.1   Common conversions and constants

J.1.1   Summary of conversions

Measurement	To convert:	Main imperial conversions [1] Multiply by: (unless otherwise stated)	To give equivalent in:	To convert:	Main metric conversions [1] Multiply by: (unless otherwise stated)	To give equivalent in:
1. Area	in²	6.4516	cm²	cm²	1.5500 × 10⁻¹	in²
	ft²	9.2903 × 10²	cm²	m²	1.0764 × 10	ft²
	yd²	8.3613 × 10⁻¹	m²		1.1960	yd²
		9.0000	ft²	are	1.1960 × 10²	yd²
	ac	4.0469 × 10³	m²		2.4711 × 10⁻²	ac
		4.0469 × 10⁻¹	ha		1.0000 × 10²	m²
		4.8400 × 10³	yd²	ha	2.4711	ac
	mi²	2.5900 × 10²	ha		1.0000 × 10⁴	m²
		2.5900	km²		1.0000 × 10⁻²	km²
		6.4000 × 10²	ac	km²	2.4711 × 10²	ac
					3.8610 × 10⁻¹	mi²
					1.0000 × 10²	ha
2. Electrical Conductivity				mS cm⁻¹	1.0000	mmho cm⁻¹
				S m⁻¹	1.0000 × 10	mS cm⁻¹
3. Length	gauge	2.500 × 10⁻¹	µm	Å	10⁻¹⁰	m
	in.	2.5400	cm	µm	10⁻⁶	m
	ft	3.0480 × 10	cm	cm	3.9370 × 10⁻¹	in.
		1.2000 × 10	in.		3.2808 × 10⁻²	ft
	yd	9.1441 × 10⁻¹	m	m	3.9370 × 10	in.
		3.0000	ft		3.2808	ft
	mi	1.6093	km		1.0936	yd
		5.280 × 10³	ft	km	6.2137 × 10⁻¹	mi
		1.760 × 10³	yd			
4. Mass	oz	2.8349 × 10	g	g	3.5274 × 10⁻²	oz
	lb	4.5359 × 10⁻¹	kg	kg	2.2046	lb
	Imperial ton (long ton)	1.0160 × 10³	kg		1.0000 × 10⁻²	quintals
		1.0160	t	quintal	1.0000 × 10²	kg
		2.2400 × 10³	lb		1.0000 × 10⁻¹	t
	US ton (short ton)	9.0718 × 10⁻¹	kg	t	2.2046 × 10⁻³	lb
		9.0718 × 10⁻¹	t		9.8421 × 10⁻¹	Imperial tons (long tons)
		2.0000 × 10³	lb		1.1023	US tons (short tons)
					1.0000 × 10³	kg
5. Plane angles [2]	degrees (°)	1.7453 × 10⁻²	radians	radians (r)	5.7296 × 10	degrees
		1.1111	grades		6.3662 × 10	grades
	grade (g)	1.5708 × 10⁻²	radians			
		9.0000 × 10⁻¹	degrees			
6. Power	ft lbf s⁻¹	1.3558	W	W	7.3756 × 10⁻¹	ft lbf s⁻¹
	hp (UK)	7.4570 × 10²	W		1.3410 × 10⁻³	hp (UK)
		1.0139	hp (metric)	hp (metric)	9.8632 × 10⁻¹	hp (UK)

Notes: See page 329.

cont

327

J.1.1   cont

7. Pressure 3/

Main imperial conversions 1/

To convert:	Multiply by: (unless otherwise stated)	To give equivalent in:
lbf in^{-2} (psi)	7.031×10^{-2}	kgf cm^{-2}
	6.8948	kPa (or kN m^{-2})
	6.8948×10^{-2}	bars
	9.71×10^{-4}	atmospheres (atm)
atmospheres (atm)	6.806×10^{-2}	cm (head of water)
	1.030×10^{3}	cm (head of water)
	1.013	bars
	1.033	kgf cm^{-2}
in. (head of mercury)	1.4693×10	lbf in^{-2} (psi)
pF	3.386×10^{-2}	bar
		cm (head of water)
	take antilog$_{10}$	(approximately)

Main metric conversions 1/

To convert:	Multiply by: (unless otherwise stated)	To give equivalent in:
cm (head of water)	1.03×10^{3}	lbf in^{-2} (psi)
	9.703×10^{-4}	atmospheres (atm)
	1.002×10^{-3}	kgf cm^{-2}
	9.833×10^{-4}	bar
	9.8×10^{-2}	kPa
	take log$_{10}$	pF (approximately)
mm (head of mercury)	3.9370×10^{-2}	in. (head of mercury)
	1.3332×10^{-3}	kPa
	1.3332×10^{-5}	bar
	1.4504×10^{-4}	lbf in^{-2} (psi)
kPa (also kN m^{-2})	2.0885×10	lbf ft^{-2} (psi)
	2.953×10^{-4}	in. (head of mercury)
	1.0000×10^{-2}	bar
	1.0197×10^{2}	kgf cm^{-2}
kgf cm^{-2}	1.4223×10	lbf in^{-2} (psi)
	9.678×10^{-1}	atmospheres (atm)
	9.39×10^{-4}	cm (head of water)
	9.804×10^{-1}	bar
	9.8066	Pa
bar	1.4504×10	lbf in^{-2} (psi)
	9.8693×10^{-1}	atmospheres (atm)
	2.953×10	in. (head of mercury)
	1.017×10^{3}	cm (head of water)
	1.0197	kgf cm^{-2}

8. Temperature

Main imperial conversions 1/

To convert:	Multiply by: (unless otherwise stated)	To give equivalent in:
°F	subtract 32 then x by 5/9	°C

Main metric conversions 1/

To convert:	Multiply by: (unless otherwise stated)	To give equivalent in:
°C	x by 9/5 then add 32	°F
°K	subtract 273.2	°C

9. Velocity

Main imperial conversions 1/

To convert:	Multiply by: (unless otherwise stated)	To give equivalent in:
ft s^{-1} (fps)	1.0973×10^{4}	cm h^{-1}
	3.048×10^{-1}	m s^{-1}
	2.6335×10^{2}	m day^{-1}
	1.0973×10^{-1}	km h^{-1}
	6.818×10^{-2}	ml h^{-1} (mph)
ml h^{-1} (mph)	4.470×10^{-1}	m s^{-1}
	1.6092	km h^{-1}
	1.4667×10	ft s^{-1}

Main metric conversions 1/

To convert:	Multiply by: (unless otherwise stated)	To give equivalent in:
cm h^{-1}	9.110×10^{-5}	ft s^{-1} (fps)
	2.778×10^{-5}	m s^{-1}
	2.400×10^{-1}	m day^{-1}
m s^{-1}	3.2808	ft s^{-1} (fps)
	2.237	ml h^{-1} (mph)
	3.6000×10^{4}	cm h^{-1}
	8.6400×10^{3}	m day^{-1}
	3.6000	km h^{-1}
m day^{-1}	3.796×10^{-4}	ft s^{-1} (fps)
	1.157×10^{-4}	m s^{-1}
	4.1667	km h^{-1}
km h^{-1}	9.113×10^{-1}	ft s^{-1} (fps)
	6.2137×10^{-1}	ml h^{-1} (mph)
	2.7778×10^{-1}	m s^{-1}

Notes: See page 329.

cont

J.1.1   cont

Measurement	Main imperial conversions [1] — To convert:	Multiply by: (unless otherwise stated)	To give equivalent in:	Main metric conversions [1] — To convert:	Multiply by: (unless otherwise stated)	To give equivalent in:
10. Volume/ capacity	in^3	1.6387×10	cm^3 or $m\ell$	cm^3 or $m\ell$	6.1023×10^{-2}	in^3
	ft^3	2.8317×10	ℓ or dm^3		1.7597×10^{-3}	pints (imp)
		2.8317×10^{-2}	m^3		2.1134×10^{-3}	liq pints (US)
		6.2289	gal (imp)		1.0×10^{-3}	ℓ or dm^3
	yd^3	7.6455×10^1	m^3	ℓ or dm^3	3.5315×10^{-2}	ft^3
	pint (imp)	5.6826×10^2	cm^3 or $m\ell$		2.1997×10^{-1}	gal (imp)
		3.4677×10	in^3		2.6417×10^{-1}	liq gal (US)
	liq pint (US)	4.7318×10^2	cm^3 or $m\ell$		1.0000×10^{-3}	m^3
	gal (imp)	4.5461	ℓ or dm^3	m^3	2.1997×10^2	gal (imp)
		4.5461×10^{-3}	m^3		2.6417×10^2	liq gal (US)
		8.0000	pint (imp)		3.5315×10	ft^3
		1.2009	liq gal (US)		1.3079	yd^3
		1.605×10^{-1}	ft^3		1.0000×10^3	ℓ
	liq gal (US)	3.7854×10^{-3}	m^3		8.106×10^{-4}	ac-ft
		3.7854	ℓ or dm^3	ha-m	97.28	ac-in.
		8.3268×10^{-1}	gal (imp)		8.110	ac-ft
	ac-in.	1.0279×10^{-2}	ha-m		1.0000×10^4	m^3
		1.0279×10^2	m^3			
	ac 6 in.	4.112×10^3	m^2 15 cm			
	ac-ft	1.233×10^{-1}	ha-m			
		1.233×10^3	m^3			
		4.3560×10^4	ft^3			
		2.713×10^5	gal (imp)			
11. Volume flow	$ft^3\,s^{-1}$ (cusec)	2.8317×10^{-1}	$m^3\,s^{-1}$ (cumec)	$m^3\,s^{-1}$ (cumec)	3.5315×10	$ft^3\,s^{-1}$ (cusec)
		2.4466×10^2	$m^3\,day^{-1}$		1.320×10^4	gal (imp) min^{-1}
		2.8317×10^1	$\ell\,s^{-1}$		7.919×10^5	gal (imp) h^{-1}
		3.7373×10^2	gal (imp) min^{-1}		7.007	ac-ft day^{-1}
	gal (imp) min^{-1}	7.577×10^{-5}	$m^3\,s^{-1}$ (cumec)		1.0000×10^3	$m^3\,day^{-1}$
		7.577×10^{-3}	$\ell\,s^{-1}$	$m^3\,day^{-1}$	8.6400×10^3	$ft^3\,s^{-1}$ (cusec)
		2.675×10^{-3}	$ft^3\,s^{-1}$ (cumec)		4.087	gal (imp) h^{-1}
	ac-ft day^{-1}	1.427×10^{-1}	$m^3\,s^{-1}$ (cumec)		8.838×10	ac-ft day^{-1}
		1.233×10^3	$m^3\,day^{-1}$		7.820×10^{-4}	$\ell\,s^{-1}$
		1.427×10^2	$\ell\,s^{-1}$		1.116×10^{-1}	$m^3\,s^{-1}$ (cumec)
		4.356×10^4	$ft^3\,s^{-1}$ (cumec)	$\ell\,s^{-1}$	1.116×10^{-4}	$ft^3\,s^{-1}$ (cumec)
		2.713×10^5	gal (imp) day^{-1}		3.5315×10^{-2}	gal (imp) min^{-1}
					1.320×10	$m^3\,s^{-1}$ (cumec)
					1.000×10^{-3}	$m^3\,s^{-1}$ (cumec)
12. Weight/area application rate	lb ac^{-1}	1.121×10^{-4}	$kg\,m^{-2}$	$kg\,m^{-2}$	8.921×10^3	lb ac^{-1}
		1.121	$kg\,ha^{-1}$		7.97×10	Imperial cwt ac^{-1}
	Imperial cwt ac^{-1}	1.255×10^{-2}	$kg\,m^{-2}$		3.982	UK tons ac^{-1}
	UK tons ac^{-1}	2.511×10^{-1}	$kg\,m^{-2}$		1.0000×10^4	$kg\,ha^{-1}$
		2.511	$t\,ha^{-1}$		8.921×10^{-1}	lb ac^{-1}

Notes: 1/ Additional conversions, including many foreign units, are given in the Economist (1980, pp 16-23) and SI units as related to soil science are discussed by Hesse (1975). See also BSI (1962), Shell International Chemical Company (1960) and J.7.3 on pp 339-340.

2/ Tables interconverting degrees and percentage slopes are given below in J.1.4.

3/ Note modern force notation: pounds weight or kilograms weight are described as pounds force (lbf) or kilograms force (kgf) rather than just lb or kg. Further pressure conversions related to soil-water potential are given in J.2.1.

J.1.2   Temperature conversions (°F and °C)

°F	°C	°F	°C	°F	°C	°F	°C	°F	°C	°F	°C
14	-10	50	10	84.2	29	118.4	48	152	66.7	184	84.4
15	-9.4	51	10.6	85	29.4	119	48.3	152.6	67	185	85
15.8	-9	51.8	11			120	48.9	153	67.2	186	85.6
16	-8.9	52	11.1	86	30	120.2	49	154	67.8	186.8	86
17	-8.3	53	11.7	87	30.6	121	49.4	154.4	68	187	86.1
17.6	-8	54	12.2	87.8	31			155	68.3	188	86.7
19	-7.2	55	12.8	88	31.1	122	50	156	68.9	188.6	87
19.4	-7	55.4	13	89	31.7	123	50.6	156.2	69	189	87.2
20	-6.7	56	13.3	89.6	32	123.8	51	157	69.4	190	87.8
21	-6.1	57	13.9	90	32.2	124	51.1			190.4	88
21.2	-6	57.2	14	91	32.8	125	51.7	158	70	191	88.3
23	-5	58	14.4	91.4	33	125.6	52	159	70.6	192	88.9
24	-4.4	59	15	92	33.3	126	52.2	159.8	71	192.2	89
24.8	-4	60	15.6	93.2	34	127	52.8	160	71.1	193	89.4
25	-3.9	60.8	16	94	34.4	127.4	53	161	71.7		
26	-3.3	61	16.1	95	35	128	53.3	161.6	72	194	90
26.6	-3	62	16.7	96	35.6	129	53.9	162	72.2	195	90.6
27	-2.8	62.6	17	96.8	36	129.2	54	163	72.8	195.8	91
28	-2.2	63	17.2	97	36.1	130	54.4	163.4	73	196	91.1
28.4	-2	64	17.8	98	36.7	131	55	164	73.3	197	91.7
29	-1.7	64.4	18	98.6	37	132	55.6	165	73.9	197.6	92
30	-1.1	65	18.3	99	37.2	132.8	56	165.2	74	198	92.2
30.2	-1	66	18.9	100	37.8	133	56.1	166	74.4	199	92.8
31	-0.6	66.2	19	100.4	38	134	56.7	167	75	199.4	93
		67	19.4	101	38.3	134.6	57	168	75.6	200	93.3
32	0			102	38.9	135	57.2	168.8	76	201	93.9
33.8	1	68	20	102.2	39	136	57.8	169	76.1	201.2	94
34	1.1	69	20.6	103	39.4	136.4	58	170	76.7	202	94.4
35	1.7	69.8	21			137	58.3	170.6	77	203	95
35.6	2	70	21.1	104	40	138	58.9	171	77.2	204	95.6
36	2.2	71	21.7	105	40.6	138.2	59	172	77.8	204.8	96
37	2.8	71.6	22	105.8	41	139	59.4	172.4	78	205	96.1
37.4	3	72	22.2	106	41.1			173	78.3	206	96.7
38	3.3	73	22.8	107	41.7	140	60	173.6	78	206.6	97
39	3.9	73.4	23	107.6	42	140	60.6	174	78.9	207	97.2
39.2	4	74	23.3	108	42.2	141.8	61	174.2	79	208	97.6
40	4.4	75	23.9	109	42.8	142	61.1	175	79.4	208.4	98
41	5	75.2	24	109.4	43	143	61.7			209	98.3
42	5.6	76	24.4	110	43.3	143.6	62	176	80	210	98.9
42.8	6	77	25	111	43.9	144	62.2	177	80.6	210.2	99
43	6.1	78	25.6	111.2	44	145	62.8	177.8	81	211	99.4
44	6.7	78.8	26	112	44.4	145.4	63	178	81.1		
44.6	7	79	26.1	113	45	146	63.3	179	81.7	212	100
45	7.2	80	26.7	114	45.6	147	63.9	179.6	82	221	105
46	7.8	80.8	27	114.8	46	147.2	64	180	82.2	230	110
46.4	8	81	27.2	115	46.1	148	64.4	181	82.8	239	115
47	8.3	82	27.8	116	46.7	149	65	181.4	83		
48	8.9	82.4	28	116.6	47	150	65.6	182	83.3		
48.2	9	83	28.3	117	47.2	150.8	66	183	83.9		
49	9.4	84	28.9	118	47.8	151	66.1	183.2	84		

To convert: to:

°F	°C	subtract 32, then multiply by 5/9
°C	°F	multiply by 9/5, then add 32

J.1.3   Double conversion table for application rate or yield

The central number refers to either the left- or the right-hand column
(eg 3 kg ha^{-1} = 2.677 lb ac^{-1}; 3 lb ac^{-1} = 3.363 kg ha^{-1})

kg ha^{-1}		lb ac^{-1}		kg ha^{-1}		lb ac^{-1}
1.121	1	0.892		22.417	20	17.844
2.242	2	1.784		33.626	30	26.765
3.363	3	2.677		44.834	40	35.687
4.483	4	3.569		56.043	50	44.609
5.604	5	4.461		67.251	60	53.531
6.725	6	5.353		78.460	70	62.453
7.846	7	6.245		89.668	80	71.374
8.967	8	7.137		100.877	90	80.296
10.088	9	8.030		112.085	100	89.218
11.209	10	8.922				

J.1.4   Conversion tables for slope data

Degrees to percentages				Percentage to degrees		Natural tangent to degrees	
Degrees	Minutes			Percent	Degrees and minutes	Ratio (tan) 1/	Degrees (rounded)
	0'	30'					
0°	0	0.87		1	0°34'	1:0.25	76
1	1.75	2.62		2	1°09'	1:0.5	63
2	3.49	4.37		3	1°43'	1:1	45
3	5.24	6.12		4	2°18'	1:1.5	34
4	6.99	7.87		5	2°52'	1:2	27
5	8.75	9.63		6	3°26'	1:2.5	22
6	10.51	11.39		7	4°00'	1:3	18
7	12.28	13.17		8	4°34'	1:3.5	16
8	14.05	14.95		9	5°09'	1:4	14
9	15.84	16.73		10	5°43'		
10	17.63	18.53		11	6°17'		
11	19.44	20.35		12	6°51'		
12	21.26	22.17		13	7°24'		
13	23.09	24.01		14	7°58'		
14	24.93	25.86		15	8°32'		
15	26.80	27.73		16	9°05'		
16	28.68	29.62		17	9°39'		
17	30.57	31.53		18	10°12'		
18	32.49	33.46		19	10°45'		
19	34.43	35.41		20	11°19'		
20	36.40	37.39		25	14°02'		
21	38.39	39.39		30	16°42'		
22	40.40	41.42		35	19°45'		
23	42.45	43.48		40	21°48'		
24	44.52	45.57		45	24°14'		
25	46.63	47.70		50	26°34'		
26	48.77	49.86		55	28°49'		
27	50.95	52.06		60	30°38'		
28	53.17	54.30		65	33°01'		
29	55.43	56.58		70	35°00'		
30	57.74	58.91		75	36°52'		
31	60.09	61.28		80	38°40'		
32	62.49	63.71		85	40°22'		
33	64.94	66.19		90	42°00'		
34	67.45	68.73		95	43°32'		
35	70.02	71.33		100	45°00'		
36	72.65	74.00					
37	75.36	76.73					
38	78.13	79.54					
39	80.98	82.43					
40	83.91	85.41					
41	86.93	88.47					
42	90.04	91.63					
43	93.25	94.90					
44	96.57	98.27					

Note: 1/ The first figure refers to the height, the second to the horizonal distance; eg on a 1:2 smooth slope there is a climb (or descent) of 5 m for every 10 m traversed horizontally along a line parallel to the maximum gradient.

1 in 2 slope

5m

10m

27° (rounded)

Percentage slope = natural tangent x 100 thus 1:2 = 50%

Annex J: Weights and measures

J.2 Soil physics

J.2.1 Conversion table for units of water potential 1/

Specific potential units (energy per unit mass)		----- Water potential units 2/ ----- (pressure or energy per unit volume)			Units based on weight	
ergs g^{-1}	joules kg^{-1}	bars	millibars	centibars or kilopascals	atmospheres	cm water 3/
1	0.000 1	0.000 001	0.001	0.000 1	0.000 000 987	0.001 017
10 000	1	0.001	10	1	0.009 87	10.17
1 000 000	100	1	1 000	100	0.987	1 017
1 000	0.1	0.001	1	0.1	0.000 987	1.017
1 013 000	101.3	1.013	1 013	101.3	1	1 030
983	0.098	0.000 983	0.983	0.098 3	0.000 970	1

Notes:
1/ The density of water is taken as 1.000 g cm^{-3}. This holds only at 4°C but is approximately correct at other temperatures.
2/ For unsaturated soils, water potential in bars is a negative value.
3/ Other equivalents include:
15 atmospheres, −15 bars, 15 340 cm water, pF 4.2 (approximately)
one-third atmosphere, −0.33 bar, 340 cm water, pF 2.5 (approximately)

Source: Adapted from Taylor and Ashcroft (1972).

J.2.2 Conversion table for sieve mesh sizes

US standard 1/		UK standard 2/		French standard 3/		IMM (obsolete)	
Mesh No	Mesh width 4/ (mm)	Mesh No	Mesh width (mm)	Mesh No	Mesh width (mm)	Mesh No	Mesh width (mm)
10	2.00	8	2.06	34	2.00	5	2.54
18	1.00	16	1.00	31	1.00	12	1.06
30	0.59	25	0.60	29	0.63		
35	0.50	30	0.50	28	0.50	30	0.42
60	0.25	60	0.25	25	0.25	50	0.25
70	0.21	72	0.21	24	0.20	60	0.21
100	0.15	100	0.15	23	0.16	80	0.16
140	0.10	150	0.10	21	0.10	120	0.10
300	0.05	300	0.05	18	0.05	200	0.06

Notes:
1/ ASTM (1981).
2/ BSI (1976).
3/ NF (1970).
4/ Note that the stated mesh width is the length of a side of a square hole. The diagonal length across the square is greater. Some particles longer than the stated width will pass through, and even some both longer and broader (but not thick), since blade-like particles can pass through edge-on to the mesh.

J.3   Soil chemistry

J.3.1   Common soil-related conversions

		To convert:	Multiply by:	To give:
1.	Soluble ions concentrations	$mmol(+)$ ℓ^{-1} me ℓ^{-1}	1 saturation % x 0.001	me ℓ^{-1} me/100 g
2.	Conductivity	EC (mS cm^{-1})	10	me ℓ^{-1} (covers range 0.1-5.0 mS cm^{-1})
		EC (mS cm^{-1})	0.36	bar or atm (osmotic pressure) (covers range 3-30 mS cm^{-1})
		EC_e (mS cm^{-1})	640 (approx)	Total soluble salts in ppm or mg ℓ^{-1} (covers range 0.1-5.0 mS cm^{-1} in water)
3.	Concentrations	mg/100 g me/100 g me ℓ^{-1} % by wt	10 10 x E.wt E.wt 10^4	ppm ppm ppm or mg ℓ^{-1} ppm
4.	Chemical concentrations (by weight)			
	Calcium	CaO Ca me/100 g Ca me/100 g $CaSO_4$	0.715 1.399 200.4 0.086	Ca CaO ppm Ca % by wt
	Magnesium	MgO Mg me/100 g Mg	0.603 1.658 121.6	Mg MgO ppm Mg
	Nitrogen	N N N $NaNO_3$ KNO_3 $(NH_4)_2SO_4$	6.067 7.218 4.717 0.165 0.138 0.212	$NaNO_3$ KNO_3 $(NH_4)_2SO_4$ N N N
	Phosphorus	P_2O_5 P	0.436 2.292	P P_2O_5
	Potassium	K_2O K me/100 g K	0.830 1.205 391.0	K K_2O ppm K
	Sodium	Na_2O Na me/100 g Na	0.742 1.348 230.0	Na Na_2O ppm Na
	Sulphur	S SO_3	2.497 0.400	SO_3 S

Annex J: Weights and measures

J.3.2 Common elements and compounds: atomic, ionic and equivalent weights

Common name	Chemical symbol	Common oxidation number or ionic charge	Atomic, molecular or ionic wt	Equivalent wt 1/
Aluminium	Al	3	27.0	9.0
Boron	B	3	10.8	3.6
Calcium	Ca	2	40.1	20.0
calcium oxide	CaO		56.1	28.0
Carbon	C	4	12.0	3.0
bicarbonate	HCO_3	−1	61.0	61.0
carbonate	CO_3	−2	60.0	30.0
Chloride	Cl	−1	35.5	35.5
Chromium	Cr	6 (or 3)	52.0	8.7
dichromate	Cr_2O_7	−2	216.0	36.0
Gypsum	$CaSO_4.2H_2O$		172.2	86.1
Hydrogen	H	1	1.0	1.0
Iron III (ferric)	Fe	3	55.8	18.6
Iron II (ferrous)	Fe	2	55.8	27.9
Magnesium	Mg	2	24.3	12.2
magnesium oxide	MgO		40.3	20.1
Manganese IV (manganic)	Mn	4	54.9	13.7
Manganese II (manganous)	Mn	2	54.0	27.5
Nitrogen	N	3	14.0	4.7
nitrate	NO_3	−1	62.0	62.0
Oxygen	O	2	16.0	8.0
Phosphorus	P	3 (or 5)	31.0	10.3
dihydrogen phosphate	H_2PO_4	−1	97.0	97.0
monohydrogen phosphate	HPO_4	−2	96.0	48.0
phosphate	PO_4	−3	95.0	31.7
Potassium	K	1	39.1	39.1
potassium oxide	K_2O		94.2	47.1
Silicon	Si	4	28.1	7.0
Sodium	Na	1	23.0	23.0
sodium hydroxide	NaOH		40.0	40.0
Sulphur	S	2	32.1	16.0
hydrogen sulphate	HSO_4	−1	97.1	97.1
sulphate	SO_4	−2	96.1	48.0
sulphide	S	−2	32.1	16.0

Note: 1/ Equivalent wt of an element = atomic wt divided by valency.

J.3.3 Preparation of common solutions

Reagent	Chemical symbol	Specific gravity (approximate figures)	% of reagent by weight and normality	Quantity needed to make 1 ℓ of 1.0 M solution 1/2/	Molarity of 1.0 N solution 2/
Acetic acid	CH_3COOH	1.05	99 (17N)	58 cm³	1.0
Ammonia solution	NH_3	0.91	25 (13N)	71 cm³	1.0
Ammonium acetate	CH_3COONH_4	–	–	77.0 g	–
Ammonium fluoride	NH_4F	–	–	37.0 g	–
Barium chloride	$BaCl_2$	–	–	208.3 g	–
Hydrochloric acid	HCl	1.18	36 (12N)	89 cm³	1.0
Nitric acid	HNO_3	1.42	70 (16N)	63 cm³	1.0
Perchloric acid	$HClO_3$	1.66	72	86 cm³	1.0
		1.54	60	103 cm³	
Potassium chloride	KCl	–	–	74.6 g	–
Potassium dichromate	$K_2Cr_2O_7$	–	–	294.2 g	0.167
Potassium hydroxide	KOH	–	–	56.1 g	1.0
Sodium acetate	CH_3COONa	–	–	82.0 g	–
Sodium bicarbonate	$NaHCO_3$	–	–	84.0 g	–
Sodium hydroxide	NaOH	–	–	40.0 g	1.0
Sulphuric acid	H_2SO_4	1.84	98 (37N)	56 cm³	0.5

Calgon solution (for dispersion in particle size analysis): 40 g sodium hexametaphosphate with 10 g anhydrous Na_2CO_3 in 5 ℓ of distilled water.

Notes: 1/ Liquid measures are approximate.
        2/ 1.0 M solution contains 1 g molecular weight of solute per litre of solution.
           1.0 N solution contains 1 g equivalent weight of solute per litre of solution.

J.4   Survey data

J.4.1   Survey site spacings and intensities

Distances (m) between		Sites per km^2	ha per site
Lines	Auger points on lines		
1 000	1 000	1	100
1 000	500	2	50
500	500	4	25
400	400	6.25	16
500	250	8	12.5
400	300	8.3	12
300	300	11.1	9
400	200	12.5	8
250	250	16	6.25
300	200	16.7	6
200	200	25	4
200	100	50	2
100	100	100	1

J.4.2   Map and AP scales: relation to actual lengths and areas (metric units)

Scale	Distance represented by 1 cm (m)	Distance representing 1 km (cm)	Area represented by 1 cm^2 (ha) 1/	Area representing 1 000 ha (cm^2)
1:1 000	10	100	0.01	10^5
1:2 500	25	40	0.0625	1.6 x 10^4
1:5 000	50	20	0.25	4 x 10^3
1:7 500	75	13	0.56	1 786
1:10 000	100	10	1.00	10^3
1:15 000	150	6.7	2.25	444
1:20 000	200	5.0	4.00	250
1:25 000	250	4.0	6.25	160
1:30 000	300	3.3	9.00	111
1:40 000	400	2.5	16.0	62.5
1:50 000	500	2.0	25.0	40
1:100 000	1 000 = 1 km	1.0	100 = 1 km^2	10
1:250 000	2.5 km	0.4	6.25 km^2	1.6
1:500 000	5 km	0.2	25 km^2	0.4
1:750 000	7.5 km	0.1	56.25 km^2	0.18
1:1 000 000	10 km	0.1	100 km^2	0.1

Note: 1/   Effectively the minimum mappable area — see Subsection 9.5.2.

Annex J: Weights and measures

J.4.3   Map and AP scales: conversion table (imperial units)

Scale	in. mi^{-1}	mi in^{-1}	mi^2 in^{-2}	ac in^{-2}
1:1 000 000	0.063 4	15.782 8	249.097	159 422.09
1:500 000	0.126 7	7.891 4	62.274	39 855.36
1:253 400	0.250 0	4.000 0	16.000	10 240.00
1:250 000	0.253 4	3.945 7	15.569	9 964.16
1:126 720	0.500 0	2.000 0	4.000	2 560.00
1:125 000	0.506 9	1.972 8	3.892	2 490.88
1:100 000	0.633 6	1.578 3	2.491 0	1 594.22
1:63 360	1.000 0	1.000 0	1.000 0	640.00
1:62 500	1.013 8	0.986 4	0.973 0	622.72
1:50 000	1.267 2	0.789 1	0.622 7	398.56
1:31 680	2.000 0	0.500 0	0.250 0	160.00
1:30 000	2.112 0	0.473 5	0.224 2	143.49
1:25 000	2.534 4	0.394 6	0.155 6	99.64
1:10 560	6.000	0.166 6	0.027 7	17.728
1:10 000	6.336	0.157 8	0.024 9	15.942
1:5 000	12.672	0.078 9	0.006 23	3.986
1:2 534	25.000	0.040 0	0.001 60	1.024
1:2 500	25.344	0.039 5	0.001 56	0.996
1:1 250	50.688	0.019 7	0.000 389	0.249

J.4.4   Aerial photograph scale relationships

	Scale	Area covered per print 1/ (km^2)	Working area per print 2/ (km^2)	No of prints per 100 km^2	Width of ground cover strip (km)	Flying height 3/	
						m	ft
Very large	1:5 000	1.3	0.8	240	1.1	760	2 500
Large	1:10 000	5.2	3.3	60	2.3	1 520	5 000
	1:20 000	21	13	15	4.6	3 040	10 000
Medium	1:25 000	33	21	10	5.7	3 800	12 500
	1:30 000	47	30	7	6.9	4 560	15 000
	1:40 000	84	53	4	9.1	6 080	20 000
Small	1:50 000	131	84	2.5	11.4	7 600	25 000
Landsat image	1:1 000 000	34 000	26 000+	± 0.005	229.0	–	–

Notes: 1/ Usual air photograph print size is about 23 cm x 23 cm (9 in. x 9 in.)
2/ The air photograph 'working area' is that covered by a single print after omitting overlap with adjacent runs and with next-but-one prints in a run, ie the area on which boundaries are drawn on alternate prints: does not apply to Landsat, all the images of which must be viewed.
3/ With camera lens of 152 mm (6 in.) focal length.
4/ Length and area relationships are given in Subsection J.4.2.

Sources: Dent and Young (1981);   White (1977).

J.4.5   Standard paper sizes

Sheet size designation 1/	Linear dimensions of sheet (in.)	(cm)	Total sheet area (cm²)	Approx area of mapped parameter (cm²) 2/
2A	46.81 x 66.22	118.9 x 168.2	20 000	8 000
A0	33.11 x 46.81	84.1 x 118.9	10 000	4 000
A1	23.39 x 33.11	59.4 x 84.1	5 000	2 000
A2	16.54 x 23.39	42.0 x 59.4	2 495	998
A3	11.69 x 16.54	29.7 x 42.0	1 247	499
A4	8.27 x 11.69	21.0 x 29.7	624	250
A5	5.83 x 8.27	14.8 x 21.0	311	124
A6	4.13 x 5.83	10.5 x 14.8	155	62
A7	2.91 x 4.13	7.4 x 10.5	78	31

Notes: 1/ International standard A designation.
         2/ This allows 15% of total sheet area for legend, and 45% for wastage, thus
             leaving 40% for map.   Actual figures will vary widely, depending on shape of
             mapped area, map scale and legend size.   Examples from recent BAI work give
             the following indications:

Map scale and paper size	---- Areas as % of total sheet area (approx) ----- Legend	Mapped areas	Wastage (borders + unmapped areas)
1:10 000 (main maps) A1	15	40	45
1:100 000 (summary map) A0	25	20	55

J.4.6   Calculation of AP scale

$$\text{Scale} = \frac{\text{Altitude}}{\text{Focal length of lens}} \text{ in the same units}$$

eg 15 000 ft divided by 0.5 ft gives 1:30 000 scale

J.5   Basic statistical formulae

Arithmetic mean, $\bar{x}$	$= \Sigma x / n$

Geometric mean $= \sqrt{x_1\ x_2\ x_3\ \dots\ x_n}$

Median   = Value that occurs as the middle term of a series (or, if
                            there are an even number of terms, the value that occurs
                            midway between the two central terms)

Mode   = Most frequently occurring term of a series

Variance $= \displaystyle\sum_{i=1}^{n} (x_i - \bar{x})^2 / n$
                            (ie variance is the mean of the sum of the squares of the
                            deviations from the arithmetic mean of the series)

Standard deviation, σ $= \sqrt{\text{Variance}}$

Coefficient of variation, v = Standard deviation/Mean, $\sigma / \bar{x}$

337

Annex J: Weights and measures

J.6   Rules of thumb

J.6.1   Soil moisture

a)   PWP — saturation percentage relationship:

4 x PWP (taken as 15–atm %) ≃ saturation percentage

Based on following table:

Soil texture 1/	No of samples	(a) PWP (15 atm %)		(b) Saturation %		Ratio (a)/(b) (mean)
		Mean	Range	Mean	Range	
Coarse	10	5.0	3.4–6.5	31.8	16.0–43.1	6.37
Medium	23	10.8	6.6–14.2	42.5	26.4–60.0	3.95
Fine	11	18.5	16.1–21.0	59.5	41.8–78.5	3.20
Organic	18	37.9	27.6–51.3	142	81.0–225	3.66

Note: 1/   NB: Designated on basis of 15–atm %.

Source: USDA (1954, p 9).

b)   PWP–FC moisture content relationship:

2 x PWP ≃ FC

c)   PWP and FC: tension and texture relationships:

Soil moisture levels		pF	atm (approx)
FC	coarse (sandy) soils	2.0	0.1
FC	medium and heavy soils	2.5	0.3
PWP	all soils	4.0	15.0

Source: FAO (1976c, p 149).

d)   Total readily available water capacity (TRAWC)

TRAWC ≃ 2/3 AWC

J.6.2   Temperature and altitude relationship

For every 100 m rise in altitude, mean temperatures fall by about 0.6°C.

J.6.3   Irrigation

a)   For surface-irrigated land, a flow of about 1 ℓ s^{-1} will irrigate 1 ha of land.

b)   Water concentration: 10^3 μS cm^{-1} ≃ 0.7 g ℓ$^{-1}$ of dissolved salts.

c)   Normal irrigation efficiency ranges:

```
surface    40–50%
sprinkler  65–75%
drip       85–90%
```

J.6.4   Bulk density

Packing density of 20 mm diameter plastic balls used in volume measurement (Subsection 6.4.2) is approximately one ball per 7.31 cm^3.   This may vary slightly between batches.

J.6.5   Hydraulic conductivity

a) A soil layer is effectively impermeable if its saturated hydraulic conductivity is ⩽ 0.1 x the value of the layer above.

b) For an 8 cm diameter hole down to 70 cm below the GWT in an auger-hole test:

Rate of rise of water (mm s^{-1}) ≈ K (m day^{-1}) where K = saturated hydraulic conductivity

J.6.6   Soil particle density

Soil particle density is usually approximately 2.65 g cm^{-3} for non-ferruginous, non-humose soils. This value can vary greatly if, in particular, there is a high content of humus (SG 1.37) or ferric oxide (SG 3.74).

J.7   Miscellaneous

J.7.1   Common mathematical and other constants

π   3.14159
e   2.71828
log e   0.43429
ln 10   2.30258

The length of one degree of latitude = 69.06 − (0.35 cos 2 ϕ) miles.

The length of one degree of longitude = 69.23 cos ϕ − (0.06 cos 3 ϕ) miles.

(ϕ = latitude)

J.7.2   Common metric unit prefixes

Prefix name	Prefix symbol	Factor multiplying basic unit
tera	T	10^{12}
giga	G	10^{9}
mega	M	10^{6}
kilo	k	10^{3}
hecto	h	10^{2}
deci	d	10^{-1}
centi	c	10^{-2}
milli	m	10^{-3}
micro	µ	10^{-6}
nano	n	10^{-9}
pico	p	10^{-12}
femto	f	10^{-15}
atto	a	10^{-18}

J.7.3   Journal of Soil Science commentary on SI units

Soil surveyors are usefully advised by the prefatory remarks made in each current issue of the Journal of Soil Science regarding SI units and SI related units. The commentary is entitled 'Units, symbols and abbreviations' and with the permission of the editor of the journal the text is reproduced below. The editional panel of the journal state that the notes are for general guidance. They are preparing (September 1983) a more detailed explanation as 'it is appreciated that difficulties may still arise'.

1.   Standard basic units

	Name	Symbol
length	metre	m
mass	kilogram	kg
time	second	s
electric current	ampere	A
thermodynamic temperature	kelvin	K
amount of substance	mole	mol
luminous intensity	candela	cd

Annex J: Weights and measures

2. Derived units

Physical quantity	Name of SI unit	Symbol of SI unit	
frequency	hertz	Hz	s^{-1}
energy	joule	J	$kg\ m^2\ s^{-2} = N\ m$
force	newton	N	$kg\ m\ s^{-2} = J\ m^{-1}$
power	watt	W	$kg\ m^2\ s^{-2} = J\ s^{-1}$
pressure	pascal	Pa	$kg\ m^{-1}\ s^{-2} = N\ m^{-2} = J\ m^{-3}$
electrical charge	coulomb	C	As
electrical potential difference	volt	V	$kg\ m^2\ s^{-3}\ A^{-1} = J\ A^{-1}\ s^{-1}$
electrical resistance	ohm	Ω	$kg\ m^2\ s^{-3}\ A^{-2} = V\ A^{-1} = S^{-1}$
electrical conductance	siemens	S	$kg^{-1}\ m^{-2}\ s^3\ A^2 = \Omega^{-1}$
electrical conductivity			$S\ m^{-1}$

3. Other SI related units

	Name	Symbol	Definition
area	hectare	ha	$10^4\ m^2$
volume	litre	l or dm^3	$10^{-3}\ m^3$
	cubic centimetre	cm^3	$10^{-6}\ m^3$
mass	tonne	t	$10^3\ kg$
pressure	bar	bar	$10^5\ Pa$

Note: dm^3 is preferred to ℓ. (BAI addendum: the symbol ℓ can be used, as in this manual, to avoid confusion between l, 1 and I).

4. Units employed in soil science (not SI)

	Name	Symbol
length	Ångstrom	Å
temperature	degree Celsius	°C
solution concentration	moles per litre	M or mol ℓ^{-1}
cation exchange	microequivalents per gram	µeq g^{-1}
	or milliequivalents per kg	meq kg^{-1}
radioactivity	Curie	Ci

5. Presentation of units

The decimal sign between digits in a number should be a point on the line, not a centred dot.

The comma for dividing figures into groups of three should be avoided; for example 252,004,700 may be written 252 004 700 or $2.520\ 047 \times 10^8$.

In any number where the decimal sign is placed before the first digit of the number a zero should always be placed before the decimal sign:

for example 0.251 and not .251

The combination of a prefix and a symbol for a unit is regarded as a single symbol and should be written with no space between the prefix and the unit; for example cm and not c m. Note that µm for the micron (one millionth of a metre), replaces µ, the symbol previously often used.

When writing the symbol for a derived unit formed from several basic units, the individual symbols should be separated by a solidus, a space or a centred dot ($\cdot$). For example, the unit for velocity, metre per second, is written m s^{-1} or $m\cdot s^{-1}$ and not ms^{-1} (ms would be a millisecond).

When a unit is raised to a power, the power refers only to the unit and not to any number preceding it, for example 2.3 cm^3 is $2.3 \times 1\ cm^3$. One, but not more than one, solidus (/) may be used instead of negative power, for example, g dm^{-3} may be written g/dm^3, and µeq g^{-1} as µeq/g.

In printed text, both numbers and SI unit symbols are printed in upright type. Algebraic symbols should be printed in italic type and can therefore be distinguished from SI symbols. Further information on the use of SI units may be obtained in the following:

Ellis G (1971). Units. Symbols and abbreviations. Royal Society of Medicine.
Hesse P R (1975). SI units and nomenclature in soil science. FAO, Rome.
Incoll L D, Long S P and Ashmore M R (1977). SI units in plant science. Current Advances in Plant Science 9, 331–343.
McGlashan M L (1971). Physicochemical quantities and units. Second Edition. Royal Institute of Chemistry.
Petersen M S (1980). Recommendations for use of SI units in hydraulic engineering. Proc Am Soc Civ Eng 106, HY12, 1981–1994.

Annex K

Remote Sensing

Contents

List of Figures and Tables

Remote Sensing

K.1   Introduction

This Annex summarises the most useful aspects of remote sensing for consultant soil surveyors, and includes brief cover of airborne and satellite imagery.   General background to theory and practice is given in Curran (1985), Lillesand and Kiefer (1987), Barrett and Curtiss (1982) and White (1977).   A more technical general text is provided by Colwell et al (1983), and a methodology for employing Landsat data in rural land use surveys in developing countries is summarised in Lock and Van Genderen (1978).   Aerial photographic interpretation for soil surveys is discussed in Soil Survey Staff (1966), Carroll et al (1977) and FAO (1967).   Thermal infra-red imagery is covered by Mason and Amos (1985) and soil moisture by Blyth (1985).   Contacts and further reading can be followed up in Carter (1986).   Eden and Parry (1986) describe applications to tropical land management.

Remote sensing techniques are based on measurements of the reflection of electromagnetic radiation from ground features such as soil, water, rock, vegetation etc.   Figure K.1 illustrates the main types and wavelengths of the radiations involved and the corresponding satellites; Table K.1 shows the main applications of various remote sensing techniques.

K.2   Remote sensing methods

K.2.1   Conventional aerial photography

This is the commonest aerial photography used in soil surveys, and is normally supplied as 23 cm x 23 cm   (9 in. x 9 in.)   black and white prints, which are used to interpret photographic tone, texture, pattern and shape as indicated in Table K.2.   (Note that these features are seldom consciously used in the interpretation, but are useful for describing how different areas have been distinguished).

Conventional black and white AP usually employs panchromatic film exposed with a yellow (Wratten No 12) filter, which eliminates haze by blocking the ultraviolet (UV) and blue wavebands (see Figure K.2).   A filter which blocks visible light but transmits UV can be used to discriminate objects otherwise undiscernible against their background, but this is a rare practice.

Conventional archive black and white AP is cheap, easy to process and familiar to users; quality and resolution are normally good.   Obtaining new AP is becoming increasingly difficult, expensive and time consuming in some countries.   For generating final products at 1:25 000 scale and above, SPOT imagery is now more cost effective.   Vegetation changes and soil moisture differences may not show clearly and, as with all AP, there are scale distortions (see Section K.3).

K.2.2   Black and white infra-red photography

Similar to conventional black and white photography, but using film sensitive in the near IR part of the spectrum.   Can be produced using a variety of filters, eg yellow for forestry and crops, or red (or IR only) for delineating water bodies.

This method has the advantages of reducing haze effects and producing much better indications of vegetation differences, soil moisture conditions and water bodies.   Many plants with similar reflectance in visible light have markedly different reflectance in the IR;   for instance broad-leaved trees appear a much lighter grey than coniferous trees on IR film.   Water is a good absorber of IR light and therefore wet soils and water bodies appear dark.

K.2.3   True colour photography

Similar to conventional black and white photography, but using colour film (which is sensitive in the same spectral range).   Usually produced with a UV filter to reduce haze.   True colour photography has the advantage of presenting a scene much as it appears to the eye, and for this reason it is often preferred by inexperienced surveyors.

Types of radiation and wavelengths used in remote sensing   Figure K.1

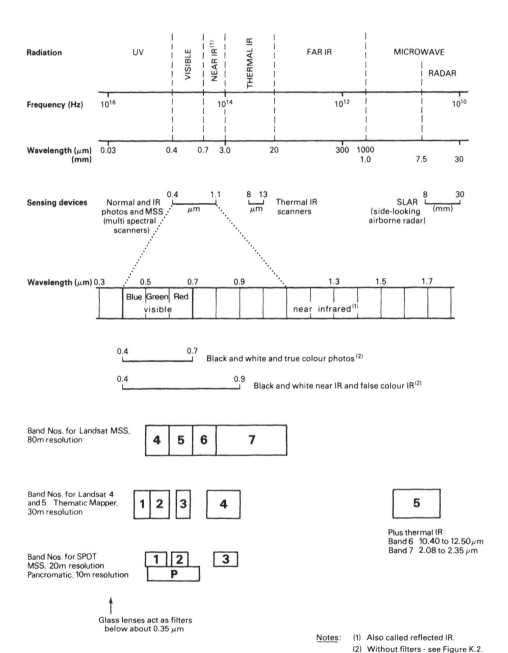

Common aerial photographic film responses and filter effects   Figure K.2

Generalised spectral response of black and white films

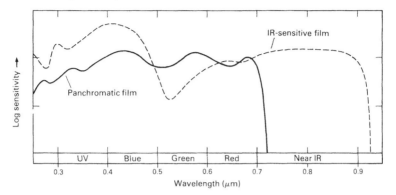

Spectral responses of colour film with ultraviolet (haze) filter

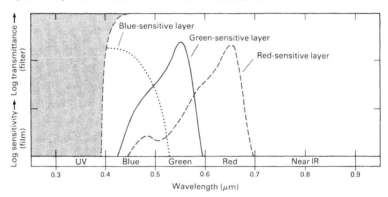

Spectral response of false colour infrared film with Wratten No. 12 (yellow) filter

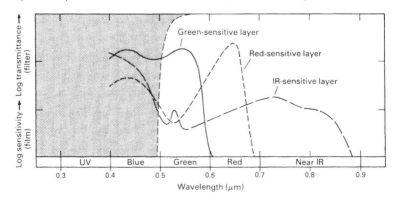

Shaded areas indicate wavelengths screened by filters

Sources: Adapted from Lillesand and Kiefer (1987) after Eastman Kodak Co. (1972a and b).

Annex K: Remote sensing

Table K.1

Summary of remote sensing techniques for soil and terrain surveys

Technique	Best scale/Resolution	Geometry	Standard Products	Processing	Interpretation	Applications	Advantages	Limitations
Airborne								
Air Photography Black and White	Very detailed to semi-detailed 1:1 000 - 1:30 000. Up to 1:120 000 possible with super wide angle lens, but not now cost effective; use satellite imagery instead.	Fair. Needs ground control.	Vertical negatives and contact prints. Uncontrolled laydowns. Obliques often available of historical events (floods etc).	Can be usefully enlarged by up to 5 x contact scale. Photomosaics; screen positives for dyeline multiple copying for field use (black and white only). Ground control-ing and stereo photogrammetric plotting for mapping.	Manual interpretation using acetate overlays is quick, easy and cheap. Stereo interpretation using hand held stereoscopes in the field. Office interpretation using desk stereoscopes and zoom transfer-scopes. Digital machines now available.	Soil surveys and terrain analysis. Topographic mapping.	Significant arch-ive data exists which is cheap to copy and utilise. Techniques are simple and can be used in the field.	Commissioning air photo-graphy is costly, time consuming and often politically sensitive. Cloud cover and the cost of ground control are serious limitations.
Air Photography Natural Colour	Detailed to semi-detailed 1:1 000 - 1:30 000. Digital or film based.						Easier to interp-ret than black and white photography. No significant extra cost for acquisition of black and white when commissioning from new.	As above plus reproduc-tion can be difficult and expensive in some countries.
Airborne Multispectral Imagery	Detailed to semi-detailed 1:1 000 - 1:30 000. Digital or film based.	Poor to Fair.	Digital and/or swath width hardcopy. Up to 11 wavebands are possible at present.	Digital: sub-scene selection, enlargement, enhancement. Fitting to base maps. Generation of photographic products.	Digital classifi-cation and manual (visual) techni-ques. Training areas can be used to calibrate larger areas.	Separation of spectral response for differing soil vegetation and terrain types.	Detailed spectral discrimination is possible. Large areas can be mapped by calibra-ting a whole image from limited ground control.	As for air photography, but is even more expensive. Digital data collection and interpretation costs can be very high and require access to sophisticated hardware and software.
Airborne Infra-red Linescan	Detailed to semi-detailed.	Poor to Fair.	Digital and/or swath width hardcopy.	Digital, and creation of photographic products.	Digital and manual classification and mapping	Thermal mapping and monitoring especially for soil moisture.	Can be flown at night. Very fine differ-ences in temper-ature can be separated.	As above plus the whole project area must be flown in the same sortie to reduce inconsistent environmental and sensor responses.
Airborne Radar (SLAR and SAAR).	Reconnaissance to Exploratory.	Very Poor.	Digital data, swath width hardcopy.	Digital, with techniques to ameliorate inconsistent responses	Digital and manual classification and mapping.	Terrain mapping in areas of persistent cloud cover and heavy vegetation.	Penetrates cloud and can also see through vegetation. Stereo cover is available on some systems.	Very expensive. Difficult to interpret due to inconsistent signatures from similar terrain depending upon viewing angle, distance, topography and vegeta-tion cover.

cont

Table K.1 cont

Satellite

Technique	Best scale/ Resolution	Geometry	Standard Products	Processing	Interpretation	Applications	Advantages	Limitations
Landsat Multispectral (MSS)	Reconnaissance to Exploratory 80 m Pixel, 3 wavebands 1:250 000 hardcopy 1:100 000 digital.	Good.	Photographic hardcopy 1:1 000 000 - 1:250 000. False colour. Digital data.	Manual using filters and colour additive viewers. Photographic: area selection, enlargement, false colouring. Digital: enlargement, destriping, contrast stretch, edge enhancement, fitting to base maps.	Manual using acetate overlays. Digital: classification, (automatic and supervised) with training areas for calibration. Band ratios.	Reconnaissance soil, terrain and vegetation surveys. Time series data for environmental change between years and season to season.	Very cost effective for initial reconnaissance studies of areas up to 30 000 km². 16 years archive of systematic coverage.	Cloud cover a persistent problem in many areas. No stereo cover. Uncertain future beyond 1989.
Landsat Thematic Mapper (TM)	Reconnaissance 30 m Pixel, 7 wavebands 1:125 000 hardcopy 1:50 000 digital.	Good.	Photographic hardcopy (3 bands) 1:1 000 000 - 1:125 000. Digital tape for all 7 bands.		As above.	Thematic mapping including soil and land systems down to 1:50 000.	7 wavebands are available including a thermal band.	As above. To fully utilise its capability digital data must be used which is expensive to acquire and requires sophisticated hard and software. A lot of data has to be discarded. Archive is limited.
SPOT Multispectral (X)	Detailed to Reconnaissance 20 m Pixel, 3 wavebands 1:50 000 hardcopy 1:25 000 digital.	Very Good.	Photographic hardcopy 1:100 000 - 1:50 000. False colour. Digital tape for all 3 bands.	Photographic: enlargement, contrast stretch. Digital: sub-scene selection, enlargement, enhancement, fitting to base maps.	Stereo coverage available. Manual using acetate overlays for field work. Digital classification.	Soil terrain mapping to 1:30 000.	Consistent systematic world coverage with ability to revisit on consecutive days on demand. Stereo coverage. More cost effective and quicker than new comparable airborne MSS data.	Cloud cover is a persistent problem in many areas. Initial costs of digital data are higher than Landsat on a unit area basis.
SPOT Pancromatic	Detailed to Reconnaissance 10 m Pixel, 1 waveband 1:50 000 hardcopy 1:25 000 digital.	Very Good.	Photographic hardcopy 1:100 000 - 1:50 000. Black and white. Digital tape.	As above.	Stereo coverage available. Manual using acetate overlays for infield annotation. Digital classification of grey scales. 3D viewing.	Base mapping to 1:25 000. Contouring.	As above. Cheaper and quicker than new black and white comparable air photography. Reproduction costs low.	As above. Limited archive at present but future is more certain than Landsat 1990 and after.

Annex K: Remote sensing

Tone	The relative shade of an area, varying from black through about 10 to 20 tones of grey to white. Smooth, dry surfaces (eg bare rock, sheet-eroded soil (unless a black soil), roads, roofs) tend to produce lighter tones; areas are darker when wet than when dry, and water usually shows up darkly unless it is very shallow/carries a high sediment load/has the sun reflected from it; conifers show up darker than broad-leaved deciduous trees in the near IR. Note that old or poor quality APs are darker towards their edges due to an optical effect called 'vignetting'; this is automatically corrected on good modern photos.
Shape	The characteristic forms of individual objects, eg palm fronds, river or channel courses, roads, houses.
Pattern	The characteristic repetition or variation of tones and shapes, and in which the individual shapes are visible, eg tree plantations (contrasted with forests); types of drainage systems (dendritic, rectangular, trellis, radial etc); termite mounds; shifting cultivation.
Texture	The characteristic smoothness or roughness of the image produced by repetition or variation of tones and shapes, but in which the individual shape elements are not visible, eg old lake basins (smooth); forests (rough); grasslands (intermediate).

Note that size of objects (hence AP scale) is important when interpreting above features.

True colour also has the apparent advantage of allowing both colour and tonal contrasts to be distinguished, giving about 100 colour/tone combinations as opposed to the 10 to 20 tones on monochromatic photographs. The increase in useful soil data is often limited, however - eg the colour of many topsoils is dark brown and colour may add little extra information. Alternatively, too many different AP units may be distinguishable on colour photos, in the sense that separate soil type or management units may not correspond to all the identifiable AP units. Matching colour photos can also be troublesome, as colour changes can occur both between runs (as atmospheric conditions change) and between batches (as film developing conditions change). Haze is also often a problem, even with a UV filter, because blue light is scattered by the atmosphere. Initial costs are only 5 to 25% above conventional black and white photographs (Dent and Young, 1981) but reprint costs are considerably higher.

K.2.4   False colour infra-red photography

Uses the same frequency range as black and white IR, and has the same haze-penetrating, moisture-indicating and vegetation-differentiating properties, plus the increased colour/tone units of colour photography. The false colours arise because the natural wavelengths are transposed to make the IR visible on the prints - so the near IR show up as red; red becomes green and green becomes blue. Blue is filtered out with a yellow filter to reduce haze effects. Table K.3 illustrates some of the main colour interpretations.

Image colour	True colour	Object
Bright red	Near IR	Dense/healthy vegetation and broad-leaved trees (chlorophyll has high IR reflectance)
Purple	Dark-blue/green	Conifers
Pink	Pale-green	Sparse grass
Yellow	Yellow/brown	Diseased or dying vegetation
Green	Brown/red	Brown-red soil
Turquoise	Brown	Smooth bare ground
Bluish or greyish green	Green	Diseased, wilted or trampled vegetation
Pale blue	Green/blue	Buildings illuminated by cloud light
White	White	Concrete, tin roofs, light-toned rocks, white-strawed cereals

After Blair-Rains (1973).

False colour IR photos are most useful in land use or vegetation mapping and crop disease identification, but are also helpful where vegetation and/or moisture status are good indicators of soil types. The photos are best taken when moisture or vegetation differences are most marked (eg after a period of drying, or when differential growth is at a maximum).

Drawbacks of false colour IR photos include the high cost of reprints and unfamilarity to users; its cost-effectiveness for soil surveys varies according to the application.

K.2.5 Side-looking airborne radar

Measures the ground reflection of a radar signal from an aircraft. Has to be side-looking because too much is reflected vertically and would saturate the receiver. The radar images are converted into black and white prints which appear as oblique views of land under harsh lighting. Land facing the aircraft transmitter appears brightest because of maximum return reflection. Water bodies reflect the signal away, and thus appear black: this 'specular reflection' also occurs with other surfaces with roughness ≼ the radar wavelength used.

Radar has the great advantage of being able to penetrate cloud, but resolution is low and the oblique view inevitably means scale and distortion errors so the prints are not suitable as map bases without cumbersome, expensive transformation.

K.2.6 Multispectral scanning

The most commonly used type of multi-spectral imagery for soil survey work has been the Landsat satellite-based information. The images are produced from a series of electronic scanners which record reflected radiation in separate spectral bands as shown in Table K.4. The latest platform, Landsat 5, was launched in March 1984. The joint French/Swedish satellite SPOT was launched in early 1986 and is now gathering sharper imagery than Landsat, down to 10 m resolution.

The data are commonly produced in the form of computer-compatible tapes (CCTs), or as 'bulk-processed' imagery - ie imagery which has not been specially processed; this is usually presented as individual black and white prints for each spectral band, or as false colour composite prints using a combination of Bands 4, 5 and 7. (Note that the false colours are not the same as in false colour IR; cf Tables K.3 and K.4). Bulk-processed photographic imagery is available at scales of 1:1 000 000, 1:500 000 and 1:250 000, or as 1:1 000 000 film negatives or positives for processing to any required scale.

Band designations and composite print colours used in Landsat imagery   Table K.4

Wavelengths (μm) and Band Nos		Interpretation/use	True colour	Colour on composite print
Landsat 1-5 Multispectral	Landsat 4-5 Thematic Mapper			
-	0.45-0.52 Band 1	Penetration into water bodies	Blue	-
0.5-0.6 Band 4	0.52-0.60 Band 2	Wavelength between the 2 chlorophyll peaks; used for vegetation and vigour differentiation	Green to yellow	Yellow (red & green)
0.6-0.7 Band 5	0.63-0.69 Band 3	Chlorophyll peak; used for vegetation and vegetation class differentiation	Yellow to red	Magenta (red & blue)
0.7-0.8 Band 6	0.76-0.90 Band 4	Responsive to vegetation biomass; used for differentiation of vegetation and moisture contents	Near IR	-
0.8-1.1 Band 7	-	Sensitive to water contents	Near IR	Cyan (blue & green)
-	1.55-1.75 Band 5	Responsive to water contents; used for differentiation of vegetation, crop water content, soil moisture	Far IR	-
-	2.08-2.35 Band 7	Responsive to rock types and hydrothermal differences	Far IR	-
-	10.40-12.50 Band 6	Thermally related phenomena	Thermal IR	-

Note: Prior to Landsat 4, Bands 1 to 3 sporadically operated a return beam vidicon (RBV) camera system, with wavelengths of respectively 0.475 to 0.575 μm, 0.580 to 0.680 μm and 0.690 to 0.830 μm. The centres of Landsat 1, 2, 3 imagery differ from those of Landsat 4, 5. Problems can be encountered with imagery for project areas which straddle scene boundaries when time series data are required.

Source: United States Geological Survey (1982).

347

Annex K: Remote sensing

The Landsat images each cover a ground area of 185 km x 185 km, with a nominal ground resolution of 80 m for MSS data and 30 m for TM. The advantages of satellite data are that repetitive cover with cartographic fidelity is available for most areas (allowing seasonal variations to be studied) and that electronic processing of CCTs allows the maximum resolution of the data to be utilised along with very rapid analysis and compilation of imagery appropriate to the application. Modern processors allow useful enlargement of Landsat MSS data to 1:100 000, TM data to 1:50 000 and SPOT data to 1:25 000. In addition, manipulation techniques such as destriping, contrast stretching, edge enlargement and classification can be rapidly carried out. Hard copy can then be produced, either as laser film-written positives or as 35 mm colour slides. Digital and photographic processing can be used to fit imagery to base maps.

Landsat imagery is very good for reconnaisance-level studies of large areas. SPOT imagery has better resolution with useable scales down to 1:25 000 and stereo cover is routinely collected. For scales below 1:25 000 air photography remains the best technique although developments in the use of remotely piloted model aeroplanes could make the acquisition of new air photography cheaper and easier in the near future. The cost effectiveness of satellite imagery is difficult to quantify and justify, as the benefits are often uncertain. The imagery can provide a lot of additional data upon which better informed decisions can be taken, but it is an additional technique which still requires fieldwork and site knowledge to be of the most use. Rough cost indicators are given in Table K.5.

Costs for Satellite Imagery 1/   Table K.5

Digital Products	Cost (£)	£ per km^2
Landsat MSS tape:		
Full 185 km x 185 km scene, 3 bands	540	0.02
Landsat TM tape		
Full scene, 185 km x 185 km 7 bands	3 000	0.09
1/4 scene, 92.5 km x 92.5 km, 7 bands	1 500	0.18
SPOT or Panchromatic MSS tape 2/		
60 km x 60 km, 3 bands, partial geometric correction	1 050	0.29
60 km x 60 km, 3 bands, full geometric correction	1 760	0.48
In addition to tape costs, usage processing charges must be included 3/		
Landsat MSS		
Geometric correction, contrast stretching,		
film writing, 3 bands, colour	750	0.03
Colour print 0.9 m x 0.9 m	140	0.01
SPOT imagery		
Geometric correction, colour or b & w	550	0.16
Film writing, colour or b & w	220	0.07
Paper print 0.9 m x 0.9 m, colour	140	0.04
Paper print 0.9 m x 0.9 m, b & w	70	0.02

Archive Photographic Products 4/	Cost (£)		
	Landsat paper print sizes (mm)		
	240 x 240	480 x 480	960 x 960
Landsat MSS - colour: scales 1:1 000 000, 1: 500 000 and 1: 250 000 5a/	35	55	140
Landsat TM - colour: 3 selected bands from the 7 available, full scene 1:1 000 000 to 1:250 000, 1/4 scene 1:500 000 to 1:125 000 5b/	35	55	140

	SPOT 600 mm x 600 mm paper prints	
	1:100 000	1:50 000
SPOT MSS - full scene 60 km x 60 km 6/	120 colour, 60 b & w	480 colour, 240 b & w

Notes: 1/ 1988 costs for purchase in UK and excluding tax.
       2/ For SPOT stereo data, two overlapping images are needed, so costs are doubled.
       3/ Some processors will give full geometric correction with partially corrected data.
       4/ These are standard, composite pictures not processed to highlight any particular features. All are 3 band colour composites.
       5/ Add generation fee if not already in hard copy archive: a/ £230, b/ £405.
       6/ Add generation fee of £560.

Costs vary according to exchange rates of the US dollar (for Landsat) and French franc (for SPOT). It is imperative that the project area coverage is fitted to the imagery scene boundaries to see how many scenes are needed. Regional receiving stations often have larger archives and cheaper prices. For digital tapes it is necessary to state which format, to suit intended and available processors.

K.3   Notes on aerial photograph usage

K.3.1   General

The main items shown on a vertical AP are illustrated in Figure K.3; the format and data provided vary slightly between survey organisations. The APs most commonly used are those taken in overlapping flight strips which are used to produce stereoscopic coverage; for full stereoscopic coverage, at least 50% overlap is needed between successive prints in a run, with about 30% overlap between runs, as shown in Figure K.4. Geometric relationships and AP effects caused by variations in relief are illustrated in Figure K.5, and basic data on print sizes and working areas of different scale photographs are given in Annex J, Tables J.4.2 and J.4.4. Figure K.6 and Table K.6 together form a test for stereoscopic vision which a team leader should use to assess the abilities of each soil surveyor on a project.

K.3.2   Scale of photography

Conventional aerial photographs always contain scale and distortion errors, the three principal types being as follows:

a) Optical errors - Arising from the optical construction of the cameras and projectors. Only points close to the principal point will be at a reasonably constant scale unless optical corrections are applied (as eg in orthophotographic map production).

b) Relief errors - Arising from changes in ground altitude over an area covered by a photo; high areas appear at a larger scale than lower areas, and radial shifts are produced on the AP print (see Figure K.5).

c) Camera tilt errors - Arising from aircraft pitching and rolling effects beyond the self-correcting capabilities of the camera mountings. Usually negligible for tilts of < 3°.

The nominal scale of an aerial photograph is the ratio of camera focal length to aircraft altitude above the ground. Both figures should be stated on the photo margin. The usual focal length is 6 in. (152 mm), thus giving a scale of 1:20 000 at 10 000 ft, for instance. Note that for every doubling of scale, four times more prints are needed to cover a given area.

In hilly areas, since the scale varies across the photo, terrain elevation should be obtained from a topographic map, either for the part of the photo being examined or for the whole photo, in which case the elevation of the principal point should be measured. In some cases it may be necessary to convert the distorted lengths and angles on APs to true values; simple geometric procedures for this are given in Lillesand and Kiefer (1987). The general formula for scale calculation is as follows:

$$\text{Print scale} = \frac{\text{Focal length}}{\text{Altitude - terrain elevation}} = \frac{f}{H-h} \text{ (see Figure K.5)}$$

eg   Photography at a nominal scale of 1:50 000 taken over land at an altitude of 1 300 ft above sea-level will have an actual scale of about 1:47 400, giving an error of over 10% in area measurements, if taken uncorrected from the photos.

An alternative method of calculating the scale of photos is to measure the distance between two widely separated points, such as road junctions, on opposite sides of the principal point, both on the photos and either on the ground or on an accurate map.

$$\text{Print scale} = \frac{\text{Distance on print}}{\text{Distance on ground}}$$

$$= \frac{\text{Distance on print}}{\text{Distance on map}} \times \text{Map scale}$$

Further details of map and AP scale conversions are given in Annex J, Subsections J.4.2 and J.4.4.

Finally, it should be noted that photographs, even of flat ground, always have some distortion (particularly at the edges) and are not suitable for accurate measurement of distances and areas. Maps produced from individual photos or laydowns should be similarly regarded. Orthophotomaps, on the other hand, combine the detail of photos with the accuracy of maps, and make an excellent mapping base. However, they require ground control in the field which can be time consuming and expensive.

Main components of a vertical aerial photograph   Figure K.3

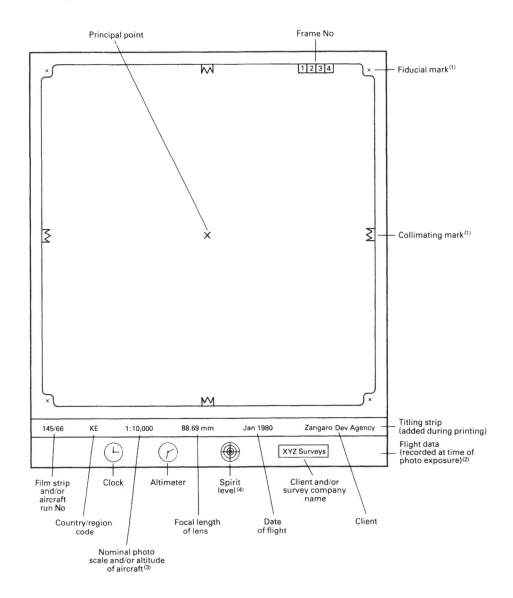

Notes: (1) For use in determining the principal point: where available, the corner
           fiducial marks should be used in preference to the collimating marks.

       (2) May include additional data, such as focal length of lens date and run number.

       (3) Altitudes usually quoted as AMSL (above mean sea-level) or AMGL
           (above mean ground level); note that AMGL is value required to calculate
           the AP scale.

       (4) This level shows camera tilt angle. 'Vertical' photographs are usually
           assumed to be those with camera tilt angles of ≤3°, and the resulting
           small scale and distortion errors are ignored.

Aerial photographic coverage for stereo interpretation   Figure K.4

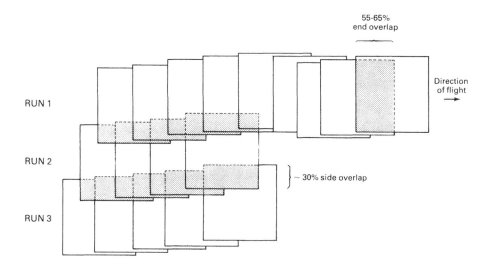

Notes: 1. Flight paths are often not exactly straight nor exactly parallel to each other.

2. Rough maps can be prepared from aerial photographic laydowns using visual alignment, but cartographic accuracy can only be achieved by using a slotted template technique or photogrammetric methods,and accurately located ground reference points (e.g. Ministry of Defence, 1971).

351

Aerial photographs: geometric relationships and relief effects   Figure K.5

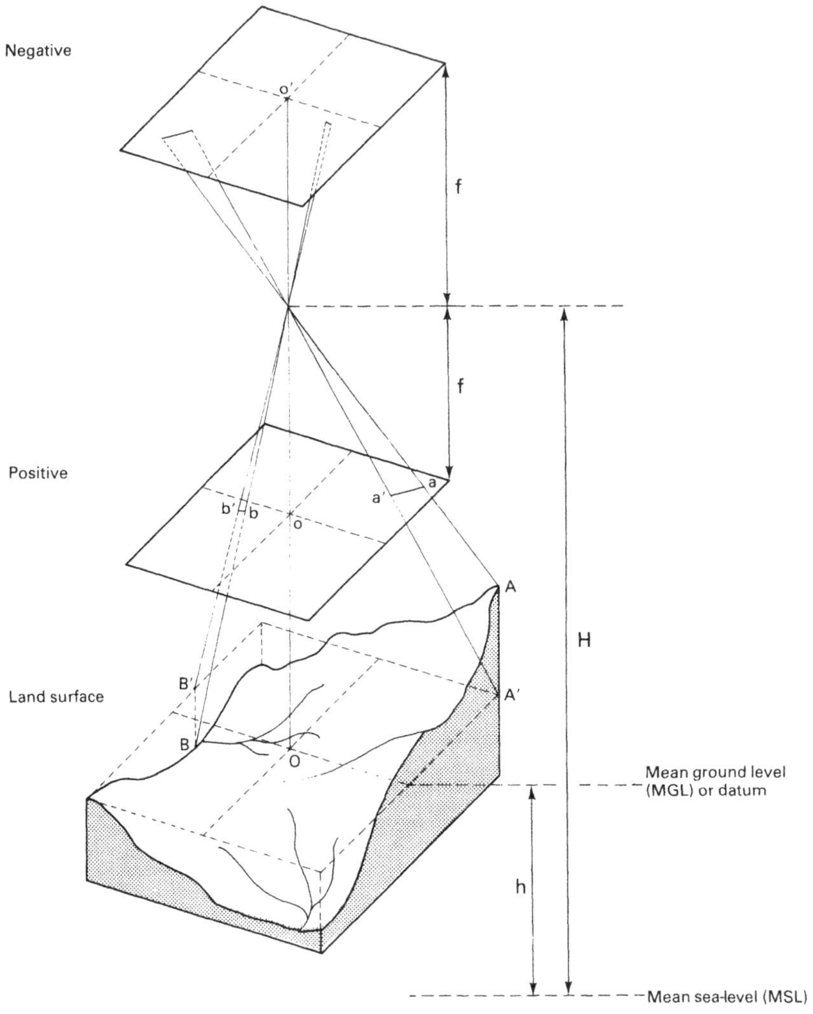

Negative

f

f

Positive

a'
a

b' b   o

Land surface

A

H

B'   A'

B   O

Mean ground level
(MGL) or datum

h

Mean sea-level (MSL)

Key   Oo' = optical axis
       o,o' = principal points
       f = focal length of camera lens
       h = height of MGL above MSL
       H = height of camera lens above MSL

Notes: 1. For calculation of AP scales from focal length (f) and camera height data, H - h gives the
           appropriate mean height value. Allowances should be made where necessary for scale
           changes due to relief variation.
        2. Lengths and most angles are affected by changes in relief, vertical lines on the ground
           suffering radial shifts as illustrated. The exceptions are angles measured through the
           principal points, which remain unaltered.

Source: Adapted from Lillesand and Kiefer (1987).

Test for stereoscopic vision

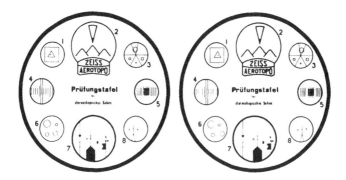

PART I

Within the rings marked 1 through 8 are designs that appear to be at different elevations. Using '1' to designate the highest elevation, write down the depth order of the designs. It is possible that two or more designs may be at the same elevation. In this case, use the same number for all designs at the same elevation.

Ring 1			**Ring 3**	
square	(2)		square	( )
marginal ring	(1)		marginal ring	( )
triangle	(3)		cross	( )
point	(4)		lower left circle	( )
			upper centre circle	( )

Ring 6			**Ring 7**	
lower left circle	( )		black flag with ball	( )
lower right circle	( )		marginal ring	( )
upper right circle	( )		black circle	( )
upper left circle	( )		arrow	( )
marginal ring	( )		tower with cross	( )
			double cross	( )
			black triangle	( )
			black rectangle	( )

PART II

Indicate the relative elevations of the rings 1 through 8.

( ) ( ) ( ) ( ) ( ) ( ) ( ) ( )
highest   lowest

PART III

Draw profiles to indicate the relative elevations of the letters in the words 'Prüfungstafel' and 'stereoskopisches Sehen'.

P R U F U N G S T A F E L   S T E R E O S K O P I S C H E S   S E H E N

Answers are given in Table K.5.

Source: Carl Zeiss (Oberkochen) Ltd.

353

Annex K: Remote sensing

Answers to stereo-vision test (Figure K.6)   Table K.6

Part 1

Ring 1		Ring 6	
square	(2)	lower left circle	(4)
marginal ring	(1)	lower right circle	(5)
triangle	(3)	upper right circle	(1)
point	(4)	upper left circle	(3)
		marginal ring	(2)

Ring 7		Ring 3	
black flag with ball	(5)	square	(4)
marginal ring	(1)	marginal ring	(2)
black circle	(4)	cross	(3)
arrow	(2)	lower left circle	(1)
tower with cross	(7)	upper centre circle	(5)
double cross	(2)		
black triangle	(3)		
black rectangle	(6)		

Part 2

(7) (6) (5) (1) (4) (2)[a] (3)[a] (8)
highest   lowest

Part 3

P R U F U N G S T A F E L   S T E R E O S K O P I S C H E S   S E H E N

---

[a] Rings 2 and 3 are at the same elevation

Source: Carl Zeiss (Oberkochen) Ltd.

354

Glossary of Soil Science and Related Terms

Glossary of Soil Science and Related Terms [1]

A horizon – The upper horizon of a mineral soil normally having the maximum organic matter content and maximum biological activity (NB: in US usage includes A2 horizons).

A2 horizon – USDA term for eluviated E horizon which has lost soil materials such as iron and aluminium oxides and silicate clays.

accelerated erosion – See erosion.

acid rocks – Rocks containing 10% or more free quartz (eg granite, rhyolite). Previously defined as rocks with ⩾ 66% silica.

acid soil – A soil with a $H^+ + Al^{3+} > OH^-$. Specifically, soils with a pH value below 7.0, but for most practical purposes a soil with a pH value about 6.5 or less. The term is usually applied to the surface layer or to the root zone unless specified otherwise.

acid sulphate soils – Soils which when wet have high contents of sulphur in reduced form; on drying soils become highly acidic (pH <3) due to production of H_2SO_4.

acidity potential – The amount of acidity (= $H^+ + Al^{3+}$) that must be neutralised to bring an acid soil to pH 7. Usually given in me/100 g soil.

acric – Of ferralsols with very low cation–exchange capacity, 1.5 me or less per 100 g clay in all or part of the B horizon within 125 cm of the soil surface (FAO Soil Legend).

acrisols – Acid, low base status soils with Bt horizons (FAO Soil Legend).

additive colour process – See colour additive process.

adsorption – The attraction of ions or compounds to the surface of a solid.

adsorption complex – The group of substances in soil capable of adsorption, consisting mainly of colloidal particles of clay and organic matter, which adsorb ions and water.

aeolian – Resulting from wind action, eg aeolian erosion, aeolian deposits.

aeration – See soil aeration.

aerial photography – See remote sensing.

aerobic – a) Having molecular oxygen as a part of the environment.
b) Growing only in the presence of molecular oxygen (such as aerobic organisms).
c) Occurring only in the presence of molecular oxygen (said of certain chemical or biochemical processes such as aerobic decomposition).

aggregate (soil) – A collection of soil particles held in a single structure such as a clod, block or prism etc.

agric horizon – Horizon immediately below the plough layer of cultivated soils containing at least 15% by volume of accumulated clay and humus.

air dry – The condition of a soil at equilibrium with the moisture in the surrounding atmosphere. The actual moisture content depends upon the relative humidity and the temperature of the surrounding atmosphere.

air porosity – The proportion (usually as a % of the bulk volume) of soil consisting of air. Usually corresponds to the volume of the large pores; that is, those drained by a tension of less than approximately 100 cm of water. See moisture tension.

albic horizon or albic E horizon – A pale, relatively coarse, eluvial horizon; clay and free iron oxides have been removed or so segregated that the colour is determined mainly by the primary sand and silt particles (USDA Soil Taxonomy and FAO Legend). A particular kind of A2 horizon.

1/ Compiled with additions and amendments from: Am Soc Agric Eng (1967); Brady (1974); Brewer (1964); Doorenbos and Kassam (1979); Fairbridge and Finkl (1979); FAO–Unesco (1974); Kilmer (1982); Lutz (1965); Reeves et al (1975); Smith (1981); Soil Sci Soc Am (1979); Stiegler (1978); USDA (1975b); Whitten and Brooks (1972). The authors' work experience has also been extensively drawn upon.

alfisols - Mineral soils that have no mollic epipedon, or oxic, or spodic horizon, but do have an argillic or natric horizon which is at least 35% base saturated. They include grey-brown podzolic, non-calcic brown and grey wooded soils in the USDA 1949 classification (USDA Soil Taxonomy).

alkali feldspar - See feldspar.

alkali soil - a) Strictly, a soil with a high degree of alkalinity (pH of 8.5 or higher).
b) A soil that contains sufficient alkali (sodium) to interfere with the growth of most crop plants, usually take as a soil with ESP ⩾ 15. Better described as a saline-sodic or sodic soil.

alkaline soil - Precisely, any soil that has a pH value > 7. Practically, a soil with a pH of > about 7.3. The term is usually applied to the surface layer or root zone but may be used to characterise a horizon or a sample.

alkalisation - The process whereby the exchangeable sodium content of a soil is increased.

allophane - A series of amorphous hydrous aluminium silicate clay minerals; CEC values of allophane clays are particularly susceptible to variations in pH and concentration of saturating solution during measurement.

alluvial fan - Deposits from a stream that divides into branching distributaries, commonly just after emerging from a constriction (ravine or canyon), an inland delta.

alluvial soil - A soil developing from recent alluvium and exhibiting essentially no horizon development or modification of the recently deposited materials.

alluvium - Mixed, unconsolidated sediments deposited by water flowing in rivers and streams or as lake currents.

alpine meadow soils - Dark soils of grassy meadows at altitudes above the timber-line. A great soil group of the hydromorphic suborder (USDA 1949 classification).

alumino-silicates - Minerals containing aluminium, silicon and oxygen as main constituents, eg microcline, $KAlSi_3O_8$.

alunee - Of acid sulphate soil, in Vietnam.

amendment - See soil amendment.

ammonia fixation - The adsorption or absorption of ammonium ions by the mineral or organic fractions of the soil in such a manner that they are relatively insoluble in water and relatively unexchangeable by the usual methods of cation exchange.

ammonification - The biochemical process whereby ammoniacal nitrogen is released from nitrogen-containing organic compounds.

amorphous - Structureless; used of non-crystalline minerals, which have no regular atomic structure.

amphibole - One of a group of common orthorhombic or monoclinic silicate minerals, mostly of magnesium, iron, calcium and sodium with a general composition -
$(MgFe^{2+},Ca)_{2-3}(MgFe^{2+},Fe^{3+},Al)5(Si,Al)$
$5(Si,Al_4)O_{11}(OH)_2$ - occurring in many igneous and metamorphic rocks. Hornblende is a common form.

amygdale - Calcareous inclusion in the vesicles of basaltic lava and similar rock.

anaerobic - a) Without molecular oxygen.
b) Living or functioning in the absence of air or free oxygen.
c) (Reactions) occurring only in the absence of molecular oxygen, such as anaerobic decomposition.

andesine - A variety of plagioclase.

andesite - A fine-grained igneous rock of intermediate composition, the volcanic equivalent of diorite; it typically contains plagioclase feldspar, amphibole, and sometimes small amounts of quartz and alkali feldspar.

andosols - Soils derived from recent volcanic deposits (FAO Soil Legend).

anion-exchange capacity - The sum total of exchangeable anions that a soil can adsorb. Expressed as me/100 g of soil (or of other adsorbing material such as clay).

anisotropic soil - Soil with (physical) properties which change with the direction of measurement of those properties (eg having different vertical and horizontal hydraulic conductivity values); cf isotropic soil.

anthropic epipedon - A thick, dark surface horizon, which is more than 50% saturated with bases, has a narrow C/N ratio and more than 250 ppm of P_2O_5 soluble in citric acid. It is formed under long-continued cultivation where large amounts of organic matter and fertilisers have been added (USDA Soil Taxonomy and FAO Legend).

anthropogenic - Man-made, the most notable soil in this category being that of puddled padi rice fields with a long history of cultivation and irrigation.

anticline - A convex geological fold; the young strata remain at the top of the succession.

antivignetting filter - See vignetting.

apatite - A naturally occurring, complex calcium phosphate which is the original source of most phosphate fertilisers. Composed of complex compounds such as $3Ca_8(PO_4):CaF_2$.

arenaceous - Sandy.

arenosols - Coarse weakly developed soils with an identifiable B horizon (FAO Soil Legend).

argillaceous - Clayey.

argillic horizon - A diagnostic illuvial B horizon characterised by an accumulation of silicate clays (USDA Soil Taxonomy and FAO Legend).

aridisols - Soils characteristic of dry places. Include red desert soils, sierozems and solonchaks in the USDA 1949 classification system (USDA Soil Taxonomy).

arroyo - Seasonal watercourse, usually dry.

association - See soil association.

Atterberg limits - See liquid limit and plastic limit.

Atterberg system - See international soil particle size classification.

auenboden - German, alluvial soil.

available nutrient - That portion of any element or compound in the soil that can be readily absorbed and assimilated by growing plants ('available' should not be confused with 'exchangeable', since the former includes eg soluble and chelated forms).

available water - The portion of water in a soil that can be readily absorbed by plant roots, ie the water held between field capacity and permanent wilting point.

azonal soils - Soils without distinct genetic horizons. A soil order in the USDA 1949 classification.

B horizon - A soil horizon usually beneath the A which is characterised by one or both of the following - (a) an accumulation of silicate clays, iron and aluminum oxides, and humus, alone or in combination; (b) a blocky or prismatic structure.

backswamp - See levee.

bahadas - In deserts, coalesced old alluvial fans.

band - A group of wavelengths, often associated with a particular colour or type of radiation.

band application - (of fertilisers): placing in a line, or band, close to a seedling, to reduce effects of soil on nutrient availability.

bar - A unit of pressure equal to 1 M dynes cm^{-2}.

bariolage - See marbling.

barkhan - Crescent-shaped desert sand dune.

basalt - A fine-grained, dark-coloured basic igneous rock consisting mainly of plagioclase feldspar and pyroxene. Basalts occur principally in lava flows and constitute over 90% of volcanic rocks.

base exchange capacity - See cation-exchange capacity.

base map - topographic sheet, either printed/ published or special ad hoc compilation, on which thematic data can be represented to produce a thematic map, showing eg (in the context of this Manual) soil boundaries, land suitability boundaries, agro-climatic parameters, recommended land use zoning.

basement - Igneous or metamorphic rocks, often Pre-Cambrian, that unconformably underlie unmetamorphosed sedimentary strata.

base saturation percentage - The proportion of the adsorption complex of a soil saturated with exchangeable cations other than hydrogen and aluminium. It is expressed as a percentage of the total cation-exchange capacity.

basic rocks - Quartz-free (usually igneous) rocks containing feldspars that are more calcic than sodic. Previously defined as rocks with 45 to 55% silica.

bauxite - An end-product of weathering, hydrated Al_2O_3, which is when pure enough an aluminium ore.

bedrock - The solid rock underlying soils and the regolith at depths ranging from zero (where exposed by erosion) to several hundred metres.

bench terrace - An embankment constructed across sloping field to allow irrigation/cultivation on a more level surface. The terrace has a steep drop on the downslope side.

bentonite - Special assembly of clay minerals (mainly smectites) formed by weathering of acid lavas and fragmented volcanic material (pyroclastic rock).

biotite - The commonest of the ferromagnesian micas, $K(Mg,Fe^{2+})_3(AlSi_3)O_{10}(OH,F)_2$, occurring in many igneous and metamorphic rocks. Usually dark in colour.

black cotton soils - Pellic vertisols of India and Africa.

black earths - An ambiguous term which can mean chernozem or vertisol(ic) soils.

bleicherde - The light-coloured, leached A2 (or E) horizon of podzol soils.

boehmite - See gibbsite.

bog soils - Marsh soils, including mucks and peats. A great soil group of the intrazonal order and hydromorphic suborder (USDA 1949 classification).

border-strip - See irrigation methods.

borovina - Freely draining, calcareous alluvial soil.

bottomland - See flood plain.

boulder - Soil or geological deposit particle size class larger than stone, more than 20 cm across: in FAO system more than 25 cm and in USDA system more than 60 cm across.

bowal - In Guinea and Ivory Coast, ironstone plateau or mesa with thick layer of hard sesquioxide sheet and/or cemented iron concretions.

braided - Of a river tract, having more than one drainage channel, with elongated islets. See geomorphic cycle. Characteristic of the old-age stage.

braunerde - See brown forest soils.

breccia - A rock composed of coarse angular fragments cemented together.

broad-base terrace - A terrace constructed across slopes to reduce erosion and runoff. Consists of a low embankment with such gentle slopes that it can be farmed.

brown earths - Soils with a mull horizon but having no horizon of accumulation of clay or sesquioxides. (Generally used as a synonym for 'brown forest soils' but sometimes for similar soils with acid reaction).

brown forest soils - Soils formed on calcium-rich parent materials under deciduous forest, and possessing a high base status but lacking a pronounced illuvial horizon. A great soil group of the intrazonal order and calcimorphic suborder (USDA 1949 classification). (A much more narrow group than the European brown forest soil or braunerde).

brown podzolic soils - Soils similar to podzols but lacking the distinct A2 (or E) horizon characteristic of the podzol group. (A zonal great soil group USDA 1949 classification).

brown soils - Soils of the temperate to cool arid regions, with a brown surface and a light-coloured transitional subsurface horizon over calcium carbonate accumulation. They develop under short grasses. A great soil group (USDA 1949 classification).

brunizems - Synonymous with prairie soils (USDA 1949 classification system).

buckshot - Small, hard concretion, commonly about 2 mm in diameter and high in FeMn.

buffer action - The processes by which changes in pH promoted by external conditions tend to be neutralised; the term is also used of various other soil constituents, notably P and K.

buffer compounds - See soil buffer compounds.

bulk density - See soil bulk density.

bulk volume - the total soil volume: includes volume of soil particles, air and water.

buried soil - Soil covered by an alluvial, loessal or other deposit, usually to a depth greater than the thickness of the solum.

burozems - Virtually synonymous with brown forest soils in Russia.

butte - See mesa.

C horizon - An unconsolidated layer (not strictly a horizon) beneath the solum that is relatively little affected by biological activity and pedogenesis and is lacking properties diagnostic of an A or B horizon. It may or may not be like the material from which the A and B have formed.

cadastral map - Map showing extent and ownership of land (and often also value for taxation purposes).

calcareous soil - Soil containing sufficient carbonate (usually $CaCO_3$, but often with $MgCO_3$) to effervesce visibly when treated with cold 0.1 M HCl.

calcic horizon - A horizon of secondary carbonate accumulation more than 15 cm in thickness, with a $CaCO_3$ equivalence of more than 15% and at least 5% more $CaCO_3$ than the C horizon (USDA Soil Taxonomy and FAO Legend).

calcite - A trigonal mineral consisting of crystalline calcium carbonate ($CaCO_3$), usually white or colourless; the principal constituent of most limestones.

calcrete - Hard caliche, lime pan.

caliche - A layer near the surface, more or less cemented by secondary carbonates of calcium or magnesium precipitated from the soil solution. It may occur as a soft thin soil horizon, as a hard thick bed just beneath the solum or as a surface layer exposed by erosion.

cambic horizon - A horizon which has been altered or changed by soil-forming processes. It usually occurs below a diagnostic surface horizon (USDA Soil Taxonomy and FAO Legend).

cambisols - Soils with a cambic B horizon as a major feature (FAO Soil Legend).

capillary conductivity - Outmoded term for hydraulic conductivity.

capillary porosity - The bulk volume of small pores which hold water in soils against a tension usually > 60 cm of water. See moisture tension; cf air porosity.

capillary water - The water held in the capillary pores of a soil (obsolete).

carapace - Ironpan (qv), hard but thin.

carbon-nitrogen ratio - The ratio of the weight of organic carbon (C) to the weight of total nitrogen (N) in a soil or in organic material. Averages about 10 in topsoils and about 8 in upper subsoils.

cat clays - Acid sulphate clays (see acid sulphate soils) with characteristic pale yellow jarosite deposits.

catena - A sequence of soils derived from similar parent material, and occurring under similar climatic conditions, but having different characteristics due to variation in relief and drainage. See also soil association, soil complex.

cation-adsorption capacity - See cation-exchange capacity.

cation exchange - The interchange between cations in solution and cations on the surface of any surface-active material such as soil colloids.

cation-exchange capacity (CEC) - The maximum amount of exchangeable cations that a soil can adsorb. Usually expressed in me/100 g of soil (or of other adsorbing material such as clay).

CCT - See computer-compatible tape.

cemented - Of a soil horizon, having a hard, brittle structure caused by cementing agents such as gypsum, calcium carbonate, or the oxides of silicon, iron and/or aluminium.

cerrado – In Brazil, level open country with distinctive savanna vegetation comprising tall grasses and low contorted trees, correlating closely with very depleted soils on old land surfaces.

characteristic curve – See soil-water characteristic.

check-basin – See irrigation methods.

chelate – Compound in which a metal ion is combined with an organic molecule by multiple chemical bonds. This can prevent, or delay, reactions of the ion within the soil but still allow uptake by plants.

cheluviation – Equivalent to podzolisation.

chernozems – Soils with a thick, nearly black or black, organic matter rich A horizon high in exchangeable calcium, underlain by a lighter-coloured transitional horizon above a zone of calcium carbonate accumulation; occur in a cool subhumid climate under a vegetation of tall and moderately tall grass prairie. A zonal great soil group (USDA 1949 classification and FAO Soil Legend).

chert – A structureless form of silica, closely related to flint, which breaks into angular fragments.

chestnut soils – Soils with a moderately thick, dark-brown A horizon over a lighter-coloured horizon that is above a zone of calcium carbonate accumulation. They develop under mixed tall and short grasses in a temperate to cool, and subhumid to semi-arid climate. (A zonal great soil group of the USDA 1949 classification, also called kastanozems in Russia).

chisel – A tillage implement with chisel-shaped units used to shatter or loosen hard, compact layers, usually in the subsoil, to depths below normal plough depth. See subsoiling.

chlorosis – A plant condition in which chlorophyll production is reduced. Chlorotic leaves range from light green through yellow to almost white.

chocolate soils – In Australia, vertic cambisols.

chroma – The relative purity, strength or saturation of a colour, directly related to the dominance of the determining wavelength of the light and inversely related to greyness; one of the three variables of colour. See Munsell colour system, hue and value.

chromic – Of soils with high chroma, ie vertisols that are not black, cambisols and luvisols with strong brown to red B horizons (FAO Soil Legend).

cinnamon soils – See vertisols; not black but, as the name implies, dusky brown.

classification – See soil classification and land classification.

clastic sediments – Sediments derived from pre-existing rock material by weathering and erosion processes; clastic sedimentary rocks include conglomerates, sandstones, siltstones and mudstones.

clay – a) A soil separate consisting of particles < 0.002 mm in equivalent diameter.
b) Soil material containing more than 40% clay, less than 45% sand and less than 40% silt (USDA system).

clay mineral – Clay-sized inorganic material (usually crystalline) found in soils.

clay skin – See cutan.

clayey – Containing large amounts of clay or having properties similar to those of clay.

claypan – A compact, slowly permeable layer in the subsoil having a much higher clay content than the overlying material, from which it is separated by a sharply defined boundary. Claypans are usually hard when dry and plastic and sticky when wet.

clod – A compact, coherent mass of soil produced artificially, usually by the activity of man as a result of ploughing, digging, etc, especially when these operations are performed on soils that are either too wet or too dry for normal tillage operation. See also ped.

coarse sand – See international and USDA soil particle size classification entries.

coarse texture – Sands, loamy sands and sandy loams but not very fine sandy loam.

cobble, cobblestone – Rounded or partially rounded rock or mineral fragment 7.5 to 25 cm in diameter (USDA).

collimating marks – Similar to fiducial marks, but less accurate; usually appear at the centre of each side of an AP.

colloid – See soil colloid.

colluvium – A deposit of rock fragments and soil material accumulated at the base of steep slopes as a result of slow weathering and gravitational action (as opposed to surface erosion effects); cf creep, solifluxion.

colour additive process – Technique (used in viewing or printing MSS data) of adding the three primary colours in various proportions to enhance or reduce contrast of different areas.

colour composite – A colour print made from several (commonly three) of the MSS bands, each of which is assigned a particular colour, so that IR reflectance can be shown.

columnar soil structure – Structure similar to prismatic structure but with rounded tops to individual peds. Often indicative of sodic soils.

command area – Area that can be irrigated under system being considered; most commonly applied to gravity-fed irrigation schemes.

compound fertiliser – A fertiliser containing two or more of the primary plant nutrients nitrogen, phosphorus and potassium.

computer-compatible tape (CCT) – Method of storing multispectral scanner data for subsequent production as a print (either with or without digital enhancement).

concretion – A chemical compound, such as calcium carbonate or iron oxide, in the form of a grain or nodule.

conductivity – See hydraulic conductivity and electrical conductivity.

conglomerate – A coarse–grained, clastic sedimentary rock formed largely of strongly cemented, rounded water-worn pebbles (cf breccia).

consistence (or consistency) – The combination of properties of soil material that determine its resistance to crushing and its ability to be moulded or changed in shape.   Described in terms such as loose, friable, firm, soft, etc.

consumptive use – The water requirement for crop growth in a given period;   includes water used by plants in transpiration and growth, plus water loss from soil by evaporation, or from intercepted precipitation;   cf ET.

contour – An imaginary line connecting points of equal elevation on a land surface.

contour terrace – A terrace aligned along, or close to, a contour.

contrast stretching – Techniques for improving contrast of images;   may be applied to photography or to computerised imagery.

controlled mosaic – A vertical aerial photographic mosaic composed to fit an accurate (or reasonably accurate) topographic base.

corned beef layer – See dry mottling.

corrugation – See irrigation methods.

creep – A colluvial process of slow mass movement of soil material down relatively steep slopes primarily under the influence of gravity, but facilitated by alternate wetting and drying and/or freezing and thawing;   cf wash, solifluxion.

crop water requirement – See evapotranspiration.

crotovina – See krotovina.

crumb – A soft, porous, more or less rounded soil structural unit 1 to 5 mm in diameter (cf granule).

crust – A surface layer ranging in thickness from a few millimetres to about 2.5 cm, that is much more compact, hard and brittle, when dry, than the material immediately beneath it.

crystal lattice – See lattice structure.

crystalline rock – A rock consisting of various minerals that have crystallised in place from magma.   See igneous rock and sedimentary rock.

cuesta – Gentle slope ending in steep drop.

cuirasse – Ironpan (qv), hard and thick.

cutan – Term for clay skins on ped and pore surfaces, which may be evidence of illuviation, proposed by Brewer (1964), whose definition reads in part: a modification of the texture, structure or fabric of natural surfaces in soil

materials due to concentration of particular soil constituents or in situ alteration of the plasma.

cyclic salt – Salt deposited by wind blowing off sea or salt lake.

Darcy's Law – Equation describing rate of flow of fluids through porous media.   See page 72).

deflation – Removal of soil particles by wind erosion.

defflocculate – a) To separate out individual particles by chemical and/or physical means, eg by Calgon.
  b) To cause the particles of the dispersed phase of a colloidal system to become suspended in the dispersion medium (cf disperse).

degradation – The changing of a soil to a more highly leached and more highly weathered condition, usually accompanied by morphological changes such as development of an A2 horizon.

degraded chernozems – Soils with a very dark-brown or black A1 horizon underlain by a dark-grey, weakly expressed A2 horizon and a brown B horizon;   formed in the forest-prairie transition of cool climates.   (A zonal great soil group of the USDA 1949 classification, also called suglinoks in Russia).

dendritic – Denoting a branching (tree-like) pattern or structure, eg a drainage pattern, or a pore system.

denitrification – The biochemical reduction of nitrate or nitrite to gaseous form of nitrogen, either as molecular nitrogen or as an oxide.

denudation – The wearing down of a land surface by the processes of weathering, mass movement, transportation and erosion.

desalinisation – Removal of salts from saline soil, usually by leaching.   Also termed desalination.

desert crust – A hard layer, containing calcium carbonate, gypsum or other binding material, exposed as the surface in desert regions.

desert pavement – Stony or gravelly land surface in desert areas arising from removal of finer particles by wind action.

desert soils – A zonal great soil group consisting of soils with a very thin, light-coloured surface horizon, which may be vesicular and is ordinarily underlain by calcareous material; formed in arid regions under sparse shrub vegetation (USDA 1949 classification).

desert varnish – A glossy sheen or coating on stones and gravel in arid regions.

desorption – The removal of sorbed material from a surface, used especially of exchangeable ions.

diagnostic horizon – Soil horizon, some or all of whose properties are used for classification purposes, especially in the USDA Soil Taxonomy system or the FAO Soil Legend.

diffusion – The transport of matter as a result of the movement of the constituent particles. The intermingling of two gases or liquids in contact with each other takes place by diffusion.

digital enhancement – Computerised treatment of data (usually from multispectral scanner) on computer-compatible tape to produce image enhancement.

diopside – See pyroxene.

diorite – A coarse-grained intermediate igneous rock consisting of salic plagioclase feldspar and ferromagnesian minerals, usually hornblende or biotite. Diorite is the plutonic equivalent of andesite.

dipslope – The gentle slope of a cuesta.

disperse – a) To break up compound particles, such as aggregates, into the individual component particles.
b) To distribute or suspend fine particles, such as clay, in or throughout a dispersion medium, such as water (cf deflocculate).

dissection – Of landscapes, see rejuvenation.

dolerite – Medium-textured igneous rock occurring mainly as dykes, sills and plugs. Dark in colour. Mineralogically and chemically the same as basalt.

dolina – Karst depression, a solution hollow.

dolomite – a) A trigonal carbonate mineral $(CaMg(CO_3)_2)$.
b) A type of limestone composed of over 15% magnesium carbonate. It occurs either as a primary precipitate from sea water (an evaporite) or by the alteration of calcite rock by magnesium-charged solutions (dolomitisation).

double layer – In colloid chemistry, the electric charges on the surface of the dispersed phase (usually negative), and those of the adjacent diffuse layer (usually positive) of ions in solution.

drift – Material of any sort deposited by geological processes in one place after having been removed from another. Glacial drift includes material moved by the glaciers and by the streams and lakes associated with them.

drip irrigation – See irrigation methods.

drumlin – A long, smooth cigar-shaped low hill of glacial till, with its long axis parallel to the direction of ice movement.

dry mottling – Typically in the C horizon of soils, mainly red-orange-yellow without much grey and with no bluish tinges; colloquially amongst soil surveyors the 'corned beef layer'.

dryland farming – The practice of crop production in low rainfall areas without irrigation.

duff – The matted, partly decomposed organic surface layer of forest soils.

dune – Accumulation of loose sand, coastal or desert.

duricrust – Term often used in Australia for silcrete.

duripan (hardpan) – An indurated horizon cemented by materials such as aluminium silicate, silica, $CaCO_3$ and iron.

dust mulch – A loose, finely granular, or powdery condition on the surface of the soil, usually produced by shallow cultivation.

dyke – A sheet-like intrusion of igneous rock, normally of intermediate grain size, that cuts discordantly through the surrounding rock.

dystric – With low base saturation, less than 50% in the upper subsoil (FAO Soil Legend).

E horizon – Equivalent of an A2 horizon, qv.

earth pillar – Column of soil, often 6 to 10 m high, capped by a boulder which protected the feature while surrounding material was eroded away.

earth resources technology satellite (ERTS) – The first of the Landsats, in 1972, was originally named ERTS-1.

EC - See electrical conductivity.

ecology - The science of interrelations of organisms and environment.

edaphology - The science that deals with the influence of soils on living things, particularly plants, and including man's use of land for plant growth.

effective rainfall - Net quantity of rainfall available for plant utilisation, ie total rainfall less losses such as surface runoff, evaporation and deep percolation.

efflata - Products of volcanic eruption.

electrical conductivity (EC) - Conductivity of a standard soil extract or paste used as a measure of salinity. ECe of a saturation extract; ECp of a saturated paste; EC1:5 of a 1:5 soil:water mixture.

eluviation - The removal of soil material in suspension or solution from a layer or layers of a soil (usually, the loss of material in solution only is described by the term leaching).

energy gradient - (As used in hydraulic conductivity): change in energy per unit distance in the direction of flow.

enstatite - See pyroxene.

entisols - Soils which have no natural genetic horizons or only the beginning of such horizons (USDA Soil Taxonomy).

eolian - See aeolian.

epipedon - A diagnostic surface horizon which includes the upper part of the soil that is darkened by organic matter, and/or the upper eluvial horizons (USDA Soil Taxonomy).

equivalent diameter –
  a) Equivalent spherical diameter – value used for irregularly shaped particles, and equal to the diameter of spherical particles with similar sedimentation behaviour.
  b) Equivalent cylindrical diameter – value used for irregular pore shapes, and equal to the diameter of cylindrical pores with similar water retention properties.

ergs – Sandy deserts.

erosion – a) The wearing away of the land surface by running water, wind, ice or other geological agents, including such processes as creep.
  b) Detachment and movement of soil or rock by water, wind, ice or gravity. The following terms are used to describe different types of water erosion:

accelerated erosion – Erosion much more rapid than normal, natural, geological erosion, primarily as a result of the influence of the activities of man or, in some cases, of animals.

gully erosion – The erosion process whereby water accumulates in narrow channels and, over short periods, removes the soil from this narrow area to considerable depths, ranging from < 50 cm to > 30 m.

natural erosion – Wearing away of the earth's surface by water, ice or other natural agents under natural environmental conditions of climate, vegetation, etc, undisturbed by man. Synonymous with geological erosion.

normal erosion – The gradual erosion of land used by man which does not greatly exceed natural erosion. See natural erosion.

rill erosion – An erosion process in which numerous small channels of only several cm in depth are formed; occurs mainly on recently cultivated soils. See rill.

scouring – Soil removal in broad rills which, on coalescing, give sheet erosion.

sheet erosion – The removal of a fairly uniform layer of soil from the land surface by runoff water.

splash erosion – The spattering of small soil particles caused by the impact of raindrops on very wet soils. The loosened and separated particles may or may not be subsequently removed by surface runoff.

ERTS – See earth resources technology satellite.

escarpment – See scarp.

esker – Sinuous, elongated glaciofluvial ridge, frequently joining kames together. See glaciofluvial deposits.

ESP – See exchangeable sodium percentage.

ET – Evapotranspiration.

etchplain – Peneplain that has been or is being subjected to rejuvenation, with the superimposition of a new or invigorated drainage pattern.

ET_m – Maximum evapotranspiration rate of a crop when soil water is not limiting; also called crop water requirement. Usually expressed as mm day^{-1} or mm per growing period.

ET_o – Reference evapotranspiration for given climate when water is not limiting. May refer to pan measurement or calculation (eg Penman). Usually expressed as mm day^{-1} or mm per growing period.

ET_{pan} – Measured evaporation from unscreened Class A pan. Usually expressed in mm day^{-1}.

eutric – With high base saturation, 50% or more in the upper subsoil (FAO Soil Legend).

evaporite – Sediment produced by evaporation of saline solutions.

evapotranspiration (ET) – The combined loss of water from a given area, and during a specified period of time, by evaporation from the soil surface and by transpiration from plants.

exchange acidity – The titratable hydrogen and aluminium that can be replaced on the adsorption complex by ions from a neutral salt solution. Usually expressed as me/100 g of soil.

exchange capacity – The total ionic charge of the adsorption complex active in the adsorption of ions. See anion-exchange and cation-exchange capacity.

exchangeable sodium percentage (ESP) – The extent to which the adsorption complex of a soil is occupied by sodium. It is expressed as follows:

$$\frac{\text{Exchangeable sodium (me/100 g soil) x 100}}{\text{Cation exchange capacity (me/100 g soil)}}$$

exposure – Area where in situ rocks are visible.

extrusive rock – Volcanic rock, usually of fine or glassy texture.

fallow – Cropland left idle in order to restore productivity, mainly through accumulation of water, nutrients or both. Summer fallow is a common stage before cereal grain in regions of limited rainfall. In bare fallow the soil is kept free of weeds and other vegetation thereby conserving nutrients and water for the next year's crop. In bush fallow, however, plants regenerate and cycle nutrients, at the same time preventing or reducing soil erosion.

false colour – Colours used to print IR (or MSS) data; wavelengths are shifted from true values to allow IR to be shown.

family – See soil family.

fan – See alluvial fan.

fayalite – See olivine.

feldspar – The most abundant mineral in igneous rocks, consisting structurally of a complex aluminium silicate framework with varying proportions of potassium, sodium, calcium and (rarely) barium. Feldspars are grouped into alkali feldspars ($KAlSi_3O_8$–$NaAlSi_3O_8$) and plagioclase feldspars ($NaAlSi_3O_8$–$CaAl_2Si_2O_8$) and are monoclinic or triclinic minerals.

felsic minerals - Light-coloured minerals, particularly feldspar and quartz.

felspar - Alternative (and pedantically, less correct) spelling of feldspar.

ferralitic (or ferrallitic) soils, French sols ferralitiques - Acid, highly weathered tropical soils, somewhat broader than the USDA category of oxisols (CCTA 1964 classification).

ferralsols - Strongly weathered tropical soils with an oxic B horizon (FAO Soil Legend, also in INEAC 1961 and CCTA 1964 systems).

ferricrete - An indurated horizon cemented by iron oxides, often with other (Mn, Al, Ti) sesquioxides.

ferrisols - Tropical soils less strongly weathered than ferralsols with well-developed blocky structure and clay skins in the B horizon: developed mainly on basic rocks (INEAC 1961 and CCTA 1964 classification).

ferrolysis - Cyclic process expounded by Brinkman (1979): reduction of Fe oxides during microbial decomposition of organic matter after water-saturation of eluvial horizons, followed by oxidation of the exchangeable ferrous iron produced.

ferromagnesian minerals - Silicate minerals rich in iron and magnesium, usually dark coloured. Includes amphibole, pyroxene, olivine, biotite and mica.

ferruginous - Denoting rocks or minerals containing iron, often resulting in a reddish colour.

ferruginous tropical soils, French sols ferrugineux tropicaux - Kaolinitic and iron-dominated tropical soils, less weathered than ferralitic soils, often falling in the USDA category of alfisols (CCTA 1964 classification).

fertiliser - Any organic or inorganic material of natural or synthetic origin which is added to a soil to supply certain elements essential to the nutrition of plants.

fertiliser requirement - The quantity of certain plant-nutrient elements needed, in addition to the amount supplied by the soil, to increase plant growth to a designated optimum.

fertility - See soil fertility.

fiducial marks - Index marks (usually four) rigidly connected with the camera lens so that lines drawn between opposite fiducial marks intersect at the principal point. Usually appear at the corners of an AP.

field capacity (FC) - The percentage of water remaining in a previously saturated soil 2 or 3 days after saturation, when free drainage has ceased. In practice, the moisture content 48 h after saturation. In many medium textured soils, value is similar to the water content in a soil at 1/3 bar equivalent suction.

fine earth fraction - The portion of a soil sample with particle diameter of 2 mm and less.

fine sand - See international and USDA soil particle size classification.

fine texture - Consisting of or containing large quantities of the fine fractions, particularly of silt and clay (includes all clay loams and clays; that is, clay loam, sandy clay loam, silty clay loam, sandy clay, silty clay and clay textural classes).

fixation - The process or processes in a soil by which certain chemical elements essential for plant growth are converted from a soluble or exchangeable form to a much less soluble or to a non-exchangeable form; for example, phosphate fixation. Contrast with nitrogen fixation.

fixed phosphorus - That phosphorus which has been changed to a less soluble form as a result of reaction with the soil.

flint - Concretionary form of silica occurring as scattered nodules and fragments of irregular shape in soft calcareous rocks.

flocculation - Opposite process to deflocculation, bonding of individual particles by chemical and/or physical means.

flood irrigation - See irrigation methods.

flood plain or floodplain - The land bordering a river or stream, built up of sediments from overflow and usually still subject to inundation.

fluorapatite - A member of the apatite group of minerals rich in fluorine. Most common mineral in rock phosphate.

fluvioglacial - See glaciofluvial.

fluvisols - Recent alluvial (or colluvial) soils (FAO Soil Legend).

forest - Large area with closed canopy of tall trees.

form line - An approximate, or uncontrolled, contour line.

forsterite - See olivine.

fragipan - Dense and brittle pan or layer in soils. The hardness is due mainly to extreme density or compactness rather than high clay content or cementation. Removed fragments are friable, but the material in place is so dense that roots cannot penetrate and water moves through it very slowly.

fulvic acid - A term of varied usage but usually referring to the mixture of organic substances remaining in solution upon acidification of a dilute alkali extract from the soil.

furrow irrigation - See irrigation methods.

gabbro - A coarse-grained basic igneous rock, the plutonic equivalent of basalt, and consisting essentially of plagioclase feldspar, pyroxene and olivine.

Gapon equation - Gives the relationship between exchangeable sodium percentage and sodium adsorption ration: often ESP lies between 0.9 and 1.4 times SAR.

gazi - See rendzinas.

genesis - See soil genesis.

geodetic survey - Mapping of such a large area that curvature of the earth has to be considered.

geomorphic cycle - Youth, maturity and old age are recognisable stages in landscape development, instanced by stream and river patterns in particular, namely the torrent tracts of youth, the broader valleys with minor rapids and waterfalls of maturity and meanders and braided tracts of old age. See also rejuvenation.

gibbsite - Major hydrous aluminium oxide found in soils: $Al(OH)_3$.

gilgai - Microrelief produced by expansion and contraction with changes in moisture. Found in soils that contain large amounts of swelling clay. In nearly level areas usually a succession of microbasins and microknolls; on sloping land microvalleys and microridges parallel to the direction of the slope.

glacial drift - Rock debris that has been transported by glaciers and deposited, either directly from the ice or from the melt-water. The debris may or may not be heterogenous.

glacial till - See till.

glaciofluvial deposits - Material moved by glaciers and subsequently sorted and deposited by streams flowing from the melting ice. The deposits are stratified and may occur in the form of outwash plains, deltas, kames, eskers and kame terraces.

gley - Conditions of poor drainage resulting in reduction of iron and other elements and in grey colours and mottles.

gleysols - Soils in which gleying is predominant (FAO Soil Legend).

glossic - Of a soil horizon tonguing into the one below or into underlying rock.

gneiss - A coarse-grained crystalline rock formed during high-grade regional metamorphism of igneous (orthogneiss) or sedimentary (paragneiss) rocks, characterised by a banded appearance and linear orientation of minerals. Composition similar to that of granite.

goethite - Kind of limonite (FeO.OH) formed usually by rapid disintegration of ferromagnesian minerals.

granite - A coarse-grained acid igneous rock consisting mainly of quartz, alkali feldspar and mica, with various accessory minerals. It occurs in intrusive bodies from crystallised magma or the granitisation (metasomatic transformation) of pre-existing rock.

granodiorite - A coarse-grained acid igneous rock containing quartz, predominantly plagioclase feldspar and ferromagnesian minerals (especially biotite or hornblende). Granodiorities are the most abundant plutonic rocks.

granule - A soil structural unit of moderately to slightly porous, approximately spherical aggregates 1 to 5 mm in diameter (cf crumb).

gravel - A soil separate consisting of particles 2 to 20 mm in equivalent diameter (international system) or 2 to 75 mm (FAO and USDA).

gravitational water - Water which moves into, through, or out of the soil under the influence of gravity.

great group - A category in the USDA Soil Taxonomy between that of the suborder and the subgroup. A 'great soil group' (USDA 1949 classification) was a broad group of soils with fundamental characteristics in common: examples are chernozems, grey-brown podzolic soils and podzols.

green manure - Plant material incorporated with the soil while green, or soon after maturity, for improving the soil.

grey-brown podzolic soils - Soils with a thin, moderately dark A1 horizon and with a greyish-brown A2 horizon underlain by a B horizon containing a high percentage of bases and an appreciable quantity of illuviated silicate clay; formed on relatively young land surfaces, mostly glacial deposits, from material relatively rich in calcium, under deciduous forests in humid temperate regions. A zonal great soil group (USDA 1949 classification).

grey desert soil - A term used in Russia, and frequently in the United States, synonymous with desert soil.

greywacke - Arenaceous rocks.

greyzems - Grey forest soils of cool temperate zones (FAO Soil Legend).

groundwater - Water that fills all the unblocked pores of underlying material below the water-table, which is the upper limit of saturation.

groundwater laterite soils - Soils characterised by hardpans or concretional horizons rich in iron and aluminium (and sometimes manganese) that have formed immediately above the water-table. A great soil group of the intrazonal order and hydromorphic suborder (USDA 1949 classification system).

groundwater podzols - Soils with an organic mat on the surface over a very thin layer of acid humus material underlain by a whitish-grey leached layer, which may be up to 90 cm thick, and is underlain by a brown, or very dark-brown, cemented hardpan layer; formed under various types of forest vegetation in cool to tropical humid climates under conditions of poor drainage. A great soil group of the intrazonal order and hydromorphic suborder (USDA 1949 classification).

grumusols - See vertisols.

gully erosion - See erosion.

gypsic horizon - A horizon of accumulation of secondary $CaSO_4$ (gypsum), more than 15 cm thick which has at least 5% more gypsum than the C or underlying stratum, and in which the product of thickness in cm and the percent gypsum is at least 150 (USDA Soil Taxonomy and FAO Legend).

H horizon - A surface horizon of organic material (see histic H horizon).

haematite - Iron oxide accessory in igneous rocks; Fe_2O_3.

half-bog soils - Soils with dark-brown or black peaty material over greyish and rust-mottled mineral soil; formed under conditions of poor drainage under forest, sedge or grass vegetation in cool to tropical humid climate. A great soil group, of the intrazonal order and hydromorphic suborder (USDA 1949 classification).

halloysite - Kaolin mineral formed usually by rapid weathering of feldspars.

halomorphic soils - Saline and alkali soils formed under imperfect drainage in arid regions and including the great soil groups solonchak or saline soils, solonetz and soloth soils. A suborder of the intrazonal soil order (USDA 1949 classification).

halophyte - Salt-tolerant plant.

hamadas - Rocky deserts.

haplic - Modal, normal, typical.

hardpan - A hardened soil layer, in the lower A or in the B horizon, caused by cementation of soil particles with organic matter or with materials such as silica, sesquioxides or calcium carbonate. The hardness does not change appreciably with changes in moisture content and pieces of the hard layer do not slake in water. See caliche, claypan and duripan.

harrowing - A secondary broadcast tillage operation which pulverises, smoothes, and firms the soil in seedbed preparation, controls weeds or incorporates material spread on the surface.

heavy metals - Metals with specific gravity > 5.

heavy soil - (Obsolete in scientific use) A soil with a high content of fine particles, particularly clay, or one with a high drawbar pull and hence difficult to cultivate.

histic H epipedon or horizon - A horizon at or near the surface, saturated with water at some season and containing a minimum of 20% organic matter if no clay is present and at least 30% organic matter if it has 50% or more of clay (USDA Soil Taxonomy and FAO Legend; cf O horizon).

histosols - Soils characterised by their high organic matter content. Bog soils and half-bog soils are included in this soil order (USDA Soil Taxonomy and FAO Soil Legend).

horizon - See soil horizon.

hornblende - An amphibole mineral common in metamorphic and igneous rocks.

hornfels - Massive, hard rock produced by thermal metamorphism.

hue - One of the three variables of colour. It is caused by light of certain wavelengths and changes with the wavelength. See Munsell colour system, chroma and value.

humic acid - A mixture of variable or indefinite composition of dark organic substances, precipitated upon acidification of a dilute alkali extract from soil.

humic gley soils - Soil of the intrazonal order and hydromorphic suborder that includes wiesenboden and related soils, such as half-bog soils, which have a thin muck or peat O_2 horizon and an A1 horizon. Developed in wet meadow and in forested swamps (USDA 1949 classification).

humification - The processes involved in the decomposition of organic matter and leading to the formation of humus.

humus - That more or less stable fraction of the soil organic matter remaining after the major portion of added plant and animal residues have decomposed. Usually it is dark coloured.

hydraulic conductivity - The rate at which a liquid (usually water) will flow through a soil mass under a given hydraulic head; the constant in Darcy's Law.

hydraulic gradient - Change in hydraulic head per unit distance in the direction of flow. See also energy and potential gradients.

hydraulic head - Energy stored in a hydraulic system, expressed as height to which water (or other fluid) rises, or would rise, above a fixed datum.

hydrolysis - Formation of an acid and a base from a salt by interaction with water: also decomposition of organic compounds by the soil solution.

hydromorphic soils - A suborder of intrazonal soils, all formed under conditions of poor drainage in marshes, swamps, seepage areas, or flats (USDA 1949 classification).

hydrous mica - Silicate clay with 2:1 lattice structure, but of indefinite chemical composition since part of the silicon in the silica tetrahedral layer has been replaced by aluminium, and a considerable, but variable, amount of potassium occurs between the crystal layers, resulting in particles larger than normal in montmorillonite and, consequently, in a lower cation-exchange capacity. A common member is illite.

hydroxyapatite - An apatite mineral rich in hydroxyl groups. A nearly insoluble calcium phosphate.

hygroscopic coefficient - The moisture content of a dry soil in equilibrium with some standard relative humidity close to saturation (usually 98%, at 1 atm pressure and at room temperature; ie approx. 27.8 bar equivalent suction), expressed as a percentage of oven-dry soil.

hysteresis - Usually applied in soils to the effects of wetting and drying on pF/moisture content relations or to acid/base effect on CEC; refers to the differences that occur in water contents (or CEC) corresponding to a particular pF (or pH) as a result of the previous wetting and drying (or increasing and decreasing pH) history of a soil.

igneous rock - Rock formed from the cooling and solidification of magma and that has not been changed appreciably since its formation. Igneous rocks are formed by solidification from a molten state, either intrusively below the earth's surface or extrusively (as lava or pyroclastic fragments).

illite - See hydrous mica.

illuvial horizon - A soil horizon of accumulation containing material translocated from an overlying layer.

illuviation - The process that is creating or has created an illuvial horizon.

image enhancement - Processing of an image to produce improved contrast between different areas or make specific features more easily visible.

impervious - Resistant to penetration by fluids or roots.

inceptisols - Soils with one or more diagnostic horizons that are thought to form rather quickly and that do not represent significant illuviation or eluviation or extreme weathering.   Soils classified as brown forest, subarctic brown forest, ando, sols bruns acides and associated humic gley and low-humic gley soils are included in this order (USDA Soil Taxonomy).

incised meanders - See rejuvenation.   Winding, twisting river reaches which are steep-banked and markedly downcut below the general valley floor level.

indurated horizon - Hard horizon, hardened eg by sesquioxide, carbonate or clay accumulation.

infiltration - The vertical movement of water into a soil, usually at the surface; subsurface movement is usually referred to as permeability or hydraulic conductivity.

infiltration rate - The rate at which water enters the soil under specified conditions.

infra-red (IR) - See false colour.

inlier - A small area of older rocks occurring within an area of younger rocks (cf outlier).

inped - Adjective for 'within a ped' describing soil pores etc.

inselberg - A steep, rounded outcrop, hill or mountain, usually of granite or gneiss, which stands out above a pediment (cf mesa).

interfluve - High ground between gullies, streams or rivers of the same drainage system (cf watershed).

intergrade - A soil with characteristics of, or properties intermediate between, two or more soil units.

intermediate rock - Rock intermediate between acid and basic rock, usually with < 10% quartz and containing a plagioclase (of the andesine-oligoclase type) and/or an alkali feldspar.   Examples are andesite and diorite.

international soil particle size classification - Size limits are in the simple progression 2 to 0.2 mm coarse sand, 0.2 to 0.02 mm fine sand, 0.02 to 0.002 mm silt and < 0.002 mm clay. See also under USDA.

interstices - Voids, pores or spaces between soil particles or soil aggregates.

intrazonal soils - Soils with more or less well-developed soil characteristics that reflect the dominating influence of some local factor, relief, parent material or age, over the normal effect of climate and vegetation (USDA 1949 classification system).

intrusive rock - Igneous rock formed from magma that has forced its way amongst pre-existing rocks.   Types of igneous intrusions include batholiths, dykes, sills etc.   Crystals are large relative to those of extrusive rocks.

ions - Atoms, or compounds, which are electrically charged as a result of the loss of electrons (to form cations) or the gain of electrons (to form anions).

ion-selective electrode - An electrochemical sensor, the potential of which (in conjunction with a suitable reference electrode) depends on the logarithm of the activity of a given ion in aqueous solution.

IR - See infra-red.

ironpan - An indurated soil horizon in which iron oxide is the principal cementing agent.

ironstone - In soil and regolith, hardened plinthite; cf laterite.

irrigation efficiency - The ratio of the water consumed by irrigated crops to the amount of water supplied.   Indicative ranges are: surface irrigation 40 to 50%; sprinkler irrigation 65 to 75% and drip irrigation 85 to 90%.

irrigation methods - The manner in which irrigation water is applied to an area.   The principal methods of applying the water include:

border-strip - The water is applied at the upper end of a strip with earth borders to confine the water to the strip.
check-basin - The water is applied rapidly to relatively level plots surrounded by levees.   The basin is a small check.
corrugation - The water is applied to small, closely spaced furrows, frequently in grain and forage crops, to confine the flow of irrigation water to one direction.
drip irrigation - See trickle irrigation.
flooding - The water is released from field ditches and allowed to flood over the land.
furrow - The water is applied to row crops in ditches made by tillage implements.
overhead or sprinkler - The water is sprayed over the soil surface through nozzles from a pressure system.
subirrigation - The water is applied in open ditches or tile lines until the water-table is raised sufficiently to wet the soil.
trickle irrigation - The water is applied from perforated piping or very low-volume emitters, with an on-off switch that may be remote controlled.

wild flooding – The water is released at high points in the field and distribution is uncontrolled.

irrigation requirement – Quantity of water that is required for crop production, not including effective rainfall.

isohumic – Of soil having an even vertical distribution of organic matter, as in some vertisols.

isomorphous substitution – Replacement of an ion in a crystal lattice by another ion, without major structural changes.

isotropic soil – Soil having the same properties in all directions (cf anisotropic soil).

jadeite – See pyroxene.

jarosite – Bright yellow compound containing S, Fe, K and Si. See sulphuric horizon.

joint – In a mass of rock, a plane of weakness and separation. Facilitates soil formation, often by tonguing, in saprolitic and weathering zones near the earth's surface.

juvenile – In the river cycle, see youth.

kabook – Murram, see laterite.

kame – An irregular mound or cone of stratified glacial drift, often elongated but less so than an esker. See glaciofluvial deposits.

kandites – The kaolin group of clay minerals.

kankar – See caliche.

kaolin – Alumino–silicate minerals with a 1:1 crystal lattice and with an approximate formula of $2Al_2O_3H_2O.2SiO_2$; kaolinite is one example.

kaolisols – Term used in Zaire by INEAC to embrace both sols ferralitiques and sols ferrugineux tropicaux.

karst – Topography characteristic of limestone country with underground river systems, sink-holes, dry valleys and solution hollows which are small basins of internal drainage.

kastanozems – Temperate steppe soils (FAO Soil Legend), also the Russian equivalent of chestnut soils.

kebir – Solonchak with a powdery surface.

kimberlite – Diamantiferous intrusive basic igneous rock, typically in 'pipe' formations.

krasnozems – See latosols.

krotovina – A former animal burrow in one soil horizon that has been filled with organic matter or material from another horizon.

kuroboku – See andosols.

labile – Of a substance which can be easily transformed and/or is readily available to plants.

lacustrine deposit – Lake sediment.

laminar flow – Flow with no cross–currents or eddies; flow of water through granular soils is usually laminar. Also called streamline or viscous flow.

land – Land is a broader term than soil. In addition to soil, its attributes include other physical characteristics such as slopes; drainage pattern; flood risks; erosion hazards; location; parcel size; and existing plant cover.

land capability – Ranking or rating of land based on the severity of land limitations (such as slope, risk of flooding, erosion hazard) for general agricultural use. See land suitability.

land characteristic – Measured or estimated land property, eg slope angle, soil texture, rainfall.

land classification – The classification of land units into classes and/or subclasses based on its suitability for general or specific purpose uses.

land element – The simplest landscape unit, with uniform lithology, form, soil and vegetation (Brink et al, 1966). See also land facet.

land evaluation – Interpretation of soils and topography for land use planning purposes.

land facet – One or more land elements that can be conveniently treated as one block for moderately extensive land use or construction and can be recognised on tone and pattern differentiation on airphotos (Brink et al, 1966). See also land system.

land form – The shape of the land, expressed geomorphologically. This is a generic term, whereas 'landform' in one word is specific, relating to eg an inselberg, a river terrace, an escarpment.

land forming – See land levelling.

land grading – See land levelling.

land levelling – Strictly levelling of land, but also used for shaping land surfaces for better water (and/or machinery) movement. Also called land forming, land shaping or land grading.

land quality – Complex attributes of land with direct effects on specific land uses; estimated or calculated from measurements of land characteristics. Examples are erosion resistance, potential crop yields.

land shaping – See land levelling.

land smoothing – Elimination of small surface irregularities without changing major contours; the final process in land levelling.

land suitability – Ranking or rating of land for a specific use or closely defined uses, taking management into account. See land capability.

land system – An area or group of areas throughout which there is a recurring pattern of topography, soils and vegetation (Christian and Stewart, 1952). One or more land facets make up a land system.

367

land use planning – Planning the uses of land that, over long periods, will best serve project aims and local aspirations, and conserve natural resources.

land utilisation type – Well-defined kind of land use, incorporating management level. Originally Dutch concept, adopted in the FAO framework for land evaluation.

landsat (more properly Landsat) – An American space satellite carrying various remote sensing devices (see MSS and RBV) for the study of land resources. To date there have been five Landsats. See also earth resources technology satellite.

laterite – An ill-defined term usually used as a general term for any horizon or formation consisting predominantly of hydrated iron and aluminium oxides. Formed by weathering under tropical conditions, especially on iron-rich rocks such as basalt. In this sense includes murram, plinthite and ironstone, although it is sometimes used for a restricted range of the less hard sesquioxidic formations.

latosols – A suborder of zonal soils including soils formed under forested, tropical, humid conditions and characterised by low silica–sesquioxide ratios of the clay fractions, low base-exchange capacity, low activity of the clay, low content of most primary minerals, low content of soluble constituents, a high degree of aggregate stability, and usually having a red colour (USDA 1949 classification). Also called krasnozems by the Russians.

lattice structure – The regular arrangement of ions in a crystalline material. Commonly used with reference to clay minerals, which consist of sheets of silica and alumina, sometimes with ions sandwiched between, and usually with some degree of isomorphous substitution.

lava – Molten rock which has flowed from a magma reservoir to the earth's surface through volcanic vents and fissures. When it solidifies basalt, andesite or rhyolite formations are the result.

leaching – The removal of materials in solution or suspension. Particularly concerns removal of soluble salts in reclamation; cf eluviation.

leaching fraction – The proportion of applied irrigation water and rainfall which drains from the soil.

leaching requirement – The proportion of applied irrigation water draining out of a profile required to keep soil salinity, or a particular ion concentration, below a specified level throughout a specified soil depth or at a particular level.

lessivé – Soil in which clay translocation has taken place to produce a Bt horizon.

leucocratic – Of rocks and minerals, light in colour.

levee (or levée) – A raised bank along the side of a river channel in its flood plain, formed by deposition at peak flow periods.

Strips of levee soil are usually well drained, contrasting with the backswamps of the flood plain further from the river.

light soil – (Obsolete in scientific use) A coarse-textured soil; a soil with a low drawbar pull and hence easy to cultivate. See coarse texture and soil texture.

lime (agricultural) – In strict chemical terms, calcium oxide. In practical terms, it is a material containing the carbonates, oxides and/or hydroxides of calcium and/or magnesium used to neutralise soil acidity.

lime requirement – The mass of calcium oxide or the equivalent of other specified liming material, required per unit area to raise the pH of a standard depth of soil to a desired value under field conditions.

limestone – A sedimentary rock composed primarily of calcite ($CaCO_3$). If dolomite ($CaCO_3$:$MgCO_3$) is present in appreciable quantities it is called a dolomitic limestone.

limonite – Group of iron oxides and hydroxides, the weathering products of iron-containing minerals.

liquid limit (LL) – The moisture content (% by weight) at which soil begins to flow.

lithic – Of a soil phase with continuous coherent hard rock at 50 cm depth or shallower (FAO Legend).

lithological discontinuity – Roman numerals are prefixed to soil horizon or layer designations in the USDA Soil Taxonomy system to indicate discontinuities in rocks or in parent materials within and, if necessary, below the solum. The uppermost material is not numbered, I being understood; the contrasting materials beneath are prefixed II, III etc.

lithosols – A great soil group of azonal soils characterised by an incomplete solum or no clearly expressed soil morphology and consisting of freshly and imperfectly weathered rock or rock fragments (USDA 1949 classification and FAO Legend).

lixivium – Obsolescent term for residual, thoroughly weathered and leached soil and regolith.

loam – The textural class name for soil having a moderate amount of sand, silt and clay. Loam soils contain 7 to 27% of clay, 23 to 50% of silt and < 52% of sand (USDA definition).

loamy – Intermediate in texture and properties between fine-textured and coarse-textured soils. Includes all textural classes with the words 'loam' or 'loamy' as a part of the class name, such as clay loam or loamy sand. See loam and soil texture.

loess – Material transported and deposited by wind and consisting of predominantly silt-sized particles with some very fine sand.

luvic – Of soils showing clay illuviation which is not, however, considered to be the dominant soil-forming process (FAO Legend).

luvisols - High base status soils with Bt horizons (FAO Legend).

luxury consumption - The intake by a plant of an essential nutrient in amounts exceeding what it needs, eg if potassium is abundant in the soil, alfalfa may take in more than is required.

lysimeter - A device such as a buried tank, used for measuring percolation and leaching losses from a column of soil under controlled conditions.

macronutrient - A chemical element necessary in large amounts (usually > 1 ppm in the plant) for the growth of plants and usually applied artifically in fertiliser or liming materials ('macro' refers to quantity and not to the essentiality of the element). For most plants N, P, K, Ca and Mg are macronutrients. See micronutrient.

magma - Molten material which exists below the solid rock of the earth's crust and fleetingly is revealed during emissions from volcanoes and fissures.

marble - Metamorphosed limestone.

marbling - Coarse mottling.

marl - Soft and unconsolidated calcium carbonate, usually mixed with varying amounts of clay or other impurities.

marsh - Periodically wet or continually flooded areas with the surface not deeply submerged. Covered dominantly with sedges, cattails, rushes, or other hydrophytic plants. Subclasses include freshwater and saline marshes.

matric potential - Amount of work that a unit quantity of water in a soil-water system is capable of performing when it moves to another system identical in all respects except that there is no soil matrix present. Numerically, it is equal to negative soil-water tension.

matrix - The fine-grained material of rock or soil in which the coarser particles are set, or the predominant material in a mixed deposit or horizon.

mature soil - A soil with well-developed soil horizons produced by the natural processes of soil formation and essentially in equilibrium with its present environment.

maturity - See geomorphic cycle.

meanders - See geomorphic cycle. Winding, twisting river reaches, characteristic of the old-age phase.

mechanical analysis - See particle size analysis.

medium sand - 0.5 to 0.25 mm in equivalent diameter (USDA soil particle size classification).

medium texture - Intermediate between fine textured and coarse textured (includes the textural classes very fine sandy loam, loam, silt loam, and silt).

melanocratic - Of rocks and minerals, dark in colour.

mellow soil - A very soft, very friable, porous soil.

melm - See pseudogley.

mesa - Isolated, steep-sided and flat-topped erosional remnant - larger than a butte (cf inselberg).

metamorphic rock - A rock that has been greatly altered from its previous condition through the action of heat (thermal metamorphism) and/or pressure. For example, marble is a metamorphic rock produced from limestone, gneiss is one produced from granite and slate is produced from shale.

metaquartzite - See quartzite.

micas - Primary monoclinic alumino-silicate minerals in which two silica layers alternate with one alumina layer, with a general formula $(K,Na,Ca)(Mg,Fe,Li,Al)_{2-8}(Al,Sl)_4O_{10}(OH,F)_2$. Micas have a perfect basal cleavage producing a flaky, layered structure and occur particularly in acid igneous rocks, in many metamorphic rocks and in derived sediments. The main varieties are biotite, phlogopite and muscovite. They separate readily into thin sheets or flakes.

micelle - Individual clay particle.

microclimate - The climate occurring over a small area, often used when local conditions (eg of elevation, exposure, vegetation cover) modify general climatic conditions of a region.

microfauna - That part of the animal population which consists of individuals too small to be clearly distinguished without the use of a microscope. Includes protozoa, nematodes, actinomycetes, algae, bacteria and fungi.

micronutrient - A chemical element necessary in only extremely small amounts (< 1 ppm in the plant) for the growth of plants. Examples are: B, Cl, Cu, Fe, Mn, Mo and Zn ('micro' refers to the amount used rather than to its essentiality). See macronutrient.

microrelief - Topography varying over short distances (a few m) and with features whose amplitude of relief is normally measurable in cm, eg gilgai.

migmatite - A very high-grade metamorphic rock in which extremes of temperature and pressure have induced partial melting so that the rock has taken on some of the characteristics of igneous texture.

mineral soil - A soil consisting predominantly of, and having its properties determined predominantly by, mineral matter. Usually contains < 20% organic matter, but may contain an organic surface layer up to 30 cm thick.

mineralisation - The conversion of an element from an organic to an inorganic form as a result of microbial decomposition.

minor element - (Obsolete) See micronutrient.

moderately coarse texture – Consisting predominantly of coarse particles. (In soil textural classification, it includes all the sandy loams except the very fine sandy loam). See coarse texture.

moderately fine texture – Consisting predominantly of intermediate-size (soil) particles or with relatively small amounts of fine or coarse particles. (In soil textural classification, it includes clay loam, sandy loam, sandy clay loam and silty clay loam). See fine texture.

mole drain – Drain formed by drawing cylinder with bullet-shaped head through soil.

mollic epipedon or horizon – A thick, dark surface layer, more than 50% base saturated (dominantly with bivalent cations), having a narrow C/N ratio (17 or less in the virgin state and 13 or less with cultivated soils), a strong soil structure, a relatively soft consistence when dry, and less than 250 ppm of P_2O_5 soluble in citric acid. This horizon characterises the mollisol soil order, but is found in other orders as well (USDA Soil Taxonomy and FAO Legend).

mollisols – Soil characterised by a mollic epipedon which is a dark, well-structured horizon that does not harden greatly on drying. Includes soils such as chernozem, prairie, chestnut in the USDA 1949 classification system (USDA Soil Taxonomy).

montmorillonite – An alumino-silicate clay mineral with a 2:1 expanding crystal lattice consisting of two silicon tetahedral layers enclosing an aluminium octahedral layer. Considerable expansion may be caused by water moving between silica layers of contiguous units.

moorboden – German, histosols.

mor – Raw humus; a surface layer of forest soils consisting of unincorporated humus, usually matted or compacted or both and distinct from the mineral soil (cf mull).

morphology – See soil morphology.

mosaic – An assembly of vertical aerial photographs or space imagery prints whose edges have been trimmed and matched to form a continuous cover of an area. See controlled and uncontrolled mosaic.

mottling – Patches of different colour or shades of colour interspersed with the dominant matrix colour.

MSS – See multispectral scanner.

muck – Highly decomposed organic material in which the original plant parts are not recognisable. Contains more mineral matter and is usually darker in colour than peat. See peat.

mulch – Any material such as straw, sawdust, leaves, plastic film, boulders, flat stones and loose soil that is spread upon the surface of the soil to protect the soil and plant roots from the effects of raindrops, soil crusting, freezing, evaporation etc.

mulch farming – A system of farming in which the organic residues are not ploughed into or otherwise mixed with the soil but are left on the surface as a mulch.

mull – A surface layer of forest soils consisting of mixed humus and mineral matter. A mull blends into the upper mineral layers without an abrupt change in soil characteristics (cf mor).

multispectral scanner system (MSS) – A passive remote sensing device which electronically detects radiation in several wavebands, scanning from side to side across the path of the aircraft or satellite.

Munsell colour system – A colour designation system that specifies the relative degrees of three variables of colour: hue, value and chroma. For example: 10YR 6/4 is a colour (of soil) with a hue = 10YR, value = 6 and chroma = 4. These notations are given standard names. See chroma, hue and value.

murram – See laterite.

muscovite – Coarse-grained white mica, $KAl_2(Si_3Al)O_{10}(OH)_2$.

nanopodzols – Shallow and humus-deficient podzols.

natric horizon – An argillic horizon which has prismatic or columnar structure and is more than 15% saturated with exchangeable sodium. This is common in solonetz and solodised solonetz soils (USDA Soil Taxonomy and FAO Legend).

natural erosion – See erosion.

natural soil – In undisturbed position, either the whole profile (see virgin soil) or lower horizons beneath a cultivated or otherwise manipulated surface.

necrotic spot – Dead part of plant; commonly used of dry discoloured patches on leaves due to effects of salinity, disease etc.

nematodes – Very small worms (usually about 0.5 to 1.5 mm long) abundant in many soils, and particularly in coarse soils; important because many of them attack and destroy plant roots.

neutral soil – A soil which has a pH around 7.0 and is neither acid nor alkaline in reaction (normally applied to surface layer, or root depth unless otherwise stated). See acid soil, alkaline soil, pH and soil reaction.

nickpoint – A place where the gradient of a river bed steepens downstream, signifying past uplift of the basin, then vigorous downcutting by the river in its lower tracts.

nitosols – Tropical soils with Bt horizon and prominent, shiny clay skins (FAO Soil Legend).

nitrification – The biochemical oxidation of ammonium to nitrate.

nitrogen fixation – The conversion of elemental nitrogen (N_2) to organic forms readily utilisable in biological processes.

nodule – See concretion.

non-calcic (or spelt noncalcic) brown soils - As brown soils (qv, USDA 1949 classification) but without calcium carbonate in the profile.

normal erosion - See erosion.

nutrient - See macronutrient and micronutrient.

nyirok - See latosols.

O horizon - Organic horizon forming the surface of some mineral soils; contains ⩾ 35% organic matter which is normally not water saturated (USDA Soil Taxonomy and FAO Soil Legend; cf histic H horizon).

obsidian - Very fine-textured (glassy, shiny) dark-coloured acid volcanic rock.

ochric epipedon or horizon - A light-coloured surface horizon generally low in organic matter which includes the eluvial layers near the surface. It is often hard and massive when dry. Although characteristic of aridisols, it is found in several of the other soil orders (USDA Soil Taxonomy and FAO Legend).

old age - See geomorphic cycle.

olivine - One of a group of orthorhombic silicate minerals, usually dark green and occurring in basic and ultrabasic igneous rocks such as peridotites; an important constituent of the mantle. Olivines vary in composition from Mg_2SiO_4 (forsterite) to Fe_2SiO_4 (fayalite).

oolith - Oolitic limestone is composed of small, subspherical rock particles, ooliths.

opportunity days - The period when land can be worked by a given tillage method, taking into account both climatic and soil conditions.

order - The highest category in the USDA Soil Taxonomy, divided into suborders. Also used for the highest category (suitable, non-suitable) in the FAO land suitability classification.

organic soil - A soil which contains a high percentage (> 20%) of organic matter throughout the solum.

orthic - Common, ordinary, typical.

orthoclase - A monoclinic potassium-rich alkali feldspar, $KAlSi_3O_8$; occurs in all igneous and many metamorphic rocks.

orthoquartzite - See quartzite.

ortstein - An indurated layer in the B horizon of podzols in which the cementing material consists of illuviated sesquioxides (mostly of iron) and organic matter.

osmotic pressure - The pressure exerted in a solution as a result of unequal concentration of salts on either side of a cell wall or semi-permeable membrane. Water tends to move from the more dilute solution through the membrane into the more concentrated solution, thus raising the pressure of the latter.

ouklip - See ferricrete.

outcrop - Strictly, the area over which a rock occurs at, or very close to, the surface; very often used as a synonym for exposure. Also any small rounded or sub-rounded rock fragments.

outlier - A small area of younger rocks occurring within an area of older rocks (cf inlier).

outwash plain - Gentle slope comprising coalesced alluvial fans and deltas of glaciofluvial origin, typically covered by irregularly stratified drift.

oven-dry soil - Soil dried at 105°C to constant weight.

overhead irrigation - See irrigation methods.

oxic horizon - A highly weathered diagnostic subsurface horizon from which most of the combined silica has been removed leaving a mixture dominated by hydrous oxide clays with some 1:1 type silicate minerals and quartz present (USDA Soil Taxonomy and FAO soil Legend).

oxisols - Soils of tropical and subtropical regions characterised by the presence of an oxic horizon. Includes latosols and some groundwater laterites (USDA Soil Taxonomy).

paddy soils, padi soils - See anthropogenic.

pakihi - Gleyed podzols.

paleosols - The lower part(s) of a polygenetic soil profile, including buried soils (qv).

pan - Horizon or layer in soil, strongly compacted, indurated, or very high in clay content. See caliche, claypan, duripan, fragipan, hardpan, ironpan and sesquioxide sheet.

parent material - The weathered mineral or organic matter from which the solum of soils is developed by pedogenic processes.

particle density - The density of soil particles (cf bulk density) which is normally about 2.65 g cm^{-3}; lower and higher particle densities occur for example in organic and iron-rich soils respectively.

particle size - The effective diameter of a soil particle.

particle size analysis - Determination of the proportions (usually as] by weight) of the different size fractions in a soil sample by sedimentation, sieving, micrometry or combinations of these methods.

paternia - Freely draining, non-calcareous, greyish alluvial soil (cf vega).

peat - Unconsolidated and undecomposed, or only slightly decomposed, organic matter accumulated under conditions of excessive moisture.

pebble - Rounded river or beach stone.

ped - A soil aggregate such as a crumb, prism, block etc, formed by natural processes (cf a clod, which is formed artificially).

pedalfer - (Obsolete) A soil order subdivision comprising soils in which sesquioxides increased relative to silica during soil formation (USDA 1938 classification).

pediment – An erosional plain of bedrock, developed between mountain and basin areas in arid zones; may be covered by alluvial or colluvial deposits, sometimes termed pedisediment.

pedocal – (Obsolete) A soil order subdivision comprising soils in which calcium accumulated during soil formation (USDA 1938 classification).

pedogenesis – The process of soil formation.

pedology – The scientific study of the soil.

pelitic – Of rocks, argillaceous.

pellic – Black, or nearly so (of vertisols).

peneplain – A former upland area eroded to a low, gently rolling surface resembling a plain.

penetrability – The ease with which a standard probe can be pushed into the soil. (May be expressed in units of distance, speed, force or work depending on the type of penetrometer used).

perched water-table – A water-table maintained above the normal GWT by an intervening, relatively impervious layer (eg compaction layer, dense clay horizon etc).

percolation – The downward movement of water through soil. Especially, the downward flow of water in saturated or nearly saturated soil at hydraulic gradients of the order of 1.0 or less (cf seepage).

peridotite – A dense ultrabasic igneous rock consisting mainly of olivine.

periglacial – At and near the edge of a present glacier snout or ice sheet; also used of the zone peripheral to the greatest extent of a former glaciation.

permanent charge – The net negative (or positive) charge of a clay particle inherent in its crystal lattice, not affected by changes in pH or by ion-exchange reactions.

permanent wilting point (PWP) – The moisture content of soil, on an oven-dry basis, at which plants wilt and fail to recover their turgidity when placed in a dark humid atmosphere. Corresponds approximately to pF 4.2.

permeability – See soil permeability.

permeation coefficient – Outmoded term for hydraulic conductivity.

petric – Of a soil phase with shallow (< 100 cm) ironstone, not continuously cemented (FAO Legend).

petrocalcic – Of a soil phase with shallow (< 100 cm) indurated caliche (FAO legend).

petroferric – Of a soil phase with shallow (< 100 cm) indurated ironstone (FAO Legend).

petrogypsic – Of a soil phase with shallow (< 100 cm) indurated highly gypseous material, usually > 60% gypsum (FAO Legend).

pF – The logarithm to base 10 of the soil moisture tension expressed as the height in cm of a column of water.

pH – The logarithm to base 10 of the reciprocal of the hydrogen-ion activity of a solution. Used as a measure of acidity/alkalinity.

pH dependent charge – That portion of the total charge of soil particles which varies with changes in pH.

phaeozems – Prairie or steppe soils; chernozem-kastanozem intergrades (FAO Soil Legend).

phase – See soil phase.

phloem – The non-woody conducting tissue in plants, chiefly concerned with the transport of nutrients.

phlogopite – Mica mineral associated mainly with limestone and dolomite.

phosphorus fixation – The essential nutrient P becomes unavailable, in optimal quantities, to plants at low and high pH. Below about pH 5.0 there is a marked enhancement of the activity of Fe, Al and Mn and in the soil, and soluble phosphates are fixed as complex, insoluble compounds of these elements. Above about pH 7.4 insoluble Ca phosphates are formed.

phreatic – Of a soil phase with groundwater between 3 and 5 m from the surface (FAO Legend).

phreatic surface – Groundwater table.

phyllite – A rock similar to slate but coarser grained.

piezometer – Fluid potential measurement device used for measuring soil-water tensions.

pisolith – Large (pea-sized) rounded concretion, 3 to 6 mm in diameter; eg pisolithic ironstone.

plaggen epipedon – Man-made surface layer more than 50 cm thick and produced by long continued manuring. Commonest in the Low Countries and northern Germany.

plagioclase – A triclinic feldspar occurring in most basic and intermediate igneous rocks and many metamorphic rocks, ranging in composition from albite ($NaAlSi_3O_8$) to anorthite ($CaAl_2Si_2O_8$).

planosols – Soils with eluviated surface horizons underlain by B horizons more strongly illuviated, cemented or compacted than associated normal soil. A great soil group of the intrazonal order and hydromorphic sub-order (USDA 1949 classification system and FAO Legend).

plasma – In soil, the material that has been or is being moved, reorganised and/or concentrated by soil-forming processes (cf skeleton grains).

plastic limit (PL) – The moisture content of a soil at which it can just be deformed without rupture.

plastic soil – A soil capable of being moulded or deformed continuously and permanently, by relatively moderate pressure, into various shapes. See consistence.

plateau – Elevated flat to undulating terrain.

platy – Consisting of soil aggregates that are developed predominately along the horizontal axes; laminated, flaky.

plinthic – Of soils containing plinthite within 125 cm of the surface (FAO Soil Legend).

plinthite – A highly weathered mixture of sesquioxides of iron and aluminium with quartz and other diluents which occurs as red mottles and which changes irreversibly to ironstone upon alternate wetting and drying. See laterite.

plough layer – The top 15 to 20 cm of soil normally inverted by ploughing with a mouldboard plough or tilled with other tillage operations.

plutonic rock – Igneous rock formed from magma; usually coarse textured.

poached soil – Puddled soil resulting from trampling by livestock.

podzols – Soils formed in humid climates, mainly cool temperate though there are tropical examples, under coniferous or mixed coniferous and deciduous forest, and characterised particularly by a highly leached, whitish-grey A2 horizon. A great soil of the zonal order (USDA 1949 classification and FAO Legend).

podzolisation – The pedological process resulting in podzols and podzolic soils.

podzoluvisols – Soils intermediate between podzols and luvisols (FAO Soil Legend).

polder – Stretch of flat land reclaimed from the sea.

polygenetic – Of soil material that has accumulated by more than one process and usually over various time spans, eg an alluvial or aeolian cover on residual or colluvial.

pore – See void.

pore-size distribution – The volume of the various size ranges of soil pores. Expressed as percentage of the bulk volume.

porosity – The percentage of the bulk volume of a soil occupied by voids or pores.

potassium fixation – The process in which exchangeable or water-soluble potassium in a soil is converted to slowly or non-available forms.

potential – See soil-water potential.

potential gradient – See energy gradient.

ppm – For solids: the proportion of a substance expressed as the number of units by weight of substance per 1 M units by weight of solid. For solutions: the number of units by weight of solute per 1 M units by weight of solution.

prairie soils – Soils formed in temperate to cool temperate, humid regions under tall grass vegetation. A zonal great soil group (USDA 1949 classification system).

pressure face – Ped face worn by contact with others, especially during moist to wet expansion periods in a shrinking and swelling soil horizon.

primary mineral – A mineral that has not been altered chemically since deposition and crystallisation from the molten state.

prismatic soil structure – A soil structure type with prism-like aggregates aligned approximately vertically.

profile – See soil profile.

psammitic – Of rocks, arenaceous.

psephitic – Of rocks, rudaceous.

pseudogley – Soil wherein surface waterlogging produces gleyed conditions, with no (or very infrequent) connection to the regional water-table.

puddled soil – A compacted soil with massive structure, usually resulting from effects of livestock or heavy machinery acting in wet conditions.

pumice – Volcanic rock that was made light and porous because of sudden release of steam and gases as it was solidifying, the material swelling up frothily.

pyrite – FeS_2, the commonest sulphide mineral.

pyroclastic – Of fragmental rocks from volcanic explosions.

pyroxene – One of a group of ferromagnesian silicate minerals of orthorhombic (orthopyroxenes) or monoclinic (clinopyroxenes) crystal structure. They occur mostly in basic and ultrabasic igneous rocks and certain metamorphic rocks. Typical compositions are $Mg_2Si_2O_6$(enstatite), $CaMgSi_2O_6$(diopside), $NaAlSi_2O_6$(jadeite).

quartz – SiO_2 (silicon dioxide or silica).

quartzite – A rock consisting almost completely of silica and formed either as a sandstone with purely siliceous cement (orthoquartzite) or from the metamorphism of a pure quartz sandstone (metaquartzite).

R horizon – A layer (not strictly a horizon) of continuous, indurated rock.

rambla – Dystrophic, gleyed alluvial soil.

rankers – Shallow soils with umbric A horizons over rock (FAO Soil Legend).

rapids – See geomorphic cycle. Characteristic of youth and to a lesser extent maturity.

RBV – See return beam vidicon.

reaction – See soil reaction.

reclamation – Processes of restoring land to productive use; applied particularly to saline soils or soils affected by industrial disturbance or contamination.

red desert soils – A zonal great soil group consisting of soils formed in warm-temperate to hot, dry regions under desert-type vegetation, mostly shrubs (USDA 1949 classification system).

red earths – Not a recommended term as it is so vague, but often used when referring to well-drained, highly weathered Fe-rich soils in the tropics.

redox potential – Abbreviation for reduction/oxidation potential, the electromotive force measured in millivolts, established at an inert electrode immersed in a solution containing equimolecular amounts of an ion or molecule in two states of oxidation. In aerated soil redox potential is positive; in waterlogged soil it is negative.

red-yellow podzolic soils – Soils formed under warm temperate to tropical, humid climates, under deciduous or coniferous forest vegetation and usually under conditions of good drainage. A zonal great soil group (USDA 1949 classification system).

regolith – The unconsolidated mantle of weathered rock and soil material on the earth's surface; loose earth materials above solid rock.

regosols – Any soils without definite genetic horizons and developing from or on deep, unconsolidated, soft mineral deposits (USDA 1949 classification system and FAO Legend).

regur – Term for vertisols of the Deccan in India.

rejuvenation – Results of geological uplift or sea-level lowering on a land area; evidence includes incision of river meanders, development of V-in-V valleys and nickpoints, retreat of scarps and dissection of plateaux.

relief – The variations in altitude of an area; cf topography.

remote sensing – The use of optical cameras or other sensing devices (eg radar, electronic scanners) mounted in aircraft or satellites to produce pictures corresponding to radiation from the ground surface.

rendzinas – Soils with brown or black friable surface horizons underlain by light-grey to pale-yellow calcareous material; developed from soft, highly calcareous parent material under grass vegetation or mixed grasses and forest in humid and semi-arid climates (USDA 1949 classification system and FAO Legend).

residual – Of soil, formed in place or with only slightly downslope displacement in surface horizons.

return beam vidicon (RBV) – A modified, electronically scanned television camera tube used in remote sensing by satellite as no film is required; usually sensitive to radiation beyond the visible range.

rhizosphere – That portion of the soil in the immediate vicinity of plant roots in which the abundance and composition of the microbial population are influenced by the presence of roots.

rhodic – Of ferralsols with a red to dusky red B horizon (FAO Soil Legend).

rhyolite – Fine-textured acid volcanic rock, mineralogically similar to granite.

rill – A small, erosional water channel with steep sides: usually only a few cm deep.

rill erosion – See erosion.

ripening – The initial pedogenetic processes that render alluvial deposits or peat suitable for agricultural use. Includes, eg in a marine polder, loss of sodium by leaching and gain (a) structure by root ramification of terrestrial plants, and (b) bearing strength for tillage.

river cycle – See geomorphic cycle.

river terrace – A near-horizontal terrace on a valley side, often paired with another on the opposite side, representing part of an older alluvial valley floor before erosion.

Roman two – Colloquial amongst soil surveyors for material below a lithological discontinuity (qv).

rubefaction, rubefication – Accumulation in well-drained soil of iron and other sesquioxides, giving a reddish colour, redder than the parent rock.

rubrozems – In Brazil, humic acrisols.

rudaceous – Of rock, coarse grained, eg breccia, conglomerate.

runoff – Rainwater discharged from an area through stream lines. Surface runoff refers to the water which is lost before entering the soil, and groundwater runoff (or seepage flow) to water which percolates through the soil.

salic horizon – A horizon at least 15 cm thick with secondary enrichment of salts more soluble in cold water than gypsum (USDA Soil Taxonomy and FAO Legend).

saline-sodic soil – A soil containing high levels of both exchangeable sodium and soluble salts. The ESP is > 15, EGe > 4 mS cm^{-1} (at 25°C), and the pH is usually 8.5 or less in the saturated soil.

saline soil – A non-sodic soil containing high levels of soluble salts. The ESP is < 15 and ECe > 4 mS cm^{-1}.

salinisation – The process of accumulation of soluble salts in soil.

saltation – Soil movement by wind or water in which particles bounce along a stream bed or the soil surface.

sand – a) A soil separate consisting of particles 2 to 0.02 mm in equivalent diameter (international system) or 2 to 0.05 mm (USDA system).
b) Soil material containing more than 85% sand (USDA), with the content of silt (USDA) plus 1.5 times the content of clay not exceeding 15%.

sandstone – Medium to coarse quartzose sedimentary rock.

saprolite - Soft, altered rock in the C horizon beneath soil proper; vestiges of the structure of the parent rock are preserved.

satellite imagery - See landsat.

saturation extract - The solution extracted from a soil at its saturation point.

savanna - Tropical grassland with scattered to fairly dense trees and shrubs, grading towards savanna woodland as the tree canopy increases.

scarify - To loosen the topsoil aggregates by means of raking the soil surface with a set of sharp teeth.

scarp or escarpment - The steep drop of a cuesta.

schist - A coarse-grained metamorphic rock in which the minerals have crystallised or recrystallised parallel to each other, formed from intermediate to basic igneous rocks.

scoria - Coarsely vesicular pumice.

scour - Of soil, to erode in shallow broad rills which on coalescing give sheet erosion.

secondary mineral - A mineral resulting from the decomposition of a primary mineral or from the reprecipitation of the decomposition products of a primary mineral (qv).

sedimentary rock - A rock formed from materials deposited from suspension or precipitated from solution.   The principal sedimentary rocks are sandstones, shales, limestones and conglomerates.

seepage - Movement (usually lateral) of water through soil (cf percolation).

self-mulching soil - A soil in which the surface layer becomes so well aggregated that it does not crust and seal under the impact of rain but instead serves as a surface mulch upon drying.

semi-desert - Area with stunted, sparse natural vegetation for most or all of the year, as against true desert where plants appear only after rainstorms which are infrequent and unpredictable.

senile - In the river cycle, see old age.

separate - See soil separate.

sericite - Fine-grained white mica with composition similar to that of muscovite.

series - See soil series.

sesquioxide - Metal oxides of the general formula X_2O_3; usually refers to Fe and Al oxides in soils, with subsidiary Mn and small amounts of Ti etc.

sesquioxide sheet - Cemented ironstone with Al and often Mn and Ti oxides also present.

SG - See specific gravity.

shale - A laminated fine-grained sedimentary rock composed mainly of clay minerals, which splits easily along bedding planes.

shear - Force acting at right angles to the direction of movement.

sheet erosion - See erosion.

sierozems - Soils with pale-greyish A horizons grading into calcareous material at a depth of 1 ft or less, and formed in temperate to cool, arid climates under a vegetation of desert plants, short grass and scattered brush.   A zonal great soil group (USDA 1949 classification system).

silcrete - An indurated horizon cemented by silica (SiO_2).

silica-sesquioxide ratio - The number of molecules of silicon dioxide (SiO_2) per molecule of aluminium oxide (Al_2O_3) plus ferric oxide (Fe_2O_3) in clay minerals or in soils.

silicates - A group of minerals constituting about 95% of the earth's crust, including the feldspars, quartz, olivines, pyroxenes, amphiboles, micas, clay minerals and garnets. They are all based around the highly stable SiO_4 tetrahedral group of atoms.

sill - A sheet formation of igneous rock which conforms to bedding or other structural planes.

silt - a) A soil separate consisting of particles 0.02 to 0.002 mm in equivalent diameter (international system) or 0.05 to 0.002 mm (USDA system).
b) Soil material containing 80% or more silt (USDA) and < 12% clay.
c) Loosely used for alluvial soil of medium to fine texture.

silting - The deposition of water-borne sediments in drains, stream channels, lakes or on flood plains, usually resulting from a decrease in the velocity of the water.

sink-hole - A funnel-shaped hollow in the ground, especially in limestone areas, down which surface water runs into an underground drainage system.

skeletal soils - See lithosols.

skeleton grains - In soil, the coarser sand, gravel and large particles not readily translocated, concentrated or internally affected by pedogenesis (cf plasma).

slaked soil - Soil with advanced structural degradation, tending to be hard when dry, but rapidly rewettable forming very loose, highly erodible soil.

slate - Metamorphosed shale.

slick spots - Small areas up to a few m in diameter of dark, dispersed organic matter arising from a high content of alkali or of exchangeable sodium.

slickensides - Polished, grooved ped surfaces in swelling and cracking clays, often lying diagonally; commonest in the subsoil of vertisols.

slow release - A fertiliser term (also known as delayed release, controlled release, controlled availability, slow-acting and metered release) used to designate a rate of dissolution much lower than for completely water-soluble compounds.   Slow release may involve either compounds which dissolve slowly or soluble compounds coated with substances slowly permeable to water.

slump - Of a soil mass, to move downhill under gravity, usually lubricated by an excess of water in the profile or just beneath it, following exceptional precipitation concentrated into a short time interval. Characteristically there is basal shearing and partial rotation of the slumped material.

smeary - See thixotropic.

smectites - The montmorillonite group of clay minerals.

smolnitza - See vertisols.

snout - The downstream edge of a glacier; may be stationary, advancing or retreating.

sodic soil - A soil that contains sufficient sodium to interfere with the growth of most crop plants, usually taken as soil with ESP ≥ 15. Sometimes confusingly referred to as alkali soil.

sodium adsorption ratio (SAR) - The following ratio of cation concentrations (in me ℓ^{-1}):
$(Na^+)/\sqrt{\{[(Ca^{2+}) + (Mg^{2+})]/2\}}$

soil - The collection of natural materials occupying parts of the earth's surface that may support plant growth, and which reflect pedogenetic processes acting over time under the associated influences of climate, relief, living organisms, parent material and the action of man.

soil aeration - The process by which air in the soil is replaced by air from the atmosphere. In a well-aerated soil, the soil air is very similar in composition to the atmosphere above the soil.   Poorly aerated soils usually contain a much higher percentage of carbon dioxide and a correspondingly lower percentage of oxygen than the atmosphere above the soil. The rate of aeration depends largely on the size, volume and continuity of pores within the soil.

soil air - Air occurring in pores and voids in a soil, usually higher in CO_2 and lower in O_2 than the surrounding air.

soil alkalinity - The degree or intensity of alkalinity of a soil, expressed by a value > 7.0 on the pH scale.

soil amendment - Any substance such as lime, sulphur, gypsum or sawdust used to alter the properties of a soil, generally to make it more productive.   Strictly speaking, fertilisers are soil amendments, but the term is used most commonly for materials other than fertilisers.

soil association - A group of defined soil units (usually soil series) occurring together in a characteristic pattern.   Soil associations very often reflect changes in relief, drainage, parent material and other charcteristics.   See also catena and soil complex.

soil buffer compounds - The clay, organic matter and compounds such as carbonates and phosphates which enable the soil to resist appreciable change in pH as external conditions change.

soil bulk density - The mass of oven-dry soil per unit bulk volume.

soil class - A group of soils having a definite range in a particular property such as acidity, degree of slope, texture, structure, land-use capability, degree of erosion, or drainage. See soil texture and soil structure.

soil classification - The systematic arrangement of soils into groups or categories on the basis of their characteristics.   Broad groupings are made on the basis of general characteristics and subdivisions on the basis of more detailed differences in specific properties.

soil colloid - Organic and inorganic matter with very small particle size and a correspondingly large surface area per unit of mass.

soil complex - A mapping unit of two or more kinds of soil, each of which occur in areas too small or too intricately mixed with other soils to be mapped separately at a given survey scale.   A more intimate and/or more random mixing of soil units than in a soil association or catena.

soil conservation - A combination of all management and land use methods which safeguard the soil against depletion or deterioration by natural or by man-induced factors (eg by erosion or structural degradation).

soil family - In soil classification one of the categories intermediate between the great soil group and the soil series.

soil fertility - The status of a soil with respect to the amount and availability to plants of elements necessary for plant growth.

soil genesis - The mode of origin of a soil with special reference to the processes responsible for the development of the solum from the unconsolidated parent material.

soil horizon - A layer of soil, usually approximately parallel to the soil surface, with distinct characteristics produced by soil-forming processes.

soil management - The sum total of all tillage operations, cropping practices and treatments conducted on or applied to a soil for the production of plants.

soil mechanics and soil engineering - Subspecialisations of soil science concerned with the application of engineering principles to problems involving the soil.

soil moisture tension - See soil-water tension.

soil monolith - A vertical section of a soil profile removed from the soil and mounted for display or study.

soil morphology - The constitution of a soil including the texture, structure, consistence, colour and other physical, chemical and biological properties of the various soil horizons and the thickness and sequence of those horizons, in the soil profile.

soil permeability - The ease with which water passes through the soil. An outmoded synonym for hydraulic conductivity of soil.

soil phase - A subdivision of a soil classification unit with characteristics that affect the use and/or management of the soil, or which do not vary sufficiently to differentiate it as a separate unit. Phases are based on characteristics such as slope, soil depth and stone content; cf soil series, soil variant.

soil profile - A vertical section of a soil.

soil reaction - The acidity or alkalinity of a soil, usually expressed as a pH.

soil salinity - The amount of soluble salts in a soil. Usually measured as a weight % or the electrical conductivity (EC) of a soil suspension or water extract.

soil separate - A size group of mineral soil particles - sand, silt or clay.

soil series - The basic unit of soil classification, being a subdivision of a soil family and consisting of soils which are essentially alike in all major characteristics except the texture of the A horizon; cf soil type, phase, variant.

soil solution - The aqueous liquid phase of the soil and its solutes consisting of ions dissociated from the surfaces of the soil particles and of other soluble materials.

soil structure - The combination or arrangement of primary soil particles into secondary particles, units or peds. The secondary units are characterised and classified on the basis of size, shape and degree of development.

soil texture - The relative proportion of the various soil particle size fractions in a soil.

soil type - The lowest unit in the US system of soil classification; a subdivision of a soil series and consisting of or describing soils that are alike in all characteristics including the texture of the A horizon. See also soil series, phase.

soil variant - Soil closely related to a given soil series but varying in at least one differentiating characteristic at the series level. Usually used for soils of too limited extent to justify establishment of a separate series; cf soil series, phase.

soil-water characteristic - The relationship between the soil-water content (by weight or volume) and the soil-water tension.

soil-water potential - A measure of the freedom of water in soil; a summation of the matric potential (qv) along with pressure, gravitational and solute potential.

soil-water tension - The equivalent negative pressure in soil water. It is equal to the equivalent pressure that must be applied to the soil water to bring it to hydraulic equilibrium, through a porous permeable wall or membrane, with a pool of water of the same composition. See matric potential.

solifluxion - Slow downhill movement of soil by alternate freezing and thawing action (cf creep).

solod - Solodised soil developed from a solonetz.

solodised soil - A soil that has been subjected to the processes responsible for the development of a solod, including leaching of soluble salts and translocation of clay and organic matter (USDA 1949 classification system).

solonchak soils - Soils with high concentrations of soluble salts, usually light coloured, without characteristic structural form, developed under halophytic plants, and occurring mostly in a subhumid or semi-arid climate. An intrazonal group of soils (USDA 1949 classification and FAO Legend).

solonetz soils - Soils with surface horizons of varying degrees of friability underlain by dark hard soil, ordinarily with columnar structure. This hard layer is usually highly alkaline. Such soils are developed under grass or shrub vegetation, mostly in subhumid or semi-arid climates. An intrazonal group of soils (USDA 1949 classification and FAO Legend).

soloth - Solod in America.

soluble sodium percentage (SSP) - The proportion of sodium ions in solution in relation to the total cation concentration, defined as follows:

$$\frac{\text{Soluble sodium concentration (me/} \ell^{-1})}{\text{Total cation concentration (me/} \ell^{-1})} \times 100$$

sols ferralitiques - See ferralitic soils.

sols ferrugineux tropicaux - See ferruginous tropical soils.

solum (plural: sola) - The upper part of the soil profile; the A and B horizons.

solution hollow - See karst.

space imagery - See landsat.

specific gravity (SG) - The ratio of the mass of a substance to the mass of an equal volume of water.

splash erosion - See erosion.

spodic horizon - A subsurface diagnostic horizon containing an illuvial accumulation of organic matter and of free sesquioxides of iron and/or aluminium (USDA Soil Taxonomy and FAO Legend).

spodosols - Soils characterised by the presence of a spodic horizon, an illuvial horizon in which active organic matter and amorphous oxides of aluminium and iron have precipitated. These soils include most podzols, brown podzolics and groundwater podzols of the 1949 classification system (USDA Soil Taxonomy).

sprinkler irrigation – See irrigation methods.

steppe – Mid-latitude grasslands consisting of level, generally treeless plains.   The term is strictly applied to such areas in Eurasia, those in North America being referred to as prairies and in South America as pampas.

sticky point – Moisture content at which soil just fails to adhere to a spatula.

Stokes' Law – Equation relating terminal settling velocity of a smooth, rigid sphere in a given fluid to the diameter of the sphere.   See page 103.

stone – A soil separate consisting of particles 2 to 20 cm in equivalent diameter (international system), 7.5 to 25 mm (FAO system) or, if cobble is used as a particle size name, 25 to 60 cm (USDA system).

stream cycle – See geomorphic cycle.

strip cropping – The practice of growing crops which require different types of tillage, such as row and sod, in alternate strips along contours or across the prevailing direction of wind.

structure – See soil structure.

stubble mulch – The stubble of crops or crop residues left essentially in place on the land as a surface cover before and during the preparation of the seedbed and at least partly during the growing of a succeeding crop.

subgroup – See great group.

subirrigation – See irrigation methods.

suborder – See great group and order.

subsoil – Soil below the plough layer, ie soil below about 20 cm.

subsoiling – Breaking of compact subsoils, without inverting them, with a special knife-like instrument (chisel) which is pulled through the soil, usually at depths of 30 to 60 cm and spacings of 60 to 150 cm.

substratum – In soil science, usually the material beneath the solum.

suglinoks – See degraded chernozems.

sulphidic – Of waterlogged soil containing 0.75% or more S.   On drainage a sulphuric horizon results (FAO Soil Legend).

sulphuric horizon – A soil layer formed as a result of artificial drainage and oxidation of sulphide-rich soil; characterised by great acidity, pH 3.5 or less, and jarosite mottling (FAO Soil Legend).

supervised – Of remote sensing, especially space imagery, that is supported by ground truth observations at key known points.

surface runoff – See runoff.

surface soil – See topsoil.

swelling clay – Clay that has an expanding lattice, normally with montmorillonite or illite dominant.

syenite – Rock similar to granite but richer in alkali feldspars.

syncline – A trough-shaped geological fold in which the younger strata remain at the top of the sequence.

syroloess – Loess soils low in humus.

takyr – Surface soil phenomenon associated often with high Na or Mg.   Soil of fine texture cracks into polygonal elements when dry and forms a platy or massive crust (FAO Soil Legend).

talus – Fragments of rock and other soil material accumulated by gravity at the foot of cliffs or steep slopes.

tasca – See rankers.

tensiometer – A device for measuring water tension in soil in situ.

tension – See soil-water tension.

terra – Latin, Italian and Portuguese for soil or land (but not the only or sometimes the main word now): terra rossa and terra fusca are respectively reddish and yellow-brown residual luvisols on limestone in southern Europe and northern Africa, terra bianca is eluviated of Fe as well as Ca, while terra roxa in Brazil falls largely within nitosols.   Spanish is tierra.

terrace – a) A level, usually narrow, plain bordering a river, lake or the sea.
b) A raised, more or less level or horizontal strip of earth usually constructed on or nearly on a contour and designed to make the land suitable for tillage and/or irrigation and to prevent accelerated erosion.

terrain – Term favoured by engineers, see land.

texture – See soil texture.

thematic map – See base map.

thermal analysis (differential thermal analysis) – A method of analysing a soil sample for constituents, based on a differential rate of heating of the unknown and standard samples when a uniform source of heat is applied.

thermal metamorphism – See metamorphic rock.

thin section – Thin (approx 0.03 mm thick) sample of rock, or resin-impregnated soil, used for microscopic studies.

thionic – High in sulphur.

thixotropic – Literally changing by touch, this term is used of soils with structure that, if broken down, rebuilds itself.   Under deformation the soil changes from a plastic solid to a mass of individual particles, with free water between them, which then within a few seconds reset to their original state.

till - a) Unstratified glacial drift deposited directly by the ice and consisting of clay, sand, gravel and boulders intermingled in any proportion.
b) To plough and prepare for seeding; to cultivate the soil.

tilth - The physical condition of soil as related to its ease of tillage, fitness as a seedbed, and its impedance to seedling emergence and root penetration.

tine - Point or prong (eg of harrow).

tirs - Term for vertisols of Morocco.

tonhauschen - See cutan.

topdressing - An application of fertiliser to a soil after the crop stand has been established.

topography - Strictly, the description of the features (natural and man-made) of an area; very often used as a synonym for relief.

toposequence - A repeating sequence of soils, slopes or vegetation correlated to relief. See also catena and soil association. In toposequences a uniform parent material is sometimes, but not always, implied.

topsoil - The layer of soil moved in cultivation.

torrents - See geomorphic cycle. Evidence of youth or rejuvenation.

trace element - See micronutrient.

transmission coefficient and transmissivity - Outmoded terms for hydraulic conductivity.

transported - Of natural soil, having moved (or currently moving) downslope or downstream.

travertine - See caliche.

trickle irrigation - See irrigation methods.

truncated - Having lost all or part of the upper soil horizon or horizons.

tufa - See caliche.

tuff - Volcanic ash usually more or less stratified and in various states of consolidation.

type - See soil type.

ultisols - Soils of humid areas characterised by the presence of either an argillic horizon or a fragipan each of which is < 35% saturated with bases. They have no spodic horizon and no oxic or natric horizons. They include soils formerly classified as red-yellow podzolics, latosols (in part) and rubrozems (USDA Soil Taxonomy).

ultrabasic - Denoting igneous rocks which are generally plutonic and contain virtually no quartz or feldspar, their main constituents being ferromagnesian silicates, oxides and sulphides. Previously defined as rocks with < 45% silica.

umbric epipedon - A thick, dark surface layer which is less than 50% saturated with bases. Andosols have umbric epipedons (USDA Soil Taxonomy and FAO Legend).

uncontrolled mosaic - A vertical aerial photographic mosaic made without correction of optical or other distortions.

unsupervised - Without ground control.

upper plastic limit - See liquid limit.

urfleinserden - Lithic rendzinas.

USDA soil particle size classes - Fine earth limits differ from the 'international' system (qv) in a widening of the silt class to 0.05 to 0.002 mm and in more than two categories of sand being recognised: there are divisions at: 2 to 1 mm very coarse sand, 1 to 0.5 mm coarse sand, 0.5 to 0.25 mm medium sand, 0.25 to 0.1 mm fine sand and 0.1 to 0.05 mm very fine sand.

value - The relative lightness or intensity of colour and approximately a function of the square root of the total amount of light. One of the three variables of colour. See Munsell colour system, hue and chroma.

variable charge soil - A soil with highly pH dependent colloid charge and hence highly pH dependent CEC.

variant - See soil variant.

varnish - See desert varnish.

varve - A distinct band representing the annual deposit in sedimentary materials regardless of origin and usually consisting of two layers, one a thick, light-coloured layer of silt and sand and the other a thin, dark layer of clay.

vascular bundle - In botany, a strand of conducting tissue comprising xylem and phloem.

vega - Freely draining, non-calcareous, brownish alluvium (cf paternia).

vegetation types - See forests, woodland, savanna, semi-desert and desert.

ventifacts - Desert pavement pebbles which have been bevelled on their windward side(s) until smooth faces result, often with sharp edges between faceted sides.

vermiculite - Clay mineral formed from material that is both micaceous and siliceous: in the process potassium and magnesium are removed.

vertic - Of cambisols and luvisols having seasonal B horizon cracks 1 cm or more wide (FAO Soil Legend).

vertical interval - Vertical distance between adjacent contour lines on map.

vertisols - Soils high in swelling clays which crack widely upon drying, resulting in shrinking, shearing and soil mass movement (USDA Soil Taxonomy and FAO Legend).

very coarse sand - 2 to 1 mm in equivalent diameter (USDA soil particle size classification).

very fine sand - 0.1 to 0.05 mm in equivalent diameter (USDA soil particle size classification).

vesicular – Of rock containing vesicles, unconnected voids with smooth walls, often round or elliptical.

vidicon – See return beam vidicon.

vignetting – A camera effect (especially with wide–angle lenses) resulting in darkening of a photographic print towards the edges; corrected by use of an antivignetting filter.

virgin soil – A soil that has not been significantly disturbed from its natural environment.

vitric – Glassy.

vivianite – Highly hydrated iron oxide, bluish as a pure mineral.

void – In soil or rock, space filled by air and/or water, between solid material. Same as pore.

volcanic rock – Ejected material from a vent or fissure in the earth's crust. See lava and pumice. Other products of eruptions include volcanic dust and volcanic ash (see tuff).

vugh – Unconnected soil void of irregular shape, with irregular walls.

vulcanism – Volcanic action.

wadi – Seasonal watercourse, usually dry.

warp – Synonym for alluvial soil, or may be restricted to freely draining alluvium.

wash – A colluvial process faster than creep (qv) or an alluvial deposition on moderate to gentle slopes influenced primarily by gravity and water action.

water – See soil water.

water duty – See irrigation requirement.

waterfall – See geomorphic cycle. Characteristic of youth and early maturity stages.

watershed – High ground between different drainage networks (cf interfluve).

water-stable aggregate – A soil aggregate stable to the action of water such as falling drops, or agitation as in wet-sieving analysis.

water-table – The upper surface of groundwater or that level below which the soil is saturated with water; locus of points in soil water at which the hydraulic pressure is equal to atmospheric pressure; see also perched water-table.

weathering – Physical and chemical changes produced in rocks, at or near the earth's surface, by weathering agents such as rain, wind, temperature changes, etc.

wet mottling – With greys and bluish tinges prominent, in a gleysol or gleyed horizon; subsidiary coloration usually yellow–orange–brown.

white micas – See muscovite and sericite.

wiesenboden – See humic gley soils.

wild flooding – See irrigation methods.

wilting point – See permanent wilting point.

windbreak – A planting of trees, shrubs or other vegetation, usually perpendicular or nearly so to the principal wind direction, to protect soil, crops, homesteads, roads, etc, against the effects of wind, such as wind erosion and the drifting of soil and snow.

woodland – In the tropics, intermediate between forest and savanna.

xanthic – Of ferralsols with a yellow to pale yellow B horizon (FAO Soil Legend).

xerophyte – Plant that grows in or on extremely dry soils or soil materials.

xerosols – Arid–zone soils with a weak ochric A horizon (FAO Soil Legend).

xylem – Wood, botanically the combination of vessels (mostly tubular) with lignified walls, and fibres, which permits aqueous solutions to be conducted to various parts of the tree or shrub.

yardangs – Fantastically shaped erosion features in Asiatic semi–deserts, where the wind has excavated long passageways (parallel to the dominant wind direction) separated by carved 'cockscomb' ridges, often capped by earth pillars.

yellow podzolic soils – See red–yellow podzolics.

yerma – Infertile, powdery soils of hot deserts.

yermosols – Arid–zone soils with a very weak ochric A horizon (FAO Soil Legend).

youth – See geomorphic cycle.

zeolite – Mineral inclusion filling voids in volcanic rock, emplaced after cooling. Zeolite crystals are mainly alumino–silicates, with affinities to kaolin.

zheltozems – In Russia, yellowish latosols, much less common than the reddish krasnozems.

zick – Synonym for solonchak.

zonal soils – Soils characteristic of a large area or zone (USDA 1949 classification system).

Annex M

Climate Classification

Contents

Annex M: Climate classification

List of Figure and Table

Climate Classification

M.1   The Köppen classification

M.1.1   Introduction   The Köppen (1923) system considers precipitation effectiveness for plant
growth as the major classification factor, and uses the appropriate seasonal values of
temperatures and precipitation in order to determine the limits of climatic groupings.
Köppen originally observed the conditions of growth required by various groups of plants,
ranging from the megatherms, which favour warm habitats, through the intermediate mesotherms,
to microtherms which thrive in a colder environment; by relating these to various
temperature limits he defined certain main climatic groups.   Other main groups are the arid
and semi-arid lands, with subgroups on a temperature basis, and those with long winters and
short summers - the polar and near-polar regions, mostly beyond the limits of tree growth,
but also defined by a mean temperature figure.

M.1.2   Main groups   The five main groups of the Köppen classification are:

A.   Tropical rainy: A hot climate with no cool season; the average temperature
                            of each month is over 18°C.

B.   Dry: Evaporation exceeds precipitation.

C.   Humid mesothermal 1/: The warmest month has a mean temperature above 10°C.   The
                            coldest month has a mean temperature between -3° and 18°C.

D.   Humid microthermal 1/: The warmest month has a mean temperature above 10°C.   The
                            coldest month has a mean temperature below -3°C.

E.   Polar: A polar type with no month averaging over 10°C.

The temperature value for the D type, for example, recognises that a 10°C average for the
warmest month coincides roughly with the poleward limits of forest; and that the -3°C mean
for the coldest month is near the equatorwards limit of frozen ground and where snow remains
on the surface for at least a month.

M.1.3   Subgroups   A second, small, letter is used to describe the rainfall distribution.   In the
A climatic regions f indicates that no month has a mean rainfall of less than 60 mm; w shows
that at least one month has less than this amount.   The values for the Af tropical rainy
climate are those which Köppen believed would support a tall tropical rain forest.   Other
small letters are used, such as m for a monsoon region which supports a rain forest despite a
short dry season.

The B climates (dry ones) are also subdivided into a BW, arid, type (W = Wüste, desert) and
BS, semi-arid or steppe type.   The BS/BW limits are identified from formulae, using
different constants combined with the rainfall and temperature values for each sub-type.
Other small letters are used as third symbols to aid description: thus BWh and BWk
(h = heiss, k = kalt) represent deserts where mean annual temperatures are over and under
18°C respectively.   Various minor characteristics of the C, D and E climates are also shown
by means of additional symbols.

M.1.4   Limitations   It must be stressed that in Köppen's form of empirical classification no
attempt is made to take account of the causes of the climate described, nor of the relations
between the location of the climatic region and those of pressure zones, air mass source
regions or other features.   Trewartha stresses the need of supplementing empirical
classifications with some description of the mode of origin of the climatic features used to
define the boundaries.

_____

1/   Note minimum of -3°C modified to 0°C by Trewartha (1968), see Table M.1.

Annex M: Climate classification

M.1.5   Climatic types   Some of the chief characteristics of various types of climate are described in Table M.1. For this purpose Trewartha's modification of Köppen's system is used. The climates are considered in groupings which are not entirely zonal; but neither are the patterns of the climates, soils and vegetation over the earth's surface. A fairly simple two-letter grouping is used, but further detail is given in some cases by the use of a third-order symbol. The map of climatic regions based on this classification (Figure M.1) shows each of these groupings.

M.2   The Thornthwaite classification 1/

During the early 1930s C W Thornthwaite put forward and modified a system of climate classification which also looks upon plants as 'meteorological instruments'. He derived a measure of the effectiveness of precipitation - a P/E Index - by dividing the total monthly precipitation (P) by the total monthly evaporation (E), and then adding the 12 values for each month. The main drawback, again, is that the actual evaporation data are not available for many parts of the world. To overcome this, Thornthwaite used observations, from various stations in the south-west USA, of temperature and precipitation, and obtained a formula which served to give the P/E Index: this formula was then used for other parts of the world. With these P/E values available he distinguished five 'humidity provinces', each with characteristic vegetation.

Humidity province	Vegetation	P/E Index
A. Wet	Rain-forest	128 and above
B. Humid	Forest	64-127
C. Subhumid	Grassland	32-63
D. Semi-arid	Steppe	16-31
E. Arid	Desert	under 16

Small letters are used for seasonal concentration of precipitation: r = abundant in all seasons; s = deficient in summer; w = deficient in winter; d = deficient in all seasons.

Thornthwaite also developed a formula to be used for calculating thermal efficiency - T/E Index - and, with values again grouped from 0 (F') to 128 and over (A'), recognised six temperature provinces:

A' - tropical	B' - mesothermal	C' - microthermal
D' - taiga	E' - tundra	F' - frost

The climate is then described by presenting the groups together. Thus the climate of the savannas of Northern Nigeria is shown as CA'w. In fact 120 possible combinations can be distinguished but, in making a world map of climatic regions, Thornthwaite chose 32 of these. Examples from a number of different regions are:

Amazon Lowlands	BA'r	Sicily	CB's
Eastern Sao Paulo	BB'r	Aquitaine	BB's
Northern Chile	EB'd	Western Britain	BC'r
Southern Chile	AC'r	East Anglia	CC'r

Thornthwaite (1948) produced a classification which considered the losses of water by evaporation from soil and plant cover as 'evapotranspiration'. Regarding the plant as 'the machinery of evaporation', he used a formula to ascertain the loss by evaporation and transpiration which would occur if water were always available to the roots. The loss is, however, calculated as a function of the temperature.

The potential evapotranspiration is then compared with the precipitation, and climatic boundaries defined on that basis. Again, the drawback is the difficulty of obtaining values for loss of moisture to the atmosphere; hence the need for the calculation in terms of temperature. The chief use of the potential evapotranspiration values, however, has been for single stations: curves drawn for monthly values can be compared with curves of precipitation, so that periods of water surplus or water deficiency can be gauged.

1/   Adapted from Money (1974).

Trewartha's modification of the Köppen system   Table M.1

Group A   Tropical rainy climates – temperature of the coolest month over 18°C

   Af   No dry season – driest month has over 60 mm.   Intertropical convergence zone (ITCZ), with Tm (or Em)
         air masses – see footnote

   Am   Short dry season – but rainfall sufficient to support rain forest (wet monsoon type)

   Aw   Dry during the period of low sun (winter) – driest month under 60 mm.   Dry Tc air in winter.   Wet
         during period of high sun – when ITCZ moves polewards and moist Tm air flows in

Group B   Dry climates – evaporation exceeds precipitation

         W – arid;   desert.   S – semi-arid;   steppe.   Boundaries between BW and BS are identified by formulae –
         eg where rainfall is evenly distributed, and r = annual rainfall (mm), and t = mean annual temperature
         (°C) the boundary between BW and BS is the isohyet r = 9.9 t + 70

   BWh   Hot desert – mean annual temperature over 18°C.   Source regions to Tc air: subtropical highs.   Dry Tc
         trade winds

   BSh   Tropical and subtropical semi-arid – a short rainy season – Tc air for most of the year

   BWk   Middle-latitude interior desert – Tc air masses in summer and Pc air masses in winter.   Large annual
         temperature range.   Persistently dry

   BSk   Middle-latitude semi-arid – dominated by dry Pc air in winter;   mainly Tc air in summer;   meagre
         rainfall, mostly in summer

         n (nebel) is used to show frequent fogs along coastlands with cool waters offshore

Group C   Humid mesothermal (moist temperature) – temperature of coldest month between 18°C and 0°C

   Cs   Subtropical, dry summers – at least three times as much rain in wettest winter month as in driest summer
         month.   Driest month < 30 mm.   Summers dominated by subtropical highs (the Cs regions lie on stable
         eastern side of the highs).   Pm air in winter, with cyclonic storms and rain

   Css   Hot summers – warmest month averages over 22°C

   Csb   Warm summers – warmest month averages under 22°C

   Ca   Humid subtropical, hot summers – warmest month over 22°C.   In summer moist Tm air from the unstable
         western side of the subtropical high over ocean.   Winters with Pc air invading, and cyclonic storms
         developing

   Caf   No dry season – driest month over 30 mm

   Caw   Dry winters – at least 10 times the rain in wettest summer month as in driest winter month

   Cb   Marine climate, cool-warm summers – warmest month under 22°C.   Mostly middle-latitude west coasts which
         receive moist Pm air and series of depressions.   Rain in all seasons

   Cbf   No dry season – the most common Cb type (Cbw describes parts of south-east Africa)

   Cc   Marine climate, short cool summers – warmest month below 22°C;   < 4 months over 10°C, rain in all seasons

Group D   Humid microthermal (rainy/snowy, cold) – temperature of coldest month under 0°C, and warmest month over 10°C

   Da   Humid continental, warm summers – warmest month over 22°C.   Precipitation in all seasons, accent on
         summer maximum;   winter snow cover.   Zone of frequent clashes between polar air and tropical air.
         Variable weather

   Db   Humid continental, cool summers – warmest month under 22°C.   As for Da, but long winter snow cover

   Dc   Subarctic climate – warmest month below 22°C, and < 4 months over 10°C.   Winters are cold, and air
         stable under Pc air mass.   Pm air sometimes gives cyclonic storms, especially in summer;   but
         precipitation is light.   Evaporation is small in winter, hence moisture remains

   Dd   Subarctic with very cold winters – coldest month below –38°C.   Very light precipitation

For the C and D groups: f – no dry season;   w – dry winter

Group E   Polar – temperature of the warmest month < 10°C

   ET   Tundra – mean temperature of warmest month above 0°C.   Pm, Pc and Arctic air masses interact, and
         cyclonic storms occur, with light precipitation, mostly in summer

   EF   Ice cap – perpetual frost;   no month with mean temperature over 0°C.   Source regions for Arctic and
         Antarctic air masses

Group H   Highland climates – undifferentiated

Note: Air mass abbreviations – E Equatorial, T Tropical, P Polar, m maritime, c continental.

Source: Adapted from Trewartha (1968).

Annex M: Climate classification

Figure M.1

World climate - Köppen system

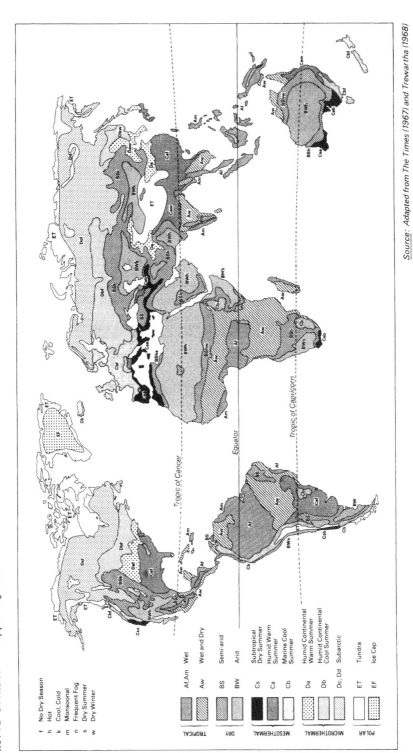

Source: Adapted from The Times (1967) and Trewartha (1968)

Properties of selected rocks 1/

Rock type	Appearance	Texture	Composition 2/	Potential contributions to soil

A. Sedimentary rocks

Breccia	Cemented, very coarse, angular fragments	Coarse	Variable	Variable, usually stony/gravelly material
Conglomerate	Cemented, very coarse, rounded fragments	Coarse	Variable	Variable, usually stony/gravelly material
Limestone	White to grey or pale reddish/brownish/yellowish	Medium to coarse	Calcite or dolomite; sometimes OM	Moderately high pH material, with high Ca and/or Mg as carbonates and/or exchangeable cations
Sandstone	As limestone, sometimes darker, often buff to pale brown	Medium to coarse	Quartz	Low pH material, low in nutrients
Shale	Variable colour, with characteristic layered structure, cleavage planes	Fine	Clay minerals, quartz; sometimes OM	Variable fine-textured material, depending on minerals present

B. Igneous rocks 3/

Andesite 4/	Light	Fine	Feldspars ($\approx$ 75%) and ferromagnesian minerals ($\approx$ 25%)	Neutral material
Basalt	Dark, smooth	Fine, sometimes glassy	Feldspars ($\approx$ 50%) and ferromagnesian minerals ($\approx$ 50%)	Fine-textured, high pH material
Diorite	Light	Coarse	As andesite	Neutral material
Gabbro	Dark, smooth	Coarse	As basalt	High pH material
Granite	Pale to speckled	Coarse	Feldspars (50%), quartz (30%) and ferromagnesian minerals (20%)	Acidic material with some more rapidly weathering pockets at higher pH
Obsidian	Dark, shiny	Very fine (glassy)	Feldspars (60%), quartz (30%) and ferromagnesian minerals (10%)	Acidic material
Rhyolite	Light	Fine to very fine	As granite	Acidic material

C. Metamorphic rocks

See rocks above using following key:

Metamorphic rock	Parent rock which has been altered
Gneiss	Granite
Marble	Limestone
Quartzite	Sandstone
Schist	Igneous rocks with much high pH to neutral soil-forming material
Slate	Shale

Notes:
1/ Illustrations of many common rocks and minerals are given in Hamilton et al (1974) and Zim and Shaffer (1963).
2/ Composition figures are rough averages; ranges can be great - eg granite can have feldspar contents of 35 to 65%.
3/ See also Figure N.2.
4/ Not to be confused with the mineral andesine.

Annex N

Tabulated Geological and Mineralogical Data

Contents

Geological time-scale

Eon	Era	Period		Epoch	Time: Ma [1]	
Phanerozoic	Cenozoic	Quaternary		Holocene	0.01	65
				Pleistocene	1.8	225
		Tertiary	Neogene	Pliocene	5	
				Miocene		570
				Oligocene	26	
			Palaeogene	Eocene	37	1000
				Palaeocene	53	
	Mesozoic	Cretaceous		late Cretaceous	65	
				early Cretaceous	100	
		Jurassic		late Jurassic	136	
				middle Jurassic	160	
				early Jurassic	176	
		Triassic			190	
	Palaeozoic	Permian			225	2000
		Carboni- ferous		Pennsylvanian	280	
				Mississippian	315	2500
		Devonian			345	
		Silurian			395	3000
		Ordovician			430	
		Cambrian			500	
Proterozoic	Pre-Cambrian	Vendian			570	
		Riphean		late Riphean	650	
				middle Riphean	900	4000
				early Riphean	1300	
		early Proterozoic			1600	
Archaean		Archaean			2500	
					4550	4550

Note: (1) Ma = millions of years.

Source: Smith (1981).

Key to composition of igneous rocks

	Low colour index			Medium colour index			High colour index			
nepheline syenite	syenite	granite	grano-diorite	diorite	gabbro	olivine gabbro	peridotite	dunite	**Coarse-grained rocks**
	micro-syenite	micro-granite	micro-grano-diorite	micro-diorite	diabase (dolerite)	olivine diabase (dolerite)			**Medium-grained rocks**
phonolite	trachyte	rhyolite	rhyolite	andesite	basalt	olivine basalt			**Fine-grained rocks**

Note: See also Table N.3.

Source: Hamilton et al. (1974).

Classification of normal (calc-alkaline) igneous rocks

Table N.2

Rock type	Amount of SiO_2 (%)	Extrusive (fine grain)	Hypabyssal (medium grain)	Plutonic (coarse grain)	Minerals 1/	SG
Acid	> 65	Rhyolite dacite	Quartz and orthoclase porphyries	Granite granodiorite	Quartz, orthoclase, Na-plagioclase, muscovite, biotite (± hornblende)	2.67 ±
Intermediate	55-65	Pitchstone andesite	Plagioclase porphyries	Diorite	Plagioclase, biotite, hornblende, quartz, orthoclase (± augite)	2.72 ±
Basic	45-55	Basalt	Dolerite	Gabbro	Ca-plagioclase, augite (± olivine, ± hornblende)	2.9-3.2
Ultrabasic	< 45	Various basic olivine basalts	Various basic dolerites	Picrite peridotite serpentinite dunite	Ca-plagioclase, olivine (± augite)	3.0-3.5

Note: 1/ See also Figure N.2.

Source: Adapted from McLean and Gribble (1979).

General identification characteristics for classification of igneous rocks

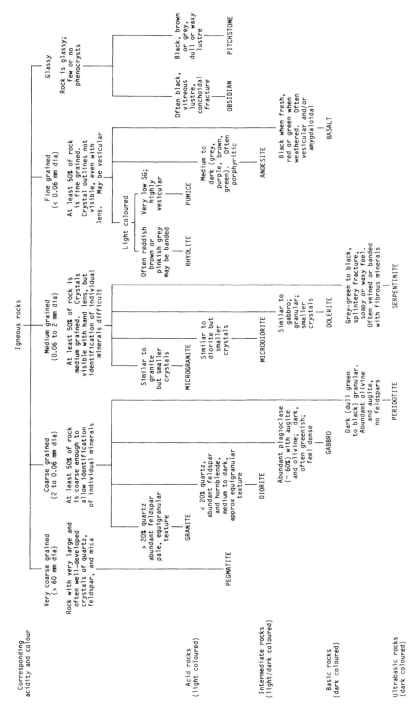

Source: Adapted from Clayton et al (1982).

General identification characteristics for common minerals in igneous rocks   Table N.4

Mineral	Identification characteristics		
Quartz	Colour	–	often colourless and transparent; it appears light grey in the hand specimen
	Hardness	–	7 scratches glass
	Fracture	–	conchoidal
	Lustre	–	vitreous
	General	–	in acid rocks quartz commonly occupies the interstices between other minerals
Orthoclase	Colour	–	commonly white to flesh-pink
	Cleavage	–	two perfect cleavages, but this may not be easily recognised in fine- and medium-grained rocks
	Fracture	–	conchoidal to uneven
	Lustre	–	vitreous, pearly parallel to cleavage
	General	–	crystals frequently appear equidimensional in the general ground mass of igneous rocks. Large well-developed crystals are common fortunately, as the mineral cannot be recognised when crystals are small or poor
Plagioclase	Colour	–	commonly white or off-white and translucent
	Cleavage	–	two good cleavages not easily recognised in small crystals
	Fracture	–	uneven
	Lustre	–	vitreous, pearly on cleavage surfaces
	General	–	crystals are commonly lath-shaped and easily recognised in coarse-grained rocks
Muscovite mica	Habit	–	crystals are commonly tabular (platy)
	Colour	–	commonly colourless or pale-grey
	Cleavage	–	perfect basal; the flakes of mica can easily be separated with a fingernail or a sharp implement
	Hardness	–	2.5 to 3 easily scratched with fingernail
	Lustre	–	vitreous, often pearly on cleavage surfaces
Biotite mica	Habit	–	crystals are commonly tabular (platy)
	Colour	–	commonly dark-brown with a distinctive coppery appearance
	Cleavage	–	perfect basal; the flakes of mica can easily be separated with a fingernail or a sharp implement
	Hardness	–	2.7 to 3.3 easily scratched with fingernail
	Lustre	–	vitreous, submetallic on cleavage surfaces
Augite and hornblende	it is often difficult to distinguish augite and hornblende in the hand specimen unless the crystals are large. The two minerals are normally identified together by the dark-green or black colour of the crystals		
Olivine	Colour	–	commonly olive-green, in some cases they may be yellowish, brownish or black
	Hardness	–	6.5 to 7
	Fracture	–	conchoidal
	Lustre	–	vitreous

Source: Clayton et al (1982).

A classification of sedimentary materials   Table N.5

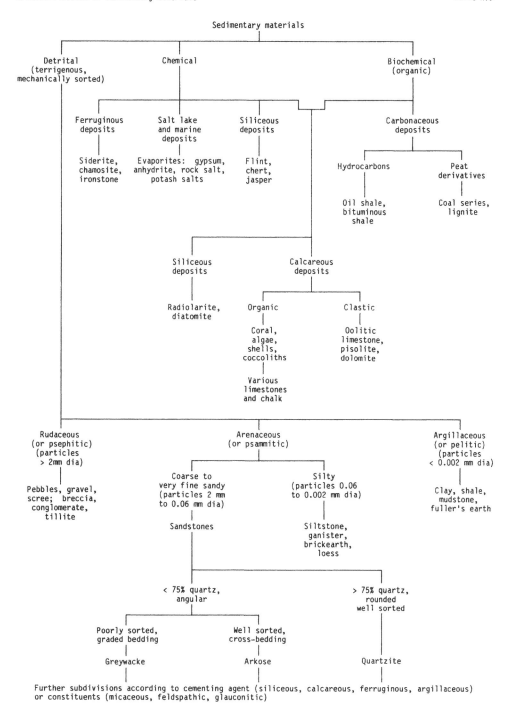

Sources: McLean and Gribble (1979), Blyth and De Freitas (1974).

Annex N: Geology and mineralogy

Classification of sedimentary rocks

Table N.6

Detrital sediments		Pyroclastic sediments	Chemical and organic sediments	
Bedded		Bedded	Bedded	Massive/bedded
Quartz, rock fragments, feldspar and other minerals	At least 50% of the rock is composed of carbonate minerals (rocks usually react with dilute HCl)	At least 50% of the grains are of fine-grained volcanic material. Rocks often composed of angular mineral or igneous rock fragments in a fine-grained matrix	Crystalline carbonate rocks depositional texture not recognisable. Fabric is non-clastic	Depositional textures often not recognisable
A. Coarse-grained rocks (2.0–0.06 mm diameter grains)				
Rock is composed of more or less rounded grains in a finer-grained matrix: CONGLOMERATE Rock is composed of angular or subangular grains in a finer-grained matrix: BRECCIA	CALCI-RUDITE	Rock is composed of: a) Rounded grains in a fine-grained matrix: AGGLOMERATE b) Angular grains in a fine-grained matrix: VOLCANIC BRECCIA	Rock is crystalline composed of calcium carbonate (> 90%) reacts with dilute HCl LIMESTONE Rock is crystalline and may show a yellowish coloration and/or the presence of voids. Reacts mildly with cold dilute HCl. Reaction increased by heating the HCl DOLOMITIC LIMESTONE Rock is crystalline and composed of magnesium carbonate (> 90%). When small chip of rock immersed in dilute HCl, there is no immediate reaction, but a slow formation of CO_2 beads on surface of chip; reaction slowly accelerates. Rate of reaction is increased by heating HCl DOLOMITE	Rock is crystalline, salty to taste and may be scratched with the fingernail HALITE (rock salt) Rock is crystalline and may be scratched with the fingernail. Grains turn into a chalky white substance when burnt for a few minutes GYPSUM
B. Medium-textured rock (0.06-0.002 mm diameter grains)				
Rock is composed of: a) mainly mineral and rock fragments SANDSTONE b) 95% quartz. The voids between the grains may be empty or filled with chemical cement QUARTZ SANDSTONE c) 75% quartz and rock fragments and up to 25% feldspar (grains commonly angular). The voids may be empty or filled with chemical cement ARKOSE d) 75% quartz and rock fragments together with 15% + fine detrital material ARGILLACEOUS SANDSTONE	CALCI-ARENITE	Rock is composed of mainly sand-sized angular mineral and rock fragments in a fine-grained matrix: TUFF		Rock is crystalline; colourless to white, frequently with a bluish tinge. It is harder than gypsum and has three orthogonal cleavages ANHYDRITE Rock is black or brownish black and has a low specific gravity (1.8–1.9). It may have a vitreous lustre and conchoidal fracture and/or breaks into pieces that are roughly cuboidal COAL
C. Find and very fine grained (< 0.002 mm diameter grains)				
Rock is composed of at least 50% fine-grained particles and feels slightly rough to the touch SILTSTONE Rock is homogeneous and fine grained. Feels smooth or only slightly rough to the touch MUDSTONE Rock has same appearance and feel as mudstone but reacts with dilute HCl CALCAREOUS MUDSTONE	CALCI-SILTITE CHALK (Bioclastic)	Rock is composed of silt-sized fragments in a fine- to very fine-grained matrix. Matrix and fragments may not always be distinguished in the hand specimen FINE-GRAINED TUFF		Rock is black or various shades of grey and breaks with a characteristic conchoidal fracture affording sharp cutting edges. The rock cannot be scratched with penknife FLINT Rock has similar appearance and hardness as flint but breaks with a more or less flat fracture CHERT
Rock is composed of at least 50% very fine grained particles and feels smooth to the touch CLAYSTONE Rock is finely laminated and/or fissile. It may be fine or very fine-grained SHALE	CALCI-LUTITE	VERY FINE-GRAINED TUFF		

Source: Adapted from Clayton et al (1982).

Classification of metamorphic rocks   Table N.7

Massive rock	Appearance
HORNFELS	Rock contains randomly orientated mineral grains (fine to coarse grained). Foliation, if present, is poorly developed. This rock type is essentially a product of thermal metamorphism associated with igneous intrusion and is generally stronger than the parent rock
MARBLE	Rock contains > 50% calcite (reacts violently with dilute HCl), is generally white in colour with a granular texture
DOLOMITIC MARBLE	Rock where the major constituent is dolomite instead of calcite (dolomite does not react immediately with dilute HCl)
GRANULITE	Rock is medium to coarse grained with a granular texture and is often banded. This rock type is associated with regional metamorphism
QUARTZITE (METAQUARTZITE)	Rock consists mainly of quartz (95%) grains which are generally randomly orientated giving rise to a granular texture

Foliated rocks	Grain size	Appearance
MIGMATITE	Coarse-grained ($\geqslant$ 2 mm)	Rock appears to be a complex intermix of metamorphic schists and gneisses and granular igneous rock. Foliations tend to be irregular and are best seen in field exposure
GNEISS		Rock contains abundant quartz and/or feldspar. Often the rock consists of alternating layers of light-coloured quartz and/or feldspar with layers of dark-coloured biotite and hornblende. Foliation is often best seen in field exposures
SCHIST		Rock consists mainly of large platy crystals of mica, showing a distinct subparallel or parallel preferred orientation. Foliation is well developed and often nodulose
PHYLLITE	Medium-grained ($\geqslant$ 0.06 mm)	Rock consists of medium- to fine-grained platy, prismatic or needle-like minerals with a preferred orientation. Foliation often slightly nodulose due to isolated larger crystals which give rise to a spotted appearance
SLATE	Fine-grained (< 0.06 mm)	Rock consists of very fine grains (individual grains cannot be recognised in hand specimen) with a preferred orientation such that the rock splits easily into thin plates

Source: Adapted from Clayton et al (1982).

Annex N: Geology and mineralogy

Main minerals contributing to soils

Mineral group and examples	Chemical formulae	Occurrence/forms/weathering/comments	Potential contributions to soil
A. Primary minerals			
Amphiboles		Easily weathered; widespread in metamorphic and igneous rocks	Ca, Mg, Na some Fe
hornblende	$NaCa_2(Mg,Fe^{2+})_4(Al,Fe^{3+})$ $(Si,Al)_8O_{22}(OH,F)_2$		
Carbonates		Easily weathered	
calcite	$CaCO_3$	Common in sediments, altered basic igneous rocks	Ca (exchangeable and as carbonate); non-acid soil materials
dolomite	$CaMg(CO_3)_2$	Common in sedimentary rocks, altered limestone	As calcite, plus Mg
Feldspars		Commonest occurring minerals; form clays when highly weathered	
albite	$NaAlSi_3O_8$	Coarse, intermediate igneous rocks with alkali feldspars	Na, clay
orthoclase	$KAlSi_3O_8$	Commonest feldspar; essential component of acid igneous rocks	K, clay
plagioclase	$Na(AlSi_3O_8)-Ca(Al_2Si_2O_8)$	Metamorphic and igneous rocks	Na, Ca, clay
Iron oxides			
goethite	$FeO(OH)$	)Most common iron oxides in igneous	Fe
haematite	Fe_2O_3	)rocks and sediments	Fe
limonite	$Fe_2O_3+FeO(OH)$ (hydrated)	)Mainly in igneous and contact	Fe
magnetite	$Fe_3O_4.SiO_2$	)metamorphic rocks; replacement)deposits	
Micas		Almost transparent sheet structures; form clays when weathered	Mg, K, clay
biotite	$K(Mg,Fe^{2+})_3(AlSi_3)O_{10}$ $(OH,F)_2$	Comparatively easily weathered; many igneous and metamorphic rocks	K, clay
muscovite	$KAl_2(AlSi_3O_{10})$ $(OH,F)_2$	Granite, pegmatite, schist, gneiss, detrital material	K, clay
Pyroxenes		Easily weathered; ultrabasic and basic volcanic and plutonic rocks	Ca, Mg, some Fe
augite	$(Ca,Mg,Fe,Al)_2(Al,Si)_2O_6$		
Silica		Most resistant to weathering; main constituents of sands	
chalcedony (flint)	SiO_2	'Low temperature' mineral; common in sediments; fibrous or ultrafine quartz	sand
quartz	SiO_2	Acid igneous rocks; many metamorphic rocks; sedimentary rocks	sand
Other primary minerals			
apatite	$Ca_5(F,Cl)(PO_4)_3$	Igneous rocks (especially pegmatites); metamorphic limestones	P
gibbsite	$Al(OH)_3$	Often an end-product of weathering in the tropics	Al
gypsum	$CaSO_4.2H_2O$	Easily weathered; evaporites, clays and limestones; associated with S	Ca, S
pyrite	FeS_2	Commonest sulphide mineral; igneous and contact-metamorphic rocks; anaerobic sediments	Fe, S
tourmaline	$Na(Mg,Fe^{2+})_3Al_6(BO_3)_3$ $(Si_6O_{18})(OH,F)_4$	Occurs in small quantities, but is only source of soil B; acidic rocks and schists, gneisses	B
zircon	$ZrSiO_4$	Very slowly weathered; more acid igneous rocks	
B. Secondary (or clay) minerals 1/			
illite	$K_{1-1.5}Al_4(Si,Al)_8O_{20}$ $(OH)_4$	Alteration of micas, alkali feldspars etc in high pH conditions	K
kaolinite 2/	$Al_4Si_4O_{10}(OH)_8$	Weathering of feldspars etc	
montmorillonite 3/	$NaMgAl_5Si_{12}O_{10}(OH)_6$	Common initial weathering stage in the tropics from basic to neutral parent material	Mg

Notes: 1/   See also Table N.10.
         2/   A member of the kandite group.
         3/   A member of the smectite group.

Major and trace elements in common rocks and minerals   Table N.9

Mineral or rock type	Major constituents	Common trace constituents
A. Sedimentary rocks		
Iron ores	Fe	P, Mo
Limestones and dolomites	Ca, Mg, Fe	Mn, Pb
Manganese ores	Mn	K, B, Co, Cu, Zn, Pb
Salt deposits	K, Na, Ca, Mg	B, I
Sandstones	Si	Various
Shales and bituminous shales	Al, Si, K	Mo, Cu, Co, B
B. Igneous rocks		
B.1 Easily weathered		
Andesine	Ca, Na, Al, Si	Cu, Mn
Anorthite	Ca, Al, Si	Cu, Mn
Apatite	Ca, P, F	Pb
Augite	Ca, Mg, Al, Si	Co, Mn, Zn, Cu, Pb
Biotite	K, Mn, Fe, Al, Si	Co, Mn, Zn, Cu
Hornblende	Mg, Fe, Ca, Al, Si	Co, Mn, Zn, Cu
Oligoclase	Na, Ca, Al, Si	Cu
Olivine	Mg, Fe, Si	Co, Mn, Zn, Cu, Mo
B.2 Moderately stable		
Albite	Na, Al, Si	Cu
Garnet	Ca, Mg, Fe, Al, Si	Mn, Cr
Ilmenite	Fe, Ti	Co, Cr
Magnetite	Fe	Zn, Co, Cr
Muscovite	K, Al, Si	
Orthoclase	K, Al, Si	Cu
Titanite	Ca, Si, Ti	
Tourmaline	Ca, Mg, Fe, B, Al, Si	
Zircon	Zr, Si	
B.3 Very stable		
Quartz	Si	

Source: Adapted from Mitchell (1964).

Clay mineral properties 1/

Mineral group and examples	Lattice structure	Mineral sheets (and main substitutions) 2/	Particle size	Stickiness	Plasticity	Shrink and swell characteristics	CEC range (me/100 g clay)
1. Kandite group 3/	1:1 (non-expanding lattice)	1 silica, 1 alumina					
kaolinite		(little or no substitution)	Small	Slight	High	None	2-15
halloysite		(little or no substitution)	Medium	Slight	High	None	2-15
2. Hydrous mica group	2:1 (slight or moderate expanding lattice)	2 silica, 1 alumina					
illite		(Al for Si in silica sheets; Mg for Fe in alumina sheet; charge balanced by interstitial K)	Medium	Medium	High	Very slight	10-40
vermiculite		(as illite, but more Al substitution, and hydrated Mg in place of interstitial K)	Fine	Medium	Very high	Medium	80-150
3. Smectite group 4/ montmorillonite	2:1 (expanding lattice)	2 silica, 1 alumina (some Al for Si in silica sheets; Fe and Mg for Al in alumina sheet; and exchangeable, hydratable cations interstitially)	Very fine	Very fine	Very high	Very great	80-150
4. Chlorite group	2:2	2 silica, 1 alumina, 1 interlayer magnesia (Al for Si in silica sheets; Al and Fe for Mg in interlayers)	Fine to medium	Medium	High	Medium to slight	Variable
5. Palygorskite or fibrous mica group attapulgite sepiolite	2:1 (modified, chain silicates; uncommon in most soils)						

Notes: 1/ Clay minerals are also referred to as secondary minerals.
      2/ Substitution = isomorphous substitution in crystal lattice.
      3/ Other members include dickite, nacrite.
      4/ Other members include nontronite, beidellite; latter low CEC 65-90 me/100 g clay.

Sources: Sopher and Baird (1978); Baver et al (1972).

Moh's scale of hardness

10 Diamond	Carbon	
9 Corundum	Alumina	
8 Topaz	Aluminium silicate	
7 Quartz	Silica	Scratches glass
6 Feldspar	Alkali silicate	Scratched by a file
5 Apatite	Calcium phosphate	
4 Fluorspar	Calcium fluoride	Scratched by a steel knife
3 Calcite	Calcium carbonate	Scratched by a copper coin
2 Gypsum	Hydrated calcium sulphate	Scratched by a fingernail
1 Talc	Hydrated magnesium silicate	

Annex N: Geology and mineralogy

Annex P

Tabulated Fertiliser Data

Contents

Composition and effects of main commercial fertilisers

Table P.1

Fertiliser	Common abbreviation (USA)	Principal fertiliser compounds	Approximate proportions of main elements (%)							Acidifying effect 1/ (kg CaO per kg N applied)	Comments
			N	P	K	Ca	Mg	S	Cl		
1. N fertilisers											
Ammonia											) Also have fungicidal
anhydrous	–	NH₃	82	–	–	–	–	–	–	-1.0	) effects; toxic
aqua	–	NH₃	15-30	–	–	–	–	–	–	-1.0	) effects on roots in) high concentrations
Ammonium chloride	–	NH₄Cl	28	–	–	–	–	–	67	-1.0	
Ammonium nitrate	AN	NH₄NO₃	33-34	–	–	–	–	–	–	-1.0	
Ammonium nitrate limestone	ANL	NH₄NO₃ CaMg(CO₃)₂	20-26	–	–	8.2	4.4	0.4	–	0 (22% N) -0.4 (26% N)	
Ammonium nitrate sulphate	ANS	NH₄NO₃ + (NH₄)₂SO₄	30	–	–	0.3	–	12	–	-2.0	
Ammonium sulphate	AS	(NH₄)₂SO₄	21	–	–	–	–	24	–	-3.0	Because of acidic reaction, often replaced by urea for tropical soils
Calcium cyanamide	–	CaCN₂	22	–	–	40	0.1	0.3	–	+1.7	Also has herbicidal effect (used against weeds)
Calcium nitrate	–	Ca(NO₃)₂	15	–	–	19	1.5	–	0.2	+1.0	
Sodium nitrate	–	NaNO₃	16	–	0.2	0.1	0.1	0.1	0.6	+1.0	S deficiency may occur where used as AS substitute, so SCU needed
Urea	–	CO(NH₂)₂	46	–	–	–	–	–	–	-1.0	
Urea formaldehyde	UF	CO(NH₂)₂ + HCHO	38	–	–	–	–	–	–	-1.0 approx	
Urea S-coated	SCU	CO(NH₂)₂ + S	35-40	–	–	–	–	7-10	–	-2.0 approx	See urea note above

General effects include: – possible changes (±) in soil pH
  – increased biological activity
  – salt damage if large quantities are used
  – possible problems caused by minor constituents and/or by-products, eg release of NH₃ from NH₄ fertilisers and its toxic
  effect on seedlings
  – effects of components other than N in commercial fertilisers

Notes: See page 402.

cont

399

Annex P: Fertiliser data

Fertiliser	Common abbreviation (USA)	Principal fertiliser compounds	Approximate proportions of main elements (%)							Acidifying effect 1/ (kg CaO per kg N applied)	Comments
			N	P	K	Ca	Mg	S	Cl		
2. P and PCa fertilisers											
Basic slag	-	Ca silico carnotite mixed	-	4-8	-	29-32	3	0.3	-		
Bone meal	-		2-4	9-12	0.2	20-25	0.4	0.1	-		
Calcium metaphosphate	CMP	Ca polyphosphate + Ca2P2O7	-	27	-	19	-	-	-		
Dicalcium phosphate	-	CaHPO4.2H2O + CaHPO4	-	17-23	-	19-29	-	-	-	} Variable, depending on reactions in soil and rate of conversion. Water-soluble types tend to acidify soil	
Fused tricalcium phosphate	FTP	Ca3(PO4)2	-	12	-	20	-	-	-		
Phosphoric acid	-	H3PO4	-	22-26	-	0.2	-	-	-		
super acid	-	H3PO4 + polyphosphoric acids	-	32-34	-	-	-	-	-		
Phosphate rock	PR	apatites	-	11-17	-	33-36	-	-	-		
Superphosphates											
ordinary (or single)	OSP	Ca(H2PO4)2.H2O + CaSO4.2H2O	-	7-9	0.2	13-20	0.2	12	-		} Used also as source of S
concentrated	CSP	Ca(H2PO4)2.H2O	-	21	-	12-14	-	0.8	-		} S deficiency may occur when used as OSP replacements
triple	TSP	Ca(H2PO4)2.H2O	-	18-22	0.3	9-14	0.3	1.4	-		
Potassium metaphosphate	-	KPO3	-	24-25	29-32	-	-	-	-		

General effects include: — pH lowering effects of water-soluble P fertiliser normally only of interest in supply of micronutrients; very low pH values
(< 2) can occur in immediate vicinity of fertiliser granules
— formation of Ca phosphate in soil can reduce pH lowering effects
— decreasing availability of heavy metals with very high phosphate levels
— effects of components other than P in commercial fertilisers

Fertiliser	Common abbreviation (USA)	Principal fertiliser compounds	N	P	K	Ca	Mg	S	Cl	Acidifying effect	Comments
3. K, NK and PK fertilisers											
Potassium bicarbonate	-	KHCO3	-	-	39	-	-	-	-		
Potassium carbonate	-	K2CO3	-	-	56	-	-	-	-		
Potassium chloride ('muriate of potash')	-	KCl	-	-	50-52	0.1	-	-	48	} Very weak acidifying effects of negligible importance	Also contains 20-30% NaCl
Potassium magnesium sulphate ('sulfomag')	-	K2SO4.2MgSO4	-	0-2	18-22	0.1-0.7	11	11-18	-		
Potassium nitrate ('nitrate of potash')	-	KNO3	13	-	37	0.3	0.3	0.2	1.2		
Potassium phosphate	-	K3PO4	-	18-22	29-45	-	-	-	-		
Potassium polyphosphate	KPP	K polyphosphates	-	26	32	-	-	-	-		
Potassium sulphate ('sulphate of potash')	-	K2SO4	-	-	40-43	-	0.1	18	-	"	For use with Cl-sensitive crops

Notes: See page 402.

Table P.1 cont

Fertiliser	Common abbreviation (USA)	Principal fertiliser compounds	N	P	K	Ca	Mg	S	Cl	Acidifying effect [1] (kg CaO per kg N applied)	Comments
4. NP fertilisers											
Ammoniated CSP	–	$NH_4H_2PO_4 + CaHPO_4$	5	20	–	12-14	0.7	–			
Ammoniated OSP	–	$NH_4H_2PO_4 + CaHPO_4 +$ apatite									
Ammoniated TSP	–	$(NH_4)_2HPO_4 +$ apatite	4	6		16	0.3	10			
Ammonium phosphate nitrate	APN	$NH_4H_2PO_4 + NH_4PO_3 + (NH_4)_2HPO_4$	9	18-22	0.3	14.0	0.3	1.4	0.3		
Ammonium phosphate sulphate	APS	$NH_4H_2PO_4 + (NH_4)_2SO_4$	30	4	–	–	–	2-15	–	Acidifying effects due to NH_4: see (1) above	
Ammonium polyphosphate	APP	$NH_4H_2PO_4 + (NH_4)_3HP_2O_7$	13-16	9-21	–	–	–	–	–		
Diammonium phosphate	DAP	$(NH_4)_2HPO_4$	12-15	24-26	–	–	–	–	–		
Monoammonium phosphate	MAP	$NH_4H_2PO_4$	18-21	20-23	–	–	–	–	–		
Nitric phosphates	NP	$CaHPO_4 + Ca(NO_3)_2 + NH_4H_2PO_4 +$ apatite	11	21	–	1.5	0.2	4.5	–		Usually replaced by APN or UAP because of cost or environmental problems
Urea ammonium phosphate	UAP	$CO(NH_2)_2 + (NH_4)_2HPO_4$	14-20	6-9	–	6-7	–	0.4	–		
Urea ammonium polyphosphate	UAPP	$CO(NH_2)_2 + (NH_4)_3HP_2O_7 + NH_4H_2PO_4$	25-34	7-15	–	–	–	–	–		
Urea phosphate	UP	$CO(NH_2)_2 + H_3PO_4$	22-30 / 18	13-19 / 20	–	–	–	–	–		
5. Mg fertilisers											
Magnesium ammonium phosphate	–	$MgNH_4HPO_4$	8	17	–	–	14	–	–		
Magnesium carbonates and limestones (dolomites)	–	$MgCO_3 \pm CaCO_3$	–	–	–	various	various	–	–		Slow-acting soil dressing
Magnesium chloride	–	$MgCl_2$	–	–	–	2	8-9	–	–		Used as foliac spray and soil dressing; care with Cl^--sensitive crops
Magnesium oxide	–	MgO	–	–	–	–	≥ 42	–	–		Slow-acting foliar nutrient
Magnesium silicate glass	–	Mg silicates	–	10	–	20	8-12	–	–		Very slow-acting soil dressing; must be finely ground
Magnesium sulphate epsom salt	–	$MgSO_4.7H_2O$	–	–	–	–	10	–	–		Readily soluble; used for foliar sprays
kieserite	–	$MgSO_4.H_2O$	–	–	–	–	17	–	–		Less soluble, but suitable for soil dressing

Notes: See page 402.

cont

Annex P: Fertiliser data

Fertiliser	Common abbreviation (USA)	Principal fertiliser compounds	Approximate proportions of main elements (%)							Acidifying effect [1] (kg CaO per kg N applied)	Comments
			N	P	K	Ca	Mg	S	Cl		
6. Ca and S fertilisers											
Calcium chloride solution	–	$CaCl_2 \cdot 6H_2O$	–	–	–	18	–	–	–		
	–	$CaCl_2$	–	–	–	10	–	–	–		
Calcium sulphate (gypsum)	–	$CaSO_4 \cdot 2H_2O$	–	–	–	23	–	18	–	Variable rise in pH	Only slightly water-soluble

Notes: [1] Methods for estimating acidity effects are given in AOAC (1975) and Finck (1982); CaO equivalent x 1.748 gives $CaCO_3$ equivalent.
Some examples of fertiliser application rates are given in Table F.6.
+ = causes pH increase
– = causes pH decrease

Sources: Terman (1982), Finck (1982), Olson et al (1971), and Tisdale and Nelson (1966).

Average composition of some common natural organic materials   Table P.2

Organic material	Composition (% by weight)						
	N	P	K	Ca	Mg	S	Cl
Activated sewage sludge	6.0	1.0	–	1.8	0.9	0.4	0.5
Blood (dried)	13.0	–	–	0.4	–	–	0.6
Bone meal (raw)	3.5	19.8	–	22.5	0.6	0.2	0.2
Bone meal (steamed)	2.0	12.2	–	23.6	0.3	0.2	–
Castor pomace	6.0	0.6	0.4	0.4	0.3	–	0.3
Cocoa meal	4.0	0.6	2.1	0.4	0.6	–	–
Cocoa shell meal	2.5	0.4	2.5	1.1	0.3	–	–
Cocoa tankage	2.5	0.6	1.0	12	–	–	–
Cottonseed meal	6.6	1.1	1.2	0.4	0.9	0.2	–
Fish scrap (acidulated)	5.7	1.3	–	6.1	0.3	1.8	0.5
Fish scrap (dried)	9.5	2.6	–	6.1	0.3	0.2	1.5
Garbage tankage	9.5	0.6	0.8	3.2	0.3	0.4	1.3
Peanut meal	7.2	0.6	1.0	0.4	0.3	0.6	0.1
Peanut hull meal	1.2	0.2	0.7	–	–	–	–
Peat	2.7	–	–	0.7	0.3	1.0	1.1
Peruvian guano	13.0	5.5	2.1	7.9	0.6	1.4	1.9
Process tankage	8.2	–	–	0.4	–	0.4	–
Soyabean meal	7.0	0.5	1.3	0.4	0.3	0.2	–
Tankage (animal)	7.0	4.3	–	11.1	0.3	0.4	0.7
Tobacco stems	1.5	0.2	4.2	3.6	0.3	0.4	1.2
Whale guano	8.5	2.6	–	6.4	0.3	–	–

Manure source	N	P	K	Organic
Dairy manure	0.7	0.1	0.5	30
Goat manure	2.8	0.6	2.4	60
Horse manure	0.7	0.1	0.4	60
Pig manure	1.0	0.3	0.7	30
Poultry manure	1.6	0.5	0.8	50
Rabbit manure	2.0	0.6	1.0	50
Sheep manure	2.0	0.4	2.1	60
Steer manure	2.0	0.2	1.6	60

Source: Fairbridge and Finkl (1979), from Tisdale and Nelson (1966); cf Table P.3.

Major nutrient contents for some organic fertilisers   Table P.3

Material	Composition (% by weight)			
	N	P	K	Moisture
Animal wastes (% of fresh material)				
Farmyard manure 1/ (range)	0.3–2.2	0.04–0.92	0.4–1.2	8–86
(mean)	0.6	0.13	0.5	76
Beef feedlot manure	0.6–3.5	0.3–0.9	0.8–2.4	5–40
Broiler litter (range)	0.4–3.6	0.1–1.7	0.3–2.0	9–75
(mean)	2.3	0.9	1.0	32
Cattle slurry	0.2–1.9	0.04–1.1	0.08–1.8	86–89
Swine slurry 2/	0.2–0.6	0.01–0.4	0.03–0.7	85–98
Urban wastes (% of dry matter)				
Activated sewage sludge	5.6	2.5	0.3	–
Digested sewage sludge	2.0	0.5	0.3	–
Miscellaneous (% of dry matter)				
Guano	16–20	5–6	2–3	–
Dried blood	10–14	–	–	–
Meat and bone meal	5–12	< 8	–	–
Bone meal	0.4–4.5	6.5–14.5	–	–
Rape cake	5	1	1	–
Spent mushroom compost	0.6–1.1	0.2–0.4	0.3–1.2	–
Peat moss	0.8	0.1	0.1	–
Cereal straw	0.6	0.1	1.2	–
Seaweed	0.6–2.0	0.1	0.9–2.1	–

Notes: 1/   Also contains (in µg g^{-1} fresh material): Cu 60; Mn 150; Zn 200.
         2/   Also contains (in µg g^{-1} fresh material): Cu 4–370; Zn 2–330.

Source: Fairbridge and Finkl (1979); cf Table P.2.

Annex P: Fertiliser data

Annex Q

Tillage Implements

Contents

Tillage Implements

Q.1.   Primary tillage

This is the first and deepest tillage operation, and forms the basis of the seedbed. The effects of badly timed and/or executed primary tillage operations are extremely difficult to correct with secondary tillage implements, and the end result is a less than optimum yield. The properties of the main primary tillage implements are summarised in Table Q.1.

Q.2   Secondary tillage

This operation is designed to refine and consolidate the seedbed to a condition suitable for successful germination and unimpeded growth. The general rule is 'a fine seedbed for fine seeds' and a seedbed, particularly in the tropics, should always be left as coarse textured as the following crop can tolerate. It is also worth remembering that a fine seedbed provides ideal conditions for weed seed germination. With wide-spaced crops such as maize and cotton the ideal is to cultivate only the strip that is going to be planted, and leave the inter-row in a rough ploughed state to withstand runoff, erosion and the germination of weeds. Far too often a seedbed is over-refined purely because it looks better; this is a waste of money, and has no agronomic value. The properties of some of the main secondary tillage implements are summarised in Table Q.2.

Whenever possible, in order to save fuel and money, and utilise tractor power more efficiently, two implements should be coupled together, eg disc harrows and spike tooth harrows, drills and harrows etc. The other advantage of doing this is in areas where soil moisture is at a premium, and seedbeds must be prepared with minimum moisture loss.

Q.3   Field operations

With the present high costs associated with operating farm machinery it is important to ensure that equipment is used in the most economic and productive way. The work pattern will vary from field to field, according to its shape, but the general rule is to operate on the longest runs possible. Having decided on the direction of operation it is no use starting in the middle and working out to each side, as eventually half the time will be spent in idle running from one end of the plot to the other. In order to reduce idle running time on the headland it is necessary to divide the field into 'lands'. In the days of horse ploughing a land was approximately 10 m (11 yd), so every 10 m the ploughman would make an 'opening', and when the field was completed there was both an 'opening' and a 'finish' every 10 m.

Nowadays, with wide implements, 'lands' are much wider, and a four-furrow one-way plough may have openings every 60 m. In operation the ploughman would 'gather' (operate in a clockwise direction) round his opening until he had ploughed 15 m on each side (a total of 30 m), having done the same at the next opening he will be left with a 30 m strip of unploughed land between the openings which is then ploughed out by 'casting' (operating in an anticlockwise direction) until he ends up with a 'finish'. Using this example, the longest 'idle' run on the headland is 30 m, and this system should be employed in the operation of all field equipment, adjusting the width of the 'land' to suit the width of the implement.

Q.4   Tyres and traction

When purchasing a tractor the selection of the correct tyre size is of vital importance, as it is the means by which a unit transmits its power to the ground. Wrong tyre equipment can severely limit the amount of usable power of a tractor.

Annex Q: Tillage implements

In general, tyres fall into two main categories – narrow section and wide section. Narrow section tyres are best for hard conditions, or hard conditions with an overlay of soft mud as the tyre can bite through the mud and gear into the hard soil underneath. Wide section tyres give flotation in soft soils, and reduce the risk of compaction. Tyre diameter is also important, and the larger the diameter the better the flotation, as more of the tyre is in contact with the ground. The ground bearing pressure, which normally coincides with tyre inflation pressure, would be approximately 2.0 kgf cm^{-2} (28 psi) at the front and 1.3 kgf cm^{-2} (18 psi) at the rear of a 2WD drive tractor.

The effects of compaction are seriously compounded, and can cause longer-lasting damage to soil structure, if smearing takes place due to tractor wheelslip and/or badly adjusted implements. Smearing, which is a compacting and slipping effect in the soil, is often caused by bad timing of operations, and working in soils that are too wet. It can also be caused by increasing the draft of an implement by wrong settings, and by poor tractor operation through wrong gear selection. A low gear at high engine revs transmits too much power to the wheels and induces slip, which can be rectified by selecting the next higher gear and reducing engine speed, whilst maintaining forward speed.

Flotation can be improved by fitting dual wheels or, for certain secondary tillage operations, by fitting cage wheels. Wheelslip can be reduced by water ballasting the rear tyres, and/or fitting wheelweights. Wheelslip can also be caused by ill-matched equipment. The case of too large an implement for the tractor power is obvious, but an implement can also be too small in the case of fully mounted implements, where the tractor is relying on increased traction from the force transferred to its back axle from the weight of the implement and its action in the soil. Note that up to about 19% wheelslip cannot be seen by the unaided eye.

Q.5   Tractor and implement compatibility

In order to control field operation costs it is important to ensure that implement size is matched to tractor type, horsepower, and the soils in which they are going to operate. If the decision is taken to increase horsepower when a tractor fleet becomes due for replacement, it is also necessary to replace the implements in order to allow full exploitation of the increased power, and to control field costs.

Most secondary tillage implements, such as disc harrows and tined cultivators, rely on a high travel speed to operate correctly, and if the implement is too wide for the tractor power, forward speed will be limited, work quality will be poor, and added expense will be incurred in carrying out further passes with the implement to achieve the desired finish. In matching implements to tractor power, and soil type, the correct balance must be made between implement width and operating speed. If in doubt, it is better to sacrifice some implement width, and achieve improved outputs through a higher operating speed.

Q.6   Machinery outputs

A simple calculation of machinery outputs can be made using the following formula:

$$\text{Machinery output (ha } h^{-1}) = \frac{S \times W \times E}{1\,000}$$

where   S = speed (km h^{-1})
         W = implement width (m)
         E = field efficiency (%)

Example:

(60 hp) tractor operating a five-furrow (35.6 cm furrow width) mouldboard plough at 8 km h^{-1}, with a field efficiency of 60% will have an output of:

$$\frac{8 \times 1.78 \times 60}{1\,000} = 0.85 \text{ ha } h^{-1}$$

Q.7   Reducing the risk of soil compaction, smearing and possible damage to soil structure

The risk of soil compaction, smearing and soil structural damage can be reduced by following these guidelines:

a)   Establish, by analysing rainfall data, the period available for seedbed preparation when the work is least likely to be interrupted by rain. Ensure that sufficient power/work units are available to complete the work programme within this period.

b)   If work is interrupted by rain, allow sufficient time for drying out before recommencing work.

c)   Ensure that the correct implements are being used for seedbed work in relation to the soil type. Use tined implements whenever possible.

d)   Train operators in the correct settings and operation of the implements.

e) In each cycle vary slightly the depth of primary tillage work to prevent the build-up of a plough pan.

f) Use 'on-land' pull implements, as opposed to 'in-furrow' type implements.

g) Do not overcultivate, and whenever possible reduce the number of passes over the land by linking different implements together.

h) In order to keep the land in good heart avoid any form of 'brute force tillage' when the soil is not in a condition to be cultivated (too wet, too dry etc).

i) With certain vegetable and cereal crops adopt a 'bed' or 'tramline' method of cultivation where tractor wheels always pass over the same, clearly defined paths.

j) In order to reduce ground bearing pressures fit dual wheels or cage wheels to the tractor. Run tractor tyres at the minimum permissible pressure. (In some operations rear tyre pressure can be reduced to 1 kg cm^{-2}, or 14 psi).

k) Never cultivate deeper than the following crop requires.

l) With transport units fit high flotation tyres, and dual axles if they are high capacity. Ensure that the rear tractor tyres and those of the trailer have the same wheel centre, and 'track' each other.

m) Avoid inducing tractor wheelslip through operating in a low gear at high engine revs.

n) When operating inter-row tillage implements fit a tine in line with the centre of the tractor wheel track, set at sufficient depth to alleviate any effect of compaction.

o) Many mouldboard ploughs can be fitted with a tine behind each body, which passes about 75 mm below the furrow bottom and reduces the risk of plough pan formation.

p) Certain field operations such as seeding, fertiliser application and spraying may have to be carried out in wet field conditions, and consideration should be given to using special low ground pressure vehicles (less than 5 psi). Aeroplanes or helicopters can also be used if fields are large enough.

Annex Q: Tillage implements

List of Tables

Primary tillage implements

Table Q.1

Implement	Variations	Soil action	Operating depth	Comments
Subsoiler	Single, double and triple shank models to match tractor horsepower. Normally fitted with angled 'wings' at the heel to increase lifting and shattering effect. Inter-shank chisel tines may be fitted to improve soil shatter action. Mole draining plug may be fitted to the heel. Slant subsoilers such as the Howard 'Paraplow' appear to give more constant depth of soil disturbance across the full implement width	Used to break up plough pan which inhibits root development and the movement of water. Also used to increase water-holding capacity, reduce run-off and the risk of erosion. Improves drainage. For the soil to heave and shatter sub-soiler must be operated when soil moisture content is low. This is not so critical with slant subsoilers	In virgin soils can be as deep as 1 000 mm, but normally just below ploughing depth, say 300 to 450 mm. In fields with subsurface drainage it is important to keep above the drainage tiles or pipes	The commonest fault is subsoiling when the soil moisture content is too high, which smears the soil, impedes the the movement of water, and causes more damage than it alleviates. Also subsoiling is often carried out as a matter of routine, whether it is required or not. Dig a few pits in the field and establish if there is a compaction problem; if there is, subsoil at the minimum depth required to alleviate it. Operating too deeply wastes time, fuel and money
Mouldboard plough	Single to 12 furrow models to suit different tractors, with furrow widths 300 to 460 mm. Plough bodies can be divided into two main groups, semi-digger which can be recognised by its long mouldboard, and digger which has a short, deep mouldboard. Either one-way or reversible models are available. The reversible plough leaves a level seedbed as there are no 'openings' or 'finishes' and is more suited to a semi-skilled operator, although the plough requires skilled 'set-ting-up'. One-way ploughs are cheaper to purchase. Different manufacturers produce a wide range of mouldboard designs including slatted mouldboards (metal strips with gaps between) which give better scouring in stiff clays	This varies from partial inversion of the furrow slice by the semi-digger base to complete inversion and deep burial of trash, weed seeds etc by the deep digger. The work of the semi-digger can be varied by the operating speed. At a slow forward speed, say 5 km h^{-1}, an unbroken furrow slice will be produced, whilst at around 9 km h^{-1} it will produce typical broken furrow work. The unbroken furrow is suited to 'weathering' (autumn ploughing for a spring seedbed), whilst broken work reduces the costs of producing an immediate seedbed	As a general guide the semi-digger base will operate to a depth of around two-thirds of its furrow width, whilst the digger base is capable of operating to a depth equal to furrow width. The normal operating depth of a semi-digger plough is around 220 mm. In Holland, the first tillage operation carried out on reclaimed land is deep digger ploughing to a depth of 1 200 mm. Similar work is carried out in certain areas of Yugoslavia. Operating depth should be altered each year to reduce the risk of plough pan formation	No longer considered the only implement to produce, economically, conditions suitable for plant growth. Badly adjusted ploughs can cause smearing at the furrow bottom, as can the tractor wheel running in the furrow, which leads to plough pan formation. In tropical soils the mouldboard plough cannot cope with hard soils, roots or stones, and with many of these soils the action of inversion is to be avoided, as there may be limited topsoil and humus. In temperate regions, with heavy clay soils, mouldboard ploughing is considered beneficial, particularly the effects of weathering in the case of autumn ploughing for a spring seedbed. Ploughing light soils often produces too 'fluffy' a seedbed, and the cost of secondary tillage work to consolidate the seedbed increases. More and more farmers are questioning the need to mouldboard plough
Disc plough	Two to six disc models either one-way, or reversible. Disc diameter varies between 700 and 900 mm. Discs can be either plain or cutaway, of which the latter gives better penetration in hard soils, and copes better with heavy surface trash. Weights can be added to the plough head-stock and frame to increase penetration in hard soils	A chopping and mixing action with only partial inversion of the furrow slice. Incomplete burial of surface trash is an advantage in areas susceptible to erosion. The height of the disc scraper dictates the degree of soil shatter. The higher the scraper the less the soil will be pulverised	Varies, but normally around 250 mm	Well accepted in the tropics and copes well with stones, roots and hard soils. In normal use has slightly higher draft than a mouldboard plough. The action of incomplete trash burial is an advantage in poor light soils. If badly adjusted can create a plough pan. Soil erosion can be caused by poor operation and management of disc ploughs, ploughing up and down slopes instead of on contours etc

cont

Table Q.1 cont

Implement	Variations	Soil action	Operating depth	Comments
Ploughing harrow	With a normal working width of 2 to 3.5 m, depending upon tractor horsepower, this implement has two gangs of discs, set at opposite diagonals to the direction of travel, similar to the offset disc harrow. Disc diameter varies between 700 and 900 mm, and discs can either be plain or cutaway. Inter-disc spacing is normally around 300 mm. An extremely heavy implement; penetration can be increased or reduced by altering the gang angle	A chopping and mixing action similar to the disc plough, with the front disc gang turn-ing the soil in the one direct-ion, and the rear gang turning in the opposite	Similar to disc plough (around 250 mm)	Used extensively in the tropics; two or three passes may be required to achieve the full working depth. If three passes are necessary, carry out the first two passes on opposite diagonals. In heavy clays with a high moisture content produces large smeared clods, which when dry are extremely difficult to break down with secondary tillage implements. In order to avoid this situation the timing of operations is extremely important
Polydisc cultivators	With a working width of around 2.5 m, this implement comprises a single gang of discs of around 650 mm diameter. The disc gang is mounted at an angle to the direction of travel, and can either be used on its own, or fitted with a seeding box	Similar to a disc harrow	Around 150 mm	At one time this was considered an extremely useful implement for establishing crops in areas of low rainfall, and moderately coarse soils. In one pass, with minimum soil dis-turbance and moisture loss, seeding could be completed. There appears to be renewed interest in this type of implement
Chisel plough	With a working width of 1.8 to 3 m, these implements normally have three rows of tines, although some models have a V-shaped main frame. A wide variety of tines is available. Some are steeply raked to reduce draft, some are fitted with 'stump jump' spring attach-ments, and others have shear bolts to reduce the risk of damage when striking an obstruc-tion. Some new models have the three rows of tines set at increasing depths, with sub-soiler units at the rear	A bursting and stirring action, with little inversion or dis-turbance of the soil layers	Between 200 and 300 mm	Ideal for penetrating hard soils (two or three passes may be required to achieve full working depth), preparing seedbeds in shallow topsoil areas, and for soil conservation as a blanket of trash remains on the surface. Much favoured by temperate climate farmers for producing a satisfactory and economical seed-bed. A fairly fast operation requiring basic operator skills. Surface weeds must be chemically controlled prior to implement use. Reduces the cost of secondary tillage operations. Requires much more emphasis in tropical agriculture, and the husbandry of poor soils
Rotary cultivator	Working widths vary between 1.2 and 3 m to match tractor power, which is normally required to be 1 hp for every 25 mm of width. This implement is driven by the tractor power-take-off unit (PTO) and comprises a central rotor to which blades or tines are bolted. The rotor and blades are surrounded by a metal shield. Different gear ratios can be selected on the implement to vary the rotor speed independent of the tractor travel speed	A chopping and soil-pulverising action close to 'brute force' tillage. The degree of soil shatter can be reduced by a slow rotor speed, increased travel speed, raising the hood at the rear of the implement, and also by reducing the number of blades or tines on the rotor	Between 50 and 200 mm	Properly operated and managed can produce a fast, fine seedbed suitable for intensive vegetable production. Good implement for paddy-field work as it helps the tractor along. Handled badly it can destroy soil structure. Never produce a fine tilth merely because it looks right. Always leave the seedbed as coarse textured as the following crop can tolerate. This implement should not be used in fields infested with perennial grass such as couch and swordgrass, as the rhizomes will be transplanted throughout the field. A costly machine to operate and maintain, frequently misused, and often the cause of plough pan formation (eg with wet soils, high rotor speed, slow travel speed)

Secondary tillage implements

Table Q.2

Implement	Variations	Soil action	Operating depth	Comments
Disc harrows	There are two main types – offset and tandem – which are available in a wide variety of working widths to suit different soils. The offset harrow has two gangs of discs set at opposite diagonals to the line of travel, and the tandem has four disc gangs, one attached to each side of a trailing and leading V-shaped frame. Disc diameter varies between 180 and 270 mm. Disc can be plain or cutaway, and the normal arrangement is to have cutaway discs on the front and plain on the rear. Disc gang angle can be increased or reduced to alter disc penetration	When operated behind a plough has a pulverising effect which reduces clod size and levels and consolidates the seedbed without dragging buried matter to the surface	The normal operating depth would be around 120 mm. In light depth would be controlled through hydraulically operated gauge wheels	By far the most popular secondary tillage implement. The smallest diameter discs with the closest inter-disc spacing normally used to 'finish off' the seedbed prior to planting. When operating in hard-baked clay soils in the tropics the degree of soil shatter, and levelling effect, will be greatly improved by attaching behind the harrow using wire ropes, a scrubber bar comprising a heavy baulk of timber, an iron bar or a heavy length of chain. As the disc harrow only refines the surface layer of soil it cannot be used on its own to produce a deeply refined seedbed for such crops as potatoes, sugarbeet etc, but should follow a tined cultivator operation
Tined cultivator	A wide range of working widths, and available with three main types of tines – rigid, sprung and spring. The spring tine is basically a rigid tine with a spring release mechanism which allows the tine to kick back if it hits an obstruction; often known as a 'stump jump' tine. A crumbler roller can be fitted to the rear to give added soil refinement and consolidation	The rigid and 'stump jump' units, which are normally used for deep cultivations, break down clods, consolidate the seedbed, and can on occasions be used as a primary tillage tool. Unsuitable for newly ploughed land as they drag buried material to the surface. The spring tine vibrates at speed and shatters clods through their natural fracture planes	The rigid and 'stump jump' tine cultivator would normally operate at a depth of about 200 mm. The spring tine unit, which is used mainly to refine the seedbed, would operate at around 100 mm	Provided that there is no danger of dragging buried material to the surface, does an excellent job of refining and consolidating a seedbed. If correctly timed, very often produces better quality work than a disc harrow on clay soils, as it shatters rather than cuts the clods
Power harrow	With operating widths ranging from 2.5 to 6 m, these implements have PTO-driven spike teeth that reciprocate, or rotate on a horizontal plane. A crumbler roller is normally fitted at the rear to give added refinement and to consolidate the seedbed	A stirring, swirling effect that pulverises the seedbed and consolidates the seedbed	From 50 to 150 mm	Ideal for producing a seedbed from ploughing in one pass. Some farmers are attaching drills behind them in order to reduce crop establishment costs. A slow and fairly expensive operation
Spike tooth harrow	Operating widths from 4.5 to 8 m. This implement is made up from 0.9 m sections coupled together, and attached at the front to a beam that connects to the tractor. The sections have diamond pattern frames to which 200 mm spike teeth are bolted	A stirring effect that pulverises the soil and consolidates the seedbed	Approximately 120 mm	The old traditional implement for finishing off the seedbed prior to planting. A fast, economical operation, which also assists levelling. As each section can flex independently of its neighbour there is good ground contour following
Turbotiller	This implement has a working width varying from 2 to 3.5 m and has a frame similar to the offset disc harrow. Attached to the frame are gangs of specially hardened cutter blades arranged so that they look like a disc with sections missing. The name 'Turbotiller' is the trade name for the Bomford and Eversbed machine; there are other manufacturers	A chopping and pulverising action that refines and consolidates the seedbed	Maximum 150 mm	A first-class implement for operating in conditions similar to the disc harrow. Has a much more positive action than the disc harrow, with better penetration, but is slightly more expensive to maintain. Not yet fully exploited in tropical agriculture

References

References

Bibliography of works cited in the text and other relevant publications: underlining of the first word in a reference (such as <u>Adams</u> below) signifies that it did not appear in the 1984 edition of the Booker Tropical Soil Manual.

<u>A</u>

Acland J (1971).   East African crops.   Longman, London.

Adams A R D and Maegraith B G (1980).   Textbook of clinical tropical diseases.   Seventh Edition. Blackwell, Oxford.

<u>Adams</u> F (Ed) (1985).   Soil acidity and liming.   Second Edition.   Am Soc of Agron, Madison, Wisconsin.

ADAS (1974).   Tests for use of concrete pipes.   ADAS/ENG/74/4.   MAFF, London.

<u>Adeoye</u> K B (1986).   Physical changes induced by rainfall in the surface layer of an alfisol, Northern Nigeria.   Geoderma 39, 59-66.

Agric Ext Lab (1969).   Report of soil analysis.   Univ California Agric Extension Service, Davis, Calif.

<u>Ahn</u> P M (1970).   West African soils.   Third Edition.   Oxford University Press.

<u>Aitken</u> J F (1983).   Relationships between yield of sugarcane and soil mapping units, and the implications for land classification.   Soil Surv and Land Eval 3 (1), 1-9.

Akehurst B C (1981).   Tobacco.   Second Edition, 750 pp.   Longman, London.

Allan J A (1980).   Remote sensing in land and land use studies.   Geography 65, 35-43.

Allan W (1965).   The African husbandman.   Oliver and Boyd, London.

Allison L E (1964).   Salinity in relation to irrigation.   Adv Agron 16, 139-180.

Am Soc Agric Eng (1962).   Measuring saturated hydraulic conductivity of soils.   Special Publ SP-SW-0262. Am Soc Agric Eng, St Joseph, Michigan.

Am Soc Agric Eng (1967).   Glossary of soil and water terms.   Special Publ SP-04-67.   Soil and Water Division.   Am Soc Agric Eng, St Joseph, Michigan.

Am Soc Agron (1967).   Irrigation of agricultural lands.   Agron Ser No 11.   Madison, Wisconsin.

Am Soc Agron (1974).   Drainage for agriculture.   Agron Ser No 17.   Madison, Wisconsin.

<u>Andriesse</u> J P (1988).   The nature and management of tropical peat soils.   Soils Bull No 59.   FAO, Rome.

AOAC (1975).   Official methods of analysis.   Twelfth Edition.   AOAC, Washington DC.

Aomine S and Jackson M L (1959).   Allophane determination in ando soils by cation exchange capacity delta value.   Soil Sci Soc Am Proc 23, 210-214.

Archer J R and Smith P D (1972).   The relation between bulk density, available water capacity, and air capacity of soils.   J Soil Sci 23, 475-480.

Arens P L (1978).   Edaphic criteria in land evaluation, pp 24-31.   <u>In</u>:   'Land Evaluation Standards for Rainfed Agriculture'.   World Soil Res Rep 49.   FAO, Rome.

411

Areola O (1982). Soil variability within land facets in areas of low, smooth relief: a case study on the Gwagwa Plains, Nigeria. Soil Survey and Land Evaluation 2(1), 9-13.

ASTM (1981). Specifications for wire cloth sieves. Standard EII. Am Soc for Testing and Materials.

Aubert G (1968). Classification des sols utilisée par les pédologues francaises. FAO World Soils Rep 32, 78-94.

Aubert G and Tavernier R (1972). Soil survey, pp 17-44. In: 'Soils of the Humid Tropics', Drosdoff M (Ed). US Nat Acad Sci, Washington DC.

Aubert H and Pinta M (1977). Trace elements in soils. Developments in Soil Science 7. Elsevier Sci Publ Co, Amsterdam.

Avella T A and Rodriguez A C (1973). Irrigation studies with the soils of the Coello District. Revista del Instituto Colombiano Agropecuario 8, 87-115.

Avery B W (1964). The soils and land use of the district around Aylesbury and Hemel Hempstead. Mem Soil Survey. HMSO, London.

Avery B W (1968). General soil classification: hierarchical and coordinate systems. Trans Ninth Int Cong Soil Sci, Adelaide 4, 169-175.

Avery B W and Bascomb C L (1974). Soil survey laboratory methods. Tech Monogr No 6. Soil Survey. Rothamsted Exp Stn, Harpenden.

Awan M (1980). The moving watertable in site drainage systems. Unpublished Ph D Thesis. Imperial College, London.

Ayers R S and Westcot D W (1976). Water quality for agriculture. Irrigation and Drainage Paper No 29. FAO, Rome.

Ayers R S and Westcot D W (1985). Water quality for agriculture. Irrigation and Drainage Paper No 29 Rev 1. FAO, Rome.

B

Bache B W (1970). Determination of pH, lime potential and aluminium hydroxide potential of acid soils. J Soil Sci 21, 28-37.

BAI (1973). Dhofar development project. Unpublished Soil Rep. BAI, London.

BAI (1979). Kano River Project Phase II. Unpublished Soil and Land Capability Rep. Vol 1 Summary and Main Report. Vol 2 Annexes. BAI, London.

BAI (1982). Ogoja agricultural development project feasibility study. Land Resour Rep. BAI, London.

Baran R, Bassereau D and Gillet N (1974). Measurement of available water and root development of an irrigated sugarcane crop in the Ivory Coast. Proc Int Soc Sugarcane Technol, XV, 2, pp 726-735.

Barber R G and Rowell D L (1972). Charge distribution and the cation exchange capacity of an iron-rich kaolinitic soil. J Soil Sci 23, 135-146.

Barber S A and Bouldin D R (1984). Roots, nutrient and water influx, and plant growth. ASA Spec Publ No 49. Am Soc of Agron, Madison, Wisconsin.

Barber S A, Walker J M and Vasey E H (1963). Mechanism for the movement of plant nutrients from the soil and fertiliser to the plant root. J Agric Food Chem 11, 204-207.

Barnes A C (1974). The sugarcane. World Crop Ser. Leonard Hill, Aylesbury.

Barnes C P et al (1971). Compaction of agricultural soils. ASAE Monogr Ser. Am Soc Agric Eng, Mich.

Barnes D, Gould B W, Bliss P I and Valentine H R (1981). Water and wastewater engineering systems. Pitman Books, London.

Barrett E C and Curtis L F (1982). Introduction to environmental remote sensing. Second Edition. Chapman and Hall, London.

Barshad I (1951). Factors affecting the molybdenum content of pasture plants 2. Effects of soluble phosphates, available nitrogen and soluble sulphates. Soil Sci 71, 387-398.

Bartelli L J et al (1966). Soil surveys and land use planning. Soil Sci Soc Am, Madison, Wisconsin.

Batey T (1971). Manganese and boron deficiency, pp 137-149. In: 'Trace Elements in Soils and Crops'. MAFF Tech Bull 21. HMSO.

Batey T and Davies D B (1971).   Soil field handbook.   ADAS Advisory Papers No 9.   MAFF, London.

Baulkwill W J (1972).   The Land Resources Division of the Overseas Development Administration.   Trop Sci
    XIV 4, 305-322.

Baver L D, Gardner W H and Gardner W R (1972).   Soil physics.   Fourth Edition, 498 pp.   Wiley, New York.

Beckett P H T (1971).   The cost-effectiveness of soil survey.   Outlook on Agric 6, 191-198.

Beckett P H T and Webster R (1971).   Soil variability - a review.   Soils and Fert 34, 1-15.

Beek K J (1978).   Land evaluation for agricultural development.   Publ 23.   ILRI, Wageningen, Neth.

Beek K J (Ed) (1987).   Quantified land evaluation.   ITC Publ No 6, Enschede, Netherlands.

Beek K J and Bennema J (1971).   Land evaluation for agricultural land use planning - an ecological method.
    Pap 1st FAO/UNDP Latin Am Semin Syst Land Wat Resour.   Mexico City.

Bell J et al (1979).   Neutron probe practice.   Inst Hydr Rep No 19, Wallingford, England.

Benton-Jones J et al (1971).   The proper way to take a plant sample for tissue analysis.   Soils and
    Crops.   Am Soc Agron Publ, 15-18.

Bergman H and Boussard J M (1976).   Guide to the economic evaluation of irrigation projects.   OECD,
    Paris.

Bernstein L (1964).   Salt tolerance of plants.   Agric Inf Bull 283.   USDA, Washington DC.

Bibby J S and Mackney D (1969).   Land use capability classification.   Tech Monogr No 1.   Soil Surv,
    Rothamsted Exp Stn, Harpenden.

Bie S W, Ulph A and Beckett P H T (1973).   Calculating the economic benefits of soil survey.   J Soil Sci
    24, 429-435.

Bingham F T (1962).   Chemical tests for available phosphorus.   Soil Sci 94, 87-95.

Black C A (Ed) (1965).   Methods of soil analysis.   Two volumes.   Agronomy No 9.   Am Soc of Agron,
    Madison, Wisconsin.

Blackburn F H (1982).   Sugarcane, 500 pp.   Longman, London.

Blaikie P (1985).   The political economy of soil erosion in developing countries, 256 pp.   Longman,
    London.

Blair-Rains A (1973).   The aerial photograph and the interpretation of vegetation.   Misc Rep 164, Land
    Resour Div ODM, Surbiton, England.

Bleeker P (1975).   Explanatory notes to the land limitation and agricultural land use potential map of
    Papua New Guinea.   Land Res Ser CSIRO Aust No 36.

Bloomfield C and Coulter J K (1973).   Genesis and management of acid sulphate soils.   Adv Agron 25,
    265-326.

Blyth F G H and De Freitas M H (1974).   Geology for engineers.   Sixth Edition.   Edward Arnold, London.

Blyth K (1985).   Remote sensing and water resources engineering.   In: Kennie T J M and Matthews M C.
    Remote sensing in civil engineering.   Surrey University Press, London.

Boekel P (1963).   The effect of organic matter on the structure of clay soils.   Neth J Agric Sci 11(4),
    250-263.

Bogdan A V (1977).   Tropical pasture and fodder plants.   Longman, London.

Bolton J L (1962).   Alfalfa.   Leonard Hill, London.

Bonneau M and Souchier B (1982).   Constituents and properties of soils (translated by V C Farmer),
    496 pp.   Academic Press, London.

Booker Agriculture International.   See under BAI.

Borden R W, Warkentin B P and Coll M (1974).   An irrigation rating for some soils in Antigua, WI.   Trop
    Agric (Trinidad) 51, 501-513.

Borst-Pauwels G W F H (1961).   Plant and Soil 14, 377-392.

Bouma J (1983).   Use of soil survey data to select measurement techniques for hydraulic conductivity.
    Agric Water Management 6, 177-190.

Bouma J, Paetzhold R F and Grossman R B (1982).   Measuring hydraulic conductivity for use in soil survey.   Soil Survey Investigations Rpt 38.   Soil Conserv Serv.   USDA, Washington DC.

Bourke D O'D (1963).   The West African millet crop and its improvement.   Sols Afr 8, 121-123.

Bouwer H (1969).   Planning and interpreting soil permeability measurements.   J Irrig and Drainage Div, ASCE 95, 391.

Bowen H J M (1979).   Environmental chemistry of the elements.   Academic Press, London.

Bower C A (1959).   The chemical amendments for improving sodium soils.   Agric Info Bull No 195.   USDA, Washington DC.

Bower C A (1961).   Prediction of the effects of irrigation waters on soils, pp 215-222.   In Proc Seminar: Salinity problems in the arid zones.   Unesco, Paris.

Bower C A and Huss R B (1948).   Rapid conductimetric method for estimating gypsum in soils.   Soil Sci 66, 199-204.

Bower C A and Maasland M (1963).   Symposium on waterlogging and salinity, W Pakistan, pp 49-61.

Boyer J (1972).   Soil potassium, pp 102-135.   In: 'Soils of the Humid Tropics', Drosdoff M (Ed).   US Nat Acad Sci, Washington DC.

Brady N C (1974).   The nature and properties of soil.   Eighth Edition.   Macmillan, New York.

Brady N C (1984).   The nature and properties of soil.   Ninth Edition.   Macmillan, New York.

Bray R H and Kurtz L T (1945).   Determination of total, organic and available forms of phosphorus in soils.   Soil Sci 59, 44.

Bregt A K, Bouma J and Jellinek M (1987).   Comparison of thematic maps derived from a soil map and from kriging of point data.   Geoderma 39, 281-290.

Breimer T and Slangen J H G (1981).   Pretreatment of soil samples before NO_3-N analysis.   Neth Jour Agric Sci 29(1), 15-22.

Bremner J M (1965).   Total nitrogen, pp 1149-1178.   In: 'Methods of Soil Analysis', Part 2, Agronomy No 9; Black C A (Ed).   Am Soc of Agron, Madison, Wisconsin.

Brewer R (1964).   Fabric and mineral analysis of soils.   Wiley, New York.

Bridges E M (1978).   World soils.   Second Edition.   Cambridge University Press, Cambridge.

Bridges E M and Davidson D A (1982).   Principles and applications of soil geography.   Longman, London.

Brink A B A, Mabbutt J A, Webster R and Beckett P H T (1966).   Report of the working group on land classification and data storage.   MEXE Rep No 940.   UK Ministry of Defence, London.

Brink A B A, Partridge T C and Williams A A B (1982).   Soil surveying for engineering, 380 pp.   Oxford University Press.

Brinkman R (1979).   Ferrolysis.   Pudoc, Wageningen, Netherlands.

Brinkman R and Smyth A J (1973).   Land evaluation for rural purposes.   ILRI Publ No 17, Wageningen, Netherlands.

Brown J C, Ambler J E, Chaney R L and Foy C D (1972).   Differential responses of plant genotypes to micronutrients, pp 389-418.   In: 'Micronutrients in Agriculture', Mortvedt J J et al (Eds).   Soil Sci Soc Am, Madison, Wisconsin.

BSI (1962, amended 1969).   Conversion factors and tables.   Part 2: Detailed Conversion Tables.   BS 350, Part 2.   British Standards Institution, London.

BSI (1975).   Methods of testing soils for civil engineering purposes.   British Standard 1377.   British Standards Institution, London.

BSI (1976).   Specifications for test sieves.   British Standard 410.   British Standards Institution, London.

BSI (1981).   Code of practice for site investigations.   British Standard 5930.   British Standards Institution, London.

Buckman H O and Brady N C (1969).   The nature and properties of soil.   Seventh Edition.   Macmillan, New York.

Bunting B T (1969).   Letter on soil classification in the United States.   Geogr J 135, 646-647.

Buol S W, Hole F D and McCracken R J (1980).   Soil genesis and classification.   Second Edition.   Iowa State University Press, Ames, Iowa.

Buringh P (1979).   Introduction to the study of soils in tropical and sub-tropical regions.   Third Edition, 146 pp.   Pudoc, Wageningen, Netherlands.

Burrough P A (1982).   Computer assistance for soil survey and land evaluation.   Soil Survey and Land Evaluation 2(2), 25-36.

Burrough P A (1986).   Principles of geographical information systems for land resource assessment, 193 pp.   Clarendon Press, Oxford.

Butler B E (1980).   Soil classification for soil survey.   Clarendon Press, Oxford.

Butler B E et al (1961).   Report on the Seventh Approximation Soil Classification System of the USDA.   Austr Soc Soil Sci Publ No 1.

C

Cadbury Ltd (1971).   Cocoa Growers Bulletin, No 7, pp 27-30.   Birmingham.

Cairncross S and Feachem R (1978).   Small water supplies.   Bull No 10.   Information and Advisory Service.   The Ross Institute of Tropical Hygiene, London.

Cairns R R and Bowser W E (1977).   Solonetzic soils and their management.   Publ 1 391, Agriculture Canada.

Campbell D (1971).   Requirements of land and water resources evaluation studies for investment purposes.   Rep AGL:TSLR/71/23.   FAO, Rome.

Canada Department of Forestry (1965).   Soil capability classification for agriculture.   Canada Land Inventory Rep No 2.   Ottawa.

Cannell R Q and Finney J R (1973).   Outlook on Agric 7, 184-189.

Carroll D M, Evans R and Bendelow V C (1977).   Air photo-interpretation for soil mapping.   Tech Monogr No 8.   Soil Surv, Rothamsted Exp Stn, Harpenden.

Carruthers I D and Clark C (1981).   The economics of irrigation.   Liverpool University Press.

Carson J (1974).   Plant root and its environment.   University Press of Virginia, Virginia.

Carter D J (1986).   The remote sensing source book.   Macarta, London.

Cate R B and Nelson L A (1965).   A rapid method for correlation of soil test analyses with plant response data.   Tech Bull 1.   ISTP Ser, North Carolina State University, Agric Exp Stn.

CBI/Kompass (1990).   Register of British industry and commerce.   Vol 1: products and services.   Vol 2: company information.   Kompass Publishers Ltd, East Grinstead.

Chan Huen Yin (1977).   A soil suitability technical grouping system for hevea.   Planters Bull 152, 135-146.   Kuala Lumpur.

Chan Huen Yin (1980).   Tropical tree crop requirements and land evaluation, p 82f.   In: 'Land Evaluation Guidelines for Rainfed Agriculture'.   World Soil Res Rep 52.   FAO, Rome.

Chang L and Burrough P A (1987).   Fuzzy reasoning: a new quantitative aid for land evaluation.   Soil Surv and Land Eval 7 (2), 69-80.

Chapman H D (1965).   Cation-exchange capacity, pp 891-901.   In: 'Methods of Soil Analysis', Part 2, Agronomy No 9; Black C A (Ed).   Am Soc of Agron, Madison, Wisconsin.

Chapman H D (1966).   Diagnostic criteria for plants and soils.   Dept of Soils and Plant Nutrition, Univ California Citrus Res Center and Agric Expl Stn, Riverside, California.

Chapman H D, Oxley J H and Curtis J H (1941).   The determination of pH at soil moisture contents approximating field conditions.   Soil Sci Soc Am Proc 5, 191-200.

Charter C F (1957).   The aims and objects of tropical soil surveys.   Soils and Fert 20, 127-128.

Chenery E M (1951).   Some aspects of the aluminium cycle.   J Soil Sci 2, 97-109.

Chenery E M (1966).   Factors limiting crop production: 4 Tea.   Span 9, 45-48.

Chijoke E O (1980).   Impact on soils of fast growing species in the lowland humid tropics.   Forestry Paper No 21.   FAO, Rome.

Child R (1953).   The selection of soils suitable for tea.   Pamphlet 5.   Tea Res Inst of East Africa, Kericho.

Child R (1973).   Coconuts.   Second Edition.   Longman, London.

Childs E C (1969).   An introduction to the physical basis of soil water phenomena.   Wiley, London.

Childs E C and Youngs E G (1974).   Soil physics twenty-five years on.   J Soil Sci 25, 399-407.

Christian C S and Stewart G A (1952).   General report on survey of Katherine-Darwin region 1946.   CSIRO Land Resource Ser No 1, Melbourne, Australia.

Christie B R (Ed) (1984).   Handbook of plant science in agriculture.   CRC Press, Boca Raton, Florida.

Clark J S (1966).   The lime potential and percent base saturation in some representative podzolic and brunizolic soils in Canada.   Soil Sci Soc Am Proc 30, 93-96.

Clarke R J and McRae R (Eds) (1988).   Coffee.   Vol 4: agronomy.   Elsevier, Barking.

Clayton C R I, Simonds N E and Matthews M C (1982).   Site investigation.   Granada, London.

Cline M G (1944).   Principles of soil sampling.   Soil Sci 58, 275-288.

Cline M G et al (1955).   Soil Survey of the Territory of Hawaii.   Soil Survey Ser 1939, No 25.   USDA, Washington DC.

Coile T S (1953).   Moisture content of small stones in soil.   Soil Sci 75, 203-207.

Coleman N T, Weed S B and McCracken R J (1959).   Cation exchange capacity and exchangeable cations in Piedmont soils in North Carolina.   Soil Sci Soc Am Proc 23, 146-149.

Colwell R N (Ed) (1983).   Manual of remote sensing.   Second Edition.   American Society of Photogrammetry, Falls Church, Virginia, USA.

Comerma J and Arias L F (1971).   A system to evaluate the land use capability of land in Venezuela.   Mimeo, Maracay, Venezuela.   Quoted by Beek, 1978.   Publ No 23.   ILRI, Wageningen, Netherlands.

Comhaire M (1966).   The role of silicon for plants.   AGRI Digest 7, 9-19.

Conacher A J and Dalrymple J B (1977).   The nine unit landsurface model: an approach to pedogeomorphic research.   Geoderma 18, 1-154.

Cooke G W (1967).   The control of soil fertility.   Crosby Lockwood, London.

Cooke G W (1975).   Fertilising for maximum yield.   Second Edition.   Crosby Lockwood, London.

Corbett W M and Tatler W (1970).   Soils in Norfolk.   Sheet TM49 (Beccles North).   Soil Survey Record No 1.   Rothamsted Exp Stn, Harpenden, Herts, UK.

Cottenie A (1980).   Soil and plant testing as a basis of fertiliser recommendations.   Soils Bull 38/2.   FAO, Rome.

Coulter J K (1972).   Soils of Malaysia.   A review of investigations on their fertility and management.   Soils Fert 35, 475-498.

Coursey D G (1967).   Yams.   Longman, London.

Cox F R and Kamprath E J (1972).   Micronutrient soil tests, pp 289-317.   In: 'Micronutrients in Agriculture', Mortvedt J J et al (Eds).   Soil Sci Soc Am, Madison, Wisconsin.

Cracknell B (1978).   ODM's experience in the commissioning of agricultural evaluations and in the feedback of the results.   Inst Brit Geog, Developing Areas Study Group, Conference on Rural Development Evaluation: cyclostyled.

Creutzberg D (1982).   Field extract of Soil Taxonomy.   Int Soil Museum, Wageningen, Netherlands.

Crops and Soils (1979).   Illustrated method for nematode sampling.   31(8), 14-15.

Cullen N A and Arnold G C (1971).   Establishment of pastures on yellow brown loams near Te Anan X Mazar and trace element responses and interactions.   NZ J Agric Res 14, 47-65.

Curran P (1985).   Principles of remote sensing.   Longman, London.

D

Dabin B and Leneuf N (1960).   Fruits 15(3), 117-127.

Dalal-Clayton D B (1988).   An historical review of soil survey and soil classification in tropical Africa.   Soil Surv and Land Eval 8 (3), 138-160.

Darcy H (1856).   Les fontaines publiques de la ville de Dijon.   Delmont, Paris.

Davidson D A (1980).   Soils and land use planning.   Longman, London.

Davies B E (Ed) (1980).   Applied soil trace elements, 482 pp.   Wiley, Chichester.

Day P R (1965).   Particle fractionation and particle size analysis, pp 545-567.   In: 'Methods of Soil Analysis', Part 1, Agronomy No 9; Black C A (Ed).   Am Soc of Agron, Madison, Wisconsin.

Deb B C (1963).   High dam soil survey project, Aswan.   FAO, Rome.

De Datta S K (1981).   Principles and practice of rice production.   Wiley, Chichester.

De Geus J G (1973).   Fertiliser guide for the tropics and subtropics.   Second Edition.   Centre d'Etude de l'Azote, Zurich.

De Gruijter J J (1977).   Numerical classification of soils and its application in survey.   Agric Res Rep 855.   Centre for Agric Publ and Documentation, Wageningen, Netherlands.

Dent D and Young A (1981).   Soil survey and land evaluation.   Allen and Unwin, London.

Dent D (1986).   Acid sulphate soils: baseline for research and development.   Int Inst for Land Recl and Improvement Publ No 39, Wageningen, Netherlands.

Dent D (1988).   Guidelines for land use planning in developing countries.   Soil Surv and Land Eval 8 (2), 67-76.

Dermott W (1967).   Wet sieving, pp 155-657.   In: 'West European Methods of Soil Structure Determination', de Boodt M (Ed).   Int Soil Sci Soc, Ghent, Belgium.

Deshaprabhu S B (Ed) (1966).   The wealth of India: raw materials.   Publ and Information Directorate, CSIR, New Dehli.

Dewis J and Freitas F C R (1970).   Physical and chemical methods of soil and water analysis.   Soils Bull No 10.   FAO, Rome.

D'Hoore J L (1964).   Soil map of Africa - explanatory monograph.   CCTA Publ 93, Lagos.

DHSS (1982).   Protect your health abroad.   Leaflet SA35.   Department of Health and Social Security, Stanmore, Middlesex.

Dieleman P J and Trafford B D (1976).   Drainage testing.   Irrigation and Drainage Paper 28.   FAO, Rome.

D'Itri F M (Ed) (1977).   Wastewater renovation and reuse.   Marcel Dekker Inc, USA.

Doering E J, Willis W O and Merrill S D (1982).   Effect of total cation concentration and sodium adsorption ratio on exchangeable sodium ratio.   Can J Soil Sci 62, 177-185.

Doggett H (1988).   Sorghum.   2nd Edition.   Longman, London.

Doll E C and Lucas R E (1973).   Testing soils for potassium, calcium and magnesium, pp 133-151.   In: 'Soil Testing and Plant Analysis', Walsh L M and Beaton J D (Eds).   Soil Sci Soc Am, Madison, Wisconsin.

Dommergues Y R and Diem H G (Eds) (1982).   Microbiology of tropical soils and plant productivity.   Developments in Plant and Soil Science No 5.   Nijhoff Junk, The Hague.

Doorenbos J and Kassam A H (1979).   Yield response to water.   Irrigation and Drainage Paper 33.   FAO, Rome.

Doorenbos J and Pruitt W O (1977).   Crop water requirements.   Irrigation and Drainage Paper 24.   FAO, Rome.

Dowson V H W and Atan A (1962).   Dates, including handling, processing and packing.   Agric Dev Pap No 72.   FAO, Rome.

Dregne H E (1976).   Soils of arid regions, 238 pp.   Elsevier, Amsterdam.

Drosdoff M (Ed) (1972).   Soils of the humid tropics.   US Nat Acad Sci, Washington DC.

Duchaufour P (1963).   Soil classification: a comparison of the American and French systems.   J Soil Sci 14, 149-155.

Duchaufour P (1982).   Pedology (translated by T R Paton), 448 pp.   George Allen and Unwin, London.

Duchaufour P (1984). Abrégé de pédologie. Masson, Paris.

Dudal R (1979). Land resources in the humid tropics. Presentation at Roy Geog Soc, London, 10 April.

Dudal R (1987). The role of pedology in meeting the increasing demands on soils. Soil Surv and Land Eval 7 (2), 101-110.

Dumm L D (1954). Drain spacing formula. Agric Eng 35, 726-730.

Dumm L D (1960). Validity and use of the transient flow concept in sub-surface drainage. Paper presented at ASE meeting, Memphis, Tennessee.

E

Eastman Kodak Co (1972a). Kodak filters for scientific and technical uses. Eastman Kodak Co, Rochester, New York.

Eastman Kodak Co (1972b). Kodak aerial films and photographic plates. Eastman Kodak Co, Rochester, New York.

Eaton F M (1935). Boron in soils and irrigation waters and its effect on plants with particular reference to the San Joaquin Valley of California. Tech Bull No 448. USDA, Washington DC.

Eaton F M (1950). Significance of carbonates in irrigation waters. Soil Sci 69, 123-133.

Eavis B W (1972). Soil physical conditions affecting seedling root growth. I. Mechanical impedance, aeration and moisture availability as influenced by bulk density and moisture in a sandy loam soil. Plant and Soil 36, 613-622.

Economist (1980). The Economist world measurement guide. The Economist Newspaper Ltd, London.

Eden M J and Parry J T (Eds) (1986). Remote sensing and tropical land management. Wiley, Chichester.

Eden T (1976). Tea. Third Edition. Longman, London.

Ellis G (1971). Units. Symbols and abbreviations. Royal Soc Medicine, London.

Elwell H A (1981). A soil loss estimation technique for Southern Africa, pp 281-292. In: 'Soil Conservation: Problems and Prospects', Morgan R P C (Ed). Wiley, Chichester.

Elwell H A and Stocking M A (1982). Developing a simple yet practical method of soil-loss estimation. Trop Agric 59(1), 43-48.

Embrechts J, Poeloengan Z and Sys C (1988). Physical land evaluation using a parametric method - application to oil palm plantations in North Sumatra, Indonesia. Soil Surv and Land Eval 8 (2), 111-122.

Emerson W W (1967). A classification of soil aggregates based on their coherence in water. Austr J Soil Res 5, 47-57.

Emerson W W, Bond R D and Dexter A (1978). Modification of soil structure. Wiley, New York.

Environmental Studies Board (1972). Water quality criteria. Natl Acad Sci and Engineering, USA.

Ernst L F (1950). A new formula for the calculation of the permeability factor with the auger hole method. TNO, Groningen, translated from Dutch by H Bouwer, Cornell Univ Ithaca, NY, 1955.

Eswaran H and Sys C (1979). The argillic horizon in low-activity-clay soils: formation and significance to classification. Pédologie 29, 176-190.

Eswaran H, Forbes T R and Laker M C (1977). Soil map parameters and classification, pp 37-57. In: 'Soil Resource Inventories', Agron Mimeo No 77-23. Department of Agronomy, Cornell Univ Ithaca, New York.

Evans H (1959). Elements other than nitrogen, potassium and phosphorus in the mineral nutrition of sugarcane. Proc Int Soc Sugarcane Technol, X, Hawaii, pp 473-508. Elsevier, Amsterdam.

Evans H (1965). Toxic sulphate soils in British Guiana and their improvement for sugarcane cultivation. Proc Int Soc Sugarcane Technol, XII, Puerto Rico, pp 695-714. Elsevier, Amsterdam.

F

Fackler E (1924). Bodenklassifikation zu Steuerzwecken. Wbl Landw Ver Bayern Vol 114.

Fairbridge R W and Finkl C W (1979). The encyclopedia of soil science, Part 1. Dowden, Hutchinson and Ross Inc, Stroudsburg, Pennsylvania.

Faniran A and Jeje L K (1983). Humid tropical geomorphology. Longman, London.

FAO (1960). Multilingual vocabulary of soil science - Jacks G V, Tavernier R and Boalch W. Land and Water Dev Div. FAO, Rome.

FAO (1963). High dam soil survey project, Aswan - Deb B C. FAO, Rome.

FAO (1966a). Guide on general and specialised equipment for soils laboratories - Golden J D et al. Soils Bull No 3. FAO, Rome.

FAO (1966b). Selection of soils for cocoa - Smyth A J. Soils Bull No 5. FAO, Rome.

FAO (1967). Aerial photo interpretation in soil survey. Soils Bull No 6. FAO, Rome.

FAO (1970a). The preparation of soil survey reports - Smyth A J. Soils Bull No 9. FAO, Rome.

FAO (1970b). Physical and chemical methods of soil and water analysis. Soils Bull No 10. FAO, Rome.

FAO (1971). Land degradation - Rauschkolb R S. Soils Bull No 13. FAO, Rome.

FAO (1972). Trace elements in soils and agriculture - Sillanp§§ M. Soils Bull No 17. FAO, Rome.

FAO (1973). Reclamation and management of calcareous soils. Soils Bull No 21. FAO, Rome.

FAO (1974). Irrigation suitability classification. Soils Bull No 22. FAO, Rome.

FAO (1975). SI units and nomenclature in soil science. Soils Bull No 28. FAO, Rome.

FAO (1976a). Framework for land evaluation. Soils Bull No 32. FAO, Rome.

FAO (1976b). Water quality for agriculture - Ayers R S and Westcot D W. Irrigation and Drainage Paper No 29. FAO, Rome.

FAO (1976c). Prognosis of salinity and alkalinity. Soils Bull No 31. FAO, Rome.

FAO (1976d). Drainage testing - Dieleman P J and Trafford B D. Irrigation and Drainage Paper No 28. FAO, Rome.

FAO (1977a). Guidelines for soil profile description (Second Edition). FAO, Rome.

FAO (1977b). Crop water requirements - Doorenbos J and Pruitt W O. Irrigation and Drainage Paper No 24. FAO, Rome.

FAO (1978). Land evaluation standards for rainfed agriculture. World Soil Res Rep 49. FAO, Rome.

FAO (1979a). Soil survey investigations for irrigation. Soils Bull No 42. FAO, Rome.

FAO (1979b). Yield response to water - Doorenbos J and Kassam A H. Irrigation and Drainage Paper No 33. FAO, Rome.

FAO (1980a). Soil and plant testing and analysis. Soils Bull No 38/1. FAO, Rome.

FAO (1980b). Soil plant testing as a basis of fertiliser recommendations - Cottenie A. Soils Bull No 38/2. FAO, Rome.

FAO (1980c). Land evaluation guidelines for rainfed agriculture. World Soil Res Rep 52. FAO, Rome.

FAO (1983). Guidelines: land evaluation for rainfed agriculture. Soils Bull No 52. FAO, Rome.

FAO (1984a). Fertilizer and plant nutrition guide. Fert and Plant Nutr Bull No 9. FAO, Rome.

FAO (1984b). Land evaluation for forestry. Forestry Pap 48. FAO, Rome.

FAO (1985a). Guidelines: land evaluation for irrigated agriculture. Soils Bull No 55. FAO, Rome.

FAO (1985b). Water quality for agriculture - Ayers R S and Westcot D W. Irrigation and Drainage Paper No 29 Rev 1. FAO, Rome.

FAO (1987). Soil and water conservation in semi-arid areas - N W Hudson. Soils Bull No 57. FAO, Rome.

FAO (1988a). Salt-affected soils and their management. Soils Bull No 39 Rev 1. FAO, Rome.

FAO (1988b). The nature and management of tropical peat soils - J P Andriesse. Soils Bull No 59. FAO, Rome.

FAO-Unesco (1973). Irrigation, drainage and salinity. An international source book. FAO-Unesco/Hutchinson, London.

FAO-Unesco (1974).   Soil map of the world, Vol 1.   Legend. Dudal R et al.   Unesco, Paris.

FAO-Unesco - ISRIC (1988).   FAO-Unesco soil map of the world.   Revised Legend.   World Soil Res Rep 60.
      FAO, Rome.

Fassbender H W (1980).   Problems connected with soil testing for phosphorus and potassium, pp 93-108.
      In: 'Soil and Plant Testing and Analysis'.   FAO, Rome.

Fauck R, Moreaux G and Thomann C (1969).   Changes in soil of Sefa (Casamance, Senegal) after 15 years of
      continuous cropping.   Agron Trop Paris 24, 263-301.

Ferguson W R (1973).   Practical laboratory planning, 147 pp.   Applied Science Publishers Ltd, Barking,
      Essex.

Finck A (1982).   Fertilisers and fertilisation.   Introduction and practical guide to crop
      fertilisation.   Verlag Chemie, Weinheim, West Germany.

Finkel H J and Nir D (1960).   Determining infiltration rates in an infiltration border.   J Geophys
      Sci 65, No 7.

Finkl C W (1982).   Soil classification.   Benchmark papers in soil science, Vol 1, 400 pp.   Hutchinson
      Ross Publishing Co, Stroudsburg, Pennsylvania.

Fleming G A (1970).   Selenium-molybdenum-gypsum trial, p 48.   In: Res Rep, Soil Div.   An Foras
      Taluntais, Dublin.

Fleming G A (1980).   Essential micronutrients I: boron and molybdenum, pp 155-197.   In: 'Applied Soil
      Trace Elements', Davies B E (Ed).   Wiley, Chichester.

Food and Agriculture Organisation of the United Nations.   See under FAO.

Forbes T R, Rossiter D and Van Wambeke A (1982).   Guidelines for evaluating the adequacy of soil
      resource inventories.   Soil Management Supp Serv Tech Monogr No 4, Washington DC.

Forsythe W M, Gavande S A and Gonzalez M A (1969).   Physical properties of soils derived from volcanic
      ash with consideration of some soils of Latin America, pp B3.1-3.6.   In: 'Panel on Soils Derived
      from Volcanic Ash of Latin America'.   IICA, Turrialba, Costa Rica.

Foster H L (1981).   The basic factors which determine inherent soil fertility in Uganda.   J Soil Sci 32,
      149-160.

Frankart R, Sys C and Verheye W (1972).   Contributions to the use of the parametric method for the
      evaluation of the classes in the different categories of the land evaluation proposed by the working
      group.   Mimeo Rep, FAO consultation on Land Evaluation.   Wageningen, Netherlands.

Freeman-Grenville G S P (1963).   The Muslim and Christian calendars.   Oxford University Press.

G

Garlick J P and Keay R W G (Eds) (1970).   Human ecology in the tropics.   Pergamon, Oxford.

Gavaud M (1986).   Les recherches réalisés par l'ORSTOM sur les sols africains.   Geoderma 13, 391-406.

Gbadegesin S and Areola O (1987).   Soil factors affecting maize yields in the Nigerian savanna and their
      relation to land suitability assessment.   Soil Surv and Land Eval 7 (3), 167-175.

Gerrard A J (1981).   Soils and landforms.   Allen and Unwin, London.

Ghadiri H and Payne D (1981).   Raindrop impact stress.   J Soil Sci 32, 41-49.

Ghosh R K (1980).   Estimation of soil-moisture characteristics from mechanical properties of soils.
      Soil Sci 130, 60-63.

Gibbon D and Pain A (1985).   Crops of the drier regions of the tropics.   Longman, London.

Gibbons F R (1961).   Some misconceptions about what soil surveys can do.   J Soil Sci 12(1), 96-100.

Giltrap D J (1982).   Computer production of soil maps, I.   Production of grid maps by interpolation.
      Geoderma 29, 295-311.

Gittinger J P (1981).   Economic analysis of agricultural projects.   John Hopkins University Press,
      Baltimore, Maryland.

Goedert W J (1983).   Management of the cerrado soils of Brazil: a review.   J Soil Sci 34, 405-428.

Golden J D, Lemos P, Carlyle R E and Freitas F C R (1966).   Guide on general and specialised equipment
      for soils laboratories.   Soils Bull No 3.   FAO, Rome.

Goldsmith P F (1977). A practical guide to the use of the universal soil loss equation, 34 pp. Unpublished BAI Tech Monogr.

Golley F B and Medina E (Eds) (1975). Tropical ecological systems. Springer, Berlin.

Greenland D J (1977). Soil drainage by intensive arable cultivation: temporary or permanent. Phil Trans Roy Soc (London), B, 281, 193-208.

Greenland D J (1979). Structural organisation of soils and crop production. In: 'Soil Physical Properties and Crop Production in the Tropics', Lal R and Greenland D J (Eds). Wiley, Chichester.

Greenland D J (Ed) (1981a). Characterisation of soils, 446 pp. Clarendon Press, Oxford.

Greenland D J (1981b). Soil management and soil degradation. J Soil Sci 32, 301-322.

Greenland D J and Lal R (Eds) (1977). Soil conservation and management in the humid tropics, p 283. Wiley, Chichester.

Greenland D J, Rimmer D and Payne D (1975). Determination of the structural stability class of English and Welsh soils. J Soil Sci 26, 294-303.

Greenwood D J (1975). Measurement of soil aeration, pp 261-272. In: 'Soil Physical Conditions and Crop Production'. MAFF Tech Bull 29, HMSO.

Gressit J L (Ed) (1982). Biogeography and ecology of Papua New Guinea. Junk, The Hague.

Griffiths E (1975). Classification of land for irrigation in New Zealand. Sci Rep NZ Soil Bur No 22.

Grigg J L (1953). Determination of available soil molybdenum. NZ Soil News 3.

Grist D H (1983). Rice. Fifth Edition, 602 pp. Longman, London.

H

Hachum A Y and Alfaro J F (1980). Rain infiltration into layered soils: prediction. Irrig and Drainage Div, Am Soc Civ Eng 106, 311-319.

Haensch G and Haberkamp D A G (1975). Dictionary of agriculture. Elsevier, Amsterdam.

Hails J R (Ed) (1977). Applied geomorphology. Elsevier, Amsterdam.

Hajrasuliha S, Baniabbassi N, Metthey J and Nielsen D R (1980). Spatial variability of soil sampling for salinity studies in South-West Iran. Irrig Sci 1, 197-208.

Hale P R and Williams B D (Eds) (1977). Liklik Buk. A rural development handbook catalogue for Papua New Guinea. English Edition. Liklik Buk Information Centre, Lae, Papua New Guinea.

Hall D G M, Reeve M J, Thomasson A J and Wright V F (1977). Water retention, porosity and density of field soils. Tech Monogr No 9. Soil Survey, Rothamsted Exp Stn, Harpenden.

Hallsworth E G (1987). Anatomy, physiology and psychology of erosion, 176 pp. Wiley, Chichester.

Hamblin A P (1982). Soil water behaviour in response to changes in soil structure. J Soil Sci 33, 375-386.

Hamilton W R, Woolley A R and Bishop A C (1974). Guide to minerals, rocks and fossils. Hamlyn, London.

Hammond L C, Pritchett W L and Chew V (1958). Soil sampling in relation to soil heterogeneity. Soil Sci Soc Am Proc 22, 548-552.

Hanks R J and Ashcroft G L (1970). Physical properties of soils. Mimeographed, Logan, Utah.

Hanna W J and Hutcheson T B (1968). Soil-plant relationships. In: 'Changing Patterns in Fertiliser Use', Nelson L A (Ed). Soil Sci Soc Am, Madison, Wisconsin.

Hansen V E, Israelsen O W and Stringham G E (1980). Irrigation principles and practices. Fourth Edition. Wiley, Chichester.

Hanson C T and Blevins R L (1979). Soil water in coarse fragments. Soil Sci Soc Am J 43, 819-820.

Hardy N, Shainberg I, Gal M and Keren R (1983). The effect of water quality and storm sequence upon infiltration rate and crust formation. J Soil Sci 34, 665-676.

Harler C R (1964). The culture and marketing of tea. Third Edition. Oxford University Press.

Harmsen K (1977). Behaviour of heavy metals in soils. Centre for Agric Publ and Documentation, Wageningen, Netherlands.

References: H - I

Harrison M N (1978). Maize in the tropics, 400 pp. Longman, London.

Harrod M F (1975). Field experience on light soils, pp 22-51. In: 'Soil Physical Conditions and Crop Production'. MAFF Tech Bull 29, HMSO, London.

Hartley C W S (1968). The soil relations and fertiliser requirements of some permanent crops in West and Central Africa, pp 155-183. In: 'The Soil Resources of Tropical Africa', Moss R P (Ed). Cambridge University Press.

Hartley C W S (1977). The oil palm. Second Edition. Longman, London.

Hartley C W S (1988). The oil palm. Third Edition. Longman, London.

Hausenbuiller R L (1985). Soil science, principles and practices. William C Brown, Dubuque, Iowa.

Haverkamp R and Vauclin M (1981). A comparative study of three forms of the Richard equation used for predicting one-dimensional infiltration in unsaturated soil. Soil Sci Soc Am J 45, 13-20.

Heald W R (1965). Calcium and magnesium, pp 999-1010. In: 'Methods of Soil Analysis', Part 2, Agronomy No 9; Black C A (Ed). Am Soc of Agron, Madison, Wisconsin.

Henery W (1975). Instrumentation and measurement for environmental sciences. Am Soc Agric Eng, Michigan.

Hesse P R (1958). Fixation of sulphur in the muds of Lake Victoria. Hydrobiologia 11, 171-181.

Hesse P R (1971). A textbook of soil chemical analysis. John Murray, London.

Hesse P R (1975). SI units and nomenclature in soil science. Soils Bull No 28. FAO, Rome.

Hewitt J S and Dexter A (1980). Effects of tillage and stubble management on the structure of a swelling soil. J Soil Sci 31, 203-215.

Hillel D (1971). Soil and water: physical principles and processes. Academic Press, New York.

Hillel D (Ed) (1980). Applications of soil physics. Academic Press, New York.

Hills R C (1970). The determination of the infiltration capacity of field soils using the cylinder infiltometer. Tech Bull No 3. British Geomorphological Res Group. Geo Abstracts, UK.

HMSO (1979). Royal Commission on environmental pollution. Seventh Report: agriculture and pollution. HMSO, London.

Hockensmith R D (1960). Water and agriculture. Pub No 62. American Association for the Advancement of Science, Washington DC.

Hockensmith R D and Steele J B (1949). Recent trends in the use of land capability classification. Proc Soil Sci Soc Am 14, 383-88.

Hodgson J M (Ed) (1974). Soil survey field handbook. Tech Monogr No 5. Soil Survey. Rothamsted Exp Stn, Harpenden.

Hodgson J M (1978). Soil sampling and soil description. Clarendon Press, Oxford.

Hogg W H (1967). Atlas of long-term irrigation needs for England and Wales. NAAS-MAFF, HMSO, London.

Holm L R G et al (1977). The world's worst weeds. Distribution and biology, 621 pp. Univ Press of Hawaii, Honolulu.

Huddleston J (1984). Development and use of soil productivity ratings in the United States. Geoderma 32, 297-317.

Hudson N W (1957). Erosion control research - progress reports on experiments at Henderson Research Station 1953-56. Rhod Agric J 54, 297-323.

Hudson N W (1971). Soil conservation. Second Edition, 1976. Batsford, London.

Hudson N W (1975). Field engineering for agricultural development. Clarendon Press, Oxford.

Hudson N W (1987). Soil and water conservation in semi-arid areas. Soils Bull No 57. FAO, Rome.

Hurd-Karrer A M (1935). J Agric Res 50, 413-427.

I

Ignatyev G M (1968). Classification of cultural and natural vegetation sites as a basis for land evaluation, pp 104-112. In: 'Land Evaluation', Stewart G A (Ed). Macmillan, Melbourne.

ILACO, BV (1981).   Agricultural compendium for rural development in the tropics and sub-tropics.
    Elsevier, Amsterdam.

ILRI (1972).   Veldboek voor land en waterdeskundigen.   ILRI, Wageningen, Netherlands.

ILRI (1972-74).   Drainage principles and applications.   Four volumes.   I (1972), II (1973), III and IV
    (1974).   Publ 16.   ILRI, Wageningen, Netherlands.

Imeson A C and Vis M (1984).   Assessing soil aggregate stability by water drop impact and ultrasonic
    dispersion.   Geoderma 34, 185-200.

Incoll L D, Long S P and Ashmore M R (1977).   SI units in plant science.   Current Advances in Plant
    Science 9, 331-343.

IRRI (1975).   Major research in upland rice.   International Rice Research Institute.   Los Banos,
    Philippines.

Irvin G (1978).   Modern cost-benefit methods.   An introduction to financial, economic and social
    appraisal of development projects.   Macmillan, London.

Isbell R F (1977).   The argillic horizon concept and its application to the classification of tropical
    soils, pp 150-157.   In: 'Proceedings of the Clamatrops Conference,' Joseph K T (Ed).   Malay Soc
    Soil Sci, Kuala Lumpur.

Isirimah N O and Ojanuga A G (1987).   Potential acid sulphate peats of the Niger Delta wetlands.   Soil
    Surv and Land Eval 7 (3), 147-156.

Islam A (1966).   Current status of soil and land capability classification in East Pakistan.   CENTO
    Conf Land Classif Non-Irrig Lands, 81-84.

Israelsen O W and Hansen V E (1962).   Irrigation principles and practices.   Third Edition.   Wiley.
    (Now in Fourth Edition; see Hansen et al, 1980).

IUPAC (1972).   Manual of symbols and terminology.   Int Union of Pure and Applied Chemistry, London.

Ive J R and Cocks K D (1987).   The value of adding searching and profiling capabilities to a land use
    planning package.   Soil Surv and Land Eval 7 (2), 87-94.

Iwata S, Tabuchi T and Warkentin B P (1987).   Soil-water interactions.   Marcel Dekker, New York.

J

Jacks G V, Tavernier R and Boalch W (1960).   Multilingual vocabulary of soil science.   Land and Water
    Dev Div.   FAO, Rome.

Jackson M L (1958).   Soil Chemical Analysis.   Prentice-Hall, New York.

Jackson M L (1965).   Free oxides, hydroxides and amorphous aluminosilicates, pp 578-603.   In: 'Methods
    of Soil Analysis', Black C A (Ed).   Am Soc of Agron, Madison, Wisconsin.

Jacob A and Uexküll H V (1963).   Fertiliser use.   Nutrition and manuring of tropical crops.   Third
    Edition.   Verlagsgesellschaft fur Ackerbau, Hanover.

James D W, Hanks R J and Jurinak J (1982).   Modern irrigated soils.   Wiley, New York.

Jansen I and Fenton T E (1978).   Computer processing of soil survey information.   J Soil Wat Conserv 33,
    188-190.

Jardine C G (1962).   Metal deficiencies in bananas.   Nature 194, 1 160-1 163.

Jarvis M G and Mackney D (1973).   Soil survey applications.   Tech Monogr 13.   Soil Survey.
    Rothamsted Exp Stn, Harpenden.

Jeffrey D W (1987).   Soil-plant relationships: an ecological approach.   Croom Helm, London.

Jenkin R N and Foale M A (1968).   An investigation of the coconut-growing potential of Christmas
    Island.   Two volumes.   Land Resources Division, Land Resource Study 4.   Tolworth, UK.

Johnson H P, Frevert R K and Evans D D (1952).   Simplified measurement and computation of soil
    permeability below the water table.   Agric Eng 33, 283-286.

Johnson W M, McClelland J E, McCaleb S B, Ulrich R and Harper W G (1960).   Classification and description
    of soil pores.   Soil Sci 89, 319-321.

Jones J B (1967).   Interpretation of plant analysis for several agronomic crops, pp 49-58.   In: 'Soil
    Testing and Plant Analysis', Part 2, Hardy G W (Ed).   Soil Sci Soc Am Spec Publ No 2, Madison, Wisc.

Jones J B (1972). Plant tissue analysis for micronutrients, pp 319-346. In: 'Micronutrients in Agriculture', Mortvedt J J et al (Eds). Soil Sci Soc Am, Madison, Wisconsin.

Jones M J and Wild A (1975). Soils of the West African savanna. Tech Comm No 55. Commonwealth Bureau of Soils, Harpenden.

Jones W O (1959). Manioc in Africa. Stanford University Press, California.

Jongerius A (1957). Morphological investigations of soil structure. Bodemkundige Studiea, No 2. Meded van der Stickting Bodemkartiering, Wageningen.

Jorquera M (Ed) (1969). Physical resource investigations for economic development - casebook of OAS field experience in Latin America. Org Amer St, Washington DC.

Joseph K T (Ed) (1977). Proceedings of the Clamatrops conference. Malay Soc Soil Sci, Kuala Lumpur.

Jumikis A R (1967). Introduction to soil mechanics. Van Nostrand, Princeton, New Jersey.

Juo A S R (1981). Chemical characteristics, pp 51-79. In: 'Characterisation of Soils', Greenland D J (Ed). Clarendon Press, Oxford.

Jurion F and Henry J (1969). Can primitive farming be modernised? INEAC, HORS Ser, 1969, 427-457. Institut National pour l'Etude Agronomique du Congo, Brussels.

K

Kalpagé F (1974). Tropical soils - classification, fertility and management. Macmillan, New Delhi.

Kasasian L (1971). Weed control in the tropics. Leonard Hill, London.

Kay D E (1973). TPI crop and product digest No 2. Root crops. Tropical Products Institute, London.

Keisling T C, Davidson J M, Weeks D L and Morrison R D (1977). Precision with which selected soil physical parameters can be estimated. Soil Sci 124(4), 241-248.

Kennie T J M and Matthews M C (1985). Remote sensing in civil engineering. Surrey University Press, London.

Kessler J and Oosterbaan R J (1974). Determining hydraulic conductivity of soils. In: 'Drainage Principles and Applications'. Publ 16, Vol 3. ILRI, Wageningen, Netherlands.

Kilmer V J (Ed) (1982). Handbook of soils and climate in agriculture. CRC Press Inc, Boca Raton, Florida.

King K F S (1966). Land capability classification and land use planning with special reference to tropical regions. 6th World For Congr. Seattle, Washington.

Kirkham C E and Powers J (1972). Advanced soil physics. Wiley-Interscience. Wiley, New York.

Klingebiel A A and Montgomery P H (1961). Land capability classification. Agric Handbook No 210. USDA Soil Conserv Serv, Washington DC.

Kohnke H (1968). Soil physics. McGraw-Hill, New York.

Kolaja V, Vrba J and Zwirnmann K H (1986). Control and management of agricultural impact on groundwater, pp 197-228. In: 'Impact of Agricultural Activities on Groundwater', Vrba J (Ed). Intl Contributions to Hydrogeology No 5.

Konanova M M (1966). Soil organic matter (translated by Novakovsky T Z and Newman A C D). Second Edition. Pergamon, Oxford.

Koolen A J and Kuipers H (1983). Agricultural soil mechanics. Springer-Verlag, New York.

Köppen W S (1923). Die Klimate der Erde. Walter der Gruyter, Berlin.

Kovda V A, Van den Berg C and Hagan R M (Eds) (1973). Irrigation, drainage and salinity: an international source book. Hutchinson, London and Unesco, Paris.

Krauskopf K B (1972). Geochemistry of micronutrients, pp 7-40. In: 'Micronutrients in Agriculture', Mortvedt J J et al (Eds). Soil Sci Soc Am, Madison, Wisconsin.

Kronberg B I and Nesbitt H W (1981). Quantification of weathering, soil geochemistry and soil fertility. J Soil Sci 32, 453-459.

Krupsky N K, Aleksandrova A M and Khiznyak S (1961). Determination of mobile aluminium in soils. 1961 Soviet Soil Sci, 1127-1130 (translation of Pochvovyedeniye 1961(10), 93-96).

Kubota J, Lazar V A, Simonson G H and Hill W W (1967). The relationship of soils to molybdenum toxicity in grazing animals in Oregon. Soil Sci Soc Am Proc 27, 679-683.

Kucera K P (1984). Accuracy of Landsat imagery and airphoto interpretation in predicting soils and land suitability for irrigation. Soil Surv and Land Eval 4 (1), 8-17.

Kumar V, Bhatia B K and Shukla U C (1981). Magnesium and zinc relationship in relation to dry matter yield and the concentration and uptake of nutrients in wheat. Soil Sci 131(3), 151-155.

Kumar V and Singh M (1980a). Sulphur and zinc interactions in relation to yield, uptake and utilisation of sulphur in soya bean. Soil Sci 130(1), 19-25.

Kumar V and Singh M (1980b). Interactions of sulphur, phosphorus and molybdenum in relation to uptake and utilisation of phosphorus by soya bean. Soil Sci 130(1), 26-31.

L

Laban P (Ed) (1981). Proceedings of the workshop on land evaluation for forestry. Publ 28. ILRI, Wageningen, Netherlands.

Lal K N and Tandon J N (1955). Effects of S/Se and Ca/B ratios on growth and sap characteristics of sugarcane. Curr Sci 24, 314-315.

Lal R (1976). Soil erosion investigations on an alfisol in Southern Nigeria. IITA Monogr No 1. IITA, Ibadan, Nigeria.

Lal R (1977a). Analysis of factors affecting rainfall erosivity and soil erodibility, pp 49-56. In: 'Soil Conservation and Management in the Humid Tropics', Greenland D J and Lal R (Eds). Wiley, Chichester.

Lal R (1977b). Soil-conserving versus soil-degrading crops and soil management, for erosion control, pp 81-86. In: 'Soil Conservation and Management in the Humid Tropics', Greenland D J and Lal R (Eds). Wiley, Chichester.

Lal R (1979). Physical characteristics of soils of the tropics: determination and management, pp 7-46. In: 'Soil Physical Properties and Crop Production in the Tropics', Lal R and Greenland D J (Eds). Wiley, Chichester.

Lal R (1981). Physical properties, pp 135-148. In: 'Characterisation of Soils', Greenland D J (Ed). Clarendon Press, Oxford.

Lal R (1987). Tropical ecology and physical edaphology. Wiley Interscience, Chichester.

Lambe P W and Whitman R V (1969). Soil mechanics. Mass Inst Technol, Massachusetts.

Landon J R (1972). A study of some common civil engineering measurements on soil. Unpublished MSc Thesis, University of Reading.

Landon J R (1988). Towards a standard field assessment of soil texture for mineral soils. Soil Surv and Land Eval 8 (3), 161-165.

Lawes D A (1966). Rainfall conservation and the yields of sorghum and groundnuts in Northern Nigeria, Samaru Res Bull No 70, Zaria.

Lee K E and Woods L G (1971). Termites and soils. Academic Press, London.

Leighty W R and Shorey E C (1930). Some carbon-nitrogen relations in soils. Soil Sci 30, 257-266.

Lillesand T M and Kiefer R W (1987). Remote sensing and image interpretation, 612 pp. Second Edition. Wiley, New York.

Lindsay W L (1972). Inorganic phase equilibria of micronutrients in soils, pp 41-58. In: 'Micronutrients in Agriculture', Mortvedt J J et al (Eds). Soil Sci Soc Am, Madison, Wisconsin.

Lipman J G and Conybeare A B (1936). Preliminary note on the inventory and balance sheet of plant nutrients in the United States. Bull New Jersey Agric Expl Stn No 607, 1-23.

Lock B F and Van Genderen J L (1978). A methodology for employing landsat data for rural land use surveys in developing countries. J Brit Interplan Soc 31, 293-304.

Lock G W (1969). Sisal. Second Edition. Longman, London.

Lof P and Van Baren H (1987). Soils of the world - wall chart. Elsevier, Amsterdam.

Loi K S, Protz R and Ross G J (1982). Relationship of clay mineral suites to parent rocks of eight soil profiles in Sarawak, Malaysia. Geoderma 27, 327-334.

Long G (1969).   Les approches de l'étude integrée des ressources naturelles en vue de l'aménagement du territoire et de la mise en valeur des terres.   Cent Etud Phyt Ecol Rep No 3191.   Montpelier.

Loveday J, Beatty H J and Norris J M (1972).   Comparison of current methods for evaluating irrigation soils.   CSIRO Division of Soils Tech Bull No 14, Canberra.

Lucas R E and Knezek B D (1972).   Climatic and soil conditions promoting micronutrient deficiencies in plants, pp 265-288.   In: 'Micronutrients in Agriculture', Mortvedt J J et al (Eds).   Soil Sci Soc Am, Madison, Wisconsin.

Lunt O R (1963).   Sensitivity of plants to exchangeable sodium percentage.   University California Rep No 5.   Agric Water Quality Res Conf Appendix B.

Lunt O R (1966).   Sodium.   In: 'Diagnostic Criteria for Plants and Soils', Chapman H D (Ed).   Department of Soils and Plant Nutrition.   University California, Riverside.

Luthin J N (1957).   Measurement of hydraulic conductivity in situ, pp 420-432.   In: 'Drainage of Agricultural Lands'.   Am Soc Agron 7.   Madison, Wisconsin.

Luthin J N and Kirkham D (1949).   A piezometer method for measuring permeability of soil in situ below a water table. Soil Sci 68, 349-358.

Lutz J F (Ed) (1965).   Glossary of soil science terms.   Soil Sci Soc Am Proc 29, 330-351.

M

Maas E V and Hoffman G J (1977).   Crop salt tolerance - current assessment.   J Irrig Drainage Div, June 1977.   Am Soc Civ Eng 103 (IRZ): 115-134.   Proceeding Paper 12993.

Maas E V (1984).   Salt tolerance of plants.   In: 'Handbook of plant science in agriculture', Christie B R (Ed).   CRC Press, Boca Raton, Florida.

Maasland M and Haskew H C (1957).   The auger hole method of measuring the hydraulic conductivity of soil and its application to the drainage design.   Proc 3rd Int Conf Irrig Drainage, 8.69-8.114.

Mackney D (1974).   Soil type and land capability.   Tech Monogr 4.   Soil Survey.   Rothamsted Exp Stn, Harpenden.

McCormack D E and Stocking M A (1986).   Soil potential ratings, I.   An alternative form of land evaluation.   Soil Surv and Land Eval 6 (2), 37-41.

McCoy D E and Donohue S J (1979).   Evaluation of commercial soil test kits for field use.   Comm Soil Sci and Pl Anal 10(4), 631-652.

McDonald L H (ed) 1982.   Agroforestry in the African humid tropics.   United Nations University, Tokyo.

McDonald R C (1975).   Soil survey in land evaluation.   Tech Rep Agric Chem Brch Dept Prim Indust Aust No 6.

McDonald R C et al (1984).   Australian soil and land survey field handbook.   Inkata Press, Melbourne.

McFarlane M J (1976).   Laterite and landscape.   Academic Press, London.

McGlashan M L (1971).   Physicochemical quantities and units.   Second Edition.   Roy Inst Chem, London.

McKeague J A et al (1984).   Tentative assessment of soil survey approaches to the characterization and interpretation of air-water properties of soils.   Geoderma 34, 69-100.

McLean A and Gribble C D (1979).   Geology for civil engineers.   Allan and Unwin, London.

McLean E O (1965).   Aluminium, pp 978-998.   In: 'Methods of Soil Analysis', Part 2, Agronomy No 9; Black C A (Ed).   Am Soc of Agron, Madison, Wisconsin.

McRae S G and Burnham C P (1981).   Land evaluation, 229 pp.   Clarendon Press, Oxford.

McVean D and Robertson V C (1969).   Ecological survey of land use and soil erosion in the West Pakistan and Azad Kashmir catchment of the River Jhelum.   J Appl Ecol 6, 77-109.

MAFF (1967).   Soil potassium and magnesium.   Tech Bull 14, HMSO.

MAFF (1971).   Trace elements in soils and crops.   Tech Bull 21, HMSO.

MAFF (1973).   Lime and liming.   Tech Bull 35, HMSO.

MAFF (1975).   Soil physical conditions and crop growth.   Tech Bull 29, HMSO.

Magdoff F R and Bartlett R J (1980).   Effect of liming acid soils on potassium availability.   Soil
    Sci 129(1), 12-14.

Malkerns Research Station (1959).   Annual report of the Research Division, Dept Agric, Swaziland.

Manrique L A (1985).   Land suitability assessment for tropical forage legumes.   Geoderma 36, 57-72.

Mason P A and Amos E M (1985).   Environmental engineering applications of thermal infra-red imagery.
    In: 'Remote Sensing in Civil Engineering'.   Kennie T J M and Matthews M C.   Surrey Univ Press.

Massoud F I (1973).   Some physical properties of highly calcareous soils and the related management
    practices.   In: 'Reclamation and Management of Calcareous Soils'.   Soils Bull No 21.   FAO, Rome.

Mauritian Sugar Industry Research Institute (1973).   Twenty-first Annual Report.   Mauritian Sugar Ind
    Res Inst, Mauritius.

Mengel K (1982).   Factors of plant nutrient availability relevant to soil testing.   Plant and Soil 64,
    129-138.

Mengel K and Busch R (1982).   The importance of the potassium buffer power on the critical potassium
    level in soils.   Soil Sci 133(1), 27-32.

Meredith R M (1965).   A review of the responses to fertiliser of the crops of northern Nigeria.   Samaru
    Misc Paper No 4, Zaria.

Metson A J (1961).   Methods of chemical analysis for soil survey samples.   New Zealand DSIR Soil Bur
    Bull 12.   Govt printer, Wellington, New Zealand.

Middleton H E (1930).   Properties of soils which influence soil erosion.   Tech Bull 178.   USDA,
    Washington DC.

Milthorpe F (1960a).   Plant-water relationships in arid and semi-arid conditions.   Unesco, Paris.

Milthorpe F (1960b).   The income and loss of water in arid and semi-arid zones.   Arid Zone Research
    Paper.   Unesco, Paris.

Ministère de la Coopération (1974).   Mémento de l'Agronome, Nouvelle (10ème) Edition.   République
    Française, Ministère de la Coopération.   Techniques Rurales en Afrique, Paris.

Ministry of Defence (1971).   Military engineering.   Vol 13, Part 13: cartography.   A/GS Trg
    Publications/3129.   Army Code No 70650.   UK Ministry of Defence, London.

Mirreh H F and Ketcheson J W (1972).   Influence of soil bulk density and matric pressure on soil
    resistance to penetration.   Can J Soil Sci 52, 477-483.

Mitchell C W (1973a).   Soil classification with particular reference to the Seventh Approximation.
    J Soil Sci 24, 411-420.

Mitchell C W (1973b).   Terrain evaluation, 221 pp.   Longman, London.

Mitchell H W (1988).   Cultivation and harvesting of the arabica coffee tree, pp 43-90.   In: 'Coffee.
    Vol 4: agronomy', Clarke R J and McRae R (Eds).   Elsevier, Barking.

Mitchell R L (1964).   Trace elements in soils, pp 320-368.   In: 'Soil Chemistry', Bear F E (Ed).
    Reinhold, New York; Chapman and Hall, London.

Mohr E C J, Van Baren F A and Van Schuylenborgh J (1972).   Tropical Soils.   Geuze, Amsterdam.

Mohrmann J C J (1966).   Operational planning of integrated surveys in developing countries.
    Publications of the ITC-Unesco Centre for Integrated Surveys, Delft, Netherlands.

Mokwunye A U and Vlek P G L (Eds) (1986).   Management of nitrogen and phosphorus fertilisers in
    Sub-Saharan Africa.   Martinus Nijhoff, Dordrecht.

Money D C (1974).   Climate, soils and vegetation.   University Tutorial Press, London.

Moore A W and Bie S W (Eds) (1977).   Uses of soil information systems.   Pudoc, Wageningen, Netherlands.

Moormann F and Van Breeman N (1978).   Rice: soil, water, land.   IRRI, Los Banos, Philippines.

Morgan R P C (1986).   Soil erosion and conservation.   Longman, London.

Morin J, Benjamini Y and Michaeli A (1981).   The dynamics of soil crusting by rainfall impact and the
    water movement in the soil profile.   J Hydr 52, 321-334.

Mortvedt J J, Giordano P M and Lindsay W L (Eds) (1972).   Micronutrients in agriculture.   Soil Sci Soc
    Am, Madison, Wisconsin.

References: M to O

Mulders M A (1987).   Remote sensing in soil science.   Elsevier, Amsterdam.

Murdoch G (1970).   Soil and land capability in Swaziland.   Min Agric, Mbabane, Swaziland.

Murdoch G (1972).   Views on land capability with bibliography 1923-70.   Misc Rep 132, Land Resour Div ODM, Surbiton, England.

Murdoch G and Lang D M (1978).   Adapting to diversity in land evaluation.   Inst Brit Geog, Developing Areas Study Group, Conference on Rural Development Evaluation: cyclostyled.

Murphy L S and Walsh L M (1972).   Correction of micronutrient deficiencies with fertilisers, pp 347-387.   In: 'Micronutrients in Agriculture', Mortvedt J J et al (Eds).   Soil Sci Soc Am, Madison, Wisconsin.

Murray D B (1959).   Deficiency symptoms of the major elements in the banana.   Trop Agric 36, 100-107.

Murthy R S, Hirekerur L R, Deshpande S B and Venkatrao B V (1982).   Benchmark soils of India.   Nat Bur Soil Surv Land Use Plann, Nagpur.

N

Nair P K R (1984).   Soil productivity aspects of agroforestry.   ICRAF, Nairobi.

Nelson F (1985).   Preliminary investigation of solifluction macrofabrics.   Catena 12, 230-233.

Nelson L A (1963).   Detailed land classification for the island of Oahu.   Bull Land Study Bur Univ Hawaii No 3.

NF (1970).   Tamis et tamisage - toiles métalliques et toiles perforées dans le tamis de controle - dimensions nominales des ouvertures.   Norme Français XII - 501-1970.   AFNOR, Paris.

Ng S K and Law W M (1971).   Pedogenesis and soil fertility in West Malaysia.   Nat Resour Res Unesco 11, 129-31.

Nicholas D J D (1961).   Determination of minor element levels in soils with the Aspergillus niger method.   Seventh Int Congr Soil Sci Trans, III, 168-182.

Nielsen D R (1987).   Emerging frontiers of soil science.   Geoderma 40, 267-273.

Nielsen D R, Biggar J W and Erh K T (1973).   Spatial variability of field-measured soil-water properties.   Hilgardia 42(7), 215-259.

Nielsen D R and Bouma J (Eds) (1985).   Spatial analysis of soil data.   Pudoc, Wageningen, Netherlands.

Nielsen D R, Jackson R D, Cary J W and Evans D D (1972).   Soil water.   Am Soc of Agron, Wisconsin.

Nieuwenhuis E (1975).   Discussion of the cost, efficiency and choice of imagery.   ITC J 1, 53-69.

Nir N (1965).   State of Israel salinity survey progress report for 1963.   State of Israel Min Agric, Water Commissioner's Office, Tel Aviv.

Norman A G (Ed) (1963).   The soybean.   Academic Press, New York.

Norman A G (1978).   Soybean physiology, agronomy and utilisation.   Academic Press, New York.

Norman M J T, Pearson C J and Searle P G E (1984).   The ecology of tropical food crops.   Cambridge University Press.

Nortcliff S (1978).   Soil variability and reconnaissance soil mapping.   J Soil Sci 29, 404-419.

Northcote K H et al (1975).   Description of Australian soils, 170 pp.   CSIRO, Melbourne.

Nye P H, Craig D and Coleman N T (1961).   Ion exchange equilibria involving aluminium.   Soil Sci Soc Am Proc 25, 14-17.

O

O'Connell D J (1975).   The measurement of apparent specific gravity of soils and its relationship to mechanical composition and plant root growth, pp 298-313.   In: 'Soil Physical Conditions and Crop Production'.   MAFF Tech Bull 29, HMSO.

Odum E P (1964).   Fundamentals of ecology.   Saunders, London.

Ollier C D (1977).   Applications of weathering studies, pp 9-50.   In: 'Applied Geomorphology', Hails J R (Ed).   Elsevier, Amsterdam.

Olsen S R (1972).   Micronutrient interactions, pp 243-264.   In: 'Micronutrients in Agriculture'.
Mortvedt J J et al (Eds).   Soil Sci Soc Am, Madison, Wisconsin.

Olsen S R and Dean L A (1965).   Phosphorus, pp 1035-1049.   In: 'Methods of Soil Analysis', Black C A
(Ed).   Agronomy No 9.   Am Soc of Agron, Madison, Wisconsin.

Olson G W (1974).   Land classification.   Search Agriculture 4(7).   Cornell University Agric Exp Stn,
Ithaca, New York.

Olson G W (1981).   Soils and the environment.   Chapman and Hall, New York.

Olson R A et al (Eds) (1971).   Fertiliser technology and use.   Soil Sci Soc Am, Madison, Wisconsin.

Onwueme I C (1978).   Tropical tuber crops: yams, cassava, sweet potatoes and cocoyams.   Wiley, New York.

Orvedal A C (1977).   Comments on interpretation potentials in relation to soil survey orders, pp 19-24.
In: 'Soil Resource Inventories'.   Agron Mimeo No 77-23.   Dept Agron, Cornell Univ Ithaca, New York.

Orvedal A C (1983).   Bibliography of soils of the tropics.   USAID, Washington, D.C.

Oster J D and Rhoades J D (1975).   Calculated drainage water compositions and salt burdens resulting from
irrigation with river waters in the Western United States.   J Environmental Quality 4, 73-79.

Oster J D and Schroer F W (1979).   Infiltration as influenced by irrigation water quality.   Soil Sci Soc
Am J 43, 444-447.

Otterman J and Gornitz V (1983).   Saltation versus soil stabilization: two processes determining the
character of surfaces in arid regions.   Catena 10, 339-362.

P

Pair C H, Hinz W H, Reid C and Frost K R (Eds) (1969).   Sprinkler irrigation.   Sprinkler Association,
Washington DC.

Panabokke C R (1967).   Soils of Ceylon and fertilizer use.   Ceylon Assoc Advanc Sci.

Parr J F and Bertrand A R (1960).   Water infiltration into soils.   Adv Agron 12, 311-363.

Patrick W H and Mahapatra I C (1968).   Transformation and availability to rice of nitrogen and phosphorus
in waterlogged soils.   Adv Agron 20, 323-360.

Pearson R W (1975).   Soil acidity and liming in humid tropics.   Cornell Intl Agric Bull 30, Ithaca, NY.

Pearson R W and Adams F (1967).   Soil acidity and liming.   Agron Ser No 12.   Am Soc of Agron,
Madison, Wisconsin.

Pedro G (1983).   Structuring of some basic pedological processes.   Geoderma 31, 289-300.

Peech M (1965).   Hydrogen-ion activity.   In: 'Methods of Soil Analysis', Part 2, Agronomy No 9;
Black C A (Ed).   Am Soc of Agron, Madison, Wisconsin.

Peech M, Alexander L T, Dean L A and Reed J F (1947).   Methods of soil analysis for soil fertility
investigations.   USDA Circ 757, Washington DC.

Penman H L (1954).   The calculation of irrigation need.   HMSO, London.

Petersen M S (1980).   Recommendations for use of SI units in hydraulic engineering.   Pro Am Soc Civ Eng
106, HY12, 1981-1994.

Petersen R G and Calvin L D (1965).   Sampling, pp 54-72.   In: 'Methods of Soil Analysis', Part 1.
Agronomy No 9; Black C A (Ed).   Am Soc of Agron, Madison, Wisconsin.

Philip J R (1954).   An infiltration equation with physical significance.   Soil Sci 77, 153.

Phillipson J (1971).   Methods of study in quantitative soil ecology.   Intl Biol Prog Handbk No 18.
Blackwell, Oxford.

Phosyn Chemicals Ltd (1980).   The important role of trace elements in plant nutrition.   Phosyn
Chemicals Ltd, Pocklington, York.

Pidgeon J D (1972).   The measurement and prediction of available water capacity of ferallitic soils in
Uganda.   J Soil Sci 23, 431-441.

Pierre W H (1965).   Plant environment and efficient water use.   Am Soc of Agron, Wisconsin.

Pitty A F (1979).   Geography and soil properties.   Methuen.

Pizer N H (1966).   Reclamation of land from the sea.   Chem Ind 791-795.

Polhamus L G (1962).   Rubber - botany, production and utilisation.   Leonard Hill, London.

Pomeroy D E (1983).   Some effects of mound-building termites on the soils of a semi-arid area of Kenya.
J Soil Sci 34, 585-575.

Ponnamperuma F N (1964).   Dynamic aspects of flooded soils and the nutrition of the rice plant,
pp 295-328.   In: 'The Mineral Nutrition of the Rice Plant', Johns Hopkins Univ Press, Baltimore, Md.

Pons L J (1970).   Acid sulphate soils (soils with cat clay phenomena) and the prediction of their origin
from pyrites muds, pp 93-107.   In: 'From Field to Laboratory'.   Publ 16, Fysich Geografisch en
Bodemkundig Laboratorium, Wageningen, Netherlands.

Pons L J and Zonneveld I S (1975).   Soil ripening and classification.   ILRI Publ No 13, Wageningen,
Netherlands.

Prasad M and Byrne E (1975).   Boron source and lime effects on the yield of three crops grown in peat.
Agron J 67, 553-556.

Pratt P F and Blair F L (1961).   A comparison of three reagents for the extraction of aluminium from
soils.   Soil Sci 91, 357-359.

Prentice A N (1972).   Cotton, with special reference to Africa.   Longman, London.

Pringle J (1975).   Soil stability measurement.   Newcastle University: cyclostyled.

Pritchard T L (1983).   Relationships of irrigation water salinity and soil salinity.   Calif Agric 37,
11-15.

Pritchett W L (1979).   Properties and management of forest soils, 512 pp.   Wiley, Chichester.

Purnell M F (1978).   Progress and problems in the application of land evaluation in FAO projects in
different countries, pp 81-88.   World Soil Resour Rep 49.   FAO, Rome.

Purnell M F (1979).   The FAO approach to land evaluation and its application to land classification for
irrigation, pp 4-8.   World Soil Resour Rep 50.   FAO, Rome.

Purnell M F (1988).   Methodology and techniques for land use planning in the tropics.   Soil Surv and
Land Eval 8 (1), 9-22.

Purseglove J W (1968).   Tropical crops: dicotyledons.   Longman, London.

Purseglove J W (1972).   Tropical crops: monocotyledons.   Longman, London.

Purseglove J W (1975).   Tropical crops: monocotyledons and dicotyledons.   Longman, London.

Pushparajah E (1983).   Use of soil survey and fertility evaluations in hevea cultivation.   Paper
presented at Fourth International Forum on Soil Taxonomy and Agrotechnology Transfer, Bangkok.

Pushparajah E and Amin L L (1977).   Soils under hevea and their management in Peninsular Malaysia,
188 pp.   Rubber Res Inst, Kuala Lumpur.

Putnam R J and Wratten S D (1984).   Principles of ecology.   Croom Helm, London and Canberra.

Puvaneswaran P and Conacher A J (1983).   Extrapolation of short-term process data to long-term landform
development: case study from South-Western Australia.   Catena 10, 321-337.

Q

Quansah C (1981).   The effect of soil type, slope, rain intensity and their interactions on splash
detachment and transport.   J Soil Sci 32, 215-224.

Quay J R (1965).   Site evaluation and soil survey.   J Am Inst Archit 43, 69-73.

Quirk J P (1957).   Effect of electrolyte concentration on soil permeability and water entry in irrigation
soils.   Intl Comm on Irrig Drainage, Third Congr 8, 114-123.

R

Ragg J M and Henderson R (1980).   A reappraisal of soil mapping in an area of southern Scotland, Part 1.
The reliability of four soil mapping units and the morphological variability of their dominant taxa.
J Soil Sci 31(3), 559-572.

Rauschkolb R S (1971).   Land degradation.   Soils Bull No 13.   FAO, Rome.

Rayner J H (1966).   Classification of soils by numerical methods.   J Soil Sci 17, 79-92.

Reinhart K B (1961).   The problem of stones in soil-moisture measurement.   Soil Soc Sci Am Proc 25, 268-270.

Reisenauer H M, Walsh L M and Hoeft R G (1973).   Testing soils for S, B, Mo and Cl, pp 173-200.   In: 'Soil Testing and Plant Analysis', Walsh L M and Beaton J D (Eds).   Soil Sci Soc Am, Madison, Wisconsin.

Renger M and Henseler K L (1971).   Investigation and numerical classification of impedance classes in Stagnogley soils of Lower Saxony.   Niedersächsischen Landesamtes fur Bodenforschung, Hanover.

Reynolds L and Simmonds D (1982).   Presentation of data in science, 209 pp.   Martinus Nijhoff, The Hague, Netherlands.

Rhoades J D (1977).   Potential for using saline agricultural drainage waters for irrigation.   Proc Water Management for Irrigation and Drainage.   ASCE, Reno, Nevada, 85-116.

Rhoades J D (1982).   Reclamation and management of salt-affected soils after drainage.   Proc First Ann W Provincial Conf Rationalisation of Water and Soil Research and Management, Alberta, Canada, 123-197.

Rhoades J D and Merrill S D (1976).   Assessing the suitability of water for irrigation: theoretical and empirical approaches, pp 69-110.   In: Soils Bull No 31.   FAO, Rome.

Richards K, Arnett R R and Ellis S (Eds) (1985).   Geomorphology and soils, 441 pp.   Allen andUnwin, London.

Richards L A (Ed) (1954).   Diagnosis and improvement of saline and alkali soils.   Handbook 60.   USDA, Washington DC.

Richardson C W, Foster G R and Wright D A (1983).   Estimation of erosion index from daily rainfall amount. Trans ASAE 26(1), 153-156.

Ridgway R B and Jayasinghe G (1986).   The Sri Lanka land information system.   Soil Surv and Land Eval 6, 20-25.

Ripple W J (Ed) (1986).   GIS for resource management: an overview.   Am Soc Photogram Rem Sens and Am Cong Surv Mapp, Falls Church, Virginia.

Riquier J (1965).   Productivité arachide en fonction des charactéristiques du sol.   ORSTOM, Tananarive, Madagascar.

Riquier J, Bramao D L and Cornet J P (1970).   A new system of soil appraisal in terms of actual and potential productivity (AGL/TESR/70/6).   FAO, Rome.

Ritchie J T (1981).   Soil water availability.   Plant and Soil 58, 327-328.

Robertson L S, Warncke D D and Baker J D (1980).   Chemical test variability within soil types.   Res Rep, Agric Exptl Stn No 390.   Dept Crop and Soil Sci, Michigan State University.

Rogoff M J (1982).   Computer display of soil survey interpretations using a geographic information system. Soil Surv and Land Eval 2 (2), 37-41.

Roose E J (1977).   Application of the universal soil loss equation of Wischmeier and Smith in West Africa, pp 177-187.   In: 'Soil Conservation and Management in the Humid Tropics', Greenland D J and Lal R (Eds).   Wiley, Chichester.

Ross Institute (1974).   Preservation of personal health in warm climates.   The Ross Institute of Tropical Hygiene, London.

Ross L, Nababsing P and Wong You Cheong (1974).   Residual effect of calcium silicate applied to sugarcane soils.   Proc Int Soc Sugarcane Technol, XV, Vol 2, 539-542.

Rubin J (1966).   Theory of rainfall uptake by soils initially drier than their field capacity and its applications.   Water Resour Res 2, 739.

Ruellan A (1984).   Les apports de la connaissance des sols intertropicaux au développement de la pédologie.   Science du Sol No 2, Paris.

Russell E W (1966).   Soil survey as a part of pre-investment.   Publications of the ITC-Unesco Centre for Integrated Surveys, Delft, Netherlands.

Russell E W (1973).   Soil conditions and plant growth.   Tenth Edition.   Longman, London.

Russell M B (1977).   Plant root systems.   McGraw-Hill, New York.

Ruthenberg H (1980).   Farming systems in the tropics.   Oxford Science Publications.   Oxford Univ Press.

S

Salter P J and Goode J E (1967).   Crop responses to water at different stages of growth.   Res Rep 2, Comm Bureau of Horticulture and Plantation crops, East Malling, Kent.

Salter P J and Williams J B (1965).   Influence of texture on the moisture characteristics of soils, II. Available water capacity and moisture release characteristics.   J Soil Sci 16, 310-317.

Salter P J and Williams J B (1967).   Influence of texture on the moisture characteristics of soils, IV. Method of estimating the available water capacities of profiles in the field.   J Soc Sci 18, 174-181.

Salter P J and Williams J B (1969).   Influence of texture on the moisture characteristics of soils, V. Relationships between particle size composition and moisture contents at the upper and lower limits of available water.   J Soil Sci 20, 126-131.

Samson J A (1980).   Tropical fruits, 250 pp.   Longman, London.

Samuels G (1979).   Silicon: a forgotten element@   Sugar J, July, 11.

Samuels G (1982).   A matter of balance.   Sugar J 45(3), 33-37.

Sanchez P A (1976).   Properties and management of soils in the tropics.   Wiley, Chichester.

Sanchez P A, Couto W and Buol S W (1982).   The fertility capability soil classification system: interpretation, applicability and modification.   Geoderma 27, 283-309.

Sanchez P A and Tergas L E (1979).   Pasture production in acid soils of the tropics.   CIAT, Cali.

Saouma E (Ed) (1975).   Ad hoc expert consultation on land evaluation.   World Soil Resour Rep 45.   FAO, Rome.

Sawhney B L (1968).   Aluminium interlayers in layer silicates.   Effect of OH/Al ratio of Al solution, time of reaction and type of structure, pp 157-163.   In: 'Clays and Clay Minerals', Bailey S W (Ed).   Pergamon Press, Oxford.

Schmugge T J, Jackson T J and McKim H L (1980).   Survey of methods for soil moisture determination. Water Resour Res 16(6), 961-979.

Schnitzer M and Khan S U (Eds) (1978).   Soil organic matter, 320 pp.   Elsevier, Amsterdam.

Schofield R K (1935).   The pF of the water in soil.   Trans Third Intl Congr Soil Sci 2, 37-48.

Schofield R K and Taylor A W (1955).   Measurements of the activities of bases in soils.   J Soil Sci 6, 137-146.

Schreier H and Zulkifli M A (1983).   Numerical assessment of soil survey data for agricultural management and planning.   Soil Surv and Land Eval 3 (2), 41-53.

Schroeder D (1984).   Soils: facts and concepts (translated by P A Gething).   German Fourth Edition. Intl Potash Inst, Bern.

Schultink G (1987).   The CRIES resource information system: computer-aided land resource evaluation for development planning and policy analysis.   Soil Surv and Land Eval 7 (1), 47-62.

Schwab G O et al (1966).   Soil and water conservation engineering.   Second Edition.   Wiley, New York.

Shaxson T F et al (1977).   A land husbandry manual.   Techniques of land use planning and physical conservation.   Land Husbandry Branch, Min Agric and Nat Resour, Malawi.

Shaxson T F (1981).   Determining erosion hazard and land use incapability - a rapid subtractive method. Soil Surv and Land Eval 1 (3), 44-50.

Shell Int Chemical Co Ltd (1960).   A pocket book of agricultural tables.   Shell Int Chem Co Ltd, London.

Sheng T C (1972).   A treatment-orientated land capability classification scheme for hilly marginal land in the humid tropics.   J Sci Res Coun Jamaica 3, 94-112.

Sherman G D (1969).   Crop growth response to application of calcium silicate to tropical soils in the Hawaiian Islands.   AGRI Digest 18, 11-19.

Shubert J R, Muth O H, Oldfield J E and Remmert L F (1961).   Fed Proc 20, 689-694.

Shuman L M (1980).   Effects of soil temperature, moisture and air-drying on extractable manganese, iron, copper and zinc.   Soil Sci 130(6), 336-343.

Sillanpää M (1959).   Comparison of some field methods of measuring hydraulic conductivity of soils. Acta Agricultura Scandanavica IX, 59-68.

Sillanpää M (1972). Trace elements in soil and agriculture. Soils Bull No 17. FAO, Rome.

Sillanpää M (1982). Micronutrients. Soils Bull No 48. FAO, Rome.

Simmonds N W (1966). Bananas. Second Edition. Longman, London.

Simonson R W (1982). Soil classification, pp 103-129. In: 'Handbook of Soils and Climate in Agriculture', Kilmer V J (Ed). CRC Press Inc, Boca Raton, Florida.

Simpson K (1983). Soil. Longman, London.

Singh M and Singh N (1979). The effect of forms of selenium on the accumulation of selenium, sulphur, and forms of nitrogen and phosphorus in forage cowpea. Soil Sci 127(5), 264-269.

Singh S S and Kanwar J S (1963). J Ind Soc Soil Sci 11, 283.

Sivarajasingham S, Alexander L T, Cady J G and Cline M G (1962). Laterite. Advances in Agronomy No 14. Academic Press, New York.

Skinner F A and Uomala P (1986). Nitrogen fixation with nonlegumes. Martinus Nijhoff, Dordrecht.

Slater C S (1957). Cylinder infiltration for determining rates of irrigation. Soil Sci Soc Am Proc 21, 457.

Slater C S (1967). Plant-water relationships. Academic Press, New York.

Smartt J (1976). Tropical pulses, 348 pp. Longman, London.

Smith D G (1981). The Cambridge encyclopedia of earth sciences. Cambridge Univ Press, Cambridge, UK.

Smith P D (1975). Soil and irrigation studies on Iran Shellcott land, Dezful, Iran. Final Report. Department of Geography, University of Durham, UK.

Smitham G M (1970). Permeable fill investigations. Appendix 5, FDEU Report 1969, 27-28.

Smyth A J (1966). Cocoa Growers Bull, No 6, 7. Cadbury Ltd, Birmingham.

Smyth A J (1970). The preparation of soil survey reports. Soils Bull No 9. FAO, Rome.

Smyth A J, Eavis B W and Williams J B (1979). Diagnostic criteria for evaluating land for irrigation. World Soil Resour Rep 50, 14-20. FAO, Rome.

Snoeck J (1988). Cultivation and harvesting of the robusta coffee tree, pp 91-128. In: 'Coffee. Vol 4: agronomy'. Clarke R J and McRae R (Eds). Elsevier, Barking.

Snyder E F (1935). Methods for determining the hydrogen-ion concentration of soils. Circ 56, USDA, Washington.

Soane B D (1975). Studies on some soil physical properties in relation to cultivations and traffic, pp 160-182. In: 'Soil Physical Conditions and Crop Production'. MAFF Tech Bull 29. HMSO.

Soil Conservation Society of America (1977). Soil erosion: prediction and control. Soil Conserv Soc Am, Ames, Iowa.

Soil Sci Soc Am (1979). Glossary of soil science terms. Soil Sci Soc Am, Madison, Wisconsin.

Soil Survey Staff (1960). Soil classification - a comprehensive system; Seventh Approximation. USDA, Washington. (Now superseded by Soil Taxonomy - see USDA, 1975b).

Soil Survey Staff (1966). Aerial-photo interpretation in classifying and mapping soils. Agric Handbook No 294. USDA, Washington DC.

Sombroek W G and Van der Weg R F (1983). Soil maps and their legends. Soil Surv and Land Eval 3 (3), 80-87.

Sopher C D and Baird J V (1978). Soils and soil management. Reston Publ Co Inc, Reston, Virginia.

Spedding D J (1969). Nature (Lond) 224, 1229-1231.

Sprague G F (Ed) (1977). Corn and corn improvement. Agron No 18. Am Soc Agron, Madison, Wisc.

Sprague H B (1964). Hunger signs in crops. Third Edition, 462 pp. Pennsylvania State University, USA.

Stakman W P (1974). Measuring soil moisture. Drainage principles and applications. Pub 16 Vol 3. Intl Inst Land Reclam and Improv, Wageningen, Netherlands.

Stewart G A (Ed) (1968). Land evaluation. Macmillan, Melbourne.

Stiegler S E (Ed) (1978). A dictionary of earth sciencies. Pan Books, London.

Stinson C H (1953). Soil Sci 75, 31-36.

Stobbs A R (1970). Soil survey procedures for development purposes. In: 'New Possibilities and Techniques for Land Use and Related Surveys', Cox I H (Ed). Geographical Publications, Berkhamsted, Herts.

Stocking M A and McCormack D E (1986). Soil potential ratings, II. An evaluation of the method in Zimbabwe. Soil Surv and Land Eval 6 (3), 73-78.

Stoops G (Ed) (1986). Soil micromorphology. Van Nostrand Reinhold, Princeton, New Jersey.

Storie R E (1933). Index for rating the agricultural value of soils. Bull Calif Agric Exp Sta No 556.

Storie R E (1964). Handbook of soil evaluation. Univ Calif Press, Berkeley.

Storie R E (1978). The Storie Index soil rating revised. Spec Publ Div Agric Sci Univ Calif No 3203.

Stout P R, Meagher W R, Pearson G A and Johnson G H (1951). Molybdenum nutrition of crop plants. I. The influence of phosphate and sulphate on the absorption of molybdenum from soils and solution cultures. Plant and Soil 3, 51-87.

Suarez D L (1981). Relation between pHc and sodium adsorption ratio (SAR) and an alternate method of estimating SAR of soil or drainage waters. Soil Sci Soc Am J 45, 469-475.

Sutton S L, Whitmore T C and Chadwick A C (1983). Tropical rain forest ecology and management. Blackwell Scientific, Oxford.

Swaine D J (1969). The trace element content of soils. Tech Comm 48. Commonwealth Agricultural Bureaux, Farnham Royal, Bucks.

Swindale L D (Ed) (1978). Soil resource data for agricultural development. University of Hawaii, Honolulu.

Sykes J (1969). Reconsideracao do conceito de umidade de murchamento permanente. Turrialba 19, 525-530.

Sys C and Frankart R (1972). Land capability in the humid tropics. Sols Afr 15, 153-175.

Sys C and Riquier J (1980). Ratings of FAO-Unesco soil units for specific crop production, pp 55-96. In: 'Report on the second FAO-UNFPA expert consultation on land resources for populations of the future.' FAO, Rome.

Sys C, Van Wambeke A, Frankart R et al (1961). La cartographie des sols au Congo, ses principes et ses méthodes. INEAC, Ser Tech 66. Institut National pour l'Etude Agronomique du Congo, Brussels.

T

Tai E A (1977). Banana, pp 441-460. In: 'Ecophysiology of Tropical Crops', Alvim P T and Kozlowski T T (Eds). Academic Press, New York.

Talsma T (1960). Comparison of field methods of measuring hydraulic conductivity. Trans Congr Irrig and Drainage. IV(6), C145-C156.

Talsma T (1968). Environmental studies of the Coleambally irrigation area and surrounding districts. Part III. Soil salinity. Water Cons and Irrig Comm NSW Bull No 2 (Land Use Series).

Tan K H (1982). Principles of soil chemistry. Dekker, New York.

Taychinov S N (1971). A method for rating soil quality. Soviet Soil Science 1, 40-49.

Taylor H M and Gardner H R (1963). Penetration of cotton seedling tap roots as influenced by bulk density, moisture content and strength of soil. Soil Sci 96, 153-156.

Taylor H M, Roberson G M and Parker J J (1966). Soil strength-root penetration relations for medium to coarse textured soil materials. Soil Sci 102, 18-22.

Taylor S A (1958). The activity of water in soils. Soil Sci 86, 83-90.

Taylor S A (1965). Managing irrigation water on the farm. Trans Am Soc Agric Eng 8(3), 433-436.

Taylor S A and Ashcroft G L (1972). Physical edaphology, 533 pp. W H Freeman and Co, San Francisco.

Taylor S A and Stewart G L (1960). Some thermodynamic properties of soil water. Soil Sci Soc Am Proc 24, 243-247.

Tea Research Institute of East Africa (1970). Teagrowers handbook, 1969. Kericho, Kenya.

Terman G L (1982). Fertiliser sources and composition, pp 295-330. In: 'Handbook of Soils and Climate in Agriculture', Kilmer V J (Ed). CRC Press Inc, Boca Raton, Florida.

Terzaghi K and Peck R E (1948). Soil mechanics in engineering practice. Wiley, Chichester.

Tessens E (1984). Clay migration in upland soils of Malaysia. J Soil Sci 35, 615-624.

Thomas D G (1970). Finger millets, pp 145-153. In: 'Agriculture in Uganda', Jameson J D (Ed), Oxford University Press.

Thomas P, Lo F K C and Hepburn A J (1976). Land capability classification of Sabah. Vol 1, Introduction and Summary. Land Resour Study No 25. Land Resour Div ODM, Tolworth, England.

Thomas R F, Blakemore L C and Kinloch D I (1980). Flow diagram keys for Soil Taxonomy. New Zealand Soil Bur Sci Rep 39.

Thompson G D (1976). Water use by sugarcane. SA Sugar J 61, 593-600 and 627-635.

Thompson H A (1978). Land evaluation for sugarcane production, pp 73-80. World Soil Res Rep 49. FAO, Rome.

Thorne D W and Thorne M D (1979). Soil water and crop production. AVI Publishing Company, Connecticut.

Thorne D W and Peterson H B (1954). Irrigated soils. Second Edition. Blakiston, New York.

Thornthwaite C W (1948). Approach towards a rational classification of climate. Geogr Rev 38, 55-94.

Thorp J and Smith G D (1949). Higher categories of soil classification: order, suborder and great soil group. Soil Sci 67, 117-126.

Times Books (1980). The Times atlas of the world. Comprehensive Edition. Sixth Edition. Times Books, in collaboration with John Bartholomew, London.

Tisdale S L and Nelson W (1966). Soil fertility and fertilisers. Second Edition. Macmillan, NY.

Tomlinson P R, Beckett P H T, Bannister P and Marsden R (1977). Simplified procedures for routine soil analyses. J Appl Ecol 14, 253-260.

Tomlinson P R (1970). Variations in the usefulness of rapid soil mapping in the Nigerian savanna. J Soil Sci 21, 162-172.

Topp G C et al (1980). The variation of in-situ measured soil water properties within soil map units. Can Jour Soil Sci 60(3), 497-510.

Tovey R and Pair C H (1963). A method for measuring water intake rate into soil for sprinkler design. Sprinkler Irrigation Association. Open Techn Conf Proc, 109-118.

Trewartha G T (1968). An introduction to climate. Fourth Edition. McGraw-Hill, New York, USA.

Troll C (1966). Seasonal climates of the earth. Springer, Berlin.

Truog E (1930). J Am Soc Agron 22, 874.

Truog E (1948). Lime in relation to availability of plant nutrients. Soil Sci 65, 1-7.

Turner D P and Sumner M E (1978). The influence of initial soil moisture content on field measured infiltration rates. Water SA, 4(1), 18-24. Soil Irrig Res Inst, Pretoria.

Turner P D (1981). Oil palm diseases and disorders. Oxford University Press.

Turner P D and Gillbanks R A (1982). Oil palm cultivation and management. Incorporated Society of Planters, Kuala Lumpur, Malaysia.

Turner R K (Ed) (1985). Land evaluation - financial, economic and ecological approaches. Soil Surv and Land Eval 5 (2), 21-33.

Twohig A (Ed) (1986). Liklik Buk. A sourcebook for development workers in Papua New Guinea. Second English Edition. Liklik Buk Information Centre, Lae, PNG.

Twyford I T (1973). Interpretation of soil analysis results. Banana Research Note No 9. Windward Islands Banana Research Scheme.

Tyurin I V, Gerasimov I P, Ivanova E N and Nosin V A (Eds) (1965). Soil survey. Israel Prog Sci Transl, Jerusalam.

U

Uehara G and Gillman G (1981). The mineralogy, chemistry and physics of tropical soils with variable charge clays, 170 pp. Westview Press, Boulder, Colorado.

Uehara G, Trangmar B B and Yost R S (1985). Spatial variability of soil properties. Pudoc, Wageningen, Netherlands.

Uexküll H R V (1986). Efficient fertilizer use in acid upland soils of the humid tropics. Fertilizer and Plant Nutrition Bulletin No 10. FAO, Rome.

Unesco (1960). Plant-water relationships in arid and semi-arid conditions - Milthorpe F. Paris.

Unesco (1971). Pedogenesis and soil fertility in West Malaysia - Ng S K and Law W W. Nat Resour Res 11, 129-131.

Unger D (1966). Soils and land use planning. Am Soc Plann Off. Philadelphia, Pennsylvania.

United States Geological Survey (1982). Landsat data users notes. Issue 23. Eros Data Centre, Sioux Falls, South Dakota.

University of California Committee of Consultants (1974). Guidelines for interpretation of water quality for agriculture. Univ Calif Davis, 13 pp.

Upton M (1973). Farm management in Africa. Oxford University Press.

USBR (1953). See USDI.

USBR (1964). Land drainage techniques and standards. Reclamation Instructions, USBR, Denver, Colorado.

USBR (1974). The earth manual. Second Edition. USBR, US Govt Printing Office, Washington DC.

USDA (1935). Methods for determining the hydrogen-ion concentration of soils - Snyder E F. Circ 56. USDA, Washington DC.

USDA (1947). Methods of soil analysis for soil fertility investigations - Peech M, Alexander L T, Dean L A and Reed J F. Circ 757. USDA, Washington DC.

USDA (1951 and supplement 1962). Soil survey manual - Kellogg C et al. Handbook 18. USDA, Washington DC.

USDA (1954a). Estimation of available phosphorus in soils by extraction with sodium bicarbonate - Olsen S R, Cole C V, Watanabe F S and Dean L A. Circ 939. USDA, Washington DC.

USDA (1954b). Diagnosis and improvements of saline and alkali soils. Richards L A (Ed). Handbook No 60. USDA, Washington DC.

USDA (1954c). Field manual for research in agricultural hydrology. Agric Handbook No 225. US Government Printing Office, Washington DC.

USDA (1961). Land capability classification - Klingbiel A A and Montgomery P H. Soil Conserv Agric Handbook No 210. USDA, Washington DC.

USDA (1965). Predicting erosion losses from cropland east of the Rocky Mountains - Wischmeier W H and Smith D D. Handbook No 282. USDA, Washington DC.

USDA (1975a). National soil survey technical work planning conference, Orlando, Florida. Report of Committee No 7 - Kinds of soil survey. Soil Conservation Service, USDA, Washington DC.

USDA (1975b). Soil taxonomy. Agric Handbk No 436. Soil Conservation Service, USDA, Washington DC.

USDA (1978). Soil potential ratings. National Soil Handbook Notice 31.

USDA (1982). Amendments to Soil Taxonomy. National Soil Taxonomy Handbook 430-VI-Issue No 1. Soil Conservation Service, USDA, Washington DC.

USDI (1953). Bureau of Reclamation manual. Vol V. Irrigated land use. USDI, Washington DC.

Utomo W H and Dexter A (1981). Soil friability. J Soil Sci 32, 203-214.

V

Vakilian M and Mahler P J (1970). Method of land evaluation applicable to Iran. Publ Soil Inst Iran No 249. Tehran.

Valdivia V S (1981). Research progress carried out in Peru on the effects of salinity on sugarcane productivity. Turrialba, 31, 237-244.

Valentine K W G (1986). Soil resource surveys for forestry. Clarendon Press, Oxford.

Valentine K W G and Chang D (1980). Map units in controlled and uncontrolled legends on some Canadian soil maps. Can J Soil Sci 60, 511-516.

Van Alphen J G and Romero R F L (1971). Gypsiferous soils. Bull 12, ILRI, Wageningen, Netherlands.

Van Barneveld G W (1977). FAO-UNDP soil surveys and land evaluations for the Second Development Programme of the Cameroon Development Corporation. Tech Rep No 7. Soil Sci Dept, Ekona.

Van Beers W F J (1974). Private communication quoted in Drainage principles and applications. Vol 3, p 129. Publ 16, III. ILRI, Wageningen, Netherlands.

Van Beers W F J (1976). The auger hole method, a field measurement of the hydraulic conductivity of soil below the water table. Fourth Edition. Bull 1, 9-32. ILRI, Wageningen, Netherlands.

Van der Kevie W (Ed) (1976). Manual for land suitability classification for agriculture in the Sudan. Tech Bull Soil Surv Admin No 21. Wad Medani, Sudan.

Van Wambeke A (1982). Calculated soil moisture and temperature regimes of South America. Soil Management Support Services, USDA, Washington DC.

Vanselow A P (1932). Equilibria of the base exchange - bentonite, permites, soil colloids and zeolites. Soil Sci 33, 95-113.

Vaughan D and Malcolm R E (1985). Soil organic matter and biological activity. Martinus Nijhoff, Dordrecht.

Veihmeyer F J and Hendrickson A M (1950). Soil and moisture in relation to plant growth. Ann Rev Plant Phys 2(2), 22.

Ven Te Chow (1959). Open-channel hydraulics. McGraw-Hill Inc, New York, USA.

Verheye W and Tessens E (1977). The latosol and podzolic soil concepts in Indonesia, pp 168-174. In: 'Proceedings of the Clamatrops Conference', Joseph K T (Ed). Malay Soc Soil Sci, Kuala Lumpur.

Verstappen H T (1983). Applied geomorphology, 437 pp. Elsevier, Amsterdam.

Vieillefon J (1967). Etudes pédohydrologiques au Togo. Docum UNDP/FAO/SF:13/TOG. FAO, Rome.

Vink A P A (1963). Planning of soil surveys in land development. Veenman en Zonen, Wageningen, Neth.

Vink A P A (1975). Land use in advancing agriculture. Springer, Berlin.

Vink A P A (1983). Landscape ecology and land use. Longman, London.

Vizier J F (1983). Etude des phénomènes d'hydromorphie dans les sols des régions tropicales à saisons contrastées. ORSTOM, Paris.

Vlek P G L (Ed) (1985). Micronutrients in tropical food crop production. Martinus Nijhoff, Dordrecht.

Vos J W E (1954). Index numbers of productivity in agriculture. Landbouwk Tijdschr 66, 451-458. The Hague.

Vrba J (Ed) (1986). Impact of agricultural activities on groundwater. Intl Contributions to Hydrogeology No 5.

W

Walker B (1983). Glossary of agricultural terms relating to soil and water. Merlin Books, Braunton.

Walker P H, Beckmann G G and Brewer R (1984). Definition and use of the term pedoderm. J Soil Sci 35, 505-510.

Wallwork J A (1976). The distribution and diversity of soil fauna. Academic Press, London.

Walsh L M and Beaton J D (Eds) (1973). Soil testing and plant analysis. Revised Edition. Soil Sci Soc Am, Madison, Wisconsin.

Walter H (1971). Ecology of tropical and sub-tropical vegetation (translated by Muller-Dombois D). Oliver and Boyd, Edinburgh.

Warkentin B P (1974). Physical properties related to clay minerals in soils of the Caribbean. Trop Ag (Trinidad) 51(2), 279-287.

Warrick A W and Nielsen D R (1980). Spatial variability of soil physical properties in the field, pp 319-344. In: 'Applications of Soil Physics', Hillel D (Ed). Academic Press, New York.

Watson J P (1977).   The use of mounds of the termite Macrotermes falciger as a soil amendment.   J Soil Sci 28, 664-672.

Webster C C and Wilson P N (1980).   Agriculture in the tropics.   Second Edition.   Longman, London.

Webster R (1968a).   Fundamental objections to the Seventh Approximation.   J Soil Sci 19, 354.

Webster R (1968b).   Soil classification in the United States: a short review of the Seventh Approximation.   Geogr J 134, 394-396.

Webster R (1977).   Quantitative and numerical methods in soil classification and survey.   Clarendon Press, Oxford.

Webster R (1985).   Quantitative spatial analysis of soil in the field.   Advances in Soil Science No 3, Springer-Verlag, New York.

Webster R and Beckett P H T (1972).   Matric suctions to which soils in South-Central England drain. J Agric Sci 78, 379-387.

Weiss E A (1983).   Oilseed crops.   Longman, London.

Werner D (1980).   Where there is no doctor.   Macmillan, London.

Western S (1978).   Soil survey contracts and quality control.   Oxford University Press, Oxford.

Westin F C and Frazee C J (1976).   Landsat data, its use in a soil survey programme.   Soil Sci Soc Am J 40, 81-87.

Whisler F D and Bouwer H (1970).   Comparison of methods for calculating vertical drainage and infiltration in soils.   J Hydrol 10, 1.

White L P (1977).   Aerial photography and remote sensing for soil survey.   Clarendon Press, Oxford.

White R E (1987).   Introduction to the principles and practice of soil science.   Second Edition. Blackwell Scientific, Oxford.

Whitehead D C (1975).   J Sci Food Agric 26, 361-367.

Whitten D G A and Brooks J R V (1972).   A dictionary of geology.   Penguin Books, London.

WHO (1982).   Vaccination certificate requirements for international travel and health advice to travellers.   World Health Organisation, Geneva, Switzerland.

Whyte R O (1976).   Land and land appraisal.   Junk B V, The Hague.

Wiggins S L (1981).   The economics of soil conservation in the Acelhuate River Basin, El Salvador. pp 399-417.   In: 'Soil Conservation - Problems and Prospects', Morgan R P C (Ed).   Wiley, Chichester.

Wilcox L V (1958).   The quality of water for irrigation use.   Inf Bull 197.   USDA, Washington.

Wilcox L V (1966).   Tables for calculating the pH_c values of waters.   Salinity laboratory, mimeo. Riverside, California.

Wild A (1987).   Russell's soil conditions and plant growth.   Eleventh Edition.   Longman, London.

Williams B G, Greenland D J, Lindstrom G R and Quirk J P (1966).   Techniques for the determination of the stability of soil aggregates.   Soil Sci 101, 157-163.

Williams C N (1975).   The agronomy of the major tropical crops.   Oxford University Press.

Williams C N and Chew W Y (1979).   Tree and field crops of the wetter regions of the tropics.   Longman, London.

Williams C N and Joseph K T (1970).   Climate, soil and crop production in the humid tropics.   Oxford University Press, Singapore.

Windward Islands Banana Research Scheme (1970).   Annual Report.

Winger R J (1956).   Field determination of hydraulic conductivity above a water table.   1956 Winter Meeting ASAE.   Chicago, Illinois.

Wischmeier W H, Johnson C B and Cross B V (1971).   A soil erodibility nomograph for farmland and construction sites.   J Soil and Water Conservation 26(5), 189-193.

Wischmeier W H and Smith D D (1965).   Predicting rainfall erosion losses from cropland east of the Rocky Mountains.   Handbook No 282.   USDA, Washington DC.

Wischmeier W H and Smith D D (1978). Predicting rainfall erosion losses. A guide to conservation planning. Agric Handbook No 537. USDA, Washington DC.

Withers B and Vipond S (1974). Irrigation design and practice. Batsford, London.

Wong I F T (1966). Soil suitability classification for dryland crops in Malaya. Proc 2nd Malay Soil Conf, 153-159. Kuala Lumpur.

Wong You Cheong (1970). The residual effect of calcium silicate applications on sugarcane. Eighteenth Annual Rep. Mauritian Sugar Ind Res Inst, Mauritius.

Wood C A R (1975). Cocoa. Third Edition. Longman, London.

Wood G (1978). BAI pioneer campsite equipment. Unpublished BAI internal document, BAI, London.

World Bank (1980). AGR Technical Note No 2. A study of tree crop farming systems in the lowland humid tropics. Vols 1 and 2. World Bank, Washington DC.

World Bank (1981). Guidelines for the use of consultants by World Bank borrowers and by the World Bank as an executing agency. World Bank, Washington DC.

World Health Organisation. See under WHO.

Wosten J H M, Bannink M H and Bouma J (1987). Land evaluation at different scales: you pay for what you get. Soil Surv and Land Eval 7 (1), 13-24.

Wright J W (1982). Land surveying for soil surveys. Clarendon Press, Oxford.

Wright M J (Ed) (1976). Plant adaptation to mineral stress in problem soils. Special Publ, Cornell Univ Agric Exp Stn. Ithaca, New York.

Wrigley G (1982). Tropical agriculture. Fourth Edition. Longman, London.

Y

Yaalon D H (Ed) (1971). Paleopedology. Intl Soc Soil Sci and Israel Univ Press, Jerusalem.

Yaalon D H (Ed) (1982). Aridic soils and geomorphic processes, 219 pp. Catena Supp No 1.

Yadav B R, Rao N H, Paliwal K V and Sarma P B (1979). Comparison of different methods for measuring soil salinity under field conditions. Soil Sci 127(6), 335-339.

Yaron B, Danfors E and Vaadia Y (1973). Arid zone irrigation. Ecol Stud Vol 5. Sporinger-Verlag, NY.

Yates R A (1964). Yield depression due to phosphate fertiliser in sugarcane. Aust J Agric Res 15(4), 537-547.

Yates R A (1978). The environment for sugarcane, pp 58-72. In: 'Land Evaluation Standards for Rainfed Agriculture'. World Soil Res Rep No 49. FAO, Rome.

Yates R A (1979). Sugarcane as an irrigated crop, pp 103-113. In: 'Land Evaluation Criteria for Irrigation'. World Soil Res Rep No 50. FAO, Rome.

Yates R A (1981). Personal communication.

Yong R N and Warkentin B P (1975). Soil properties and behaviour. Developments in geotechnical engineering No 5. Elsevier Scientific Publ Co, Amsterdam.

Yoshino M M (Ed) (1984). Climate and agricultural land use in monsoon Asia. University of Tokyo Press.

Young A (1973). Soil survey procedures in land development planning. Geogr J 139(1), 53-64.

Young A (1976). Tropical soils and soil survey. Cambridge University Press.

Young A (1985). The potential of agroforestry as a practical means of sustaining soil fertility. Working Paper 34. ICRAF, Nairobi.

Young A (1987). Distinctive features of land use planning for agroforestry. Soil Surv and Land Eval 7 (3), 133-140.

Young A and Brown P (1962). The physical environment of Northern Nyasaland. Gvt Printer, Zomba, Malawi.

Young A and Goldsmith P F (1977). Soil survey and land evaluation in developing countries: a case study in Malawi. Geogr J 143, 407-431.

Yu T R (Ed) (1983). The physical chemistry of paddy soils. Science Press, Beijing.

Yuan T L and Fiskell J G A (1958). Aluminium studies. I. Soil and plant analysis of aluminium by modification of the aluminon method. J Agric Food Chem 7, 115-117.

Z

Zachar D (1982). Soil erosion, 518 pp. Elsevier, Amsterdam.

Zachmann D W, Chateau P C D and Klute A (1982). Simultaneous approximation of water capacity and soil hydraulic conductivity by parameter identification. Soil Sci 134, 157-163.

Zim H S and Shaffer P R (1963). Rocks and minerals. Golden Press, New York.

Zimmerman J D (1966). Irrigation. Wiley, New York.

Zimmerman R (1982). Environmental impact of forestry. Conservation Guide No 7. FAO, Rome.

Zonneveld I S (1966). Plant ecology in integrated surveys of the natural environment. Publ ITC-Unesco Cent Integr Surv. Delft.

Zwarich M A and Shaykewich C F (1969). An evaluation of several methods of measuring bulk densities of soils. Can J Soil Sci 49, 241-245.

Zwerman P and Prundeanu J (1958). Certain characteristics of land in relation to the tendency of farmers to establish conservation practices. Agron J 50, 438-440.

Index of Authors and Other Personal Names

443

Appendix I

FAO-Unesco Soil Map of the World
Revised Legend

Contents

FAO-Unesco Soil Map of the World Revised Legend

I.1   Introduction

In the mid 1980's an exhaustive appraisal of the 1974 FAO-Unesco legend took place, based on its international and national use over the previous decade. This has now resulted in a Revised Legend (FAO-Unesco 1988) whose main contents are summarised in this Appendix (Tables I.1 to I.9). Since maps using the old legend will continue to be used for many years, the summary given in the first edition of this manual is retained in this edition, pp 33-37 and 231-239.

I.2   The main changes 1974 to 1988

Appendix 4 of the Revised Legend summarises the changes made; among the more important are the following:

a)   Level 1 major soil groupings added to the 1988 legend:
Alisols, anthrosols, calcisols, gypsisols, leptosols, lixisols and plinthosols.

b)   1974 Level 1 major soil groupings deleted:
Lithosols; rendzinas and rankers (now grouped within the leptosols); yermosols and xerosols.

c)   Level 2 soil unit changes:
All orthic units changed to haplic; plinthic units grouped under plinthosols.
Other changes are summarised in Table I.1.

d)   1988 diagnostic horizons:

Modifications: albic E, cambic B, natric B.

Replacements: argillic B (now argic B), oxic B (now ferralic B).

Additions: argic B (was argillic B), ferralic B (was oxic B), fimic A, petrocalcic (was petrocalcic phase), petrogypsic (was petrogypsic phase).

Deletions: none (only replacements).

e)   1988 diagnostic properties:

Modifications: abrupt textural change; ferralic properties; ferric properties; plinthite.

Replacements: exchange complex dominated by amorphous materials (now andic properties); continuous coherent hard rock (now continuous hard rock); high salinity (now salic properties); hydromorphic properties (now gleyic and stagnic properties).

Additions: andic properties (was exchange complex dominated by amorphous materials); calcareous materials; continuous hard rock ('coherent' deleted); fluvic properties (was recent alluvial deposits); geric properties; gleyic properties (was part of hydromorphic properties); gypsiferous materials; nitic properties; organic soil materials; salic properties (was high salinity); sodic properties; stagnic properties (was part of hydromorphic properties); strongly humic materials.

Deletions: albic material; aridic moisture regime; gilgai microrelief (now gilgai phase); high organic matter in the B horizon; takyric features (now takyric phase); thin iron pan (now placic phase). Other deletions covered under 'Replacements'.

Appendix I: FAO-Unesco soil map of the world revised legend

f) Level 3 units: Soil Sub-Units.

The Revised Legend has added a further tier of units to the 1974 Legend, Level Sub-Units for specific needs at national or regional level. These Sub-Units, involvir combinations of properties of Soil Units, are potentially very numerous, and only broa guidelines are given. Note, in particular, that:

i) Soil Sub-Units must be clearly defined and should not overlap, or conflict with, th Level 1 Major Soil Groupings or Level 2 Soil Units.

ii) Because the number of letters usable for Soil Sub-Unit suffixes are limited, th same letters can have different meanings from one Sub-Unit to another (see example below).

iii) If Soil Sub-Units are to be used, care must be taken to ensure that they ar compatible with local use; further details are due to be published by FAO-Unesco ar any future global guidelines should also be adhered to.

Specifically, the Soil Sub-Units allow accommodation of the following in more detaile soil studies:

iv) Intergrades between Major Soil Groupings and/or Soil Units, eg

Niti-haplic acrisols: AChn	haplic acrisols showing nitic properties within the B above 125 cm, and so an intergrade between hapli acrisols and haplic nitisols.
Calci-mollic solonetz: SNmk	mollic solonetz with a calcic horizon within 125 cm, and so an intergrade between calcic and mollic solonetz.

v) Specification of important soil characteristics which are shown as phases i small-scale work, eg

Fragi-ferric luvisols: LVFy	luvisols having a fragipan within 125 cm.

vi) Characteristics additional to those used in the Level 2 Soil Unit definitions, eg

Grumi-eutric vertisols: VReg	eutric vertisols which, when dry, have a strong fine granular structure in the top 18 cm.
Mazi-eutric vertisols: VRem	eutric vertisols which, when dry have a hard, massive structure in the top 18 cm.

vii) Characteristics more detailed than those in the Level 2 Soil Unit definitions, eg

Hyper-calcaric cambisols: CMch	calcaric cambisols in which the calcareous material has more than 40% $CaCO_3$ equivalent.
Epi-gleyic podzols: PZge	gleyic podzols which show gleyic properties within the top 50 cm.

Two specifiers or intergrades (or a combination of both) can also be use occasionally, eg ando-mollieutric cambisols.

Appendix I: FAO-Unesco soil map of the world revised legend

List of Tables

Summary of 1974 - 1988 changes in FAO-Unesco Soil Units (Level 2)   Table I.1

Major Soil Grouping	New or new name	Changes to Level 2 Soil Units (1974 equivalent)	Amended	Deleted
Acrisols	haplic	(orthic)	gleyic, humic	orthic
Alisols	all new	-	-	-
Andosols	haplic gelic, gleyic umbric	(ochric) - (humic)	- - -	humic, ochric
Anthrosols	all new	-	-	-
Arenosols	calcaric, gleyic, haplic	-	albic, luvic	-
Calcisols	all new	-	-	-
Cambisols	calcaric humic	(calcic) -	-	calcic
Chernozems	gleyic	-	luvic	-
Ferralsols	geric haplic umbric	(acric) (orthic) -	plinthic, humic	acric, orthic
Fluvisols	mollic, salic, umbric	-	-	-
Gleysols	andic, thionic umbric	- (humic)	-	humic, plinthic
Greyzems	haplic	(orthic)	gleyic	orthic
Gypsisols	all new	-	-	-
Histosols	fibric, folic, gelic, terric, thionic	-	-	eutric, dystric
Kastanozems	gypsic	-	calcic	-
Leptosols	all new	-	-	-
Lixisols	all new	-	-	-
Luvisols	haplic stagnic	(orthic) -	gleyic	orthic, plinthic
Nitisols	haplic mollic, rhodic umbric	(eutric, dystric) - (humic)	-	eutric, dystric humic
Phaeozems	-	-	luvic, gleyic	-
Planosols	umbric	(humic)	-	humic, solodic
Plinthosols	all new	-	-	-
Podzols	cambic carbic gelic haplic	(leptic) (humic) - (orthic)	gleyic	humic, leptic orthic, placic
Podzoluvisols	gelic, stagnic	-	eutric, dystric gleyic	-
Regosols	gypsic, umbric	-	-	-
Solonchaks	calcic, gelic gypsic, sodic haplic	- - (orthic)	mollic, gleyic	orthic, takyric
Solonetz	calcic, gypsic haplic stagnic	- (orthic) -	mollic, gleyic	orthic
Vertisols	calcic, eutric dystric, gypsic	-	-	pellic, chromic

Source: FAO-Unesco (1988, pp 103 to 109).

Appendix I: FAO-Unesco soil map of the world revised legend

Major Soil Grouping (Level 1)	Symbols 1/ (1974)	(1988)	Revised description 2/
Acrisols	A	AC	Illuvial soils with low base saturation, low activity clay. More leached than luvisols, lower BSP than lixisols, lower CEC than alisols, less weathered than ferralsols. B horizon CEC < 24 me/100g clay and BSP < 50 in some part (cf nitisols, planosols)
Alisols	1/	AL	Illuvial soils with low base saturation, high activity clay. B horizon CEC < 24 me/100g clay and BSP < 50 in some part (cf acrisols, lixisols, luvisols, nitisols, planosols)
Andosols	T	AN	Soils with > 35 cm of andic materials (recent volcanic deposits)
Anthrosols	1/	AT	Soils profoundly modified by human activities
Arenosols	Q	AR	Sandy, generally weakly developed soils, but excluding soils with fluvic properties (fluvisols) and recent volcanic soils (andosols)
Calcisols	1/	CL	Soils with carbonate accumulations or horizons. Lacking salic and gleyic properties. Excluding soils with characteristics of vertisols and planosols
Cambisols	B	CM	Weathered soils without translocation of soil material; a cambic B horizon; is the only diagnostic B horizon. Excluding soils with characteristics of vertisols and andosols
Chernozems	C	CH	'Black earths' of the temperate steppes. Deep mollic A horizon rich in organic matter, plus calcic horizon and/or soft lime. No features of vertisols, planosols or andosols
Ferralsols	F	FR	Strongly weathered soils of the humid tropics with high sesquioxide contents. Ferralic horizon not overlain by argic or natric B horizons
Fluvisols	J	FL	Recently deposited soils. Depositional rather than pedogenetic profiles. Includes fluviatile, marine, lacustrine and colluvial sediments
Gleysols	G	GL	Unconsolidated soils, gleyic properties dominate top 50 cm. Excluding coarse-textured and alluvial soils and soils with characteristics of vertisols or arenosols
Greyzems	M	GR	'Grey forest soils' of cool temperate zones. Mollic A and argic B horizons (cf kastanozems, phaeozems). No diagnostic features of planosols
Gypsisols	1/	GY	Soils with a gypsic and/or petrogypsic horizon. Lacking salic and gleyic properties. Excluding soils with characteristics of vertisols and planosols
Histosols	O	HS	Soils high in fresh or partly decomposed organic matter
Kastanozems	K	KS	'Chestnut' or 'brown temperate steppe' soils. Similar to chernozems but with shallower mollic and cambic horizons May contain calcic/gypsic horizons (cf phaeozems, greyzems)
Leptosols	1/	LP	Soils < 50 cm deep or very stony unconsolidated material (< 15% fine earth above 125 cm)
Lixisols	1/	LX	Weathered illuvial soils with high base saturation. B horizon CEC < 24 me/100g clay in part, and BSP ≥ 50 throughout (cf acrisols, alisols, luvisols, nitisols, podzoluvisols)
Luvisols	L	LV	Illuvial soils with high base saturation, high activity clay. B horizon CEC ≥ 24 me/100g clay and BSP ≥ 50 throughout (cf acrisols, alisols, lixisols, nitisols, podzoluvisols)
Nitisols	N	NT	Illuvial soils with merging argic B horizons and nitic properties (including shiny ped faces). Usually formed on basic rocks (cf acrisols, podzoluvisols, luvisols)
Phaeozems	H	PH	'Prairie soils'. Lighter than chernozems, chernozem-kastanozem intergrade. Mollic A horizon and BSP ≥ 50 above 125 cm. Lacking features of vertisols, planosols and andosols (cf greyzems, nitisols)
Planosols	W	PL	Soils with an albic E horizon with hydromorphic properties abruptly overlying a slowly permeable B horizon (not a natric or spodic B horizon) above 125 cm
Plinthosols	1/	PT	Soils with ≥ 25% plinthite in a horizon ≥ 15 cm thick in top 50 cm (or within 200 cm if underlying an albic E horizon or a horizon with gleyic/stagnic properties above 100 cm)
Podzols	P	PZ	Soils with a spodic B usually below a strongly bleached horizon. (Note: some lowland tropical podzols (bleached sands) are classified as albic arenosols - Young, 1976, p 245)
Podzoluvisols	D	PD	Soils intermediate between podzols and luvisols. Argic B horizon with irregular upper boundary due to albic E tonguing or nodule formation
Regosols	R	RG	Weakly developed soils from unconsolidated (but not coarse-textured) materials; lacking properties of leptosols, fluvisols, andosols and arenosols
Solonchaks	Z	SC	Saline soils showing salic properties. Excluding soils with properties of fluvisols
Solonetz	S	SN	Sodic soils having a natric B horizon. Alkaline reaction
Vertisols	V	VR	Dark montmorillonite-rich clays with characteristic shrinking/swelling properties

Note: 1/ Denotes that soil name only appears in 1988 classification
2/ Descriptions highly abbreviated; see full definitions in FAO-Unesco (1988, pp 35 to 55 and 66 to 79)

Table I.3

Key to occurrence of revised FAO-Unesco soil units (Level 2)

Occurrence of Soil Units (Level 1 Units, columns):
AC Acrisols · AL Alisols · AN Andosols · AT Anthrosols · AR Arenosols · CL Calcisols · CM Cambisols · CH Chernozems · FR Ferralsols · FL Fluvisols · GL Gleysols · GR Greyzems · GY Gypsisols · HS Histosols · KS Kastanozems · LP Leptosols · LX Lixisols · LV Luvisols · NT Nitisols · PH Phaeozems · PL Planosols · PT Plinthosols · PZ Podzols · PD Podzoluvisols · RG Regosols · SC Solonchaks · SN Solonetz · VR Vertisols

Level 2 Soil Unit Adjective	Symbol 2/	Connotation	Major properties 3/
albic 1/	a	Strongly bleached	Albic E horizon above 125 cm
andic 1/	a	Soil name in Japan	Andic properties, smeary consistence
aric 1/	a	Plough layer	Horizon remnants due to deep ploughing
calcaric	c	CaCO3 present	Calcareous at least > 20 and < 50 cm
calcic	c	CaCO3 accumulation	Calcic horizon and/or CaCO3 above 125 cm
cambic 1/	b	Changed horizon	Cambic B horizon above 125 cm
carbic 1/	x	(Organic) carbon	High dispersed OM in spodic B
chromic	x	Bright colours	High chromas in B horizon
cumulic 1/	c	Added sediments	Man-made sediments > 50 cm thick
dystric	d	Low fertility	Low BSP (<50) at least > 20 and < 50 cm
eutric	e	High/moderate fertility	High BSP (≥50) at least > 20 and < 50 cm
ferralic 1/	o	High sesquioxides	CEC < 24 cmol(+) per kg clay in subsoil
ferric	f	Fe mottling/accumulation	Ferric properties
fibric 1/	f	Fibrous organic matter	Weakly decomposed OM; very poor drainage
fimic 1/	f	Added manure	Man-made manured surface > 50 cm thick
folic 1/	i	Intact organic matter	Well drained OM
gelic	i	Permafrost present	Permafrost above 200 cm
geric 1/	g	Strongly weathered	Bases + Al ≤ 1.5 me/100g clay in B
gleyic	g	Gleying features	Gleyic properties above 100 cm
glossic 1/	w	Tonguing features	A horizon tonguing into B or C
gypsic 1/	j	Gypsum accumulation	Gypsic horizon above 125 cm
haplic 1/	h	Simple	Simple, normal horizon sequence
humic 1/	u	Organic matter	OM in B horizon
lithic	s	Rock present	Continuous hard rock within 10 cm
luvic	w	Lessivé	Argic B horizon
mollic	m	Soft, well-structured	Mollic A horizon
petric 1/	m	Hard layer	Petrocalcic/petrogypsic horizon
plinthic 1/	v	Brick-like clay	Plinthite above 125 cm
rendzic 1/	k	Shallow, stony	Mollic A horizon over calcareous material
rhodic	r	Red	Red to dusky red B horizon
salic 1/	z	(Soluble) salts	High ECe
sodic	n	Sodium accumulation	≥ 15% exch Na (or ≥ 50% exch Na + Mg) at least > 20 and < 50 cm
stagnic 1/	s	Stagnant water	Stagnic properties
terric 1/	s	Humified organic matter	Highly decomposed OM to ≥ 35cm; imperfect to poor drainage
thionic	t	Sulphur accumulation	Sulfuric horizon and/or sulfidic material above 125 cm
umbric	u	Dark coloured	Umbric A horizon (or dystric H horizon)
urbic 1/	u	Urban	Waste materials to ≥ 50 cm
vertic	z	Inverted	Vertic properties
vitric 1/	z	Glassy material	Andic but not smeary
xanthic 1/	x	Yellow	Yellow to pale yellow ferralic B horizon

Notes: 1/ Indicates new or amended in 1988 Revised Legend
2/ These symbols are used to designate Level 2 soil units, they should not be confused with horizon suffixes (Table 1.6)
3/ Highly abbreviated; definitions are given in FAO-Unesco (1988) pp 35 to 55 and 66 to 79

Appendix I: FAO-Unesco soil map of the world revised legend

Cartographic representation of revised FAO-Unesco mapping legend   Table I.4

Property and order referred to	Description of map symbol, as used on the World Soil Map sheets at 1:5 000 000
1. Soil association	Symbol of dominant Soil Unit or, if leptosols predominate, the symbol (eg LPe) followed by those of up to two associated Units (See Table I.2). Associated Units should total ⩾ 20% of the mapped area; important soils occupying < 20% are added as inclusions
2. Sub-Unit specifier	Symbol of Sub-Unit (see Table I.3)
3. Description of association	Given by a number corresponding to that of the description on the accompanying map legend

4. Soil texture	Textural class 1/ in top 30 cm of dominant soil referred to by $\overline{a}$ number, preceded by a dash:
	    1: Coarse < 15% clay, and > 65% sand
	    2: Medium < 35% clay, and < 70% sand
	                (or ⩽ 85% sand if clay ⩾ 15%)
	    3: Fine > 35% clay
	Where textures cannot be separated, up to three symbols may be used, separated by an oblique

5. Relief	Dominant slope class 1/ indicated by a small letter following the texture symbol:
	    a: Level to gently undulating 0 to 8%
	    b: Rolling to hilly 8 to 30%
	    c: Steeply dissected to mountainous < 30%
	Two symbols may be used in areas of complex relief

Examples of mapping symbols 2/

a)	FRx 1 - 2ab	Xanthic ferralsols, medium textured, level to rolling, described as Unit 1 on legend
b)	PLe 5 - 2/3b	Eutric planosols, medium and fine textured, rolling, and gleyic luvisols (minor occurrence, so not shown in symbol), described as Unit 5 on legend
c)	LPe-LVx-ANh 3 - c	Eutric leptosols, chromic luvisols and haplic andosols, steeply dissected (no information on texture); Unit 3 on legend

Notes: 1/   These classes are too broad for most development work, particularly irrigation planning.
           2/   For Level 3 Soil Sub-Unit symbols see notes in Section I.1 (f).

Source: Adapted from FAO-Unesco (1988, pp 93 to 96).

Simplified key to diagnostic horizons of the revised FAO-Unesco system   Table I.5

Diagnostic horizon and symbol 1/		Characteristics 2/
Albic E	E	Bleached, usually sandy material, lacking clay and free iron oxides
Argic B	Bt	Horizon with higher clay content than overlying horizon (eg by illuvial deposition)
Calcic	Ak, Bk, Ck	Secondary carbonate accumulation, with $CaCO_3$ equivalent $\geqslant$ 15% and 5% more than underlying horizons
Cambic B	Bw	In situ altered B - most B horizons not meeting criteria for argic, natric, spodic or oxic
Ferralic B	Bws	Highly weathered SL or finer texture. Low CEC, illuvial clay and weatherable minerals
Fimic A	Af 3/	Man-made manurial horizon
Gypsic	Ay, By, Cy	Secondary $CaSO_4$ accumulation $\geqslant$ 5% more than underlying horizon
Histic H	H	High OM and peaty
Mollic A	Aa 3/	'Fertile earth' topsoil, well structured and dark with moderately high OM and base saturation > 50% (cf umbric A)
Natric B	Btn	High clay (as Bt) and Na (or Na + Mg) accumulation. Exchangeable Na usually > 15% and has columnar structure
Ochric A	Ao 3/	A horizon of dry areas. Pale, low OM and/or thin or hard and massive. Excluding finely stratified material,eg alluvium
Petrocalcic	Amk, Bmk, Cmk	Continuous, indurated calcic horizon; may have silica too
Petrogypsic	Amy, Bmy, Cmy	Cemented gypsic horizon; $CaSO_4$ usually > 60%
Spodic B	Bh, Bs or Bhs	OM and/or sesquioxide-rich horizon of podzols
Sulphuric	Bd 3/	Oxidised sulphide-rich materials; pH < 3.5 and jarosite mottles
Umbric A	Ae 3/	'Infertile earth' topsoil, with moderately high OM and base saturation < 50% (cf mollic A) but excluding fimic horizons

Notes: 1/ Horizon symbols are also used as qualitative designations and so do not always indicate that criteria for diagnostic horizons are met. Unless otherwise stated, most diagnostic horizons are > 15 cm thick and their upper limits lie within the top 125 cm of soil. If two or more occur, the upper diagnostic B horizon is used for classification.
2/ Highly abbreviated; see FAO-Unesco, 1988 (p.20f) and cf Table C.15.
3/ No standard FAO symbols exist; these are suggested forms for Booker Tate use.

Horizon symbol suffixes

b = Buried or bisequal
c = Concretions
g = Mottling reflecting oxidation changes
h = OM accumulation
i = Permafrost
j = Jarosite
k = $CaCO_3$ accumulation
m = Strongly cemented
n = Na accumulation
p = Disturbed by tillage
q = Silica
r = Strong reduction due to groundwater
s = Sesquioxide accumulation
t = Illuvial clay accumulation
u = (Only used when separating two horizons without suffixes; Au1, Au2, etc)
w = Alteration in situ (clay, colour, structure)
x = Fragipan occurence
y = $CaSO_4$ accumulation
z = Accumulation of salts more soluble than gypsum

Master horizons and layers

H Organic horizons, waterlogged
O Organic horizons, well drained
A Mineral surface horizons with OM or soil formation
E Eluvial horizons
B Mineral subsurface horizons with accummulation or pedological alterations
C Unconsolidated mineral material
R Continuous indurated rock

Vertical differentiation of similar horizons is denoted by suffixed numbers:
eg: (from surface) Bt1 - Bt2 - Bt3

Lithological discontinuities are denoted by prefixed numbers:
eg: Bt1 - Bt2 - 2Bt3

Source: FAO-Unesco (1988, pp 20 to 27 and p 35), FAO (1977a).

Appendix I: FAO-Unesco soil map of the world revised legend

Major diagnostic property	Characteristics 1/
Abrupt textural change	Lower horizon has significantly higher clay content than that of upper horizon (change occurs within distance of < 5 cm)
Albic material	Deleted, see 'albic E horizon' in Table I.5
Andic properties	Recent volcanic deposits: high Al and P, low bulk density; and/or high volcanic glass content
Calcareous material	Strong effervescence with 10% HCl, or > 2% $CaCO_3$ equivalent
Continuous hard rock	Underlying material continuous except for few cracks with displacement ≤ 10 cm. Sufficiently hard and coherent when moist to prevent hand digging by spade. Excludes subsurface horizons (eg duripan and the petro- horizons)
Ferralic properties	B horizons with CEC of < 24 me/100 g clay
Ferric properties	Red mottles and/or concretions
Fluvic properties	Fluviatile, marine and lacustrine sediments. Regular additions of fresh materials and irregular OM content and/or stratification in ≥ 25% of soil volume
Geric properties	Highly weathered. Extractable bases + Al ≤ 1.5 me/100g, and pH (KCl) ≥ 5.0 or pH (KCl) - (pH H_2O) ≥ - 0.1
Gleyic properties	Reduced conditions caused by groundwater saturation; capillary fringe reaches surface; evident Fe reduction and segregation; > 95% of matrix N, GY, G or B Munsell colours, unless low or inert Fe colouring. cf Stagnic properties
Gypsiferous material	Soil material with ≥ 5% gypsum
Humic	See 'strongly humic'
Hydromorphic properties	Deleted, see 'gleyic properties'
Interfingering	Penetrations by an albic E horizon into an underlying argic or natric B, but each insufficiently wide to constitute tonguing (qv)
Nitic properties	Soil material of ≥ 30% clay, possessing easily broken moderate to strong angular blocky structures with shiny ped faces
Organic soil materials	High organic matter, measured by organic C contents
Plinthite	Iron-rich clay with quartz, commonly red mottles. Irreversible change to ironstone on drying. Low organic matter
Salic properties	Top 30 cm of soil with EC_e > 15 mS cm^{-1} or, if pH > 8.5, with EC_e > 4 mS cm^{-1}
Slickensides	Polished and grooved surfaces produced by sliding of soil masses
Smeary consistence	Soil material that changes under pressure from a plastic solid into a liquified state (when it smears between the fingers) and back to the solid condition
Sodic properties	ESP > 15, or exch Na + exch Mg > 50%
Soft powdery lime	Soft carbonates, oxides and/or hydroxides of calcium and/or magnesium soft enough to be cut by a finger nail
Stagnic properties	Reduced conditions caused by surface water saturation; mottling and/or Fe-Mn concretions or low chromas. cf Gleyic properties
Strongly humic	Soil material with > 1.4% organic C
Sulphidic materials	Waterlogged soil with ≥ 0.75% sulphur and less than three times as much $CaCO_3$ equivalent as sulphur
Takyric features	Deleted, see 'takyric phase' in Table I.7
Tonguing	Penetration by albic E into argic B; depths of 'tongues' must be greater than widths and be > 15% of the mass of the upper part of the argic B
Vertic properties	Of clay soils with vertisol features but not qualifying as true vertisols
Weatherable minerals	Minerals which are relatively unstable in a humid climate, including: quartz, 1:1 lattice clays, 2:1 lattice clays except Al-interlayered chlorite
Yermic properties	Deleted, see 'yermic phase' in Table I.7

1/ Highly abbreviated; see FAO-Unesco, 1988 (p 28f) for details.
    Unless otherwise stated, the upper limit of the diagnostic properties lie within the top 125 cm of soil.

Revised FAO-Unesco soil phases

Table I.7

Phase name	Description
Anthraquic	Stagnic properties within 50 cm of soil surface due to surface water stagnation after long irrigation, particularly of rice
Duripan	Duripan (silica cementation) within 100 cm of soil surface
Fragipan	Fragipan (hard, seemingly cemented, loamy) with upper level within 100 cm of soil surface. May be 15 to 200 cm thick
Gelundic	Soils with surface polygon formation due to frost heaving
Gilgai	Microbasin and microknoll (or microvalley and microridge) relief associated with clayey soils, mainly vertisols
Inundic	Standing or flowing water present on surface for more than 10 days in growing period
Lithic	Continuous, hard rock within 50 cm of soil surface
Petroferric	Iron-rich, essentially continuous and indurated horizon within 100 cm of soil surface (cf skeletic phase)
Phreatic	GWT between 3 and 5 m from soil surface
Placic	Thin iron pan, generally 2 to 10 mm thick, within 100 cm
Rudic	Gravel, stones, boulders or rock outcrops sufficient to make mechanised agriculture impractical
Salic	$EC_e > 4$ mS cm^{-1} within 100 cm of soil surface. (Not used for solonchaks)
Skeletic	$\geq$ 25 cm thick layer of $\geq$ 40% by volume of hardened, coarse fragments/oxidic nodules/plinthite/ironstone within 50 cm of soil surface. Not continuously cemented (cf petroferric phase)
Sodic	ESP > 6 within 100 cm of soil surface. (Not used for soils with natric B or sodic properties)
Takyric	Fine-textured soils whose surfaces when dry form a massive or platy crust cracked into polygons
Yermic	Soils with < 0.6% organic C in top 18 cm (or < 0.25% organic C if coarser than SL) plus features of arid conditions

Abbreviated from FAO-Unesco (1988, pp 60f).

Appendix I: FAO-Unesco soil map of the world revised legend

Major soil grouping and symbol	Main diagnostic horizon(s) and other properties of main soil grouping 3/	Other properties in (or lacking in) some units 3/
Acrisols AC	Bt B horizon: CEC < 24 me/100g clay and BSP < 50 in some part of top 125 cm	Gleyic (< 100 cm) properties; Ao ± ferric properties; plinthite; Aa or Ae and strongly humic Not soils NT, PD, PL
Alisols AL	Bt B horizon: CEC ≥ 24 me/100g clay and BSP < 50 in some part of top 125 cm	Stagnic(50 cm) or gleyic (< 100 cm) properties; Ao ± ferric properties; Aa or Ae and strongly humic Not soils NT, PD, PL
Andosols AN	Aa or Ae, or Ao over Bw No other diagnostic horizons Andic materials from surface to ≥ 35 cm No salic properties No gleyic properties < 50 cm	Aa, Ae, or Ao and Bw, ± smeary consistence ± average texture finer than ZL in top 100 cm; gleyic properties < 100 cm Not soil VR
Anthrosols AT	Pronounced evidence of human action	Af; deep cultivation; accumulation of fine sediments; wastes, refuse etc
Arenosols AR	Coarser than SL to ≥ 100 cm, unconsolidated Ao and E only diagnostic horizons permitted No fluvic or andic properties	Ferralic properties; some properties of Bw or Bt; presence of calcareous material between 20 and 50 cm; gleyic properties < 100 cm; BSP > or < 50
Calcisols CL	One or more of: Bk, Bmk or soft lime < 125 cm Ao, Bw, Bt only other diagnostic horizons permitted No salic properties No gleyic properties < 100 cm	Ao, Bw, Bt Coarse, unconsolidated materials Not soils PL, VR
Cambisols CM	Ae or Ao, and Bw No other diagnostic horizons No salic properties No gleyic properties < 50 cm	Gleyic properties (only > 50cm); calcareous between 20 and 50 cm; ferralic or vertic properties; strong brown/red (chromic); BSP > or < 50 Not soils AN or VR
Chernozems CH	Aa with moist chroma ≤ 2 to ≥ 15 cm Bk and/or soft lime < 125 cm No Bn No salic properties No gleyic properties < 50 cm if Bt absent No uncoated silt and quartz grains on ped surfaces	Bk; Bt; tonguing of A horizon into Bw or C; Bt plus gleyic properties < 100 cm Not soils AN, PL or VR
Ferralsols FR	Ferralic B (Bws)	Ao with pale yellow (xanthic) or dusky red (rhodic) (or neither) Bws; Aa or Ae and strongly humic; plinthite; geric properties
Fluvisols FL	Aa, Ae or Ao H or Bd or sulfidic materials < 125 cm	Calcareous material between 20 and 50 cm salic properties; BSP > or < 50
Gleysols GL	A (all types) Gleyic properties < 50 cm No fluvic or salic properties Not coarse textured No plinthite 125 cm	H, Bw, Bk and/or By, Bd; andic properties; BSP > or < 50 Not soils AR, VR
Greyzems GR	Aa with moist chroma ≤ 2 to ≥ 15 cm Bt Uncoated silt and quartz grains on ped surfaces	Gleyic properties within 100 cm Not soil PL
Gypsisols GY	By and/or Bmy Ao, Bw, Bt, Bk, Bmk only other diagnostic horizons permitted No salic properties No gleyic properties < 100 cm	Bk, Bt, Bmy Not soils PL, VR
Histosols HS	H or O with ≥ 40 cm (cumulatively) of organic soil materials in top 80 cm (or ≥ 60 cm if OM mainly sphagnum or moss or has BD < 0.1)	Well drained; highly decomposed or raw OM; Bd or sulfidic materials above 125 cm

Notes: See page 457   cont

456

Table I.8 cont

Major soil grouping and symbol	Main diagnostic horizon(s) and other properties of main soil grouping 3/	Other properties in (or lacking in) some units 3/
Kastanozems KS	Aa and moist chroma > 2 to ⩾ 15 cm Bk and/or By and/or soft lime < 125 cm No Bn No salic properties No gleyic properties < 50 cm if Bt absent	Bk, Bt, By Not soils AN, PL, VR
Leptosols LP	< 30 cm effective soil depth, or < 20% fine earth to 75 cm Aa, Ae or Ao with or without Bw	Calcareous material; hard rock < 10 cm BSP > or < 50
Lixisols LX	No Aa Bt B horizon: CEC < 24 me/100 g clay in some part and BSP ⩾ 50 throughout top 125 cm	E; ferric, stagnic (< 50 cm) or gleyic (< 100 cm) properties; plinthite Not soils NT, PD, PL
Luvisols LV	No Aa Bt B horizon: CEC ⩾ 24 me/100g clay and BSP ⩾ 50 throughout top 125 cm	E; strong red/brown (chromic); Bk,and/or soft lime < 125 cm; ferric, vertic, stagnic (< 50 cm) or gleyic (< 100 cm) properties Not soils NT, PD, PL
Nitisols NT	Bt that has clay content within 20% of maximum in top 150 cm Gradual/diffuse A/B boundary Nitic properties within top 125 cm No tonguing of A; no ferric or vertic properties No plinthite above 125 cm	Ao ± red/dusky red (rhodic); Aa or Ae and strongly humic Not soil PD
Phaeozems PH	Aa CEC > 16 me/100 clay and BSP ⩾ 50 above 125 cm No Bk, no By, no soft lime above 125 cm; no Bn No Bws; no salic properties No gleyic properties < 50 cm if Bt absent No uncoated silt and quartz grains if Aa has moist chroma ⩽ 2 to ⩾ 15 cm	Calcareous material 20 to 50 cm; Bt; Bt plus stagnic (< 50 cm) or gleyic (< 100 cm) properties Not soils AN, NT, PL, VR
Planosols PL	E with stagnic properties in part, over slowly permeable horizon above 125 cm (not Bn or spodic B)	Ao with BSP > or < 50; Aa; Ae; eutric or dystric H
Plinthosols PT	⩾ 25% by volume of plinthite in horizon ⩾ 15 cm thick above 50 cm (or above 125 cm if underlying E or horizon with gleyic or stagnic properties in top 100 cm)	E; Aa, Ae or dystric H and strongly humic Ao and BSP > or < 50
Podzols PZ	Spodic B	E; stagnic (< 50 cm) or gleyic (< 100 cm) properties; various free iron to carbon ratios
Podzoluvisols PD	Bt with irregular upper boundary of B due to deep tonguing of E or to Fe nodules > 2 cm diameter No Aa	Stagnic (< 50 cm) or gleyic (< 100 cm) properties BSP > or < 50
Regosols RG	Ao or Ae only diagnostic horizons Unconsolidated materials, but not coarse-textured or stony No fluvic properties No gleyic properties < 50 cm No salic properties	Calcareous or gypsiferous material between 20 and 50 cm; BSP > or < 50 Not soils AN or VR
Solonchaks SC	Salic properties No diagnostic horizons other than those on right No fluvic properties	A or H; Bw; Bk, By or soft lime < 125 cm; sodic properties; gleyic properties < 100 cm
Solonetz SN	Bn	Aa, Ao; Bk, By or soft lime < 125 cm; gleyic properties < 100 cm; stagnic properties < 50 cm
Vertisols VR	Clay ⩾ 30% and cracks ⩾ 1 cm wide to ⩾ 50 cm At depths 25 to 100 cm slickenslides, or parallelipiped structure, with or without gilgai	Bk, By or soft lime < 125 cm; BSP > or < 50

Notes: 1/   Highly abbreviated;   full definitions should always be checked.
         2/   Permafrost criteria have been omitted;   depth criteria indicate distance below soil surface.
         3/   Note that some symbols used here are not in the FAO-Unesco system - for these and other abbreviations see Tables I.5 and I.6.

Source: FAO-Unesco (1988, pp 36 to 55 and 66 to 79).

Table I.9

Simplified flow diagram of revised FAO-Unesco key to major soil groupings

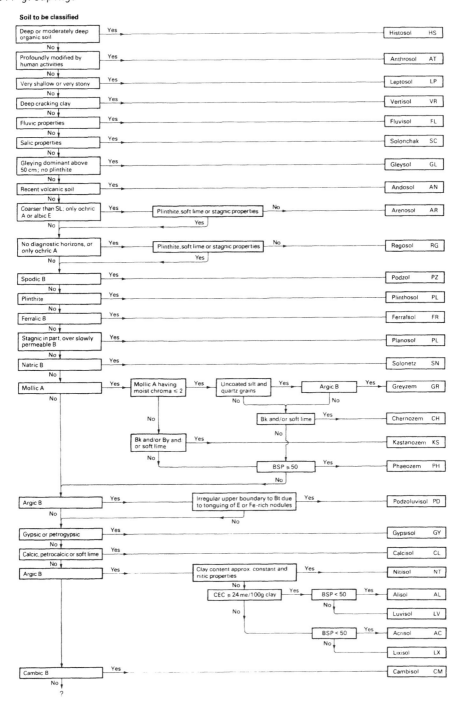

Notes: 1. Soil legend highly abbreviated; does not include, for example, most depth criteria nor criteria for buried soils, nor exceptions (e.g. recent alluvial soils with salic properties are solonchaks, not fluvisols). Details in FAO-Unesco (1988, pp66 to 79) should always be checked.

2. See Table I.5 for horizon symbols.

Appendix II

Soil and Water Salinity Update

Contents

oil and Water Salinity Update

.1   Sample calculation of leaching requirement 1/

A maize crop is irrigated by furrow irrigation in a uniform loam soil under the following conditions:-

Electrical conductivity of river water, EC_w	=	1.2 mS cm^{-1}
Crop evapotranspiration	=	800 mm/season
Salinity limit for 90% yield of maize 2/	=	2.5 mS cm^{-1}
Salinity limit for 100% yield of maize 2/	=	1.7 mS cm^{-1}

(a) Leaching requirement for 90% yield potential:

$$LR = \frac{EC_w}{5EC_e - EC_w} = \frac{1.2}{5(1.7) - 1.2} = 0.10 \quad \text{(see Section 8.7)}$$

Actual amount of water (AW) required to satisfy crop ET and leaching requirements is:-

$$AW = \frac{ET}{1 - LR} = \frac{800}{1 - 0.10} = 890 \text{ mm/season}$$

(b) Leaching requirement for 100% yield potential:

$$LR = \frac{1.2}{5(2.5) - 1.2} = 0.16$$

Actual amount of water required to satisfy crop ET and leaching requirement is:

$$AW = \frac{800}{1 - 0.16} = 952 \text{ mm/season}$$

With a field application efficiency of, say 65%, the actual water application to satisfy the 800 mm/season ET demand = 800/0.65 = 1 230 mm/season. Thus, provided that enough of the losses in excess of ET are deep percolation losses, then no additional leaching to control salinity is necessary.

.2   Adjusted SAR: updated calculation method 3/

The adjusted SAR procedure presented previously (see page 164) is no longer recommended. Oster and Rhoades (1975), Oster and Schroer (1979) and Suarez (1981) carefully evaluated that procedure and concluded that it overstates the sodium hazard. They suggest that, if used, the value obtained by that method should be multiplied by 0.5 to evaluate more correctly the effects of HCO_3 on calcium precipitation. The method currently recommended by FAO (1985b) is the newer adj RNa procedure of Suarez (1981), although the old SAR procedure is also acceptable (see page 164).

The new method is similar to the old SAR calculation, but adjusts the calcium concentration of the irrigation water to the expected equilibrium value following an irrigation, and includes the effects of carbon dioxide (CO_2), of bicarbonate (HCO_3) and of salinity (EC_w) upon the calcium originally present in the applied water but now a part of the soil-water. The procedure assumes a soil source of calcium - from soil lime ($CaCO_3$) or other soil minerals such as silicates - and no precipitation of magnesium.

Notes: 1/ Adapted from FAO (1985b, p 26).
           2/ From Table 8.3.
           3/ Adapted from FAO (1985b).

Appendix II: Soil and water salinity update

List of Tables

Appendix II: Soil and water salinity update

The new term for this is adj RNa (adjusted Sodium Adsorption Ratio). It can be used to predict more correctly potential infiltration problems due to relatively high sodium (or low calcium) in irrigation water supplies (Suarez 1981; Rhoades 1982) and can be substituted for SAR in Table 8.9. The equation for adj RNa of the surface soil is:

$$adj\ RNa = \frac{Na}{\sqrt{\dfrac{Ca_x + Mg}{2}}}$$

where: Na = Sodium in the irrigation water (me/ℓ)
Ca_x = A modified calcium value taken from Table II.1 in me/ℓ
Mg = Magnesium in the irrigation water (me/ℓ)

To use the Ca_x table (Table II.1), first determine the HCO_3 to Ca ratio and EC_w from the water analysis, using HCO_3 and Ca in me/ℓ and the water salinity (EC_w) in mS cm^{-1}. An appropriate range of calculated HCO_3/Ca ratios appears on the left side of the table and the range of EC_w across the top. Find the HCO_3/Ca ratio that falls nearest to the calculated HCO_3/Ca value for the subject water and read across to the EC_w column that most closely approximates the EC_w for the water being evaluated. The Ca_x value shown represents the me/ℓ of Ca that is expected to remain in solution in the soil water at equilibrium. Table II.2 compares the three calculation procedures for SAR, adj SAR and adj RNa.

Calcium concentration in near-surface soil-water following irrigation   Table II.1

Ratio of HCO3:Ca in applied water	Calcium concentration (Ca_x) in me/ℓ for salinity of applied water (EC_w) in mS cm^{-1}											
	0.1	0.2	0.3	0.5	0.7	1.0	1.5	2.0	3.0	4.0	6.0	8.0
.05	13.20	13.61	13.92	14.40	14.79	15.26	15.91	16.43	17.28	17.97	19.07	19.94
.10	8.31	8.57	8.77	9.07	9.31	9.62	10.02	10.35	10.89	11.32	12.01	12.56
.15	6.34	6.54	6.69	6.92	7.11	7.34	7.65	7.90	8.31	8.64	9.17	9.58
.20	5.24	5.40	5.52	5.71	5.87	6.06	6.31	6.52	6.86	7.13	7.57	7.91
.25	4.51	4.65	4.76	4.92	5.06	5.22	5.44	5.62	5.91	6.15	6.52	6.82
.30	4.00	4.12	4.21	4.36	4.48	4.62	4.82	4.98	5.24	5.44	5.77	6.04
.35	3.61	3.72	3.80	3.94	4.04	4.17	4.35	4.49	4.72	4.91	5.21	5.45
.40	3.30	3.40	3.48	3.60	3.70	3.82	3.98	4.11	4.32	4.49	4.77	4.98
.45	3.05	3.14	3.22	3.33	3.42	3.53	3.68	3.80	4.00	4.15	4.41	4.61
.50	2.84	2.93	3.00	3.10	3.19	3.29	3.43	3.54	3.72	3.87	4.11	4.30
.75	2.17	2.24	2.29	2.37	2.43	2.51	2.62	2.70	2.84	2.95	3.14	3.28
1.00	1.79	1.85	1.89	1.96	2.01	2.09	2.16	2.23	2.35	2.44	2.59	2.71
1.25	1.54	1.59	1.63	1.68	1.73	1.78	1.86	1.92	2.02	2.10	2.23	2.33
1.50	1.37	1.14	1.44	1.49	1.53	1.58	1.65	1.70	1.79	1.86	1.97	2.07
1.75	1.23	1.27	1.30	1.35	1.38	1.43	1.49	1.54	1.62	1.68	1.78	1.86
2.00	1.13	1.16	1.19	1.23	1.26	1.31	1.36	1.40	1.48	1.54	1.63	1.70
2.25	1.04	1.08	1.10	1.14	1.17	1.21	1.26	1.30	1.37	1.42	1.51	1.58
2.50	0.97	1.00	1.02	1.06	1.09	1.12	1.17	1.21	1.27	1.32	1.40	1.47
3.00	0.85	0.89	0.91	0.94	0.96	1.00	1.04	1.07	1.13	1.17	1.24	1.30
3.50	0.78	0.80	0.82	0.85	0.87	0.90	0.94	0.97	1.02	1.06	1.12	1.17
4.00	0.71	0.73	0.75	0.78	0.80	0.82	0.86	0.88	0.93	0.97	1.03	1.07
4.50	0.66	0.68	0.69	0.72	0.74	0.76	0.79	0.82	0.86	0.90	0.95	0.99
5.00	0.61	0.63	0.65	0.67	0.69	0.71	0.74	0.76	0.80	0.83	0.88	0.93
7.00	0.49	0.50	0.52	0.53	0.55	0.57	0.59	0.61	0.64	0.67	0.71	0.74
10.00	0.39	0.40	0.41	0.42	0.43	0.45	0.47	0.48	0.51	0.53	0.56	0.58
20.00	0.24	0.25	0.26	0.26	0.27	0.28	0.29	0.30	0.32	0.33	0.35	0.37
30.00	0.18	0.19	0.20	0.20	0.21	0.21	0.22	0.23	0.24	0.25	0.27	0.28

Notes: 1/   Assumes a soil source of calcium from lime ($CaCO_3$) or silicates; no precipitation of magnesium, and partial pressure of CO_2 near the soil surface (PCO_2) of 0.0007 atm.

          2/ Ca_x represents Ca in the applied irrigation water but modified due to salinity of the applied water (EC_w), its HCO_3/Ca ratio (HCO_3 and Ca in me/ℓ) and the estimated partial pressure of CO_2 in the surface few mm of soil.

Source: FAO (1985b) adapted from Suarez (1981).

Comparison of methods to calculate the sodium hazard of water   Table II.2

Results of water analysis:

EC_w	=	1.15 mS cm^{-1}	CO_3	=	0.42 me/ℓ
Ca	=	2.32 me/ℓ	HCO_3	=	3.66 me/ℓ
Mg	=	1.44 me/ℓ			
Na	=	7.73 me/ℓ			

Cation Sum	=	11.49 me/ℓ	Anion Sum	=	4.08 me/ℓ

1. The original method of Sodium Adsorption Ratio (SAR) calculation:

$$SAR = \frac{Na}{\sqrt{\dfrac{Ca + Mg}{2}}} = \frac{7.73}{\sqrt{\dfrac{2.32 + 1.44}{2}}} = 5.64$$

2. The adjusted Sodium Adsorption Ratio (adj SAR) from the procedure given in Ayers and Westcot (1976), see page 164:

$$adj\ SAR = SAR\ [1 + (8.4 - pHc)]$$

$$where\ pHC = (pK_2 - pK_c) + p(Ca + Mg) + p(Alk)$$

$$
\begin{aligned}
(pK_2 - pK_c) &= 2.3 \\
p(Ca + Mg) &= 2.7 \\
p(Alk) &= \underline{2.4} \\
pHc &= 7.4
\end{aligned}
$$

$$adj\ SAR = 5.64\ [1 + (8.4 - 7.40)] \qquad = 11.3$$

3. The adjusted Sodium Adsorption Ratio (adj RNa), see page 460:

$$HCO_3/Ca = 1.76\ \text{(from water analysis above)}$$
$$Ca_x = 1.43\ me/\ell\ \text{(from Table II.1)}$$

$$adj\ RNa = \frac{Na}{\sqrt{\dfrac{Ca_x + Mg}{2}}} = \frac{7.73}{\sqrt{\dfrac{1.43 + 1.44}{2}}} = 6.45$$

Source: Adapted from FAO (1985b).

Appendix II: Soil and water salinity update

Relative salt tolerances of selected crops 1/   Table II.3

Tolerant	Moderately tolerant	Moderately sensitive	Sensitive
	Approximate threshold salinity at which yield loss starts		
ECe = 6 to 10 mS cm^{-1}	ECe = 3 to 6 mS cm^{-1}	ECe = 1.3 to 3.0 mS cm^{-1}	ECe = 1.3 mS cm^{-1}

FIBRE, SEED AND SUGAR CROPS

Tolerant	Moderately tolerant	Moderately sensitive	Sensitive
Barley (Hordeum vulgare)	Cowpea (Vigna unguiculata)	Broadbean (Vicia faba)	Bean (Phaseolus vulgaris)
Cotton (Gossypium hirsutum)	Oats (Avena sativa)	Castorbean (Ricinus communis)	Guayule (Parthenium argentatum)
Jojoba (Simmondsia chinensis)	Rye (Secale cereale)	Maize (Zea mays)	Sesame (Sesamum indicum)
Sugarbeet (Beta vulgaris)	Safflower (carthamus tinctorius)	Flax (Linum usitatissimum)	
	Sorghum (Sorghum bicolor)	Millet, foxtail (Setaria italica)	
	Soybean (Glycine max)	Groundnut/Peanut (Arachis hypogaea)	
	Triticale (X Triticosecale)	Rice, paddy (Oryza sativa)	
	Wheat (Triticum aestivum)	Sugarcane (Saccarum officinarum)	
	Wheat, durum (Triticum turgidum)	Sunflower (Helianthus annuus)	

GRASSES AND FORAGE CROPS

Tolerant	Moderately tolerant	Moderately sensitive	Sensitive
Alkali grass, nuttall (Puccinellia airoides)	Barley (forage) (Hordeum vulgare)	Orchard grass (Dactylis glomerata)	Alfalfa (Medicago sativa)
Alkali sacaton (Sporobolus airoides)	Brome, mountain (Bromus marginatus)	Rye (forage) (Secale cereale)	Bentgrass (Agrostis stolonifera palustris)
Bermuda grass (Cynodon dactylon)	Canary grass, reed (Phalaris, arundinacea)	Sesbania (Sesbania exaltata)	Bluestem, Angleton (Dichanthium aristatum)
Kallar grass (Diplachne fusca)	Clover, Hubam (Melilotus alba)	Siratro (Macroptilium atropurpureum)	Brome, smooth (Bromus inermis)
Saltgrass, desert (Distichlis stricta)	Clover, sweet (Melilotus)	Sphaerophysa (Sphaerophysa salsula)	Buffelgrass (Cenchrus ciliaris)
Wheatgrass, fairway crested (Agropyron elongatum)	Fescue, meadow (Festuca pratensis)	Timothy (Phleum pratense)	Burnet (Poterium sanguisorba)
Wildrye, Altai (Elymus angustus)	Harding grass (Phalaris tuberosa)	Trefoil, big (Lotus uliginosus)	Clover, alsike (Trifolium hydridum)
Wildrye, Russian (Elymus junceus)	Panic grass, blue (Panicum antidotale)	Vetch, common (Vicia angustifolia)	Clover, bersim (Trifolium alexandrinum)
	Rape (Brassica napus)		Clover, ladino (Trifolium repens)
	Rescue grass (Bromus unioloides)		Clover, red (Trifolium pratense)
	Rhodes grass (Chloris gayana)		Clover, strawberry (Trifolium fragiferum)
	Ryegrass, Italian (Lolium italicum multiflorum)		Clover, white Dutch (Trifolium repens)
	Ryegrass, perennial (Lolium perenne)		Corn (forage) (maize) (Zea mays)
	Sudan grass (Sorghum sudanense)		Cowpea (forage) (Vigna unguiculata)
	Trefoil, birdsfoot (Lotus corniculatus)		Dallis grass (Paspalum dilatatum)
	Wheat (forage) (Triticum aestivum)		Foxtail, meadow (Alopecurus pratensis)
	Wheatgrass, standard crested (Agropyron sibiricium)		Grama, blue (Bouteloua gracilis)
	Wheatgrass, slender (Agropyron trachycaulum)		Lovegrass (Eragrostis sp.)
	Wheatgrass, western (Agropyron smithii)		Milkvetch, cicer (Astragalus cicer)
	Wildrye, beardless (Elymus triticoides)		Oatgrass, tall (Arrhenatherum, Danthonia)
	Wildrye, Canadian (Elymus canadensis)		Oats (forage) (Avena sativa)

Note: See page 463

cont

rant	Moderately tolerant	Moderately sensitive	Sensitive
	Approximate threshold salinity at which yield loss starts		
= 6 to 10 mS cm^{-1}	ECe = 3 to 6 mS cm^{-1}	ECe = 1.3 to 3.0 mS cm^{-1}	ECe = 1.3 mS cm^{-1}

T, NUT AND VEGETABLE CROPS

ragus sparagus officinalis) palm hoenix dactylifera)	Artichoke (Helianthus tuberosus) Beet, red (Beta vulgaris) Fig (Ficus carica) Jujube (Zizyphus jujuba) Olive (Olea europaea) Papaya (Carica papaya) Pineapple (Ananas comosus) Pomegranate (Punica granatum) Squash, zucchini (Cucurbita pepo melopepo)	Broccoli (Brassica oleracea botrytis) Brussels sprouts (B.oleracea gemmifera) Cabbage (B.oleracea capitata) Cauliflower (B.oleracea botrytis) Celery (Apium gravolens) Corn, sweet (Zea mays) Cucumber (Cucumis sativus) Eggplant (Solanum melongena esculentum) Grape (Vitis sp.) Kale (Brassica oleracea acephala) Kohlrabi (B.oleracea gongylode) Lettuce (Latuca sativa) Muskmelon (Cucumis melo) Pepper (Capsicum annum) Potato (Solanum tuberosum) Pumpkin (Cucurbita peop pepo) Radish (Raphanus sativus) Spinach (Spinacia oleracea) Squash, scallop (Cucurbita pepo melopepo) Sweet Potato (Ipomoea batatas) Tomato (Lycopesicon) Turnip (Brassica rapa) Watermelon (Citrullus lanatus)	Almond (Prunus dulcis) Apple (Malus sylvestris) Apricot (Prunus armeniaca) Avocado (Persea americana) Banana (Musa sp.) Bean (Phaseolus vulgaris) Blackberry (Rubus sp.) Boysenberry (Rubus ursinus) Carrot (Daucus carota) Cherimoya (Annona cherimola) Cherry, sweet (Prunus avium) Currant (Ribes sp.) Gooseberry (Ribes sp.) Grapefruit (Citrus paradisi) Lemon (Citrus limon) Lime (Citrus aurantiifolia) Loquat (Eriobotrya japonica) Mango (Mangifera indica) Okra (Abelmoschus esculentus) Onion (Allium cepa) Orange (Citrus sinensis) Parsnip (Pastinaca sativa) Passionfruit (Passiflora edulis) Peach (Prunus persica) Pear (Pyrus communis) Persimmon (Diospyros virginiana) Plum: Prune (Prunus domestica) Pummelo (Citrus maxima) Raspberry (Rubus idaeus) Rose apple (Sysgium jambos) Sapote, white (Casimiroa edulis) Strawberry (Fragaria sp.) Tangerine, mandarin (Citrus reticulata)

e: 1/ These data serve only as a guide to the relative tolerances among crops. Absolute tolerances vary with
       climate, soil conditions and cultural practices. Detailed tolerances can be found in Table 8.3 and
       Maas (1984).

rce: Adapted from FAO (1985b), from Maas (1984); cf Tables 8.2 to 8.4.

Appendix II: Soil and water salinity update

Common laboratory determinations for irrigation water analysis   Table II.4

Parameter	Symbol	Unit	Usual range in irrigation water
SALINITY			
Salt Content			
Electrical Conductivity	EC_w	mS cm^{-1}	0 - 3
(or)			
Total Dissolved Solids	TDS	mg ℓ^{-1}	0 - 2000
Cations and Anions			
Calcium	Ca^{++}	me ℓ^{-1}	0 - 20
Magnesium	Mg^{++}	me ℓ^{-1}	0 - 5
Sodium	Na^{+}	me ℓ^{-1}	0 - 40
Carbonate	CO_3^{--}	me ℓ^{-1}	0 - 0.1
Bicarbonate	HCO_3^{-}	me ℓ^{-1}	0 - 10
Chloride	Cl^{-}	me ℓ^{-1}	0 - 30
Sulphate	SO_4^{--}	me ℓ^{-1}	0 - 20
NUTRIENTS 1/			
Nitrate-Nitrogen	NO_3-N	mg ℓ^{-1}	0 - 10
Ammonium-Nitrogen	NH_4-N	mg ℓ^{-1}	0 - 5
Phosphate-Phosphorus	PO_4-P	mg ℓ^{-1}	0 - 2
Potassium	K^{+}	mg ℓ^{-1}	0 - 2
MISCELLANEOUS			
Boron	B	mg ℓ^{-1}	0 - 2
Acidity/Basicity	pH		6.0 - 8.5
Sodium Adsorption Ratio	SAR		0 - 15

Note: 1/ NO_3-N means analysis for NO_3 but reported in terms of chemically equivalent nitrogen.   Similarly, for NH_4-N, analysis for NH_4 but reported in terms of chemically equivalent elemental nitrogen.   The total nitrogen available to plants is the sum of the equivalent elemental nitrogen. The same reporting method is used for phosphorus.

Source: FAO (1985b, p 10)

Subject Index

Entries in the Glossary, Annex L, are not repeated below. Readers are advised to look for items both in this list and in Annex L, pp 335-380.

Note also that soil, plant nutrient elements, climate and altitude are mentioned throughout Annex F, especially pp 292-308 on soil suitability for individual crops, but these numerous allusions are not cross-referenced here.

A

A horizon 234,453
abrupt textural change 232,234,447,454
accumulated infiltration rate 64,217
acetic acid 334
acid rock 388-389
acid sulphate soil 143,152,238
acidifying effect of fertiliser 399-402
acidity see pH
acric soil 36,238,449
acrisols 35-36,232,235,238,241,246,448-451,456
ADAS field texture classes 230
additive ratings 56
adjusted sodium absorption ratio 164,459-461
aerial photographs 5-6,19,341-344,346-354
  interpretation 8,24,344,346
  processing 344,346
  scale relationships 336-337,349
Africa, land management in 2
Africa, Soil Map of (CCTA) 33,231,233
agglomerate 392
agric horizon 249
air drying 107,109
air-filled porosity 83: see also soil porosity
albic horizon 232,249,447,453
albic soil 36,232,234,238-239,449,451,454
albite 394-395
alfalfa see lucerne
alfisols 239-241,244-246,248
alisols 447,449-451,456
alkali soil 157
alkalinity see pH
allophane 118,122,132
alluvial soil 232,447: see also fluvisols
aluminium 102,113,115-116,122,128,132-134,137,334
  in water 172
ammonia 174,334,464
ammonium acetate 118,120,123,126,334
ammonium sulphate 399
amphibole 394: see also hornblende
amygdaloidal rock 389
anaerobism 83,97
andepts 238,246
andesine 395
andesite 385,387-389
andic soil 447,449,451,454
andosols 35-36,232,236,238,241,246-247,449-451,456
anhydrite 391-392
animal nutrition 144,151,154-155
anion exchange resin 137
anorthite 395

anthraquic soil 455
anthropic epipedon 249
anthrosols 447,449-451,456
apatite 394-395,397,400-401
apples 150,159-160,283,463
aquic soil 238,248
arabica coffee 273,282,296-297
arable (USBR definition) 47,49,257,261
area nomenclature 182
arenosols 35-36,232,236,238,241,246,449-451,456
argic horizon 447,453
argillic horizon 232,234,249,447
aric soil 451
arid climate 232,234,382
  (dry climate of Köppen 381,383-384)
aridic soil moisture regime 447
aridisols 239-241,244-246
arkose 391-392
arsenic in water 172
artichokes 463
asparagus 150,159,463
associations see soil associations
Astragallus 155
atomic weights 334
attapulgite 396
auger site record sheet 319
auger-hole method (bulk density) 78
auger-hole method (hydraulic conductivity) 72-73, 218-226
augite 388-390,394-395
available water capacity 61,91-95,230,338
  interpretation of results 96
average infiltration rate 65,216
avocadoes 129,150,159-160,463
azonal soil 246

B

B horizon 234,453
background data 12,15,18-19
bacterial activity 117
Balmoral series 68
bananas 160,270-271,280,284,287,289,292-293,463
bands for Landsat imagery 347
barium chloride 118,120,334
barium in soil 144
barley 100,129-130,150-151,159-162,280,283,289,462
basalt 385,387-389
base saturation 113,122-123,131
basic infiltration rate 63,67-70
basic rock 388-389
basic slag 155-400

465

Printed and bound by CPI Group (UK) Ltd, Croydon, CR0 4YY

01/11/2024

01782630-0018